Lecture Notes in Artificial Intelligence 10037

Subseries of Lecture Notes in Computer Science

More information about this series at http://www.springer.com/series/1244

Giovanni Adorni · Stefano Cagnoni
Marco Gori · Marco Maratea (Eds.)

AI*IA 2016:
Advances in
Artificial Intelligence

XVth International Conference
of the Italian Association for Artificial Intelligence
Genova, Italy, November 29 – December 1, 2016
Proceedings

 Springer

Editors
Giovanni Adorni
University of Genoa
Genova
Italy

Stefano Cagnoni
University of Parma
Parma
Italy

Marco Gori
University of Siena
Siena
Italy

Marco Maratea
University of Genoa
Genova
Italy

ISSN 0302-9743   ISSN 1611-3349   (electronic)
Lecture Notes in Artificial Intelligence
ISBN 978-3-319-49129-5   ISBN 978-3-319-49130-1   (eBook)
DOI 10.1007/978-3-319-49130-1

Library of Congress Control Number: 2016956538

LNCS Sublibrary: SL7 – Artificial Intelligence

Printed on acid-free paper

This Springer imprint is published by Springer Nature
The registered company is Springer International Publishing AG
The registered company address is: Gewerbestrasse 11, 6330 Cham, Switzerland

Preface

This volume collects the contributions presented at the XV Conference of the Italian Association for Artificial Intelligence (AI*IA 2016). The conference was held in Genova, Italy, from November 28 to December 1, 2016. The conference is organized by AI*IA (the Italian Association for Artificial Intelligence), and it is held annually.

The conference received 53 submissions. Each paper was carefully reviewed by at least three members of the Program Committee, and finally 39 papers were accepted for publication in these proceedings.

Following the 2013 and 2015 editions of the conference, we adopted a "social" format for the presentations: The papers were made available to conference participants in advance. Each paper was shortly presented at the conference, then assigned a time slot and a reserved table where the authors were available for discussing their work with the interested audience. The aim of this format is to foster discussion and facilitate idea exchange, community creation, and collaboration.

AI*IA 2016 featured exciting keynotes by Pietro Leo, Executive Architect, IBM Italy, CTO for Big Data Analytics; Giorgio Metta, Vice Scientific Director of IIT; and Dan Roth, University of Illinois at Urbana-Champaign.

The conference program also included seven workshops: the Second Italian Workshop on Artificial Intelligence for Ambient Assisted Living (AAL 2016), the Third Workshop on Artificial Intelligence and Robotics (AIRO 2016), the 10th Italian Workshop on Artificial Intelligence for Cultural Heritage (AI*CH 2016), the 5th Italian Workshop on Machine Learning and Data Mining (MLDM.it 2016), the 23rd RCRA International Workshop on Experimental Evaluation of Algorithms for Solving Problems with Combinatorial Explosion (RCRA 2016), and the First Italian Workshop on Deep Understanding and Reasoning: A Challenge for Next-Generation Intelligent Agents (URANIA 2016), plus a Doctoral Consortium.

The chairs wish to thank the Program Committee members and the reviewers for their careful work in selecting the best papers, the chairs of the workshops and of the Doctoral Consortium for organizing the corresponding events, as well as Angelo Ferrando, Frosina Koceva, and Laura Pandolfo for their help during the organization of the conference.

September 2016
Giovanni Adorni
Stefano Cagnoni
Marco Gori
Marco Maratea

Organization

AI*IA 2016 was organized by AI*IA (in Italian, Associazione Italiana per l'Intelligenza Artificiale), in cooperation with the Department of Informatics, Bioengineering, Robotics and Systems Engineering and the Polytechnic School of the University of Genoa (Italy).

Executive Committee

General Chair

Giovanni Adorni   Università di Genova, Italy

Program Chairs

Stefano Cagnoni   Università di Parma, Italy
Marco Gori   Università di Siena, Italy
Marco Maratea   Università di Genova, Italy

Doctoral Consortium Chairs

Ilaria Torre   Università di Genova, Italy
Viviana Mascardi   Università di Genova, Italy

Local Chairs

Alessio Merlo   Università di Genova, Italy
Simone Torsani   Università di Genova, Italy

Program Committee

Matteo Baldoni   Università di Torino, Italy
Stefania Bandini   Università Milano-Bicocca, Italy
Roberto Basili   Università di Roma Tor Vergata, Italy
Nicola Basilico   Università di Milano, Italy
Federico Bergenti   Università di Parma, Italy
Stefano Bistarelli   Università di Perugia, Italy
Luciana Bordoni   ENEA, Italy
Francesco Buccafurri   Università Mediterranea di Reggio Calabria, Italy
Amedeo Cappelli   CNR, Italy
Luigia Carlucci Aiello   Sapienza Università di Roma, Italy
Amedeo Cesta   CNR, Italy
Antonio Chella   Università di Palermo, Italy
Carlo Combi   Università di Verona, Italy
Gabriella Cortellessa   CNR, Italy
Stefania Costantini   Università dell'Aquila, Italy

Giuseppe De Giacomo	Sapienza Università di Roma, Italy
Nicola Di Mauro	Università di Bari, Italy
Francesco Donini	Università della Tuscia, Italy
Agostino Dovier	Università di Udine, Italy
Floriana Esposito	Università di Bari, Italy
Stefano Ferilli	Università di Bari, Italy
Salvatore Gaglio	Università di Palermo, Italy
Patrick Gallinari	University of Paris 6, France
Marco Gavanelli	Università di Ferrara, Italy
Georg Gottlob	Oxford University, UK
Nicola Guarino	CNR, Italy
Luca Iocchi	Sapienza Università di Roma, Italy
Evelina Lamma	Università di Ferrara, Italy
Nicola Leone	Università della Calabria, Italy
Chendong Li	Dell, USA
Francesca Alessandra Lisi	Università di Bari, Italy
Sara Manzoni	Università Milano-Bicocca, Italy
Paola Mello	Università di Bologna, Italy
Alessio Micheli	Università di Pisa, Italy
Alfredo Milani	Università di Perugia, Italy
Michela Milano	Università di Bologna, Italy
Stefania Montani	Università del Piemonte Orientale, Italy
Alessandro Moschitti	Università di Trento, Italy
Angelo Oddi	CNR, Italy
Andrea Omicini	Università di Bologna, Italy
Maria Teresa Pazienza	Università di Roma Tor Vergata, Italy
Alberto Pettorossi	Università di Roma Tor Vergata, Italy
Roberto Pirrone	Università di Palermo, Italy
Piero Poccianti	Consorzio Operativo Gruppo MPS, Italy
Gian Luca Pozzato	Università di Torino, Italy
Luca Pulina	Università di Sassari, Italy
Daniele P. Radicioni	Università di Torino, Italy
Francesco Ricca	Università della Calabria, Italy
Fabrizio Riguzzi	Università di Ferrara, Italy
Andrea Roli	Università di Bologna, Italy
Salvatore Ruggieri	Università di Pisa, Italy
Fabio Sartori	Università Milan-Bicocca, Italy
Ken Satoh	National Institute of Informatics and Sokendai, Japan
Andrea Schaerf	Università di Udine, Italy
Floriano Scioscia	Politecnico di Bari, Italy
Giovanni Semeraro	Università di Bari, Italy
Roberto Serra	Università di Modena e Reggio Emilia, Italy
Giovanni Squillero	Politecnico di Torino, Italy
Pietro Torasso	Università di Torino, Italy
Leonardo Vanneschi	Università Milan-Bicocca, Italy
Eloisa Vargiu	Euracat, Spain

Marco Villani   Università di Modena e Reggio Emilia, Italy
Giuseppe Vizzari   Università Milano-Bicocca, Italy

Additional Reviewers

Giovanni Amendola	Alessandra De Paola	Marco Morana
Giuliano Armano	Luca Di Gaspero	Paolo Naggar
Marco Alberti	Antonino Fiannaca	Mirco Nanni
Eneko Agirre	Claudio Gallicchio	Laura Pandolfo
Cristina Baroglio	Laura Giordano	Giovanni Pani
Ludovico Boratto	Michele Gorgoglione	Fabio Patrizi
Annalina Caputo	Hiroyuki Kido	Andrea Pazienza
Federico Capuzzimati	Yves Lesperance	Zohreh Shams
Martine Ceberio	Tingting Li	Maria Simi
Giuseppe Cota	Marco Lippi	Francesco Santini
Rosario Culmone	Giuseppe Loseto	Pierfrancesco Veltri
Bernardo Cuteri	Marco Manna	Gennaro Vessio
Berardina Nadja De Carolis	Federico Meschini	Riccardo Zese
	Marco Montali	

Sponsoring Institutions

AI*IA 2016 was partially funded by the *Artificial Intelligence* journal, by the Department of Informatics, Bioengineering, Robotics and Systems Engineering, and the Polytechnic School, by the Istituto Nazionale di Alta Matematica "F. Severi" — Gruppo Nazionale per il Calcolo Scientifico, by Camlin Italy, by the city of Genoa, and by the Italian Association for Artificial Intelligence.

Contents

Optimization and Evolutionary Algorithms

Understanding Characteristics of Evolved Instances for State-of-the-Art
Inexact TSP Solvers with Maximum Performance Difference   3
  Jakob Bossek and Heike Trautmann

On-line Dynamic Station Redeployments in Bike-Sharing Systems   13
  Carlo Manna

Optimized Word-Size Time Series Representation Method Using a Genetic
Algorithm with a Flexible Encoding Scheme   26
  Muhammad Marwan Muhammad Fuad

Efficient Search of Relevant Structures in Complex Systems   35
  Laura Sani, Michele Amoretti, Emilio Vicari, Monica Mordonini,
  Riccardo Pecori, Andrea Roli, Marco Villani, Stefano Cagnoni,
  and Roberto Serra

Classification, Pattern Recognition, and Computer Vision

Flat and Hierarchical Classifiers for Detecting Emotion in Tweets   51
  Giulio Angiani, Stefano Cagnoni, Natalia Chuzhikova, Paolo Fornacciari,
  Monica Mordonini, and Michele Tomaiuolo

Spam Filtering Using Regularized Neural Networks with Rectified
Linear Units   65
  Aliaksandr Barushka and Petr Hájek

User Mood Tracking for Opinion Analysis on Twitter   76
  Giuseppe Castellucci, Danilo Croce, Diego De Cao, and Roberto Basili

Using Random Forests for the Estimation of Multiple Users' Visual
Focus of Attention from Head Pose   89
  Silvia Rossi, Enrico Leone, and Mariacarla Staffa

Multi-agent Systems

An Analytic Study of Opinion Dynamics in Multi-agent Systems
with Additive Random Noise   105
  Stefania Monica and Federico Bergenti

Combining Avoidance and Imitation to Improve Multi-agent
Pedestrian Simulation 118
    Luca Crociani, Giuseppe Vizzari, and Stefania Bandini

Knowledge Representation and Reasoning

Educational Concept Maps for Personalized Learning Path Generation. 135
    Giovanni Adorni and Frosina Koceva

Answer Set Enumeration via Assumption Literals 149
    Mario Alviano and Carmine Dodaro

On the Application of Answer Set Programming to the Conference
Paper Assignment Problem. 164
    Giovanni Amendola, Carmine Dodaro, Nicola Leone,
    and Francesco Ricca

A Subdivision Approach to the Solution of Polynomial Constraints
over Finite Domains Using the Modified Bernstein Form. 179
    Federico Bergenti, Stefania Monica, and Gianfranco Rossi

$\mathcal{I}$-DLV: The New Intelligent Grounder of DLV 192
    Francesco Calimeri, Davide Fuscà, Simona Perri, and Jessica Zangari

Abducing Compliance of Incomplete Event Logs 208
    Federico Chesani, Riccardo De Masellis, Chiara Di Francescomarino,
    Chiara Ghidini, Paola Mello, Marco Montali, and Sergio Tessaris

Boosting the Development of ASP-Based Applications in Mobile
and General Scenarios. 223
    Francesco Calimeri, Davide Fuscà, Stefano Germano, Simona Perri,
    and Jessica Zangari

Relationships and Events: Towards a General Theory of Reification
and Truthmaking. 237
    Nicola Guarino and Giancarlo Guizzardi

A Self-Adaptive Context-Aware Group Recommender System 250
    Reza Khoshkangini, Maria Silvia Pini, and Francesca Rossi

A Model+Solver Approach to Concept Learning. 266
    Francesca Alessandra Lisi

Machine Learning

A Comparative Study of Inductive and Transductive Learning
with Feedforward Neural Networks. 283
    Monica Bianchini, Anas Belahcen, and Franco Scarselli

Structural Knowledge Extraction from Mobility Data.   294
Pietro Cottone, Salvatore Gaglio, Giuseppe Lo Re, Marco Ortolani,
and Gabriele Pergola

Predicting Process Behavior in WoMan   308
Stefano Ferilli, Floriana Esposito, Domenico Redavid,
and Sergio Angelastro

On-line Learning on Temporal Manifolds.   321
Marco Maggini and Alessandro Rossi

Learning and Reasoning with Logic Tensor Networks   334
Luciano Serafini and Artur S. d'Avila Garcez

Semantic Web and Description Logics

Probabilistic Logical Inference on the Web.   351
Marco Alberti, Giuseppe Cota, Fabrizio Riguzzi, and Riccardo Zese

Probabilistic Hybrid Knowledge Bases Under the Distribution Semantics   364
Marco Alberti, Evelina Lamma, Fabrizio Riguzzi, and Riccardo Zese

Context-Awareness for Multi-sensor Data Fusion in Smart Environments. . . .   377
Alessandra De Paola, Pierluca Ferraro, Salvatore Gaglio,
and Giuseppe Lo Re

Reasoning about Multiple Aspects in Rational Closure for DLs   392
Valentina Gliozzi

A Framework for Automatic Population of Ontology-Based
Digital Libraries   406
Laura Pandolfo, Luca Pulina, and Giovanni Adorni

Reasoning About Surprising Scenarios in Description Logics of Typicality. . .   418
Gian Luca Pozzato

Natural Language Processing

A Resource-Driven Approach for Anchoring Linguistic Resources
to Conceptual Spaces.   435
Antonio Lieto, Enrico Mensa, and Daniele P. Radicioni

Analysis of the Impact of Machine Translation Evaluation Metrics
for Semantic Textual Similarity.   450
Simone Magnolini, Ngoc Phuoc An Vo, and Octavian Popescu

QuASIt: A Cognitive Inspired Approach to Question Answering
for the Italian Language.................................... 464
   Arianna Pipitone, Giuseppe Tirone, and Roberto Pirrone

Spoken Language Understanding for Service Robotics in Italian.......... 477
   Andrea Vanzo, Danilo Croce, Giuseppe Castellucci, Roberto Basili,
   and Daniele Nardi

Planning and Scheduling

DARDIS: Distributed And Randomized DIspatching and Scheduling....... 493
   Thomas Bridi, Michele Lombardi, Andrea Bartolini, Luca Benini,
   and Michela Milano

Steps in Assessing a Timeline-Based Planner 508
   Alessandro Umbrico, Amedeo Cesta, Marta Cialdea Mayer,
   and Andrea Orlandini

Formal Verification

Learning for Verification in Embedded Systems: A Case Study 525
   Ali Khalili, Massimo Narizzano, and Armando Tacchella

Learning in Physical Domains: Mating Safety Requirements
and Costly Sampling..................................... 539
   Francesco Leofante and Armando Tacchella

Author Index 553

Optimization and Evolutionary Algorithms

Understanding Characteristics of Evolved Instances for State-of-the-Art Inexact TSP Solvers with Maximum Performance Difference

Jakob Bossek[(✉)] and Heike Trautmann

Information Systems and Statistics Group, University of Münster, Münster, Germany
{bossek,trautmann}@wi.uni-muenster.de

Abstract. State of the Art inexact solvers of the NP-hard Traveling Salesperson Problem (TSP) are known to mostly yield high-quality solutions in reasonable computation times. With the purpose of understanding different levels of instance difficulties, instances for the current State of the Art heuristic TSP solvers LKH+restart and EAX+restart are presented which are evolved using a sophisticated evolutionary algorithm. More specifically, the performance differences of the respective solvers are maximized resulting in instances which are easier to solve for one solver and much more difficult for the other. Focusing on both optimization directions, instance features are identified which characterize both types of instances and increase the understanding of solver performance differences.

Keywords: Transportation · Metaheuristics · Combinatorial optimization · TSP · Instance hardness

1 Introduction

In the Traveling Salesperson Problem (TSP) we aim to find a minimal cost roundtrip tour in an edge-weighted graph, which visits each node exactly once and returns to the starting node. A plethora of algorithmic approaches for this famous NP-hard combinatorial problem was developed in the past decades. Inexact solvers for the TSP are known to produce high-quality solutions in reasonable time compared to exact solvers such as Concorde [1]. Recently, the EAX solver [2] was shown to be competitive to the well-known State of the Art LKH algorithm [3], more specifically respective restart variants LKH+restart and EAX+restart even improve the original versions [4] on the Euclidean TSP. However, there is no single inexact solver which operates best on all possible problem instances regarding solution quality. In this work, we investigate performance differences of the current State of the Art TSP solvers on specifically evolved instances.

Efficient algorithm selection approaches [5] in this field are conducted in a feature- and instance-based fashion. TSP features, e.g. in [6–9][1], are computed on benchmark instances and related to algorithm performance allowing for

---

[1] These feature sets are available in the R-package salesperson [10].

© Springer International Publishing AG 2016
G. Adorni et al. (Eds.): AI*IA 2016, LNAI 10037, pp. 3–12, 2016.
DOI: 10.1007/978-3-319-49130-1_1

constructing algorithm selection models for unseen instances based on machine learning techniques.

Understanding which instance characteristics pose a specific level of difficulty onto high-performing TSP solvers is an active research field, see e.g. [4,11]. In this paper, we specifically address LKH+restart compared to EAX+restart as the two current State of the Art TSP solvers with potential for improving their standalone application by means of a portfolio approach [4]. We are specifically interested in instances on which both solvers exhibit maximum performance difference, i.e., which are much harder to solve for one of the solvers, while we focus both directions. Thus, the performance ratio is used as fitness function of a sophisticated evolutionary algorithm for evolving instances which was already used in a similar fashion for single solvers in [7,8,12,13]. Two variants of solver performance are contrasted. The classical mean par10 score is supplemented by focussing on the median solver runtime over a fixed number of runs diminishing the influence of timeouts in individual runs. Moreover, the influence of rounding instances to a grid structure is analysed systematically. Additionally, we contrast characteristics of instances which are much harder or much easier for one solver w.r.t. the other.

Section 2 details the evolutionary algorithm. Experimental results are then presented in Sect. 3. Conclusions and an outlook on future research are given in Sect. 4.

2   EA for Evolving Instances

Algorithm 1 reflects the process of the evolutionary algorithm in terms of pseudocode. The core parameter of the EA is the kind of fitness function used. As the EA aims at generating instances with maximum performance difference of two solvers, we define the fitness function as the performance ratio $P_{(A,B)}(I)$ for a pair of solvers A and B, i.e.

$$P_{(A,B)}(I) = \frac{P_A(I)}{P_B(I)}$$

on a specific instance I, where $P_A(I)$ and $P_B(I)$ are the solver performances of solver A and B on instance I. Solver performance in our scenario is either determined by the standard indicator *penalized average runtime* or by the *penalized median runtime*. The former repeatedly measures the runtime of the solver on an instance until the optimal tour (pre-computed by Concorde) has been found and computes the arithmetic mean subsequently. In case the cutoff time *timeLimit* is reached, ten times the cutoff time is used for further computations as a penalty. However, inside the EA, the actual cutoff time is used ensuring that the probability of removal of such a solution at later algorithm stages is not unreasonably low. The evaluation at the final generation uses the classical par10 score with the full penalty. The median score is much more insensitive to outliers and maximum ratio in medians is much harder to obtain.

Algorithm 1. Evolving EA

1: **function** EA(fitnessFun, popSize, instSize, generations, timeLimit, cells, rnd=true)
2:     poolSize = ⌊ popSize / 2 ⌋
3:     population = GENERATERANDOMINSTANCES(popSize, instSize)   ▷ in $[0,1]^2$
4:     **while** stopping condition not met **do**
5:         **for** $i = 1 \rightarrow$ popSize **do**
6:             fitness[i] = FITNESSFUN(population[i])
7:         **end for**
8:         matingPool = CREATEMATINGPOOL
9:   ▷ 2-tournament-selection
10:         offspring[1] = GETBESTFROM
11:             CURRENTPOPULATION   ▷ 1-elitism
12:         **for** $i = 2 \rightarrow$ popSize **do**
13:             Choose $p1$ and $p2$ randomly from the
14:             mating pool
15:             offspring[i] = APPLYVARIATIONS($p1$, $p2$)
16:             Rescale offsspring to $[0,1]^2$ by dimension
17:             **if** rnd **then**
18:                 Round each point to nearest cell grid
19:             **end if**
20:         **end for**
21:         population = offspring
22:     **end while**
23: **end function**

The initial population of size *popSize* is randomly generated in $[0,1]^2$ for instances of size *instSize* and the performance ratio is computed. Distances are scaled by multiplying with a factor of 100 and afterwards rounded to the nearest integer. This step is neccassary since EAX expects integer distances. The EA is then run for a fixed number of *generations* and the evolutionary loop is executed as follows: The mating pool is formed by 2-tournament selection supplemented by the best solution of the current population (1-elitism). Two different mutation operators are applied to each combination of randomly drawn instance pairs of the mating pool. Uniform mutation replacing coordinates of selected nodes with new randomly chosen coordinates is applied with a very low probability possibly followed by gaussian mutation adding normal noise to the selected point coordinates. Therefore, global as well as local changes can come into effect. In the current version the EA does not use any recombination operator. A single EA generation ends after rescaling the instance to completely cover $[0,1]^2$ and, if *rnd* = *true*, rounding the points to the nearest cell grid. The latter relates to important relevant structures in practice such as the structural design of circuit boards.

3   Experiments

3.1   Experimental Setup

In total 200 TSP instances were evolved. For all four considered optimization directions, i.e. $P_{(LKH,EAX)}$, $P_{(EAX,LKH)}$, $P_{(LKH+restart,EAX+restart)}$ and $P_{(EAX+restart,LKH+restart)}$, in each case 25 instances were generated with activated and deactivated rounding respectively. Based on preliminary experiments and experimental results of [8,12] the EA parameters were set as follows: $timeLimit = 120$, $popSize = 30$, $generations = 5000$, $uniformMutationRate = 0.05$, $normalMutationRate = 0.1$, $normalMutationSD = 0.025$ and $cells = 100$. We used the reference implementation LKH 2.0.7 based on the former implementation 1.3 [14], the original EAX implementation as well as specific restart variants as described in [4]. The solvers were repeatedly evaluated, three times inside the EA due to a limited computational budget but ten times for final evaluations. As described in Sect. 2 either the par10 score or the median score was computed for the final instances.

For comparison and practical validation, performance ratios of the respective solvers on TSPLIB instances[2] of comparable size, i.e. $200 \leq instSize \leq 400$ were computed for both kinds of performance measures. Moreover, 100 random instances in $[0,1]^2$ were generated while the same rounding strategy of the distance matrix was applied as used inside the EA for the evolved instances. All experiments were run on the parallel linux computer cluster PALMA at University of Münster, consisting of 3528 processor cores in total. The utilized compute nodes are 2,6 GHz machines with 2 hexacore Intel Westmere processors, totally 12 cores per node and 2 GB main memory per core.

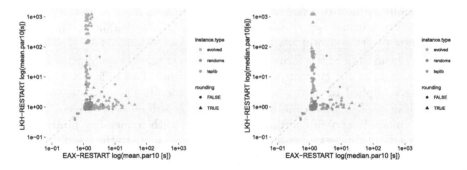

Fig. 1. Average (left) and median (right) par10 scores (log-scale) of LKH+restart and EAX+restart on evolved, random and TSPLIB instances. A specific symbol visualizes whether instances were rounded to a grid structure (rnd) or not (nrnd).

---

[2] TSPLIB-Instances: a280, gil262, kroA200, kroB200, lin318, pr226, pr264, pr299, rd400, ts225, tsp225.

3.2 Results

Figures 1 and 2 give an overview about the performance scores of the considered solver pairs, i.e., both for the original as well as the restart variants. Evolved instances are visualized together with random and TSPLIB instances.

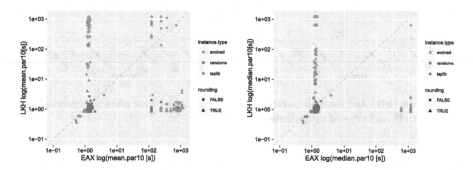

Fig. 2. Average (left) and median (right) par10 scores (log-scale) of LKH and EAX on evolved, random and TSPLIB instances. A specific symbol visualizes whether instances were rounded to a grid structure (rnd) or not (nrnd).

It becomes obvious that in both pairings the presented EA successfully generated instances with much higher performance differences of both solvers than usually present in random as well as TSPLIB instances. Whether the instance was rounded to a grid structure inside the EA does not have a structural influence on the relation of the performance scores. Moreover, we see that generating easier instances for LKH+restart compared to the EAX+restart is a much harder task than in case of considering the opposite direction. Specifically, EAX+restart timeouts did not occur here. On the contrary, for variants without restart, this effect cannot be observed. In addition, in some cases, the EA did not converge in that instances are of similar difficulty for both solvers. This behaviour, however, is due to a part of solver runs resulting in timeouts as reflected by the location of the points in case the median scores are considered (Fig. 2, lower right part). In general, evaluating with median scores diminishes the influence of timeouts. Maximum median scores can only be obtained in case at least fifty percent of solver runs on a specific instance result in a timeout. Therefore, there could be potential of using this kind of performance measure inside the EA. Results are presented further down.

The previous observations are reflected in Fig. 3 as well. Here, boxplots of performance scores on the evolved instances for each solver depending on the optimization direction as well as the rounding activations are given. Supplementary to Figs. 1 and 2 *all* solvers have been evaluated on the respective instances. Not surprisingly, instances specifically generated for the restart variants do not result in such extreme performance differences for the classical variants and the other way round. However, the basic tendency can be observed here as well.

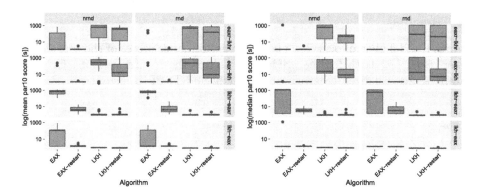

Fig. 3. Mean (left) and median par10 scores (right) of the four solvers depending on rounding (rnd) and type of optimization (log-scale).

Evaluating with median scores shows that especially the pairing (LKH,EAX), i.e. generating easier instances for LKH, does not show the desired performance effects. Figure 4 explicitly provides boxplots of the performance ratios $P_{(A,B)}$ and thus summarizes all effects previously listed, in particular the huge differences in performance ratios compared to random and TSPLIB instances.

Understanding Characteristics of Extreme Instances. Next we try to understand which structural TSP characteristics, termed *TSP instance features*, of the evolved instances are suitable to distinguish between easy and hard to solve instances respectively. Those features could be used in algorithm selection scenarios to select the best solver out of a portfolio of solvers. A classification approach was used in order to separate the respective groups of the most extreme instances by means of well-known established TSP features as introduced in [6,8]. The R-package salesperson [10] was used for this purpose. However, we did not consider expensive features such as local search, clustering distance based features as well as branch and cut techniques in order to avoid much computational overhead, especially regarding the possible influence on future algorithm selection models as e.g. in [4].

In case of the restart variants, the ten most extreme instances w.r.t. performance ratios on mean par10 scores are selected, i.e. the ten best EA results of both optimization directions for the solver pairings. A random forest was used to distinguish between both instance sets combined with a feature selection approach based on leave-one-out cross-validation and an outer resampling resulting in a median misclassification error of zero and a respective mean misclassification error of 0.3 regarding all folds. Thus, we are able to separate both instance sets in a satisfying way including an indication which features are of crucial importance here. Figure 5 shows the respective feature importance plot created by means of the R-package flacco [15]. Red dots reflect features which are used in at least 70 % of the folds, orange labeled features at least in 30 % and black ones at least once. In this regard the median distance of the minimum spanning tree is

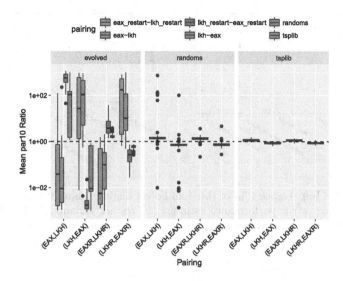

Fig. 4. Performance ratios based on mean par10 scores of the four considered algorithms on all considered instance sets.

identified as the crucial feature separating both instance classes. The results conicide with the results of [8] where the mean distance of the minimum spanning tree was identified as a separating feature between easy and hard instances for TSP approximation algorithms. This result is promising with respect to future work in the algorithm selection context: The computation of minimum spanning tree characteristics is an computationally cheap task and we strive for cheap feature, since wasting a lot of runtime for the feature computation before actually solving the TSP itself is senseless.

The same analysis was conducted for the original solver variants. However, as evolved instances are much denser in the lower right and upper left corner in Fig. 2 than in the restart case, we only selected the respective five most extreme instances. In this case different features play a key role in explaining solver performance differences including nearest-neighbor based features as visualized in Fig. 5. Again, the median misclassification error vanishes while the mean misclassification error is 0.2, i.e. only two out of the ten instances are misclassified.

Median Scores as EA Internal Performance Measures. In order to investigate possible potential of using the median par10 score inside the EA for performance evaluation, we conducted a smaller experiment focussed on the restart variants with inactive rounding as these solvers in our view are most interesting. This lead to fifty evolved instances, i.e. 25 for each optimization direction.

Figure 6 gives an overview about the resulting median par10 scores on the newly evolved instances together with the random instance and TSPLIB results. We see that the EA is not successful in improving the performance ratio of both

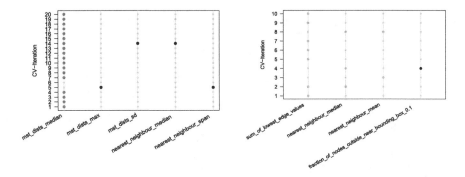

Fig. 5. Variable Importance Plot of Random Forest distinguishing the (left): ten most extreme instances w.r.t. performance ratio for the restart variants, (right): five most extreme instances w.r.t. performance ratio for the original algorithm variants.

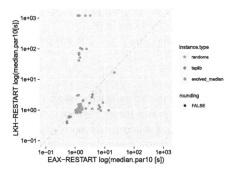

Fig. 6. Median par10 scores (log-scale) of LKH+restart and EAX+restart on evolved (median score fitness), random and TSPLIB instances.

solvers even further compared to the resulting median evaluation on the instances originally generated inside the EA using mean par10 scores. The same is true for comparing the mean par10 scores on both scenarios (see Fig. 7). However, slight improvements are visible in case easier instances for LKH+restart are evolved. Most probably the median alone does not provide enough differentiation between varying solver results over the repetitions.

However, in our view an adequate combination of mean and median scores inside the EA fitness function is promising in order to get deeper insights into solver variance on the considered instances. We will investigate this issue in further studies together with increasing the number of solver repetitions along the evolutionary loop.

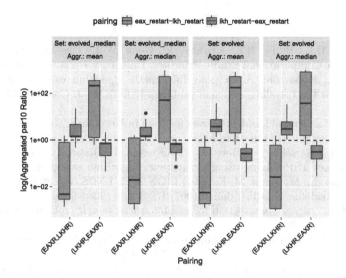

Fig. 7. Comparison of mean and median par10 ratios (log-scale) of instance sets evolved for LKH+restart and EAX+restart either by using the mean par10 or the median score inside the EA as fitness function.

4   Conclusions

This work focusses on the two current State of the Art inexact TSP solvers LKH and EAX together with their respective restart variants. In order to increase understanding of performance differences of both solvers, a sophisticated evolutionary algorithm was used to evolve instances which lead to maximum performance difference of both solvers on the specific instances. Both directions are analyzed, i.e. we generated instances which are easier for solver A but much harder for solver B as well as the opposite case. In this regard we observed substantial differences in solver performance ratios compared to random or TSPLIB instances on the evolved instances. By feature-based analysis of the most extreme instances in terms of performance ratio crucial features are identified for both solver pairings which are indicated to have an influence on solver-specific problem difficulty. Moreover, we contrasted the classical mean par10 score with a respective median version to even increase the challenge of evolving instances with high solver performance differences.

Future studies will focus on generalizing the results to higher instance sizes and on designing a more sophisticated fitness function inside the EA to even increase solver performance differences on the evolved instances.

Acknowledgments. The authors acknowledge support by the European Research Center for Information Systems (ERCIS).

References

1. Applegate, D.L., Bixby, R.E., Chvatal, V., Cook, W.J.: The Traveling Salesman Problem: A Computational Study. Princeton University Press, Princeton (2007)
2. Nagata, Y., Kobayashi, S.: A powerful genetic algorithm using edge assembly crossover for the traveling salesman problem. INFORMS J. Comput. **25**, 346–363 (2013)
3. Helsgaun, K.: General k-opt submoves for the Lin-Kernighan TSP heuristic. Math. Program. Comput. **1**, 119–163 (2009)
4. Kotthoff, L., Kerschke, P., Hoos, H., Trautmann, H.: Improving the state of the art in inexact TSP solving using per-instance algorithm selection. In: Dhaenens, C., Jourdan, L., Marmion, M.-E. (eds.) LION 2015. LNCS, vol. 8994, pp. 202–217. Springer, Heidelberg (2015). doi:10.1007/978-3-319-19084-6_18
5. Kotthoff, L.: Algorithm selection for combinatorial search problems: a survey. AI Mag. **35**, 48–60 (2014)
6. Hutter, F., Xu, L., Hoos, H.H., Leyton-Brown, K.: Algorithm runtime prediction: methods & evaluation. Artif. Intell. **206**, 79–111 (2014)
7. Smith-Miles, K., van Hemert, J.: Discovering the suitability of optimisation algorithms by learning from evolved instances. Ann. Math. Artif. Intell. **61**, 87–104 (2011)
8. Mersmann, O., Bischl, B., Trautmann, H., Wagner, M., Bossek, J., Neumann, F.: A novel feature-based approach to characterize algorithm performance for the traveling salesperson problem. Ann. Math. Artif. Intell. **69**, 1–32 (2013)
9. Pihera, J., Musliu, N.: Application of machine learning to algorithm selection for TSP. In: Fogel, D., et al. (eds.) Proceedings of the IEEE 26th International Conference on Tools with Artificial Intelligence (ICTAI). IEEE press (2014)
10. Bossek, J.: salesperson: Computation of Instance Feature Sets and R Interface to the State-of-the-Art Solvers for the Traveling Salesperson Problem. R package version 1.0 (2015)
11. Fischer, T., Stützle, T., Hoos, H.H., Merz, P.: An analysis of the hardness of TSP instances for two high-performance algorithms. In: Proceedings of the 6th Metaheuristics International Conference, Vienna, Austria, pp. 361–367 (2005)
12. Mersmann, O., Bischl, B., Bossek, J., Trautmann, H., Markus, W., Neumann, F.: Local search and the traveling salesman problem: a feature-based characterization of problem hardness. In: Hamadi, Y., Schoenauer, M. (eds.) LION 6. LNCS, vol. 7219, pp. 115–129. Springer, Heidelberg (2012)
13. Nallaperuma, S., Wagner, M., Neumann, F., Bischl, B., Mersmann, O., Trautmann, H.: A feature-based comparison of local search and the christofides algorithm for the travelling salesperson problem. In: Foundations of Genetic Algorithms (FOGA) (2013) (accepted)
14. Lacoste, J.D., Hoos, H.H., Stützle, T.: On the empirical time complexity of state-of-the-art inexact tsp solvers. Optimization Letters (to appear)
15. Kerschke, P., Dagefoerde, J.: flacco: Feature-Based Landscape Analysis of Continuous and Constraint Optimization Problems. R package version 1.1 (2015)

On-line Dynamic Station Redeployments in Bike-Sharing Systems

Carlo Manna[(✉)]

Insight Research Centre for Data Analytics, University College Cork, Cork, Ireland
carlo.manna@insight-centre.org

Abstract. Bike-sharing has seen great development during recent years, both in Europe and globally. However, these systems are far from perfect. The uncertainty of the customer demand often leads to an unbalanced distribution of bicycles over the time and space (congestion and/or starvation), resulting both in a loss of customers and a poor customer experience. In order to improve those aspects, we propose a dynamic bike-sharing system, which combines the standard fixed base stations with movable stations (using trucks), which will able to be dynamically re-allocated according to the upcoming forecasted customer demand during the day in real-time. The purpose of this paper is to investigate whether using moveable stations in designing the bike-sharing system has a significant positive effect on the system performance. To that end, we contribute an on-line stochastic optimization formulation to address the redeployment of the moveable stations during the day, to better match the upcoming customer demand. Finally, we demonstrate the utility of our approach with numerical experiments using data provided by bike-sharing companies.

Keywords: On-line combinatorial optimization · Uncertainty · Smart cities

1 Introduction

Bike-sharing systems (BSS) are in place in several cities in the world, and are an increasingly important support for multi-modal transport systems [1]. BSS are widely adopted with 747 active systems, a fleet of over 772,000 bicycles and 235 systems in planning or under construction [2]. A BSS typically has a number of base stations scattered throughout a city. At the beginning of the day, each station is stocked with a pre-determined number of bikes. Users with a membership card can pick up and return bikes from any designated station, each of which has a finite number of docks. At the end of the work day, trucks are used to move bikes around so as to return to some pre-determined configuration at the beginning of the day. Due to the individual movement of customers according to their needs, there is often congestion (more than required) or starvation (fewer than required) of bikes at certain base stations. According to CapitalBikeShare Company [3], in a city like Washington, at a minimum, there are around 100

© Springer International Publishing AG 2016
G. Adorni et al. (Eds.): AI*IA 2016, LNAI 10037, pp. 13–25, 2016.
DOI: 10.1007/978-3-319-49130-1_2

cases of empty stations and 100 cases of full stations per day and at a maximum there are about 750 cases of empty stations and 330 cases of full stations per day. As demonstrated in [4] this can result in a significant loss of customer activity. Such loss in demand can have two undesirable outcomes: (a) loss in revenue; (b) increase in carbon emissions, as people resort to less environmentally-friendly modes of transport. To address such a problem, the majority of the proposed solutions aim to find more efficient methods for dynamic rebalancing of the number of bicycles on the base stations taking into account the uncertainty of the customer demand or predicting the customer demand at each station [5,6]. Basically, by using trucks a number of bicycles are transferred from one station to another to accomplish the upcoming demand. This operation takes place ones a day (or more in some specific situations).

The aim of this paper is to study a totally different approach, in which the fixed base stations are augmented by a number of movable stations (using trucks) with the dual purpose of both: (1) dynamically adding/re-allocating dock stations in city areas to match customer demand in real-time and, (2) a dynamic re-balancing of the number of bicycles in some particular fixed station where a redeployment of a docking station is unnecessary.

Particularly, we consider a problem in which the day time is partitioned in time intervals of equal length. We suppose that for each of those time periods, the probability distributions of the travel demands between different locations are known. At start of the day, we compute the best possible locations for the dock stations in each time period, taking into account the stochastic nature of the demand, with the aim to maximize the number of customers. This is an on-line problem. That means, although the solution is a sequence of decisions for each time period, only the immediate decision is actually taken (i.e. the station allocations for the incoming time period). While this time period, a new computation is performed, using more updated travel demand predictions, and a new decision for the next period is taken from the re-computed decision sequence. This carry on until the termination of the time horizon.

The main advantage of such a system is that the bike stations configuration is not fixed, but it can change adaptively with the travel demand day by day and, in each day, it can change during a number of time periods, in the respect of specific time constraints (i.e. a new configuration must be computed in advance with enough time to allow the repositioning of the stations). Finally, updating the decisions each time periods, allow reducing the uncertainty in the predictions. This because the majority of the travel demand prediction techniques are based on auto-regressive models, which make use of the most recent known data to predict future outcomes [7].

The key distinction from existing research on bike sharing is that we consider the dynamic redeployment of bicycle stations (instead of just rebalancing the number of bicycles in the existing stations). This approach from one side does not exclude the possibility to rebalancing as in the existing research. However this extends such a research to a novel approach to the BSS by which is possible to dynamically change the configuration of the dock stations in real-time to maximize the potential customer demand. In doing this, while numerous models exist in literature, in this paper the potential customer demand is based on a primary concept: the *distance decay*. That is "the attenuation of a pattern or a process with distance" [8]. In other words, people are less willing to use a facility allocated too far from them. This is a focal concept in a variety of modelling contexts, such as transporation, migration and location theory [9].

Specifically, our key contributions are as follows:

- A mixed and linear programming (MILP) formulation to maximize the expected demand assigned to the moveable bicycle stations while simultaneously address the distance decay;
- An on-line stochastic optimization formulation to address the dock stations re-allocation problem during the day to accomplish the estimated customer demand in real-time;
- A potentially novel approach to the design of a bike-sharing system, along with numerical results for an initial investigation.

Extensive numerical simulations using datasets of two bike-sharing companies, namely Capital Bikeshare (Washington, DC) [10] and Hubway (Boston, MA) [11] show that the proposed approach can improve the customer usage of the bike-sharing system and that the computation time is reasonably fast to be used in real-time.

2   Related Work

Although bike sharing systems are relatively new, they have been studied extensively in the literature. For this reason, we only focus on threads of research that are of relevance to this paper. However, on the best of our knowledge, there is no any previous work for on-line stochastic redeployment of moveable stations in bike-sharing.

The first thread of research focus on the bicycles rebalancing between the stations. Particularly, [12–14] focus on the problem of finding routes at the end of the day for a set of carriers to achieve the desired configuration of bikes across the base stations. They have provided scalable exact and approximate approaches to this problem by either abstracting base stations into mega stations or by employing insights from inventory management or by using variable neighbourhood search based heuristics. Those works assume there is only one fixed redeployment of bikes that happens at the end of the day. In contrast, [15] predict the stochastic demand from user trip data of Singapore metro system using poisson distribution and provide an optimization model that suggests the best location of the stations and a dynamic bicycles redeployment for the model to minimize the number of unsatisfied customers. However, they assume that redeployment of bikes from one station to another is always possible without considering the routing of carriers, which is a major cost driver for the bike-sharing company. In [16] they overcome this problem, developing a mixed integer programming formulation which includes the carrier routes into the optimization model. Finally, other relevant works have been proposed in [17] to deal with unmet demand in rush hours. They provide a myopic redeployment policy by considering the current demand. They employed Dantzig-Wolfe and Benders decomposition techniques to make the decision problem faster. [18] also provides a myopic online decisions based on assessment of demand for the next 30 min.

The second thread of research is complementary to the work presented in this paper is on demand prediction and analysis. [19] provides a service level analysis of the BSS using a dualbounded joint-chance constraints where they predict the near future demands for a short period of time. Finally, in [20], the BSS is represented as a dual markovian waiting system to predict the actual demand.

As we already highlighted, all the aforementioned works differ from the one proposed in this paper as we consider a dynamic re-allocation of a certain number of bicycle stations during the day. This lead to a formulation of the problem which is different from all the previous provided in literature.

3   Problem Description

In this section we formally describe the bike-sharing system with dynamic redeployment. It is compactly described using the following tuple: $\langle T, A, K, \mathbf{S}, \mathbf{P}, \mathbf{D}, \mathbf{X}, \hat{\mathbf{X}}, \delta, l \rangle$, where A represents the set of areas for which the demand has to be covered, K represents the set of possible locations for the dock stations with $K \subseteq A$, T is the time horizon. $\mathbf{S}$ is a binary vector representing the totality of decisions on the allocated stations, with S_k^t denotes the decision on whether or not enabling a dock station in $k \in K$ at time $t \in T$. Furthermore, $\mathbf{P}$ is a binary vector representing the distribution of the service coverage between areas and dock stations. In particular, $P_{a,k}^t$ denotes whether or not the area $a \in A$ is served by the station in $k \in K$, at time $t \in T$. $\mathbf{D}$ is a vector of the distances (or travel time) between all areas and station locations, with $d_{a,a'}$ $(d_{a,k})$, the distance/travel time between the area a and a' (between the area a and the station in k). $\mathbf{X}$ is a vector of the probability distribution of the potential customer demand between different areas, with $x_{a,a'}^t$ denoting the potential travel demand from the area a to the area a' (with $(a, a') \in A$), at time $t \in T$. The potential customer demand at time t denotes the maximum number of possible customers to be served. This is expressed in form of probability distribution. Similarly, $\hat{\mathbf{X}}$ is a vector of the *expected* covered customer demand between different areas, with $\hat{x}_{a,a'}^t$ denoting the expected travel demand from the area a to the area a' (with $(a, a') \in A$), at time $t \in T$. Finally $\delta \in [0, 1]$ is a distance decay parameter [8], through which we take into account the aforementioned *distance decay* concept. Finally l is the maximum number of moveable stations. The expected covered customer demand denotes the number of customers which is expected to use the bike-sharing system. Hence, given a potential demand $x_{a,a'}^t$, it holds $\hat{x}_{a,a'}^t \leqslant x_{a,a'}^t$ for each $a \in A$ and $t \in T$. $\hat{\mathbf{X}}$ depends on various factors, such as the total distance/travel time between the customer location and the dock station, and from the distance decay parameter δ.

Given the potential customer demand $\mathbf{X}$ instantiated from a known probability distribution, at each time step $t \in T$, the goal is to maximize a profit function $W(\cdot)$ of the overall expected covered demand $\hat{\mathbf{X}}$ over the total time horizon $\mathbf{T}$, subject on some problem specific constraints $C(\cdot)$ (both are specified in the following paragraph). This is achieved through finding the best possible sequence of decisions $\langle S^1, \ldots, S^T \rangle$, concern the locations of the dock stations over the entire time horizon $\mathbf{T}$.

Finally, as reported in the relevant literature [21], in defining the solution approach, we have made the following assumptions:

1. the expected covered demand decrease with the total distance travelled on foot (i.e. the distance between the starting point and the pick up station, and the distance between the return station and the arrival point), given a predefined decay parameter δ.
2. each area can be served by only one station.

The assumption (1) states that the distance travelled on bicycle between the stations does not have any negative impact on the demand. This because the potential demand $\mathbf{X}$ is already estimated considering this factor. Conversely, the distance travelled on foot does have a negative impact on the service usage. The assumption (2) states that the totality of the demand in an area $a \in A$, considering a either starting or arrival point, can be served by only one station located at $k \in K$. This hypothesis is one of the most used in many facility location models, which assumes that the customer always use the closest facility.

4   Solution Approach

We propose an on-line stochastic combinatorial optimization approach for the system described in the last section. Consequently, we first outline the solution approach for the deterministic case (in which we consider no uncertainty into the customer demand) and then, we remove such hypothesis and extend this formulation to the on-line stochastic case.

4.1   Deterministic Case

The deterministic model is based on a single scenario. That means considering $\mathbf{X}$ as a single per-determined scenario values for the potential customer demand between different areas. The solution is expressed as a sequence of decisions to be taken at each time step. We do not need to recompute the solution at each time step, since the demand remain the same in each period. Hence the problem does not have any on-line feature.

To address the deterministic case we propose a MILP formulation with the following decision variables:

$$S_k^t = \begin{cases} 1 \text{ if a station is located in } k \text{ at time } t \\ 0 \text{ otherwise} \end{cases} \tag{1}$$

$$P_{ak}^t = \begin{cases} 1 \text{ if the area } a \text{ is served by station in } k \text{ at time } t \\ 0 \text{ otherwise} \end{cases} \tag{2}$$

The proposed MILP formulation is defined as follow:

$$\max \sum_{a,t} \hat{x}_{a,a'}^t \tag{3}$$

s.t.:

$$\hat{x}_{a,a'}^t = x_{a,a'}^t \left[1 - \delta \left(\sum_k d_{a,k} P_{a,k}^t + \sum_k d_{a',k'} P_{a',k'}^t \right) \right] \quad \forall\, a,t \tag{4}$$

$$\sum_k P_{a,k}^t = 1 \qquad\qquad \forall\, a,t \tag{5}$$

$$\hat{x}_{a,a'}^t \geqslant 0 \qquad\qquad \forall\, a,t \tag{6}$$

$$P_{a,k}^t - S_k^t \leqslant 0 \qquad\qquad \forall\, a,k,t \tag{7}$$

$$2 \leqslant \sum_k S_k^t \leqslant l \qquad\qquad \forall\, t \tag{8}$$

$$S_k^t \in \{0,1\} \qquad\qquad \forall\, k,t \tag{9}$$

$$P_{a,k}^t \in \{0,1\} \qquad\qquad \forall\, a,k,t \tag{10}$$

The objective (3) maximizes the total expected covered demand over the entire time horizon. Constraint (4) specifies the interaction between the expected covered demand with the potential demand and the station locations. The constraint (4) explicit the condition 1 reported in the previous section, which assumes the expected covered demand decreases linearly with the total distance travelled on foot. Further, beyond a certain distance (which scales with δ), the expected covered demand becomes 0. This according to a primary concept in the central place theory, namely *coverage range*, which denotes the maximum distance (or travel time) a user is willing to overcome to utilize a service facility

[21]. Constraint (5) specifies that the demand may only be assigned to one station at each time step, as mentioned in the condition 2 in the previous section. Constraint (6) ensures the expected covered demand be a non negative value. Constraint (7) limits assignment to open stations only, while constraint (8) specifies that the number of open stations, at each time step, must be in a predefined range (from a minimum of 2 up to l). Finally, constraints (9) and (10) are integrality conditions.

Among many possible models describing the relationships between expected covered demand and distance, we have chosen the one reported in (4) as the linear nature of such relation allows us to propose a linear formulation for the optimization problem.

4.2   On Line Stochastic Case

In this section we remove the deterministic hypothesis and introduce the on-line stochastic optimization model. We use a similar approach as in [22], with some more additional ideas to address the peculiarity of the BSS system. The general approach is to evaluate many scenarios draw from the probability distribution $\mathbf{X}$ of the potential customer demand. Then, each of those scenarios represents a deterministic problem as the one detailed in the previous section. Hence, we solve each scenario deterministically, using the MILP formulation reported in (1)–(10). Each solution represents the best dock stations configuration to maximize the (3), over the entire time horizon for each scenario. Once we have the best solution for each scenario, we need to combine all these solutions in order to find the best decision over all scenarios. In other words, we need to choose only one solution among the totality of the all computed solutions.

To that end, we propose the following heuristic: let $\mathbf{S}_i$ be the solution for the scenario i with $i = 1 \ldots, N$, over the entire time horizon, with S_i^t the dock station configuration at time t, and $S_i^t \subseteq \mathbf{S}_i$ for each i and $t = 1 \ldots, T$. Finally, s_{ik}^t denotes the decision on whether or not enabling a dock station in $k \in K$ at time $t \in T$ for the scenario i. s_{ik}^t represents a single element of the station configuration S_i^t. We define $V(t, k)$ a matrix of the location $k \in K$ and the time t, with t rows and k columns. The value of each cell $v(t, k)$ is a non negative integer.

We initialize each element of $v(t, k) = 0$. During the deterministic optimization step, for each time that is $s_{ik}^t = 1$, we increment the corresponding element $v(t, k)$ such that $v(t, k) = v(t, k) + 1$

Put in other words, we select the time t at which a location k has been chosen as dock station, and included in a deterministic solution $\mathbf{S}_i$ for the scenario i and increment those elements $v(t, k)$ by 1. We continue to update $V(t, k)$ after optimizing each sampled scenario i.

At the end of the sampling/optimization step, the final value of the all elements $v(t, k)$ denote the total number of times those choices have been selected by the totality of the solutions.

Then, we can evaluate each deterministic solution $\mathbf{S}_i$ individually by means of $V(t, k)$.

Algorithm 1. On line stochastic optimization procedure

1: **for** t from 1 to T **do**
2:     $0 \leftarrow V(k,t)$
3:     **for** i from 1 to N **do**
4:         $\omega_i \leftarrow$ **sample**$(\mathbf{X})$
5:         $S_i \leftarrow$ **solve**(ω_i)
6:         store S_i in Γ_t
7:         **for all** (t,k): $s_{ik}^t = 1$ **do**
8:             $v(t,k) \leftarrow v(t,k) + 1$
9:     $S_{best} \leftarrow \operatorname{argmax}_{S_i} \sum_{(t,k):s_{ik}^t=1} v(t,k)$

We choose as best solution $\mathbf{S}_{best}$ among all the plans $\mathbf{S}_i$ as follow:

$$S_{best} = \operatorname*{argmax}_{S_i} \sum_{(t,k):s_{ik}^t=1} v(t,k) \tag{11}$$

The on-line stochastic optimization procedure is reported in the box Algorithm 1. At each time step $t \in H$ (line 1), we initialize the matrix $V(t,k)$ (line 2). Then, we sample N different scenarios ω_i from the probability distribution $\mathbf{X}$ (line 4) and solve each of those deterministically using the MILP formulation (1)–(10) (line 5) finding the solution $\mathbf{S}_i$ and finally storing it in Γ_t. At the end of each deterministic optimization procedure, we update each element $v(t,k)$ of the matrix $V(t,k)$ according to the decisions included in the solution $\mathbf{S}_i$ (lines 7-8). Finally (line 9), once each deterministic scenario has been solved, we choose the best solution S_{best} according to the Eq. (11).

Notice that, in the spirit of the proposed on-line formulation, at each time step t, only the decisions $S_{best}^{t+1} \subseteq \mathbf{S}_{best}$ for the immediate next time step $t+1$ is taken.

5   Numerical Experiments

We evaluate our approach with respect to run-time and demand growth for the BSS on synthetic data based on real world data sets. In particular, for each generated instance, we have compared a BSS schema with fixed dock stations with the one with moveable stations.

5.1   Data Preparation

For the numerical experiments we have used data sets provided by bike-sharing companies [10,11]. These data sets contain numerous attributes. For the purpose of this paper, among all the attributes provided in the data, we used the followings: (1) Customer trip records and (2) Geographical locations of base stations.

In order to generate different instances, we assumed the demand follows a poisson distribution as in the model provided by [15]. We divided each day in 4 time slots (from 8 : 00 up to 24 : 00). For each time slot, we learn the parameter λ that governs the poisson distribution from real data. Then, we assumed such probability distribution as the demand $\mathbf{X}$ for each instance. Notice that we also learn a single parameter λ for the entire day and assumed this as our demand $\mathbf{X}$ for the fixed dock stations case.

5.2   Experimental Setting

We generated 3 different instances, and for each of those we have considered different case studies. Each case study consider a different combination of the following parameters: *(i)* number of areas $\mathbf{A}$ for which the demand has to be covered, *(ii)* number of total dock-stations (considering the sum of the fixed and the moveable stations) and *(iii)* the maximum number of moveable stations l. Basically, in each case study we consider a scenario with a fixed number of settled dock-stations and a number of moveable stations during the entire time horizon. In this, the location for the settled dock-stations were chosen randomly among the station locations provided by the solution found running the deterministic optimization problem (1)–(10) for the corresponding scenario.

We start from a smaller case in which we have 12 areas and a total of 6 stations, up to the biggest case of 40 areas and 20 total dock stations.

We have computed the fixed dock stations case using the deterministic model (1–10), and the Algorithm 1 for the moveable stations case. In this case the time horizon has 4 time slots of 4 h each (from 8 : 00 up to 24 : 00). Notice that, as the stations are moveable in each time slot, these stations are not available to the customers while moving. Hence, we have considered such stations only available in the first 2 h of each time slot, while moving (and not available) in the remain time. For this reason, to make the simulation as realistic as possible, the computation time never has to exceed the 2 h limit for each run. Hence, for any case study the number of samples N in the Algorithm 1 was 1000, except for the biggest case (40 areas and 20 total dock stations), where to limit the run time we set $N = 500$.

Finally, throughout the experiments the distance decay parameter δ is 0.1, following a general case reported in similar works related to facility location planning [21].

5.3   Results

The algorithms were written using ILOG CPLEX Optimization Studio V12.6 incorporated within C++ code on a 2.70 GHz Intel Core i7 machine with 8 GB RAM on a Windows 7 operating system.

The results for each case study are reported throughout the Tables 1, 2, 3, 4, and 5. We can see that for each instance and for each case study, using a dynamic re-deployment of the dock-stations bring to a significant increasing in term of customer demand. Although the results seem do not show some general rule, we can notice that the more the system growth in complexity (more areas to be covered and more moveable stations), the more the performance of the system

Table 1. Case study with **12** areas and **6** dock stations in total and with a set of moveable stations from **1** to **3**. The stochastic optimization procedure has been computed using **N = 1000** samples.

Instance	Moveable stations	Demand growth	Runtime (sec)
1	1	12.15 %	320
2	1	9.49 %	378
3	1	12.16 %	316
1	2	12.75 %	371
2	2	8.51 %	372
3	2	13.23 %	368
1	3	12.81 %	398
2	3	10.13 %	436
3	3	14.06 %	377

Table 2. Case study with **18** areas and **9** dock stations in total, and with a set of moveable stations from **3** to **5**. The stochastic optimization procedure has been computed using **N = 1000** samples.

Instance	Moveable stations	Demand growth	Runtime (sec)
1	3	13.61 %	961
2	3	8.75 %	999
3	3	15.53 %	989
1	4	13.81 %	1012
2	4	8.96 %	1069
3	4	16.81 %	1021
1	5	14.94 %	986
2	5	8.90 %	1107
3	5	16.87 %	1023

increase in term of customer demand. In particular, comparing the results from Table 1 (smallest case) and the one in the Table 4 we can see an improvement of the demand of a minimum of 9.26 % (which refers to the instance 2, using 3 moveable stations for the case in Table 1 and 8 moveable stations for the case in Table 4). Also, we can see that the maximum improvement (12.36 %) in the customer demand is achieved for the instance 1, with 3 moveable stations in the case study reported in Table 1 and with 8 moveable stations in the case reported in Table 4. According to this, the improvement seems more related

Table 3. Case study with **24** areas and **12** dock stations in total, and with a set of moveable stations from **4** to **6**. The stochastic optimization procedure has been computed using **N = 1000** samples.

Instance	Moveable stations	Demand growth	Runtime (sec)
1	4	17.52 %	2483
2	4	13.20 %	2519
3	4	21.06 %	2403
1	5	17.36 %	2528
2	5	13.79 %	2521
3	5	20.88 %	2422
1	6	18.16 %	2555
2	6	13.94 %	2506
3	6	20.99 %	2467

Table 4. Case study with **30** areas and **15** dock stations in total, and with a set of moveable stations from **6** to **8**. The stochastic optimization procedure has been computed using **N = 1000** samples.

Instance	Moveable stations	Demand growth	Runtime (sec)
1	6	23.86 %	4433
2	6	19.36 %	4592
3	6	23.61 %	4391
1	7	24.76 %	4504
2	7	19.21 %	4669
3	7	24.12 %	4394
1	8	25.17 %	4473
2	8	19.39 %	4872
3	8	25.72 %	4512

to the implementation of the dynamic station re-deployment on larger systems then related with the number of moveable platforms. Finally, for the largest case reported in Table 5 the result confirm such tendency, although the improvement does not seems too significant. However, it is difficult to say from this results if this may depends on the few number of samples used by the Algorithm 1, or by different reasons.

Table 5. Case study with **40** areas and **20** dock stations in total, and with a set of moveable stations from **8** to **10**. The stochastic optimization procedure has been computed using **N = 500** samples.

Instance	Moveable stations	Demand growth	Runtime (sec)
1	8	25.27 %	6609
2	8	20.14 %	6846
3	8	25.94 %	6531
1	9	25.85 %	6742
2	9	20.19 %	6930
3	9	25.88 %	6557
1	10	25.94 %	6858
2	10	20.97 %	7029
3	10	25.90 %	6704

6   Conclusions

In this paper, we investigated the problem of on-line dynamic redeployment in bike-sharing systems. We proposed an on-line stochastic optimization formulation to address this problem. Then, we carried out numerical experiments using data provided by bike-sharing companies to show the effectiveness of our approach.

When we consider the on-line station redeployments, the results show a significant improvement in the customer demand which go from a minimum around 8 % up to a maximum of a 25 %. We also found that the more the system growth in complexity (more areas to be served and more stations) the more the improvement in customer demand becomes significant.

As future works there are many points to deal with. First of all, in order to generalize the results, we need to formulate models for which the relationship between the expected customer demand and the distance between the facilities is non linear. This will lead to a different optimization formulation. Further, we need to integrate into the proposed formulation, the dynamic rebalancing of the bicycles and eventually, the cost for the truck routings into the optimization formulation. Finally, we need to test such an approach on bigger instances (300 areas with 100 − 150 stations in total) which can be significant for large cities such as Barcellona, Madrid or Dublin [23].

References

1. Shaheen, S., Guzman, S., Zhang, H.: Bikesharing in europe, the americas, and asia. Transp. Res. Rec. J. Transp. Res. Board **2143**, 159–167 (2010)
2. Meddin: The bike-sharing world map (2015). http://www.bikesharingworld.com/
3. Capitalbikeshare (2015). www.capitalbikeshare.com/home

4. Fricker, C., Gast, N.: Incentives and redistribution in homogeneous bike-sharing systems with stations of finite capacity. EURO J. Transp. Log. **5**(3), 261–291 (2016)
5. O'Mahony, E., Shmoys, D.: Data analysis and optimization for (citi)bike sharing. In: AAAI Conference on Artificial Intelligence (2015)
6. Boucher, X., Brissaud, D., Tran, T.D., Ovtracht, N., dArcier, B.F.: 7th industrial product-service systems conference - pss, industry transformation for sustainability and business modeling bike sharing system using built environment factors. Procedia CIRP **30**, 293–298 (2015)
7. Bowman, J., Ben-Akiva, M.: Activity-based disaggregate travel demand model system with activity schedules. Transportation research part a: policy and practice **35**(1), 1–28 (2001)
8. Gregory, D., Johnston, R., Pratt, G., Watts, M., Whatmore, S.: The Dictionary of Human Geography, 5th edn. Wiley, Hoboken (2011)
9. Fotheringham, A.S.: Spatial Interaction Models: Formulations and Applications. Springer, Heidelberg (1988)
10. Various 1 (2016). http://www.capitalbikeshare.com/system-data
11. Various 2 (2016). http://hubwaydatachallenge.org/trip-history-data
12. Raviv, T., Kolka, O.: Optimal inventory management of a bike-sharing station. IIE Trans. **45**(10), 1077–1093 (2013)
13. Raviv, T., Tzur, M., Forma, I.A.: Static repositioning in a bike-sharing system: models and solution approaches. EURO J. Transp. Log. **2**(3), 187–229 (2013)
14. Rainer-Harbach, M., Papazek, P., Hu, B., Raidl, G.R.: Balancing bicycle sharing systems: a variable neighborhood search approach. In: Middendorf, M., Blum, C. (eds.) EvoCOP 2013. LNCS, vol. 7832, pp. 121–132. Springer, Heidelberg (2013). doi:10.1007/978-3-642-37198-1_11
15. Shu, J., Chou, M.C., Liu, Q., Teo, C.P., Wang, I.L.: Models for effective deployment and redistribution of bicycles within public bicycle-sharing systems. Oper. Res. **61**(6), 1346–1359 (2013)
16. Ghosh, S., Varakantham, P., Adulyasak, Y., Jaillet, P.: Dynamic redeployment to counter congestion or starvation in vehicle sharing systems. In: International Conference on Automated Planning and Scheduling (2015)
17. Contardo, C., Morency, C., Rousseau, L.M.: Balancing a dynamic public bike-sharing system (2012)
18. Pfrommer, J., Warrington, J., Schildbach, G., Morari, M.: Dynamic vehicle redistribution and online price incentives in shared mobility systems. IEEE Trans. Intell. Transp. Syst. **15**(4), 1567–1578 (2014)
19. Nair, R., Miller-Hooks, E.: Fleet management for vehicle sharing operations. Transp. Sci. **45**(4), 524–540 (2011)
20. Leurent, F.: Modelling a vehicle-sharing station as a dual waiting system: stochastic framework and stationary analysis. 19 p., November 2012
21. Farhan, B., Murray, A.T.: Distance decay and coverage in facility location planning. Ann. Reg. Sci. **40**(2), 279–295 (2006)
22. Hentenryck, P.V., Bent, R.: Online Stochastic Combinatorial Optimization. The MIT Press, Cambridge (2006)
23. Dublin bike sharing map (2016). http://www.dublinbikes.ie/all-stations/station-map

Optimized Word-Size Time Series Representation Method Using a Genetic Algorithm with a Flexible Encoding Scheme

Muhammad Marwan Muhammad Fuad[✉]

Aarhus University, MOMA, Palle Juul-Jensens Boulevard 99, 8200 Aarhus N, Denmark
marwan.fuad@clin.au.dk

Abstract. Performing time series mining tasks directly on raw data is inefficient, therefore these data require representation methods that transform them into low-dimension spaces where they can be managed more efficiently. Owing to its simplicity, the piecewise aggregate approximation is a popular time series representation method. But this method uses a uniform word-size for all the segments in the time series, which reduces the quality of the representation. Although some alternatives use representations with different word-sizes in a way that reflects the various information contents of different segments, such methods apply a complicated representation scheme, as it uses a different representation for each time series in the dataset. In this paper we present two modifications of the original piecewise aggregate approximation. The novelty of these modifications is that they use different word-sizes, which allows for a flexible representation that reflects the level of activity in each segment, yet these new medications address this problem on a dataset-level, which simplifies establishing a lower bounding distance. The word-sizes are determined through an optimization process. The experiments we conducted on a variety of time series datasets validate the two new modifications.

Keywords: Genetic algorithm · Time series representation · Word-size

1 Introduction

A large number of medical, financial and economic applications produce measurements that can be recorded over a period of time. This organized temporary information is known by the name of *time series*. Examples of time series include seismic data, stock prices, weather forecast, and many others.

Time series data are ubiquitous. It is estimated that much of today's data come in the form of time series [1]. Some of these time series may contain trillions of observations. The size of the dataset itself can also be very large. All this raises various challenges concerning storing and processing these data.

Time series data are high-dimensional, which makes their representation a challenging task. Luckily, time series data are also highly correlated that many dimensionality reduction techniques can produce a faithful representation of the data.

© Springer International Publishing AG 2016
G. Adorni et al. (Eds.): AI*IA 2016, LNAI 10037, pp. 26–34, 2016.
DOI: 10.1007/978-3-319-49130-1_3

The purpose of time series mining is to extract knowledge from time series databases, which are usually very large. The major tasks of time series mining are: query-by-content, anomaly detection, motif discovery, prediction, clustering, classification, and segmentation [2].

Classification is one of the main tasks in data mining. The purpose of classification is to assign an unlabeled data object to one or more of predefined classes.

Time series mining has developed several representation methods, also called dimensionality reduction techniques, as this is the main objective of this process, which, in turn, yields data that require a smaller storage space and a shorter processing time. Examples of the most common dimensionality reduction techniques include *Discrete Fourier Transform* (DFT) [3, 4], *Discrete Wavelet Transform* (DWT) [5], *Singular Value Decomposition* (SVD) [6], *Adaptive Piecewise Constant Approximation* (APCA) [7], *Piecewise Aggregate Approximation* (PAA) [8, 9], *Piecewise Linear Approximation* (PLA) [10], and *Chebyshev Polynomials* (CP) [11].

PAA divides each time series into equal-sized words and represents each word in the high-dimension space (raw data) with the average of the data in that word.

Bio-inspired optimization algorithms have been widely applied to different data mining tasks such as classification, clustering, and feature selection [12, 13, 14].

In this paper we apply two modifications to the PAA representation, based on the genetic algorithms, which reduce the dimensionality of the time series using words of different sizes. The first modification uses a representation with a fixed number of words. The second one uses a variable number of words. In the two cases, as we mentioned earlier, the word size itself is changeable. We compare the two modifications with the original PAA and we see that they both give better results in a classification task compared with the original PAA.

The rest of this paper is organized as follows; in Sect. 2 we present a background on time series mining. The two new modifications are introduced in Sect. 3 and their performances are evaluated in Sect. 4. We conclude this paper with Sect. 5.

2   Background

Today's technologies generate large volumes of data. Data mining aims to extract knowledge hidden in these data. Data mining encompasses several tasks the main of which are [1, 15–17]:

- **Data Preparation or Data Pre-processing:** Most of the raw data are unprepared, noisy, or incomplete. All this makes data mining systems fail to process these data properly. For this reason, a preparation stage is required before handling these data. This stage may include different processes such as data cleansing, normalizing, handling outliers, completion of missing values, and deciding which attributes to keep and which ones to discard.
- **Prediction:** This task includes a kind of estimation, except that it concerns values that are beyond the range of already observed data. This is equivalent to extrapolation in numerical analysis.

- **Query-by-content:** In this task the algorithm searches for all the objects in the database that are similar to a given pattern.
- **Clustering:** This is a process in which the algorithm groups the data objects into classes of similar objects. Clustering is different from classification in that in clustering we do not have target variables. Instead, clustering is a process of partitioning the data space into groups of similar objects.
- **Classification:** In classification we have categorical variables which represent classes. The task is to assign class labels to the dataset according to a model learned from a learning stage on a training data, where the classes are known (supervised learning). When given new data, the algorithm aims to classifying these data based on the model acquired during the training stage, and later this model is applied to a testing dataset.

Classification is in fact one of the main tasks of data mining and it is particularly relevant to the experimental section of this paper. There are a number of classification models, the most popular of which is *k-nearest-neighbor* (*k*NN). In this model the object is classified based on the *k* closest objects in its neighborhood. A special case of particular interest is when $k = 1$.

Performance of classification algorithms can be evaluated using different methods. One of the widely used ones is *leave-one-out cross-validation* (LOOCV) (also known by *N-fold cross-validation*, or *jack-knifing*), where the dataset is divided into as many parts as there are instances, each instance effectively forming a test set of one. N classifiers are generated, each from N − 1 instances, and each is used to classify a single test instance. The classification error is then the total number of misclassified instances divided by the total number of instances [15].   □

Data mining has been applied to different data types. Of these type, times series stand out as one of most widespread data types. Time series are particularly subject to noise, scaling, and outliers. These problems are usually handled by pre-processing the data [2].

Due to the high-dimensionality of time series data, most classic machine learning and data mining algorithms do not work well on time series [18]. The major motivation of representation methods is to reduce the high dimensionality of these data. This is achieved mainly by taking advantage of another characteristic of time series, which is their high feature correlation.

Representation methods follow the GEMINI framework [19]. GEMINI reduces the dimensionality of the time series by converting them from a point in an *n*-dimension space into a point in an *N*-dimensional space, where $N \ll n$. If the similarity measure defined on the reduced space is a lower bound of the original similarity measure then the similarity search returns no false dismissals. A post-processing sequential scan on the candidate answer set is performed to filter out all the false alarms and return the final answer set.

Other representation methods use several low-dimensional spaces corresponding to different resolution levels [20–22].

One of the most popular time series representation methods is *Piecewise Aggregate Approximation* – PAA [8, 9]. PAA divides a time series *S* of *n*-dimensions into equal-sized segments (words) and maps each segment to a point of a lower *N*-dimensional space, where each point in the reduced space is the mean of value of the data points falling within that segment. The similarity measure given in the following equation:

$$d^N(S, T) = \sqrt{\frac{n}{N}} \sqrt{\sum_{i=1}^{N} \left(\overline{S}_I - \overline{t}_I \right)^2} \tag{1}$$

is defined on the lower-dimensional space. This similarity measure is a lower bound of the Euclidean distance defined on the original space. In Fig. 1 we shows an example of PAA for $n = 12, N = 3$.

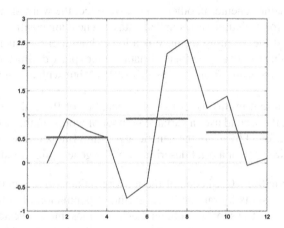

Fig. 1. PAA representation

It is worth mentioning that the authors of [8] use a compression ratio of 1:4 (i.e. every 4 points in the original time series are represented by 1 point in the reduced space).

In [7] the authors of the original PAA presented APCA as an extension of PAA. Whereas PAA uses frames of equal lengths, APCA relaxes this condition and represents each time series in the dataset by a set of constant value segments of varying lengths such that their individual reconstruction errors are minimal. The intuition behind this idea is that different regions of the time series contain different amounts of information. So while regions of high activity contain high fluctuations, other regions of low activity show a flat behavior, so a representation method with high fidelity should reflect this difference in behavior.

In general, finding the optimal piecewise polynomial representation of a time series requires $O(Nn^2)$. However, the authors of [7] utilize an algorithm which produces an approximation in $O(nlog(n))$ [18].

3   Genetic Algorithm Based Piecewise Aggregate Approximation

PAA has two main advantages; the first is it simplicity, the second is that the same dimensionality reduction scheme is applied to all the time series in the dataset. This guarantees that a lower-bounding distance can be easily established. However, because of the uniformity of representation, PAA is not able to distinguish regions with high information-content from those with low-information content. This is, as we saw in the previous section, was the motivation behind APCA. But APCA overcame this problem at the expense of a complicated representation scheme, but also, and this is our main concern, by using a rationale that is based on individual time series. This directly results in the difficulty of introducing a lower-bounding distance.

In this paper we try to modify PAA to allow it to reflect different activity in different regions, by using segments with different lengths, i.e. word-sizes. However, and this the main characteristic of our method, all the time series in the dataset are segmented according to the same scheme. In other words, we select the word-sizes for the whole dataset instead of for each time series in that dataset. The main positive consequence of this strategy is that the resulting distance is a lower bound of the Euclidean distance. Besides, determining the word-size for each dataset takes place during the training stage, so the method has the same simplicity as that of PAA but with a more faithful representation.

We actually present in this paper two modifications of PAA, the first uses a fixed number of words, the second uses a variable number of words. In the two modifications the word-sizes are found by using the genetic algorithm. But before we introduce these two medications we present a brief description of the genetic algorithm.

The Genetic Algorithm: GA is one of the most prominent global optimization algorithms. Classic GA starts by randomly generating a population of chromosomes that represent possible solutions to the problem. Each chromosome is a vector whose length is equal to the number of parameters, denoted by *nbp*. The fitness function of each chromosome is evaluated in order to determine the chromosomes that are fit enough to survive and possibly mate. In the *selection* step a percentage *sRate* of chromosomes is selected for mating. *Crossover* is the next step in which the offspring of two parents are produced to enrich the population with fitter chromosomes. *Mutation*, which is a random alteration of a certain percentage *mRate* of chromosomes, enables GA to explore the search space. In the next generation the fitness function of the offspring is calculated and the above steps repeat for a number of generations *nGen*.

Classic GA, which encodes solutions using chromosomes with a predefined length, may not be appropriate for some problems. Another more flexible encoding scheme, introduced in [23] as a variant of classifier systems, uses chromosomes of variable lengths. Later, this concept was used to solve different optimization problems where the number of parameters is not fixed.   □

As we mentioned earlier, we present two modifications of PAA, the first one, which we refer to by GA-PAA, has chromosomes with a fixed number of parameters $nbp = \frac{n}{4}$. Each of these parameters represents a possible word-size. The optimization process searches for the optimal combination of word-sizes, for that dataset, which optimizes a fitness function of a certain time series data mining task, under the constraint that the traditional compression ratio with which PAA is usually applied (i.e. 1:4, see Sect. 2) is preserved.

The second modification is denoted by VCL-GA-PAA. This version relaxes the above constraint of keeping the compression ratio of 1:4 and, theoretically, allows for any compression ratio. In other word, the value of *nbp* is variable and is part of the optimization process. However in practice, and in order for GA to converge, we constraint the value by $2 \leq nbp \leq \frac{n}{2}$.

The fitness function we choose is the classification error. In other words, the optimization process searches for the optimal word-sizes (for GA-PAA and VCL-GA-PAA) that minimize the classification error.

Our motivation for choosing the classification error as a fitness function is because classification is the most popular data mining task [18]. However, we could choose another data mining task, or we could even apply multi-objective optimization.

4   Performance Evaluation

We compared the performance of GA-PAA and VCL-GA-PAA with that of PAA in a 1NN classification task on 20 time series datasets of different sizes and dimensions available at UCR Time Series Classification Archive [24], which makes up between 90 % and 100 % of all publicly available, labeled time series data sets in the world [25]. Each dataset in the archive consists of a training set and a testing set.

The length of the time series on which we conducted our experiments varies between 24 (ItalyPowerDemand) and 1639 (CinC_ECG_torso). The size of the training sets varied between 16 (DiatomSizeReduction) and 100 (Trace), and (ECG200). The size of the testing sets varies between 20 (BirdChicken), (BeetleFly), and 1380 (CinC_ECG_torso). The number of classes varies between 2 (ItalyPowerDemand), (BeetleFly), (ECG200), (Herring), (Wine), (ECGFive-Days), (SonyAIBORobotSurface), (BirdChicken), and 7 (Lighting7).

The experiments consist of two stages; in the training stage each of GA-PAA and VCL-GA-PAA, and for each dataset, is trained on the training set to obtain the optimal word-sizes that yield the minimum classification error. In the testing stage those word-sizes are applied to the corresponding datasets to obtain the classification error on that dataset.

In Table 1 we present the results of our experiments. The best result (the minimum classification error) for each dataset is shown in bold-underlined printing in yellow-shaded cells.

As we can see from the table, both GA-PAA and VCL-GA-PAA outperform PAA for all the datasets tested. We also see that out of the 20 datasets VCL-GA-PAA outperformed GA-PAA for 13 datasets, whereas GA-PAA outperformed VCL-GA-PAA for 2 datasets, and for 5 datasets they both yielded the same classification error.

As for analyzing the compression ratio, we see that for the 13 datasets where VCL-GA-PAA gave a smaller classification error the compression ratio was 1:3 for 2 datasets (ECGFiveDays) and (Car) and 1:2 for 3 datasets (ItalyPowerDemand), (Meat), and (Gun_Point). Also in the case of the 5 datasets for which both VCL-GA-PAA and GA-PAA give the same results, the compression ratio was 1:3 for one dataset (OliveOil) and 1:2 for another (ECG200). Such compression ratios (1:3 or 1:2) are not of practical

Table 1. The 1NN classification error (1NN - CE) and the corresponding compression ratios (CR) of PAA, GA-PAA, and VCL-GA-PAA

dataset	PAA		GA-PAA		VCL-GA-PAA	
	1NN - CE	CR	1NN - CE	CR	1NN - CE	CR
CBF	0.050		0.050		**0.039**	1:16
CinC_ECG_torso	0.104		0.102		**0.068**	1:11
DiatomSizeReduction	0.065		0.049		**0.046**	1:13
ECG200	0.130		**0.090**		**0.090**	1:2
ECGFiveDays	0.146		0.044		**0.039**	1:3
FaceFour	0.205		0.148		**0.068**	1:19
Lighting7	0.397		**0.274**		**0.274**	1:17
OliveOil	0.133		**0.100**		**0.100**	1:3
SonyAIBORobotSurface	0.258		0.186		**0.151**	1:10
Meat	0.067	1:4	0.050	1:4	**0.033**	1:12
Wine	0.370		**0.222**		0.278	1:3
Herring	0.484		**0.391**		**0.391**	1:5
BirdChicken	0.450		0.300		**0.150**	1:25
BeetleFly	0.250		**0.150**		**0.150**	1:13
ArrowHead	0.206		**0.160**		0.166	1:2
Beef	0.333		0.267		**0.200**	1:11
Trace	0.250		0.170		**0.090**	1:7
Gun_Point	0.093		0.060		**0.053**	1:2
ItalyPowerDemand	0.068		0.047		**0.029**	1:2
Car	0.267		0.233		**0.183**	1:3

interest. However, it is important to mention here that the experiments were designed so that VCL-GA-PAA returns the best results when $2 \leq nbp \leq \frac{n}{2}$. In other words, the fact that VCL-GA-PAA outperformed or gave the same result as GA-PAA for such low compression ratios (1:3 or 1:2) does not necessarily mean that VCL-GA-PAA will not still outperform or give the same result as GA-PAA for higher compression ratios. It is just the way the experiments were designed.

On the other hand, we see that some of the compression ratios of VCL-GA-PAA were high or even very high for some datasets (1:19 for (FaceFour) and 1:25 for (Bird-Chicken)). In fact, out of the 20 datasets tested, the compression ratio was higher than 1:8 for 10 datasets when VCL-GA-PAA was applied. This shows the high performance of VCL-GA-PAA.

5   Conclusion

In this paper we presented two modifications of PAA, a popular time series representation method, which are based on the genetic algorithm. The two modifications divide the time series into segments of different word-sizes, unlike the original PAA which uses a uniform word-size. The main advantage of this segmentation scheme that we used is to reflex the different information content of different segments. Although other methods also proposed a representation with different word-sizes, the novelty of our modifications is: 1- The optimal word-sizes are determined on a dataset-level, with the advantage this strategy brings by easily establishing a lower bounding distance 2- The optimal word-sizes are obtained during the training stage through an optimization process, so these modifications have the same simplicity of the original PAA, which is its main advantage.

We presented two modifications of PAA, the first, GA-PAA uses the same compression ratio as the original PAA. Its main difference is that the word-sizes are of variable lengths. The second modification, VCL-GA-PAA uses variable compression ratios. Consequently, the time series of a certain dataset can be represented by a data-dependent number of words each of which has a different word-size.

The experiments of a classification task we conducted on a variety of time series datasets show that the two modifications GA-PAA and VCL-GA-PAA outperform PAA. The highly-flexible representation scheme of VCL-GA-PAA was particularly competitive as this representation yielded smaller classification errors even for higher compression ratios.

We believe that the main interesting feature that VCL-GA-PAA offers is the possibility to completely customize the representation by setting, a priori, a predefined compression ratio for a certain application and then run the optimization process to obtain the optimal word-sizes that correspond to that compression ratio.

References

1. Larose, D.T.: Discovering Knowledge in Data: An Introduction To Data Mining. Wiley, New York (2005)
2. Esling, P., Agon, C.: Time-series data mining. ACM Comput. Surv. (CSUR) **45**(1), 12 (2012)
3. Agrawal, R., Faloutsos, C., Swami, A.: Efficient similarity search in sequence databases. In: Proceedings of the 4th Conference on Foundations of Data Organization and Algorithms (1993)
4. Agrawal, R., Lin, K.I., Sawhney, H.S., Shim, K.: Fast similarity search in the presence of noise, scaling, and translation in time-series databases. In: Proceedings of the 21st International Conference on Very Large Databases, Zurich, Switzerland, pp. 490–501 (1995)
5. Chan, K.P., Fu, A.W-C.: Efficient time series matching by wavelets. In: Proceedings of 15th International Conference on Data Engineering (1999)
6. Korn, F., Jagadish, H., Faloutsos, C.: Efficiently supporting ad hoc queries in large datasets of time sequences. In: Proceedings of SIGMOD 1997, Tucson, AZ, pp. 289–300 (1997)
7. Keogh, E., Chakrabarti, K., Pazzani, M., Mehrotra, S.: Locally adaptive dimensionality reduction for similarity search in large time series databases. In: SIGMOD, pp 151–162 (2001)
8. Keogh, E., Chakrabarti, K., Pazzani, M., Mehrotra, S.: Dimensionality reduction for fast similarity search in large time series databases. J. Know. Inform. Syst. **3**, 263–286 (2000)
9. Yi, B.K., Faloutsos, C.: Fast time sequence indexing for arbitrary Lp norms. In: Proceedings of the 26th International Conference on Very Large Databases, Cairo, Egypt (2000)

10. Morinaka, Y., Yoshikawa, M., Amagasa, T., Uemura, S.: The L-index: an indexing structure for efficient subsequence matching in time sequence databases. In: Proceedings of 5th Pacific Asia Conference on Knowledge Discovery and Data Mining, pp. 51–60 (2001)

11. Cai, Y., Ng, R.: Indexing spatio-temporal trajectories with Chebyshev polynomials. In: SIGMOD (2004)

12. Muhammad Fuad, M.M.: Differential evolution-based weighted combination of distance metrics for k-means clustering. In: Dediu, A.-H., Lozano, M., Martín-Vide, C. (eds.) TPNC 2014. LNCS, vol. 8890, pp. 193–204. Springer, Heidelberg (2014). doi: 10.1007/978-3-319-13749-0_17

13. Muhammad Fuad, M.M.: Differential evolution versus genetic algorithms: towards symbolic aggregate approximation of non-normalized time series. In: Sixteenth International Database Engineering & Applications Symposium, IDEAS 2012, Prague, Czech Republic, 8–10 August, 2012. BytePress/ACM (2012)

14. Muhammad Fuad, M.M.: Multi-objective optimization for clustering microarray gene expression data - a comparative study. In: Jezic, G., Howlett, R.J., Jain, L.C. (eds.). SIST, vol. 38, pp. 123–133Springer, Heidelberg (2015). doi:10.1007/978-3-319-19728-9_10

15. Bramer, M.: Principles of Data Mining. Springer, London (2007)

16. Gorunescu, F.: Data mining: Concepts, Models and Techniques. Blue Publishing House, Cluj-Napoca (2006)

17. Mörchen, F.: Time series knowledge mining. Ph.D. thesis, Philipps-University Marburg, Germany, Görich & Weiershäuser, Marburg, Germany (2006)

18. Maimon, O., Rokach, L.: Data Mining and Knowledge Discovery Handbook. Springer, New York (2005)

19. Faloutsos, C., Ranganathan, M., Manolopoulos, Y.: Fast subsequence matching in time-series databases. In: Proceedings of ACM SIGMOD Conference of Minneapolis (1994)

20. Muhammad Fuad, M.M., Marteau P.F.: Multi-resolution approach to time series retrieval. In: Fourteenth International Database Engineering & Applications Symposium, IDEAS 2010, Montreal, QC, Canada (2010)

21. Muhammad Fuad, M.M., Marteau P.F.: Speeding-up the similarity search in time series databases by coupling dimensionality reduction techniques with a fast-and-dirty filter. In: Fourth IEEE International Conference on Semantic Computing, ICSC 2010, Carnegie Mellon University, Pittsburgh, PA, USA (2010)

22. Muhammad Fuad, M.M., Marteau, P.F.: Fast retrieval of time series by combining a multiresolution filter with a representation technique. In: The International Conference on Advanced Data Mining and Applications, ADMA 2010, ChongQing, China, November 21 (2010)

23. Smith, S. F.: A learning system based on genetic adaptive algorithms. Doctoral dissertation, Department of Computer Science, University of Pittsburgh, PA (1980)

24. Chen,Y., Keogh, E., Hu, B., Begum, N., Bagnall, A., Mueen, A., Batista, G.: The UCR time series classification archive (2015). www.cs.ucr.edu/~eamonn/time_series_data

25. Ding, H., Trajcevski, G., Scheuermann, P., Wang, X., Keogh, E.: Querying and mining of time series data: experimental comparison of representations and distance measures. In: Proceedings of the 34th VLDB (2008)

Efficient Search of Relevant Structures in Complex Systems

Laura Sani[1], Michele Amoretti[1(✉)], Emilio Vicari[1], Monica Mordonini[1],
Riccardo Pecori[1,4], Andrea Roli[2], Marco Villani[3], Stefano Cagnoni[1],
and Roberto Serra[3]

[1] Dip. di Ingegneria dell'Informazione, Università degli Studi di Parma, Parma, Italy
`michele.amoretti@unipr.it`
[2] Dip. di Informatica - Scienza e Ingegneria,
Università di Bologna, Sede di Cesena, Italy
[3] Dip. di Scienze Fisiche, Informatiche e Matematiche,
Università degli Studi di Modena e Reggio Emilia, Modena, Italy
[4] SMARTest Research Centre, Università eCAMPUS, Novedrate, CO, Italy

Abstract. In a previous work, Villani et al. introduced a method to identify candidate emergent dynamical structures in complex systems. Such a method detects subsets (clusters) of the system elements which behave in a coherent and coordinated way while loosely interacting with the remainder of the system. Such clusters are assessed in terms of an index that can be associated to each subset, called Dynamical Cluster Index (DCI). When large systems are analyzed, the "curse of dimensionality" makes it impossible to compute the DCI for every possible cluster, even using massively parallel hardware such as GPUs.

In this paper, we propose an efficient metaheuristic for searching relevant dynamical structures, which hybridizes an evolutionary algorithm with local search and obtains results comparable to an exhaustive search in a much shorter time. The effectiveness of the method we propose has been evaluated on a set of Boolean models of real-world systems.

Keywords: Complex systems · Hybrid metaheuristics · Local search

1 Introduction

The study of complex systems is related to the analysis of collective behaviors and emerging properties of systems whose components are usually well-known. Measuring the complexity of a composite system is a challenging task; dozens of measures of complexity have been proposed, several of which are based on information theory [1]. Detecting clusters of elements that interact strongly with one another is even more challenging, especially when the only information available is the evolution of their states in time.

The problem of finding groups of system elements that have a tighter dynamical interaction among themselves than with the rest of the system is a typical

© Springer International Publishing AG 2016
G. Adorni et al. (Eds.): AI*IA 2016, LNAI 10037, pp. 35–48, 2016.
DOI: 10.1007/978-3-319-49130-1_4

issue in data analysis; notable examples are the identification of functional neuronal regions in the brain and the detection of specific groups of genes ruling the dynamics of a genetic network.

The method proposed by Villani et al. [2] identifies emergent dynamical structures in complex systems, also referred to as relevant sets (RSs) in the following. To do so, it assesses the relevance of each possible subset of the system variables, computing a quantitative index, denoted as Dynamical Cluster Index (DCI). Therefore, to fully describe a dynamical system based on the DCI it would be necessary to compute such an index for all possible subsets of the system variables. Unfortunately, their number increases exponentially with the number of variables, soon reaching unrealistic requirements for computation resources. As a consequence, to extract relevant DCI information about a system by observing its status over time, it is absolutely necessary to design efficient strategies, which can limit the extension of the search by quickly identifying the most promising subsets.

In this paper, we propose HyReSS (Hybrid Relevant Set Search), a hybrid metaheuristic for searching relevant sets within dynamical systems, based on the hybridization of an evolutionary algorithm with local search strategies. In the tests we have made on data describing both real and synthetic systems, HyReSS has been shown to be very efficient and to produce results comparable to an exhaustive search in a much shorter time.

The paper is organized as follows. In Sect. 2, we discuss previous related work. In Sect. 3, we describe the DCI-based approach. In Sect. 4, we present the evolutionary metaheuristic. In Sect. 5, we report some experimental results. Finally, in the last section, we conclude the paper summarizing our achievements and discussing future research directions.

2   Related Work

Several measures of complexity are based on information theory [1], which is convenient since any dynamically changing phenomenon can be characterized in terms of the information it carries. Hence, these measures can be applied to the analysis of any dynamical system. A widely-known information-theoretic framework by Gershenson and Fernandez [3] allows one to characterize systems in terms of emergence, self-organization, complexity and homeostasis. Such a framework has been applied, for example, to characterize adpative peer-to-peer systems [4], communications systems [5] and agroecosystems [6].

The DCI method [7,8] is an extension of the Functional Cluster Index (CI) introduced by Edelman and Tononi in 1994 and 1998 [9,10] to detect functional groups of brain regions. In our previous work, we extended the CI domain to non-stationary dynamical regimes, in order to apply the method to a broad range of systems, including abstract models of gene regulatory networks and simulated social [11], chemical [2], and biological [8] systems.

Genetic Algorithms (GAs) are popular search and optimization techniques, particularly effective when little knowledge is available about the function to optimize. However, some studies [12,13] show that they are not well-suited to fine-tune searches in complex spaces and that their hybridization with local search methods, often referred to as memetic algorithms (MAs) [12], can greatly improve their performance and efficiency. Nevertheless, basic GAs and MAs are designed to find absolute optima and therefore are not capable of maintaining the diversity of solutions during evolution, which is essential when multimodal functions are analyzed, and the goal is to find as many local optima as possible.

To compensate for this shortcoming, various techniques, commonly known as *niching methods*, have been described in the literature, that maintain population diversity during the search process and allow the search to explore many peaks in parallel. Most niching methods, however, often require that problem-specific parameters, strictly related to the features of the search space, be set *a priori* to perform well. This is documented, for example, in [14–16], that describe applications to mechatronics, image processing, and multimodal optimization, respectively.

Among the most renowned niching algorithms we can recall Fitness Sharing [17], Sequential Niching [18], Deterministic Crowding [19], and restricted tournament selection [20]. In this work we have used a modified version of deterministic crowding, because that method does not require *a priori* setting of problem-related parameters, such as the similarity radius, and its complexity is low, since it scales as $O(n)$ with the number of dimensions of the search space. This is probably the main reason why the usage of deterministic crowding is still often reported in the recent literature [21–23].

3 Approach

Many complex systems, both natural and artificial, can be represented by networks of interacting nodes. Nevertheless, it is often difficult to find neat correspondences between the dynamics expressed by these systems and their network description. In addition, network descriptions may be adequate only in case of binary relationships. In the case of systems characterized by non-linear interactions among their parts, the dynamic relationships among variables might not be entirely described by the topology alone, which does not represent the actual dynamical interactions among the elements. In contrast, many of these systems can be described effectively in terms of coordinated dynamical behavior of groups of elements; relevant examples are Boolean networks [24], chemical or biological reaction systems [2] and functional connectivity graphs in neuroscience [25,26]. Furthermore, in several cases, the interactions among the system elements are not known; it is therefore necessary to deduce some hints about the system organization by observing the behavior of its dynamically relevant parts.

The goal of the work described in this paper is to identify groups of variables that are good candidates for being relevant subsets, in order to describe the organization of a dynamical system. We suppose that *(i)* the system variables express some dynamical behavior (i.e., there exists at least a subset of the observed states of the system within which they change their value), *(ii)* there exist one or more subsets where these variables are acting (at least partially) in a coordinated way, and *(iii)* the variables of each subset have weaker interactions with the other variables or RSs than among themselves. The outcome of the analysis is essentially a list of possibly overlapping subsets, ranked according to some criteria, which provide clues for understanding the system organization.

The approach we use (*Dynamical Cluster Index*, or DCI, method) has been previously presented by some of the authors of this paper. Here we briefly summarize the method, pointing to the relevant literature for further details [7,8]. The DCI method relies on information theoretical measures, related with the Shannon Entropy [27]. Given the observational nature of our data, the probabilities are estimated by the relative frequencies of their values. Let us now consider a system U composed of K variables (*e.g.*, agents, chemicals, genes, artificial entities) and suppose that S_k is a subset composed of k elements, with $k < K$. The $DCI(S_k)$ value is defined as the ratio between the *integration I* of S_k and the *mutual information M* between S_k and the rest of the system:

$$DCI(S_k) = \frac{I(S_k)}{M(S_k; U \setminus S_k)} \tag{1}$$

where $I(S_k)$ measures the statistical independence of the k elements in S_k (the lower $I(S_k)$, the more independent the elements) while $M(S_k; U \setminus S_k)$ measures the mutual dependence between the subset S_k and the rest of the system $U \setminus S_k$. In formulas:

$$I(S_k) = \sum_{s \in S_k} H(s) - H(S_k) \tag{2}$$

$$M(S_k; U \setminus S_k) = H(S_k) + H(U \setminus S_k) - H(S_k, U \setminus S_k) \tag{3}$$

where $H(X)$ is the entropy or the joint entropy, depending on X being a single random variable or a set of random variables.

Any subset of the system variables (Candidate Relevant Set - CRS - in the following) having $M = 0$ does not communicate with the rest of the system: it constitutes a separate system and its variables can be excluded from the analysis. The DCI scales with the size of the CRS, as already pointed out in [9], so it needs to be normalized by dividing each member of the quotient in Eq. 1 by its average value in a reference system where no dynamical structures are present. Following our previous works [7,8,28] our reference is a homogeneous system composed of the same number of variables and described by the same number of observations as the system under analysis. The values of the observations for the homogeneous system are generated randomly, according to the uni-variate distributions of each variable that could be estimated from the real observations if all variables were independent. Formally:

$$C'(S) = \frac{I(S)}{\langle I_h \rangle} / \frac{M(S; U \setminus S)}{\langle M_h \rangle} \tag{4}$$

Finally, in order to assess the significance of the normalized DCI values, a statistical index T_c can be computed [9]:

$$T_c(S) = \frac{C'(S) - \langle C'_h \rangle}{\sigma(C'_h)} = \frac{C(S) - \langle C_h \rangle}{\sigma(C_h)} \tag{5}$$

where $\langle C_h \rangle$, $\sigma(C_h)$, $\langle C'_h \rangle$ and $\sigma(C'_h)$ are, respectively, the average and the standard deviation of the DCI indices and of the normalized cluster indices from a homogeneous system with the same size as S_k.

The CRSs can be ranked according to their T_c: in both cases the analysis returns a huge set of candidates, most of which are a subset (or superset) of other CRSs. In order to identify the most relevant information, in [7] a post-processing sieving algorithm has been proposed, able to reduce the list of CRSs to the most representative ones. The algorithm is based on the consideration that if CRS A is a proper subset of CRS B and ranks higher than CRS B, then CRS A should be considered more relevant than CRS B. Therefore, the algorithm keeps only those CRSs that are not included in or do not include any other CRS with higher T_c. This "sieving" action stops when no more eliminations are possible: the remaining groups of variables are the proper RSs. This procedure can also be extended to the identification of hierarchical relations among RSs: this topic is the subject of ongoing work.

In this paper, we focus onto a particularly critical issue, i.e., how to efficiently detect the highest-ranked CRSs according to their T_c. Indeed, the number of CRSs increases exponentially with the system's dimension, the number of CRSs of size k in a system of size K being $\binom{K}{k}$. However, to characterize a dynamical system of interest, one does not need to know the T_c index of all possible CRSs, but only to detect the CRSs for which the T_c is highest. To do so, we developed HyReSS, a hybrid metaheuristic described in the following section, postponing to future investigations the use of its results to detect the RSs and their hierarchy.

4   HyReSS: A Hybrid Metaheuristic for RS Detection

HyReSS hybridizes a basic genetic algorithm with local search strategies that are driven by statistics, computed at runtime, on the results that the algorithm is obtaining.

A genetic algorithm is first run to draw the search towards the basins of attraction of the main local maxima in the search space. Then, the results are improved by performing a series of local searches to explore those regions more finely and extensively.

The method can be subdivided into five main cascaded steps:

1. Genetic algorithm;
2. CRS relevance-based local search;
3. CRS frequency-based local search;
4. Group cardinality-based local search;
5. Merging.

4.1 Genetic Algorithm

The first evolutionary phase is a genetic algorithm based on the Deterministic Crowding (DC) algorithm, one of the most efficient and commonly used niching techniques. As specified above, HyReSS does not search a single CRS, but the set $\mathcal{B}$ of the N_{best} highest-T_c CRSs.

Each individual corresponds to one CRS and is a binary string of size N, where each bit set to 1 denotes the inclusion in the CRS of the corresponding variable, out of the N that describe the system. A list (termed "best-CRS memory" in the following) is created to store the best individuals that have been found along with their fitness values. At the end of the run, it should contain all CRSs in $\mathcal{B}$.

The initial population, of size p, is obtained by generating random individuals according to a pre-set distribution of cardinality (pairs, triplets, etc.). This kind of generation aims to create a sample that is as diversified as possible (avoiding repetitions) as well as representative of the whole search space.

The fitness function to be maximized corresponds directly to the T_c and is implemented through a CUDA[1] kernel that can compute in parallel the fitness values of large blocks of individuals.

Evolution proceeds by selecting $p/2$ random pairs of individuals and creating p children by means of a single-point crossover. After crossover, each child possibly replaces the most similar parent of lower fitness. To safeguard genetic diversity, a parent is only replaced if the child is not already present in the population.

This evolutionary process is iterated until the population is no more able to evolve, *i.e.*, the new generation remains equal to the previous one. When that happens, new random parents are generated.

Mutation (implemented as bit flips) is applied with a low probability (P_{mut}) after each mating.

The termination condition for this evolutionary phase is reached when the number of evaluations of the fitness function exceeds a threshold α_f or new parents have been generated for α_p times.

The implemented algorithm is elitist, since a child is inserted in the new population only if its fitness is better than the fitness of the parent it substitutes. Therefore, the overall fitness of the population increases monotonically with the algorithm iterations. After the end of the evolutionary algorithm, the N_{gBest} fittest individuals are selected to seed the subsequent phases.

4.2 Variable Relevance-Based Search

While running the genetic algorithm, a relevance coefficient RC_i is computed for each variable i of the system under examination. RC_i is higher if variable i is frequently included in high-fitness CRSs.

---

[1] https://developer.nvidia.com.

At the end of each generation t of the GA, a fitness threshold is set, separating high-fitness CRSs from low-fitness ones, and corresponding to a certain percentile β of the whole fitness range.

$$\tau(t) = minFitness + (maxFitness - minFitness) * \beta \qquad (6)$$

A presence coefficient (PC_i) and an absence coefficient (AC_i) are defined, for variable i, as the sum of the fitness values of the CRSs having fitness greater than τ, in which the variable has been present or absent, respectively, cumulated over the generations and normalized with respect to the number of generations in which the corresponding CRSs have been included.

Based on these two coefficients, the ratio $R_{ap,i} = AC_i/PC_i$ is computed. The variable is classified as relevant if PC_i is greater than a threshold (the γ_{th} percentile of the full range of PC_i values) and $R_{ap,i}$ is lower than a certain threshold δ.

The corresponding local search procedure performs a recombination of the most relevant variables with other, randomly chosen, ones. As a first step, all possible subsets (simple combinations) of the most relevant variables are computed, excluding the subsets of cardinality 0 and 1. Then, for each subset dimension, the individual with the highest fitness is selected. Such individuals are the basis for generating new CRSs, by forcing the presence or absence of relevant/irrelevant variables and by randomly adding other variables into the RCSs. Every newly generated individual is evaluated and, should its fitness be higher, replaces the lowest-fitness individual in the best-CRS memory.

At the end of this phase a local search is performed in the neighborhood of the best individual of the best-CRS memory, which is updated in case new individuals with appropriate fitness are obtained.

4.3 Variable Frequency-Based Search

In this phase, the same procedure used to generate new individuals and to explore the neighborhood of the best one is repeated, based on a different criterion.

We consider the frequency with which each variable has been included in the CRSs evaluated in the previous phases and use this value to identify two classes of variables, which are assigned higher probability of being included in the newly generated CRSs:

– variables with frequency much lower than the average;
– variables with frequency much greater than the average.

In fact, variables of the former kind may have been previously "neglected", thus it may be worth verifying whether they are able to generate good individuals, while variables of the latter kind are likely to have been selected very frequently in the evolutionary process because they actually have a significant relevance.

4.4   Group Cardinality-Based Search

During the previous phases, HyReSS records the frequency with which groups of each possible cardinality $(2, \ldots, N-1)$ have occurred. These indices are then normalized according to the a priori probability of occurrence of such groups, given by the corresponding binomial coefficient $\binom{N}{c}$ where N is the total number of variables and c the cardinality of the group.

New CRSs are then generated using a procedure driven by such indices, such that cardinalities having lower values have higher probability of occurring and are possibly stored into the best-CRS memory according to their fitness.

4.5   Merging

In this phase a limited pool of variables is selected by considering all variables that are included in the highest-fitness CRSs in the best-CRS memory. In practice, a size θ for the pool is set; then, the best individuals are progressively OR-ed bitwise, in decreasing order of fitness starting from the best two CRSs, until the result of the bitwise OR contains θ bits set to 1 or all the CRSs have been processed. A final exhaustive search over all the possible CRSs that comprise the selected variables, is made, and the best-CRS memory is updated accordingly.

5   Experimental Results

In this section we illustrate three examples of dynamical systems we have used as benchmarks for HyReSS. The first one is a deterministic simulation of a chemical system called *Catalytic Reaction System (CatRS)*, described by 26 variables. The second one is a stochastic artificial system reproducing a *Leaders & Followers (LF)* behavior, featuring 28 variables. These examples have been analyzed using both exhaustive search and HyReSS. The third example, denoted as *Green Community Network (GCN)*, features 137 variables, a size for which an exhaustive search is not feasible on a standard computer. Thus, it was analyzed only by HyReSS. However, we could compare its results with those provided by field experts.

In all our test, we have performed 10 independent runs of HyReSS, to take the stochastic nature of the tool properly into account. We evaluated the results of HyReSS, when possible, by comparing the list of highest-T_c subsets it produced with the results of an exhaustive search. To let results be comparable, we relied on the same homogeneous system to compute normalized DCI values in both approaches. Tests were run on a Linux server equipped with a 1.6 GHz Intel I7 CPU, 6 GB of RAM and a GeForce GTX 680 GPU by NVIDIA. The parameters regulating the behavior of HyReSS were set as reported in Table 1.

Results are summarized in Table 2 and are discussed in the following subsections.

Table 1. HyReSS parameter settings. The parameters are defined in Sect. 4.

System	P_{mut}	p	α_f	α_p	β	γ	δ	θ
CatRS	.1	16384	163840	3	.75	.75	.3	15
LF	.1	16384	163840	3	.75	.75	.3	15
GCN (56 vars.)	.1	25600	256000	3	.75	.75	.3	15
GCN (137 vars.)	.1	50176	501760	3	.75	.75	.3	15

Table 2. Summary of HyReSS performances and comparison with an exhaustive search (ES), when possible.

System	N. Variables	N. Samples	Time (ES) [s]	Time (HyReSS) [s]	Speedup
CatRS	26	751	180	24	7.5
LF	28	150	300	19	15.8
GCN	56	124	n.a	71	n.a.
GCN	137	124	n.a	258	n.a

5.1 Catalytic Reaction System

The set of observations comes from the simulation of an open well-stirred chemostat (CSTR) with a constant incoming flux of feed molecules (empty ellipses in Fig. 1) and a continuous outgoing flux of all molecular species proportionally to their concentration. Six catalyzed reactions produce six new chemical species (pattern-filled ellipses in Fig. 1) and are divided in two dynamical arrangements, a linear chain and a circle. The system asymptotic behavior is a fixed point: we perturbed each single produced chemical species, in order to allow the variables to change their concentrations over time and thus highlight their dynamic relationships (for details, see [2]). In this work, we encoded each species' trajectory as a binary variable, the 0 and 1 symbols meaning "concentration is changing" and "concentration is not changing" respectively.

As the system has "only" 26 variables, it has been possible to perform an exhaustive search to be used as reference, which took about 180 s. Producing almost identical results (the resulting error rate is less than 0.02^2), the average running time of HyReSS was 24 s.

5.2 Leaders and Followers

The model is an abstract representation of a basic leader-followers (LF) model: it consists of an array of n binary variables $X = [x_1, x_2, \ldots, x_n]$, which could

---

[2] In one of the 10 runs, HyReSS failed to detect 1 of the first 50 RSs detected by the exhaustive search.

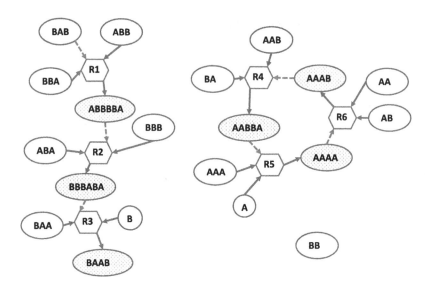

Fig. 1. The simulated CatRS. Circular nodes represent chemical species, while the white ones represent the species injected in the CSTR and the pattern filled ones those produced by the reactions. The diamond shapes represent reactions, where incoming arrows go from substrates to reactions and outgoing arrows go from reactions to products. Dashed lines indicate the catalytic activities.

represent, for example, the opinion of n people in favor or against a given proposal. The model generates independent observations on the basis of the following rules:

- the variables are divided into four groups:
  - $G1 = \{A0, A1, A2, A3\}$
  - $G2 = \{A7, A8, A9\}$
  - $G3 = \{A12, A13, A14, A15, A16, A17, A18, A19, A20\}$
  - $G4 = \{A22, A23, A24, A25, A26, A27\}$
- the remaining variables $A4, A5, A6, A10, A11$ and $A21$ assume the value 0 or 1 with identical probability;
- variables $A0, A7, A12, A22$ and $A23$ are the leaders of their respective groups; at each step they randomly assume value 0 respectively with probability 0.4, 0.3, 0.3, 0.3 and 0.6, 1 otherwise;
- the other variables (*i.e.*, the followers) copy or negate the values of their leaders, with the exception of the followers belonging to group $G4$ computing the OR or AND function of their two leaders.

Given these rules, the system comprises only well-defined and non-interacting groups, dynamically separate with respect to the other independent random variables. However, its stochasticity could occasionally support the emergence of spurious relationships, which make the automatic detection of groups non-trivial.

The LF case we have considered features 28 variables. An exhaustive search takes about 300 s. HyReSS completes its run in 19 s, on average, always providing the same results as the exhaustive search, considering the 50 highest-T_c subsets as a reference.

5.3   Green Community Network

In this case, the data come from a real situation and show the participation (*i.e.*, the presence or absence) of 137 people in a series of 124 meetings, held during a project (the so-called Green Community project [29]) which involved four mountain communities and focused on studies addressing energy efficiency and renewable energy production. The full original data set was multimodal, by far broader and more complex: some of us however (during the "MD" project, within which the DCI methodology was first proposed) extracted this simplified data set to verify whether the DCI analysis would be able to evidence the formation of specific dynamics among subsets of participants, despite the apparently simplicity of the information carried by the observations.

We have considered two versions of the GCN. The first one includes all 137 variables, whereas the second one has only 56 variables, representing people who attended more than one meeting. Both cases have too many variables to perform an exhaustive search.

The real situation is complex, with a part of the sociologically significant groups composed by smaller but very active (and sometimes partially overlapping) subgroups, having very often T_c values higher than the values computed for larger groups which include them: as a consequence the complete analysis would require the application of the sieving algorithm, as already mentioned in Sect. 3. In this work, however, we are focusing on the search of the highest-ranked CRSs, a step fundamental for the correct detection of the sociologically significant groups. In this regard, HyReSS is quite effective, (i) finding almost all the expected highest-ranked CRSs, (ii) identifying unknown groups *a posteriori* certified by the human experts and (iii) highlighting the presence of groups judged "interesting and requiring further investigations" by the human experts. HyReSS achieved these results in a very efficient way, with average running time of 71 s for 56 variables and 258 s for 137 variables.

6   Conclusion

In this paper we have presented HyReSS, an ad-hoc hybrid metaheuristic, tailored to the problem of finding the candidate Relevant Subsets of variables that describe a dynamical system. In developing our search algorithm, we have combined GAs' capacity of providing a good tradeoff between convergence speed and exploration, with local searches which refine and extend results, when correctly seeded. Using the deterministic crowding algorithm as a basis for the GA we guarantee that a large number of local maxima are taken into consideration in the early stages of HyReSS. The subsequent local search stages extend the results of the GA (stochastic) search more systematically to those CRSs that are most likely to have high fitness values, according to a few rules essentially derived from common sense.

In the benchmarks we took into consideration, HyReSS was very fast on an absolute scale, thanks to the GPU implementation of the fitness function. Even more importantly, at least for the smaller-size systems for which the comparison was possible, it could provide the same results as an exhaustive search based on the same parallel code, performing much fewer fitness evaluations and, consequently, in a significantly shorter time. The results obtained on the larger problems, on which the speedup with respect to an exhaustive search is virtually incommensurable and for which a ground truth is therefore not available, were qualitatively aligned with the expectations of a domain expert who analyzed the data.

The availability of an efficient algorithm will allow us to extend our research on the detection of candidate RSs to dynamical systems of much larger sizes than previously possible. At the same time, it will allow for devising more complex analyses, by which we aim to detect also hierarchical relationships among RSs.

From an algorithmic viewpoint, we expect to be able to further optimize HyReSS by fully parallelizing the search, whose GPU implementation is currently limited to the evaluation of the fitness function, which, as usually happens, is the most computation-intensive module within the algorithm. The modular structure of HyReSS will allow us to perform a detailed analysis of the algorithm in order to highlight which stage is most responsible for the algorithm performance and possibly design some optimized variants accordingly. Finally, we will also study the dependence of the algorithm on its parameters, to further improve its effectiveness and, possibly, to devise some self-adapting mechanisms to automatically fit their values to the system under investigation.

Acknowledgments. The authors thank the UE project "MD – Emergence by Design", Pr.ref. 284625 7th FWP-FET program for providing the data, which where in turn kindly provided by the Green Community project, sponsored by the National Association for Municipalities and Mountain Communities (UNCEM).

References

1. Prokopenko, M., Boschetti, F., Ryan, A.J.: An information-theoretic primer on complexity, self-organization, and emergence. Complexity **15**(1), 11–28 (2009)
2. Villani, M., Filisetti, A., Benedettini, S., Roli, A., Lane, D., Serra, R.: The detection of intermediate-level emergent structures and patterns. In: Miglino, O., et al. (eds.) Advances in Artificial Life, ECAL 2013, pp. 372–378. The MIT Press, Cambridge (2013)
3. Gershenson, C., Fernandez, N.: Complexity and information: measuring emergence, self-organization, and homeostasis at multiple scales. Complex **18**(2), 29–44 (2012)
4. Amoretti, M., Gershenson, C.: Measuring the complexity of adaptive peer-to-peer systems. Peer-to-Peer Netw. Appl. **9**(6), 1031–1046 (2016)
5. Febres, G., Jaff, K.: Calculating entropy at different scales among diverse communication systems. Complexity **21**(S1), 330–353 (2016)

6. Marull, J., Font, C., Padr, R., Tello, E., Panazzolo, A.: Energy landscape integrated analysis: a proposal for measuring complexity in internal agroecosystem processes (barcelona metropolitan region, 18602000). Ecol. Indic. **66**, 30–46 (2016)
7. Filisetti, A., Villani, M., Roli, A., Fiorucci, M., Serra, R.: Exploring the organisation of complex systems through the dynamical interactions among their relevant subsets. In: Andrews, P. et al., (eds.) Proceedings of the European Conference on Artificial Life 2015, ECAL 2015, pp. 286–293. The MIT Press (2015)
8. Villani, M., Roli, A., Filisetti, A., Fiorucci, M., Poli, I., Serra, R.: The search for candidate relevant subsets of variables in complex systems. Artif. Life **21**(4), 412–431 (2015)
9. Tononi, G., McIntosh, A., Russel, D., Edelman, G.: Functional clustering: identifying strongly interactive brain regions in neuroimaging data. Neuroimage **7**, 133–149 (1998)
10. Tononi, G., Sporns, O., Edelman, G.M.: A measure for brain complexity: relating functional segregation and integration in the nervous system. Proc. Natl. Acad. Sci. **91**(11), 5033–5037 (1994)
11. Filisetti, A., Villani, M., Roli, A., Fiorucci, M., Poli, I., Serra, R.: On some properties of information theoretical measures for the study of complex systems. In: Pizzuti, C., Spezzano, G. (eds.) WIVACE 2014. CCIS, vol. 445, pp. 140–150. Springer, Heidelberg (2014)
12. Chen, X., Ong, Y.S., Lim, M.H., Tan, K.C.: A multi-facet survey on memetic computation. IEEE Trans. Evol. Comput. **15**(5), 591–607 (2011)
13. Hu, X.M., Zhang, J., Yu, Y., Chung, H.S.H., Li, Y.L., Shi, Y.H., Luo, X.N.: Hybrid genetic algorithm using a forward encoding scheme for lifetime maximization of wireless sensor networks. IEEE Trans. Evol. Comput. **14**(5), 766–781 (2010)
14. Behbahani, S., de Silva, C.W.: Niching genetic scheme with bond graphs for topology and parameter optimization of a mechatronic system. IEEE/ASME Trans. Mechatron. **19**(1), 269–277 (2014)
15. Chang, D., Zhao, Y., Zheng, C.: A real-valued quantum genetic niching clustering algorithm and its application to color image segmentation. In: International Conference on Intelligent Computation and Bio-Medical Instrumentation (ICBMI), pp. 144–147, December 2011
16. Pereira, M.W., Neto, G.S., Roisenberg, M.: A topological niching covariance matrix adaptation for multimodal optimization. In: IEEE Congress on Evolutionary Computation (CEC), pp. 2562–2569, July 2014
17. Goldberg, D.E., Richardson, J.: Genetic algorithms with sharing for multimodal function optimization. In: Proceedings of the Second International Conference on Genetic Algorithms on Genetic Algorithms and Their Application, pp. 41–49. L. Erlbaum Associates Inc., Hillsdale (1987)
18. Beasley, D., Bull, D.R., Martin, R.R.: A sequential niche technique for multimodal function optimization. Evol. Comput. **1**(2), 101–125 (1993)
19. Manner, R., Mahfoud, S., Mahfoud, S.W.: Crowding and preselection revisited. In: Parallel Problem Solving From Nature, North-Holland, pp. 27–36 (1992)
20. Harik, G.R.: Finding multimodal solutions using restricted tournament selection. In: Proceedings of the 6th International Conference on Genetic Algorithms, pp. 24–31. Morgan Kaufmann Publishers Inc., San Francisco (1995)
21. Lacy, S.E., Lones, M.A., Smith, S.L.: Forming classifier ensembles with multimodal evolutionary algorithms. In: IEEE Congress on Evolutionary Computation (CEC), pp. 723–729 (2015)

22. Will, A., Bustos, J., Bocco, M., Gotay, J., Lamelas, C.: On the use of niching genetic algorithms for variable selection in solar radiation estimation. Renew. Energy **50**, 168–176 (2013)
23. Yannibelli, V., Amandi, A.: A deterministic crowding evolutionary algorithm to form learning teams in a collaborative learning context. Expert Syst. Appl. **39**(10), 8584–8592 (2012)
24. Villani, M., Barbieri, A., Serra, R.: A dynamical model of genetic networks for cell differentiation. PloS one **6**(3), e17703 (2011)
25. Shalizi, C.R., Camperi, M.F., Klinkner, K.L.: Discovering Functional Communities in Dynamical Networks. In: Airoldi, E., Blei, D.M., Fienberg, S.E., Goldenberg, A., Xing, E.P., Zheng, A.X. (eds.) ICML 2006. LNCS, vol. 4503, pp. 140–157. Springer, Heidelberg (2007)
26. Sporns, O., Tononi, G., Edelman, G.: Theoretical neuroanatomy: Relating anatomical and functional connectivity in graphs and cortical connection matrices. Cereb. Cortex **10**(2), 127–141 (2000)
27. Cover, T., Thomas, A.: Elements of Information Theory, 2nd edn. Wiley-Interscience, New York (2006)
28. Villani, M., Carra, P., Roli, A., Filisetti, A., Serra, R.: On the robustness of the detection of relevant sets in complex dynamical systems. In: Rossi, F., Mavelli, F., Stano, P., Caivano, D. (eds.) WIVACE 2015. CCIS, vol. 587, pp. 15–28. Springer, Heidelberg (2016)
29. Anzoise, V., Sardo, S.: Dynamic systems and the role of evaluation: the case of the green communities project. Eval. Program Plan. **54**, 162–172 (2016)

Classification, Pattern Recognition, and Computer Vision

Flat and Hierarchical Classifiers for Detecting Emotion in Tweets

Giulio Angiani, Stefano Cagnoni, Natalia Chuzhikova, Paolo Fornacciari,
Monica Mordonini, and Michele Tomaiuolo(✉)

Dipartimento di Ingegneria dell'Informazione, Università degli Studi di Parma,
Parco Area delle Scienze 181/A, 43124 Parma, Italy
{angiani,cagnoni,fornacciari,monica,tomamic}@ce.unipr.it,
michele.tomaiuolo@unipr.it

Abstract. Social media are more and more frequently used by people
to express their feelings in the form of short messages. This has raised
interest in emotion detection, with a wide range of applications among
which the assessment of users' moods in a community is perhaps the most
relevant. This paper proposes a comparison between two approaches to
emotion classification in tweets, taking into account six basic emotions.
Additionally, it proposes a completely automated way of creating a reliable training set, usually a tedious task performed manually by humans.
In this work, training datasets have been first collected from the web and
then automatically filtered to exclude ambiguous cases, using an iterative
procedure. Test datasets have been similarly collected from the web, but
annotated manually. Two approaches have then been compared. The first
is based on a direct application of a single "flat" seven-output classifier,
which directly assigns one of the emotions to the input tweet, or classifies
it as "objective", when it appears not to express any emotion. The other
approach is based on a three-level hierarchy of four specialized classifiers,
which reflect a priori relationships between the target emotions. In the
first level, a binary classifier isolates subjective (expressing emotions)
from objective tweets. In the second, another binary classifier labels subjective tweets as positive or negative. Finally, in the third, one ternary
classifier labels positive tweets as expressing joy, love, or surprise, while
another classifies negative tweets as expressing anger, fear, or sadness.
Our tests show that the a priori domain knowledge embedded into the
hierarchical classifier makes it significantly more accurate than the flat
classifier.

Keywords: Social media · Emotions detection · Hierarchical
classification

1   Introduction

In recent years, people's interest in public opinion has dramatically increased.

Opinions have become key influencers of our behavior; because of this, people do not just ask friends or acquaintances for advice when they need to take

© Springer International Publishing AG 2016
G. Adorni et al. (Eds.): AI*IA 2016, LNAI 10037, pp. 51–64, 2016.
DOI: 10.1007/978-3-319-49130-1_5

a decision, for example, about buying a new smartphone. They rather rely on the many reviews from other people they can find on the Internet. This does not apply only to individuals but also to companies and large corporations. Sentiment analysis and opinion mining have spread from computer science to management and social sciences, due to their importance to business and society as a whole. These studies are possible thanks to the availability of a huge amount of text documents and messages expressing users' opinions on a particular issue. All this valuable information is scattered over the Web throughout social networks such as Twitter and Facebook, as well as over forums, reviews and blogs.

Automatic classification is a generic task, which can be adapted to various kinds of media [1,2]. Our research aims, firstly, at recognizing the emotions expressed in the texts classified as subjective, going beyond our previous approaches to the analysis of Twitter posts (tweets), where a simple hierarchy of classifiers, mimicking the hierarchy of sentiment classes, discriminated between objectivity and subjectivity and, in this latter case, determined the polarity (positive/negative attitude) of a tweet [3]. The general goal of our study is to use sentiment detection to understand users' moods and, possibly, discover potential correlations between their moods and the structure of the communities to which they belong within a social network [4]. In view of this, polarity is not enough to capture the opinion dynamics of a social network: we need more nuances of the emotions expressed therein. To do this, emotion detection is performed based on Parrott's tree-structured model of emotions [5]. Parrott identifies six primary emotions (anger, fear, sadness, joy, love, and surprise); then, for each of them, a set of secondary emotions and, finally, in the last level, a set of tertiary emotions. For instance, a (partial) hierarchy for *Sadness* could include secondary emotions like *Suffering* and *Disappointment*, as well as tertiary emotions like *Agony, Anguish, Hurt* for *Suffering* and *Dismay, Displeasure* for *Disappointment*.

Secondly, our research aims to significantly increase the size of the training sets for our classifiers using a data collection method which does not require any manual tagging of data, limiting the need for reference manually-tagged data only to the collection of an appropriate test set.

Accordingly, in the work described in this paper, we have first collected large training datasets from Twitter using a robust automatic labeling procedure, and a smaller test set, manually tagged, to guarantee the highest possible reliability of data used to assess the performance of the classifiers derived from those training sets. Then, we have compared the results obtained by a single "flat" seven-output classifier to those obtained by a three-level hierarchy of four classifiers, that reflects apriori knowledge on the domain. In the first level of our hierarchy, a binary classifier isolates subjective from objective tweets; in the second, another binary classifier labels subjective tweets as positive or negative and, in the third, one ternary classifier labels positive tweets as expressing joy, love, or surprise, while another classifies negative tweets as expressing anger, fear, or sadness.

The paper is structured as follows. Section 2 offers a brief overview of related work. Section 3 describes the methodology used in this work. Section 4 describes the experimental setup and summarizes and compares the results obtained by the flat and the hierarchical classifiers. Section 5, concludes the paper discussing and analyzing the results achieved.

2   Related Work

Although widely studied, sentiment analysis still offers several challenges to computer scientists. Recent and comprehensive surveys of sentiment analysis and of the main related data analysis techniques can be found in [6, 7]. Emotional states like joy, fear, anger, and surprise are encountered in everyday life and social media are more and more frequently used to express one's feelings. Thus, one of the main and most frequently tackled challenges is the study of the mood of a network of users and of its components (see, for example, [8–10]). In particular, emotion detection in social media is becoming increasingly important in business and social life [11–15].

As for the tools used, hierarchical classifiers are widely applied to large and heterogeneous data collections [16–18]. Essentially, the use of a hierarchy tries to decompose a classification problem into sub-problems, each of which is smaller than the original one, to obtain efficient learning and representation [19, 20]. Moreover, a hierarchical approach has the advantage of being modular and customizable, with respect to single classifiers, without any loss of representation power: Mitchell [21] has proved that the same feature sets can be used to represent data in both approaches.

In emotion detection from text, the hierarchical classification considers the existing relationship between subjectivity, polarity and the emotion expressed by a text. In [22], the authors show that a hierarchical classifier performs better on highly imbalanced data from a training set composed of web postings which has been manually annotated. They report an accuracy of 65.5 % for a three-level classifier versus 62.2 % for a flat one. In our experiments, we performed a similar comparison on short messages coming directly from Twitter channels, without relying on any kind of manual tagging.

Emotions in tweets are detected according to a different approach in [23]. In that work, polarity and emotion are concurrently detected (using, respectively, SentiWordNet [24] and NRC Hashtag Emotion Lexicon [25]). The result is expressed as a combination of the two partial scores and improves the whole accuracy from 37.3 % and 39.2 % obtained, respectively by independent sentiment analysis and emotion analysis, to 52.6 % for the combined approach. However this approach does not embed the a-priori knowledge on the problem as effectively as a hierarchical approach, while limiting the chances to build a modular customizable system.

3   Methodology

A common approach to sentiment analysis includes two main classification stages, represented in Fig. 1:

1. Subdivision of texts according to the principles of objectivity/subjectivity. An objective assertion only shows some truth and facts about the world, while a subjective proposition expresses the author's attitude toward the subject of the discussion.
2. Determination of the polarity of the text. If a text is classified as subjective, it is regarded as expressing feelings of a certain polarity (positive or negative).

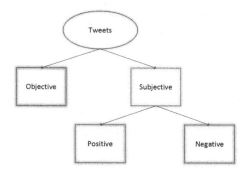

Fig. 1. Basic classification.

As mentioned earlier, the purpose of this paper is to improve the existing basic classification of tweets. Within this context, improving the classification should be considered as an extension of the basic model in the direction of specifying the emotions which characterize subjective tweets, based on Parrott's socio-psychological model. According to it, all human feelings are divided into six major states (three positive and three negative):

– positive feelings of love, joy, surprise
– negative feelings of fear, sadness, anger

We take into consideration flat and hierarchical classifiers, which are shown in Figs. 2 and 3, respectively. Hierarchical classification is based on the consistent application of multiple classifiers, organized in a tree-like structure. In our case, a first step uses a binary classifier that determines the subjectivity/objectivity of a tweet. The second step further processes all instances that have been identified as subjective. It uses another binary classifier that determines the polarity (positivity/negativity) of a tweet. Depending on the polarity assessed at the previous level, the third step classifies the specific emotion expressed in the text (love, joy or surprise for positive tweets; fear, sadness or anger for negative tweets).

To limit the need for human intervention in the definition of the training data, and thus allow for the collection of larger datasets, we devised a strategy

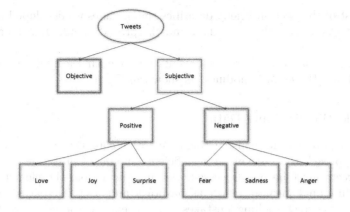

Fig. 2. Hierarchical classification.

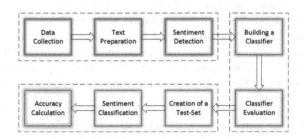

Fig. 3. Flat classification.

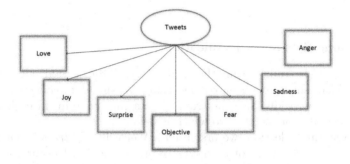

Fig. 4. Application structure.

to completely automate the collection of one training set for the construction of the flat multi-classifier, and other four for training the classifiers comprised in our hierarchical model.

We used manual labeling only in building the test sets, since the reliability of such data is absolutely critical for the evaluation of results of the classifiers taken into consideration.

The rest of this section briefly describes the modules we developed to implement our method, as well as the data and the procedure adopted to create the training sets.

The project has been developed using Java within the Eclipse IDE. It is structured into three main modules, as shown in Fig. 4.

3.1   Collecting Training Data

The main requirement for constructing an emotion classifier based on a machine learning approach is to download a sufficient amount of posts for the training phase. Tweets must be pre-processed, clearing them from the elements which have no emotional meaning, such as hashtags and user references. It is also important to correct spelling mistakes, and to encode special characters and emoticons appropriately as text tokens. Each sample of the training set represents a tweet and is composed by the processed text and an emotion class used as a label.

3.2   Training Sets and Classification

Our classifiers were trained using the "Naive Bayes Multinomial" algorithm provided by Weka. They have been trained using training data collected automatically and systematically into the training sets that contain, in each line, one tweet and the label of the class to which it belongs.

For the feature selection, we first used a Weka filter (*StringToWordVector*) to turn a string into a set of attributes representing word occurrences. After that, an optimal set of attributes (N-grams) was selected using the *Information Gain* algorithm provided by Weka, which estimates the worth of a feature by measuring the information gain with respect to the class.

For the hierarchical classifier, four training sets have been created, with data labeled according to the task of each of the four classifiers: "OBJ" or "SUB" for the objectivity classifier, "POS" or "NEG" for the polarity classifier. For the lower-level emotion classifiers (one for the tweets labeled as positive, one for those labeled as negative by the higher-level classifiers) the following labels/classes have been considered:

– Classes of positive polarity: LOVE, JOY, SURPRISE
– Classes of negative polarity: ANGER, SADNESS, FEAR

For the flat classifier, a multiclass file was created, with each sample labeled according to one of seven classes, six representing the emotions taken into account and the seventh for the objective tweets. Namely: LOVE, JOY, SURPRISE, ANGER, SADNESS, FEAR, OBJ.

The channels used to obtain emotive tweets (according to the corresponding hashtag) involve emotions that Parrott identified as either primary, secondary, or tertiary. Parrot's taxonomy of basic human feelings, including only the primary and secondary level, is presented in Table 1.

Table 1. Primary and secondary emotions of Parrott's socio-psychological model.

Primary emotion	Secondary emotion
Love	Affection, Lust/Sexual desire, Longing
Joy	Cheerfulness, Zest, Contentment, Pride, Optimism, Relief
Surprise	Surprise
Anger	Irritability, Exasperation, Rage, Disgust, Envy, Torment
Sadness	Suffering, Sadness, Disappointment, Shame, Neglect
Fear	Horror, Nervousness

The training set defining the objectivity or subjectivity of a tweet has been downloaded from the SemEval3 public repository.[1]

3.3 Classifying Data

We used the function library provided by Weka to develop a Java application for assessing the quality of our classifiers. The application supports both classification models taken into consideration for processing a test set using the Weka classifier models trained with the data described above, labels data and assesses the classifiers' accuracy by comparing the labels assigned to the test data by the classifiers to the actual ones, reported in the test set.

4 Results

In this section we present the results of our research. We first describe the experimental setup and, in particular, the procedure we followed to collect the data sets for training and testing the classifiers, as well as the preliminary tests we made to evaluate the quality of data and to determine the optimal number of features as well as the size of the N-grams used as features.

Finally, we compare the flat and the hierarchical classifiers on the basis of the accuracy they could achieve on the test set.

4.1 Collecting Data

Our training sets were built in a completely automated way, without human intervention.

- **Raw training set (Training Set 1).** Our raw training set (in the following called TS1) consists of about 10,000 tweets: we collected about 1500 tweets for each emotion and as many objective tweets. For the six nuances of emotions, we gathered data coming from several Twitter channels, following Parrott's classifications. Thus, the selection of channels was made methodically, without

---

[1] https://www.cs.york.ac.uk/semeval-2013/task2/index.php?id=data.html.

human evaluation. For each emotion, we used all the three levels of Parrott's model: for example, to extract tweets expressing *sadness* we downloaded data from the channel related to the primary emotion, *#Sadness*, but also from those related to secondary (*#Suffering, #Disappointment, #Shame, ...*) and tertiary emotions (*#Agony, #Anguish, #Hurt* for *Suffering*; *#Dismay, #Displeasure* for *Disappointment*, and so on). The objective (neutral) tweets were selected from the data set used for the SemEval competition[2].

– **Refined training set (Training Set 2).** Since the raw training set contains tweets obtained directly from Twitter channels, it may certainly contain spurious data. Thus, we adopted an automatic process to select only the most appropriate tweets. We filtered TS1 to remove the most ambiguous cases, and obtained a second training set (in the following called TS2) of about 1000 tweets for each of the six primary emotions. The filtering process was based on six binary classifiers, one for each emotion. The training set for each of them was balanced and considered two classes: the "positive" class included all raw tweets automatically downloaded from sources related to the emotion associated to the classifier; the "negative" class included tweets coming, in equal parts, from the other five emotions and from the set of objective tweets. Finally, TS2 included only the tweets which could be classified correctly by the binary classifier, in order for the tweets we used for training the main classifiers (i.e., those in TS2) to be as prototypical as possible.

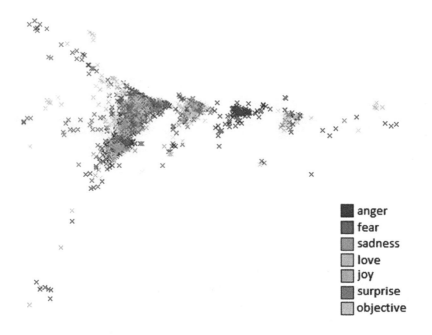

Fig. 5. Visualization of the first two PCA components for the test set.

[2] https://en.wikipedia.org/wiki/SemEval.

– **Test Set.** Tweets for the test set were downloaded in the same way as those for the training set, but they were manually annotated. They consist of 700 tweets, 100 for each of the six emotions in addition to 100 objective tweets. Even if, obviously, a representation sufficiently relevant for a correct classification would require a much larger number of features, we plot their first two components obtained by Principal Component Analysis (PCA) in Fig. 5 to give a first rough idea of the distribution of the tweets in the feature space. Objective tweets (yellow) and tweets related with sadness (green) are quite clearly separated from the others even in this minimal representation. Instead, other emotions are much closer and significantly overlapped, especially those related with surprise (violet). This could actually be justified considering that secondary and tertiary emotions can play a very significant role in recognizing this emotion, since it can be equally associated with both positive and negative events.

4.2   Optimizing the Parameters of Classifiers

For each classifier (four for the hierarchical and one for the flat approach), a systematic preliminary analysis was performed to optimize some relevant parameters that affect the training phase. We selected a grid of configurations and then used cross-validation to estimate the quality of classifiers configured according to it. In particular, we searched for the optimal length of N-grams to be used as features. Figure 6 shows the case of the flat classifier, but the other cases are similar. It can be observed that accuracy nearly peaks at N-gram = 2. Longer sequences increase the complexity of the training phase, without producing any significant improvement of the results. We also analysed the dependence of the performance on the number of features selected, using Weka's Information Gain algorithm. In Fig. 6 one can observe that its increase does not provide a

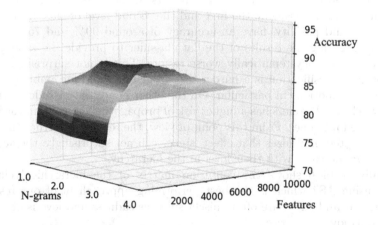

Fig. 6. Optimization of the system parameters.

Table 2. Parameter optimization results.

Classifier	N-Gram (max)	Features
Flat	2	1500
Sub/Obj	2	2500
Pos/Neg	2	2500
Anger/Sadness/Fear	2	1550
Love/Joy/Surprise	2	1500

monotonic improvement of the classifier quality. Instead, it has a peak at around 1500 features.

Table 2 shows the results of the parameter optimization step. In particular, the N-Gram (max) value is 2 for all our classifiers (unigram and bigram are considered). The last column shows the number of features that optimizes the performance of the classifiers.

4.3 Accuracy

Tables 3 and 4 report the results obtained on the test set by the seven-output flat classifier and by the hierarchical one. They display the accuracy of each approach, when trained on TS1 or on TS2, in order to assess the effect of the refinement step. These tables confirm the advantage of hierarchical classification, which intrinsically exploits a priori domain knowledge embedded into the whole classifier structure. They also show that some emotions (e.g., sadness) are classified rather well, while others are harder to classify (e.g., anger). For each class the best results in terms of precision, recall and F-measure have been emphasized. These results show that the best results have been obtained using the filtered training set (TS2).

Table 5 reports in detail the partial results of the three classification levels of the hierarchical classifiers: the first and the second level of classification, i.e., subjectivity and polarity, have an accuracy of around 90 % and 75 %, respectively. Aggregating the results of the flat classifier to provide the same partial responses provides systematically worse results. This is not surprising, since a seven-output classification is a harder task in general, and for ambiguous (and often mixed) emotions in particular. On the other hand, the cascaded structure of a hierarchical classifiers has a higher risk of propagating errors from the higher levels to the lower ones. From this point of view, the results show that the structure we adopted minimizes that effect, since, still not surprisingly, the accuracy of classifiers increases with their level in the hierarchy.

Finally, Table 6 shows the confusion matrix of the hierarchical classifier trained using TS2 which is the best performing approach in of our research. Notably, fear and anger are often misclassified as sadness and love is often misclassified as joy or surprise

Table 3. Accuracy of flat classifier, using alternatively the two training sets.

	Flat					
	TS1			TS2		
	Prec	Rec	F_M	Prec	Rec	F_M
Objective	**0.70**	0.69	0.69	0.58	0.73	0.65
Anger	**0.36**	0.18	0.24	**0.36**	0.23	0.28
Fear	0.39	0.22	0.28	**0.50**	0.13	0.21
Sadness	0.29	**0.68**	0.41	0.31	0.66	0.42
Love	0.45	0.29	0.35	**0.50**	0.32	**0.39**
Joy	0.32	0.34	0.33	0.36	**0.42**	**0.39**
Surprise	0.46	0.39	0.42	**0.50**	0.42	0.46
Total accuracy	40,48 %			42,33 %		

Table 4. Accuracy of hierarchical classifier, using alternatively the two training sets.

	Hierarchical					
	TS1			TS2		
	Prec	Rec	F_M	Prec	Rec	F_M
Objective	0.66	0.87	**0.75**	0.60	**0.88**	0.71
Anger	0.31	0.27	0.28	0.34	**0.30**	**0.32**
Fear	0.40	0.22	0.28	0.46	**0.23**	**0.31**
Sadness	0.34	0.56	0.43	**0.38**	0.60	**0.46**
Love	0.40	**0.34**	0.37	0.43	0.28	0.34
Joy	**0.39**	0.34	0.37	0.39	0.38	0.39
Surprise	0.47	0.41	0.44	0.50	**0.45**	**0.48**
Total accuracy	43,61 %			**45,17 %**		

Table 5. Accuracy of the intermediate results of hierarchical classification, based on TS1 and TS2.

	TS1	TS2
Objective/Subjective	91.90 %	90.06 %
Positive/Negative	73,36 %	75,54 %
Final classification	43.61 %	45.17 %

Table 6. Confusion matrix of the hierarchical classification based on TS2.

->	Objective	Fear	Anger	Sadness	Love	Joy	Surprise
Objective	**88**	3	2	3	2	1	1
Fear	15	**22**	15	23	1	13	5
Anger	15	7	**30**	38	4	5	1
Sadness	8	1	13	**63**	1	10	8
Love	5	8	8	11	**29**	20	21
Joy	8	4	7	16	17	**39**	10
Surprise	7	2	12	12	12	11	**47**

5   Conclusion

Emotion detection in social media is becoming a task of increasing importance, since it can provide a richer view of a user's opinion and feelings about a certain topic. It can also pave the way to detecting provocatorial and anti-social behaviors. In this work, we have analyzed the problem of automatic classification of tweets, according to their emotional value. We referred to Parrott's model of six primary emotions: anger, fear, sadness, joy, love, and surprise. In particular, we compared a flat classifier to a hierarchical one. Our tests show that the domain knowledge embedded into the hierarchical classifier makes it more accurate than the flat classifier. Also, our results prove that the process of automatic construction of training sets is viable, at least for sentiment analysis and emotion classification, since our automatic filtering of training data makes it possible to create training sets that improve the quality of the final classifier with respect to a "blind" collection of raw data based only on the hashtags. The results we have obtained are comparable with those found in similar works [26].

References

1. Ugolotti, R., Sassi, F., Mordonini, M., Cagnoni, S.: Multi-sensor system for detection and classification of human activities. J. Ambient Intell. Humaniz. Comput. **4**, 27–41 (2013)
2. Matrella, G., Parada, G., Mordonini, M., Cagnoni, S.: A video-based fall detector sensor well suited for a data-fusion approach. Assistive Technol. Adap. Equipment Inclusive Environ. Assistive Technol. Res. Ser. **25**, 327–331 (2009)
3. Fornacciari, P., Mordonini, M., Tomauiolo, M.: A case-study for sentiment analysis on Twitter. In: Workshop dagli Oggetti agli Agenti, WOA 2015 (2015)
4. Fornacciari, P., Mordonini, M., Tomauiolo, M.: Social network and sentiment analysis on Twitter: towards a combined approach. In: 1st International Workshop on Knowledge Discovery on the WEB, KDWeb 2015 (2015)
5. Parrott, W.G.: Emotions in Social Psychology: Essential Readings. Psychology Press, Philadelphia (2001)
6. Liu, B.: Sentiment Analysis and Opinion Mining: Synthesis Lectures on Human Language Technologies, vol. 5, pp. 1–167 (2012)

7. Mohammad, S.M.: Sentiment analysis: detecting valence, emotions, and other affectual states from text. Emotion Measurement (2015)
8. Mislove, A., Lehmann, S., Ahn, Y.Y., Onnela, J.P., Rosenquist, J.N.: Pulse of the nation: US mood throughout the day inferred from Twitter. Northeastern University (2010)
9. Healey, C., Ramaswamy, S.: Visualizing Twitter sentiment (2010). Accessed 17 June 2016
10. Allisio, L., Mussa, V., Bosco, C., Patti, V., Ruffo, G.: Felicittà: Visualizing and estimating happiness in Italian cities from geotagged tweets. In: ESSEM@ AI* IA, pp. 95–106 (2013)
11. Kao, E.C., Liu, C.C., Yang, T.H., Hsieh, C.T., Soo, V.W.: Towards text-based emotion detection a survey and possible improvements. In: 2009 International Conference on Information Management and Engineering, ICIME 2009, pp. 70–74. IEEE (2009)
12. Strapparava, C., Valitutti, A., et al.: WordNet affect: an affective extension of WordNet. In: LREC, vol. 4, pp. 1083–1086 (2004)
13. Strapparava, C., Mihalcea, R.: Learning to identify emotions in text. In: Proceedings of the 2008 ACM Symposium on Applied Computing, pp. 1556–1560. ACM (2008)
14. Poria, S., Cambria, E., Winterstein, G., Huang, G.B.: Sentic patterns: dependency-based rules for concept-level sentiment analysis. Knowl. Based Syst. **69**, 45–63 (2014)
15. Franchi, E., Poggi, A., Tomaiuolo, M.: Social media for online collaboration in firms and organizations. Int. J. Inf. Syst. Model. Des. (IJISMD) **7**, 18–31 (2016)
16. Dumais, S., Chen, H.: Hierarchical classification of web content. In: Proceedings of the 23rd Annual International ACM SIGIR Conference on Research and Development in Information Retrieval, pp. 256–263. ACM (2000)
17. Silla Jr., C.N., Freitas, A.A.: A survey of hierarchical classification across different application domains. Data Min. Knowl. Disc. **22**, 31–72 (2011)
18. Addis, A., Armano, G., Vargiu, E.: A progressive filtering approach to hierarchical text categorization. Commun. SIWN **5**, 28–32 (2008)
19. Koller, D., Sahami, M.: Hierarchically classifying documents using very few words. In: Proceedings of the 14th International Conference on Machine Learning (1997)
20. Armano, G., Mascia, F., Vargiu, E.: Using taxonomic domain knowledge in text categorization tasks. International Journal of Intelligent Control and Systems (2007). Special Issue on Distributed Intelligent Systems
21. Mitchell, T.: Conditions for the equivalence of hierarchical and flat Bayesian classifier. Technical report, Center for Automated Learning and Discovery, Carnegie-Mellon University (1998)
22. Ghazi, D., Inkpen, D., Szpakowicz, S.: Hierarchical approach to emotion recognition and classification in texts. In: Farzindar, A., Kešelj, V. (eds.) AI 2010. LNCS (LNAI), vol. 6085, pp. 40–50. Springer, Heidelberg (2010). doi:10.1007/978-3-642-13059-5_7
23. Al-Hajjar, D., Syed, A.Z.: Applying sentiment and emotion analysis on brand tweets for digital marketing. In: 2015 IEEE Jordan Conference on Applied Electrical Engineering and Computing Technologies (AEECT), pp. 1–6. IEEE (2015)

24. Esuli, A., Sebastiani, F.: SentiWordNet: a publicly available lexical resource for opinion mining. In: Proceedings of LREC, vol. 6, pp. 417–422. Citeseer (2006)
25. Mohammad, S.M., Kiritchenko, S.: Using hashtags to capture fine emotion categories from tweets. Comput. Intell. **31**, 301–326 (2015)
26. Ghazi, D., Inkpen, D., Szpakowicz, S.: Hierarchical versus flat classification of emotions in text. In: Proceedings of the NAACL HLT 2010 Workshop on Computational Approaches to Analysis and Generation of Emotion in Text, pp. 140–146. Association for Computational Linguistics (2010)

Spam Filtering Using Regularized Neural Networks with Rectified Linear Units

Aliaksandr Barushka and Petr Hájek[(✉)]

Institute of System Engineering and Informatics,
aculty of Economics and Administration,
University of Pardubice, Pardubice, Czech Republic
{Aliaksandr.Barushka,Petr.Hajek}@upce.cz

Abstract. The rapid growth of unsolicited and unwanted messages has inspired the development of many anti-spam methods. Machine-learning methods such as Naïve Bayes (NB), support vector machines (SVMs) or neural networks (NNs) have been particularly effective in categorizing spam /non-spam messages. They automatically construct word lists and their weights usually in a bag-of-words fashion. However, traditional multilayer perceptron (MLP) NNs usually suffer from slow optimization convergence to a poor local minimum and overfitting issues. To overcome this problem, we use a regularized NN with rectified linear units (RANN-ReL) for spam filtering. We compare its performance on three benchmark spam datasets (Enron, SpamAssassin, and SMS spam collection) with four machine algorithms commonly used in text classification, namely NB, SVM, MLP, and k-NN. We show that the RANN-ReL outperforms other methods in terms of classification accuracy, false negative and false positive rates. Notably, it classifies well both major (legitimate) and minor (spam) classes.

Keywords: Spam filter · Email · Sms · Neural network · Regularization · Rectified linear unit

1 Introduction

Spam is defined as unsolicited and unwanted messages sent electronically by a sender having no current relationship with the recipient [1]. Email spam is a subset of electronic spam involving nearly identical messages sent to numerous recipients by email. Clicking on links in spam email may send users to phishing web sites or sites that are hosting malware. Spam email may also include malware as scripts or other executable file attachments. By contrast, SMS spam is typically transmitted over a mobile network [2].

Recent studies have shown that on average 80 % of emails is spam, with significant differences in spam rates among countries (see e.g. the Global Spam Map). As a result, serious negative effects on the worldwide economy have been observed [3, 4], including lower productivity or the cost with delivering spam and viruses/phishing attacks.

© Springer International Publishing AG 2016
G. Adorni et al. (Eds.): AI*IA 2016, LNAI 10037, pp. 65–75, 2016.
DOI: 10.1007/978-3-319-49130-1_6

It is hardly possible to find who the first person to send spam was. The idea behind it is very simple. If the spammer sends a message to millions of people and only one person replies back it makes this activity profitable. This is true for both email and SMS spam because of the availability of unlimited pre-pay SMS packages. In order to increase the cost of sending spam, a highly accurate spam filter is necessary [5].

Over the years, various anti-spam techniques have been developed. Machine-learning methods have been particularly effective in detecting spams, including techniques such as Naïve Bayes (NB) classifiers [6, 7], decision trees [8], support vector machines (SVMs) [9], k-nearest neighbour algorithm (k-NN) [10], or multilayer perceptron neural network (MLP) [11]. These techniques treat spam filtering as a binary classification problem, categorizing the incoming message as either spam or non-spam. This is a challenging task because both misclassifying a legitimate message to spam and misclassifying a spam to non-spam brings costs [12]. The main idea behind machine-learning approaches is to automatically construct word lists and their weights based on a classification of messages. However, spammers usually attempt to decrease the probability of being detected as spam by using legitimate words [5].

The current state of the art in spam filtering has been recently surveyed for both email [13, 14] and SMS spam filtering [2]. The surveys conclude that Bayesian approaches remain highly popular with researchers, whereas neural networks (NNs) are significantly under-researched in this field. In contrast to Bayesian approaches, NNs (and SVMs) are more computationally expensive, limiting their maximum potential application in online spam filtering [14]. However, NNs have recently shown promising potentials for textual classification, especially when equipped with advanced techniques such as rectified linear units and dropout regularization [15]. Thus, the main limitations of traditional NNs can be addressed, namely slow optimization convergence to a poor local minimum and overfitting issues.

To perform spam filtering, we therefore use a regularized NN with rectified linear units (RANN-ReL) [16] and compare it with four machine learning algorithms commonly used in text classification [17], namely NB, SVM, MLP, and k-NN classifiers. We demonstrate that the RANN-ReL outperforms other methods on three benchmark spam datasets, namely Enron, SpamAssassin and SMS spam collection.

The remainder of this paper is organised in the following way. Section 2 briefly reviews related literature. Section 3 presents the spam datasets used for the automatic filtering. Section 4 introduces the NN with dropout regularization and rectified linear units. In addition, methods used for comparative analysis are presented. The experiments are performed in Sect. 5. Section 6 discusses the obtained results and concludes the paper.

2 Machine Learning Techniques in Spam Filtering

The negative economical impacts of spam drive some countries to adopt legislation, for details see [3, 18]. However, this approach has a serious drawback that spam messages can be sent from multiple countries [19]. Moreover, it is not easy to track the actual senders of spam [13]. In addition to legislation, some authors have proposed changes in protocols and operation models [3]. Other non-machine-learning approaches include rule-, signature- and hash-based filtering, whitelists and blacklists, and traffic analysis [14]. Another way to tackle the issue is to apply machine learning spam filters, which automatically identify whether the message is spam or not based on the content of the message. Once message identified as spam it can be either moved to spam folder or deleted. In contrast to other text classification tasks, spam filters have to address several specific issues [20]: (1) skewed and changing class distributions; (2) unequal and uncertain misclassification costs of spam and legitimate messages; (3) complex text patterns; and (4) concept drift.

Spam filtering using machine learning approaches starts with text pre-processing [13]. First, tokenization is performed to extract the words (multi-words) in the messages. Next, the initial set of words is usually reduced by stemming, lemmatization and stop-words removal. Bag-of-words (BoW) (also known as the vector-space model) is a common approach to represent the weights of the pre-processed words. Specifically, binary representation, term frequency (tf) and term frequency–inverse document frequency (tf-idf) are popular weighting schemes. A feature selection algorithm may be further applied to detect words with the highest relevance. These algorithms include term and document frequencies, information gain, χ^2 statistic, or evolutionary algorithms used in filters or wrappers [21]. Finally, machine learning classifiers are applied on the pre-processed dataset.

The first classifiers of spam/legitimate messages employed the NB algorithms mainly due to their simplicity and computational efficiency [6, 7, 22]. SVM is another popular classification algorithm used in spam filtering. Drucker et al. [9] reported that, when compared with decision trees, SVMs are more robust to both different datasets and pre-processing techniques. The comparative study by [23] demonstrated that SVM, Adaboost and logistic regression methods are superior to NB and k-NN approaches. Furthermore, Koprinska et al. [24] showed that random forest outperforms traditional decision trees, SVMs and NB and, in addition, the performance of the spam filter can be boosted by using semi-supervised co-training paradigm. Another comparative study was conducted by [25], demonstrating that SVM performed better than NB and k-NN classifiers. Recent studies also confirm the superiority of SVMs over NB, decision trees and MLP [26].

Several interesting research directions in spam filtering have recently emerged such as the utilization of social networks and unlabelled training data. Shen and Li [5] exploited the social relationships among email correspondents and their (dis)interests to automatically detect spam. Another issue being addressed in spam filtering is the limited availability of labelled training data. Laorden et al. [4] proposed an anomaly based spam filtering system using a data reduction algorithm to the labelled dataset. Thus, processing time was reduced while maintaining high detection rates.

3   Spam Datasets

An important aspect when benchmarking different spam classifiers is the dataset used for evaluating its performance. To examine the performance of the chosen methods, the following spam datasets were used: (1) Enron[1], (2) SpamAssassin[2], and (3) SMS[3] dataset.

Enron spam dataset [7] is a popular dataset with spam and ham email messages. This spam dataset has been used in a number of studies, see [13] for an overview. This dataset contains a total of 5172 emails, including 3672 legitimate and 1500 spam emails.

SpamAssasin spam dataset is another popular corpus which has been used as a benchmark in many studies. This dataset contains 3252 emails, of which 2751 are legitimate and 501 are spam emails. Note that this dataset is, when compared with the Enron spam dataset, more imbalanced with 84.6 % legitimate emails.

SMS spam dataset [27] was chosen in order to diversify spam corpuses. Unlike Enron and SpamAssassin datasets, SMS spam dataset includes 4822 legitimate and 746 spam SMS messages, this is a total of 5568 messages.

Before being able to classify legitimate and spam messages, we performed data pre-pocessing. First, all words were converted to lower case letters and tokenization was performed. To represent messages, we further removed stop-words using the Rainbow stop-word handler. These words do not provide any semantic information and add noise to the model [4]. Snowball stemmer was used as a stemming algorithm. To represent the weights of the pre-processed words, we used *tf.idf* as the most common BoW approach. In this scheme, weights w_{ij} are calculated as follows:

$$w_{ij} = \left(1 + \log(tf_{ij})\right) \times \log(N/df_i) \tag{1}$$

where N denotes the total number of messages, tf_{ij} is the frequency of the i-th word in the j-th message, and df_i denotes the number of messages with at least one occurrence of the i-th term. To select the most relevant words, we ranked them according to their *tf. idf*. For our experiments, we used the top 200, 1000 and 2000 words in a BoW fashion.

4   Methods

In this section, we introduce a NN with dropout regularization and rectified linear units (RANN-ReL). To demonstrate the effectiveness of the RANN-ReL, we compared the results with four methods commonly used in spam classification tasks, namely NB, SVM, MLP, and k-NN classifier. Therefore, we also provide a brief description of these methods.

---

[1] http://csmining.org/index.php/enron-spam-datasets.html.

[2] http://csmining.org/index.php/spam-assassin-datasets.html.

[3] https://archive.ics.uci.edu/ml/datasets/SMS+Spam+Collection.

Neural Network with Dropout Regularization and Rectified Linear Units. Complex tasks require a large number of hidden units to model them accurately. However, such a complex adaptation on training data may lead to overfitting, preventing a high accuracy on testing data. The overfitting issue can be effectively addressed by using dropout regularization. In the fully connected layers of feed-forward NN, dropout regularization randomly sets a given proportion (usually a half) of the activations to zero during training. Thus, hidden units that activate the same output are omitted.

Commonly used sigmoidal units are reported to suffer from the vanishing gradient problem, often accompanied with slow optimization convergence to a poor local minimum [28]. Rectified linear (ReL) unit tackles this problem so that when it is activated above 0, its partial derivative is 1. The ReL function can be defined as follows:

$$h_i = \max(w_i^T x, 0) = \begin{cases} w_i^T x & \text{if } w_i^T x > 0 \\ 0 & \text{otherwise} \end{cases}, \tag{2}$$

where w_i is the weight vector of the i-th hidden unit, and x is the input vector. The ReL function is therefore one-sided and does not enforce a sign symmetry or anti-symmetry. On the other hand, the main disadvantage of using the ReL is the fact that this function allows a NN to easily get sparse representation. It also leads to a less intensive computation because there is no need to compute the exponential function in activations and sparsity can be exploited. The combination of dropout regularization with ReL units has shown promising synergistic effects in [29, 30].

Naïve Bayes. The NB classifier represents a probability-based approach. The NB has become a popular method in spam filtering due to its simplicity. It uses information learnt from training data to compute the probability that the message is spam or legitimate given the words appearing in the message. However, it is based on the assumption that feature values are independent given the class, which is often not fulfilled in text classification tasks. Overall, the NB learning is relatively easy to implement and accommodates discrete features reasonably well. It has also been reported to be quite as accurate or robust as some other machine learning methods.

k-NN. Another simple machine learning method is the k-NN classifier. It is considered an example-based classifier, where training data are used for comparison rather than an explicit class representation. There is basically no training phase. A new message is classified based on the k most similar messages (mostly using the Euclidean distance). Moreover, finding the nearest neighbor(s) can be speed up using indexing.

Multilayer Perceptron. The MLP is a classifier based on the feed-forward artificial neural network. It consists of multiple layers of nodes, where each layer is fully connected to the next layer in the network. Nodes in the input layer represent the input data. All other nodes maps inputs to the outputs by performing linear combination of the inputs with the node's weights and applying an activation function. The backpropagation algorithm is a commonly used algorithm to train the MLP [11].

Support Vector Machines. SVM has been reported to be an effective classifier in spam filtering due to its ability to handle high-dimensional data. It finds the optimal separating hyperplane that gives the maximum margin between two classes. A sub-set of the training data (so-called support vectors) are used to define the decision boundary. Sequential Minimal Optimization (SMO) is a frequently used optimization technique to find the parameters of the separating hyperplane. The SMO decomposes the overall quadratic programming problem into sub-problems, using Osuna's theorem to ensure convergence. In case of non-linear classification, kernel functions are used to map the problem from its original feature space to a new feature space where linear separability is ensured.

5 Experimental Results

In this study, the RANN-ReL was trained using gradient descent algorithm with the following parameters:

- input layer dropout rate = 0.2,
- hidden layer dropout rate = 0.5,
- number of hidden units = {10, 20, 50, 100, 200},
- learning rate = {0.05, 0.10},
- the number of iterations = 1000.

The structure and parameters of the RANN-ReL, the same as for the remaining methods, were found using grid search procedure.

The SVMs was trained by the SMO algorithm. In the experiments, we examined the following parameters of the SVMs:

- kernel function = polynomial,
- the level of polynomial function = {1, 2},
- complexity parameter $C = \{2^0, 2^1, 2^2, ..., 2^8\}$.

The MLP was trained using the backpropagation algorithm with the following parameters:

- the number of neurons in the hidden layer = {10, 20, 50, 100, 200},
- learning rate = {0.05, 0.10, 0.30},
- momentum = 0.2,
- the number of iterations = 1000.

Finally, we used k-NN classifier with the Euclidean distance function and the number of neighbours set to 3.

To estimate the generalization performance of the classifiers, we used 10-fold cross-validation on the three spam datasets. The overall performance estimate is represented by the average and std. dev. over the 10 classifiers. To evaluate the performance, we used common measures in spam filtering, namely Accuracy, FP (false positive) and FN (false negative) rates. FP are legitimate messages that are mistakenly regarded as spam, whereas false negatives FN are spam messages that are not detected.

Table 1. Classification results on Enron, SpamAssassin and SMS datasets

Dataset	NB	SVM	MLP	k-NN	RANN-ReL
Acc [%]					
Enron	91.35 ± 1.18	96.55 ± 0.78*	95.90 ± 2.95	91.85 ± 1.31	**98.33 ± 0.57***
SpamAssassin	96.61 ± 1.43	**99.85 ± 0.22***	99.52 ± 0.39*	98.72 ± 0.56*	99.80 ± 0.22*
SMS	95.07 ± 0.77	97.58 ± 0.68*	95.77 ± 3.78	93.36 ± 0.92	**98.54 ± 0.43***
FN rate [%]					
Enron	12.04 ± 1.65	2.93 ± 0.87	3.21 ± 2.88	9.34 ± 1.63	**1.72 ± 0.74**
SpamAssassin	6.01 ± 1.30	**0.09 ± 0.17**	0.32 ± 0.35	0.33 ± 0.32	0.17 ± 0.22
SMS	12.44 ± 3.63	12.11 ± 4.42	28.98 ± 29.55	44.46 ± 6.46	**8.67 ± 3.19**
FP rate [%]					
Enron	**0.34 ± 0.51**	4.73 ± 1.64	6.28 ± 11.18	5.21 ± 1.98	1.63 ± 0.99
SpamAssassin	21.46 ± 6.09	0.50 ± 1.11	1.34 ± 1.75	6.15 ± 2.94	**0.40 ± 0.90**
SMS	3.77 ± 0.79	0.62 ± 0.76	0.40 ± 0.45	0.79 ± 0.38	**0.34 ± 0.25**

* significantly higher at $p = 0.05$.

Table 2. Classification results on Enron, SpamAssassin and SMS datasets for 200 words

Dataset	NB	SVM	MLP	k-NN	RANN-ReL
Acc [%]					
Enron	89.02 ± 1.28	94.77 ± 1.03*	94.21 ± 1.80	91.85 ± 1.31	**96.20 ± 0.85***
SpamAssassin	96.61 ± 1.43	99.24 ± 0.44*	99.30 ± 0.51*	98.72 ± 0.56*	**99.51 ± 0.35***
SMS	93.59 ± 0.96	95.32 ± 0.84*	95.22 ± 1.33	93.36 ± 0.92	**96.09 ± 0.76***
FN rate [%]					
Enron	11.35 ± 1.78	4.82 ± 1.16	5.45 ± 2.92	9.34 ± 1.63	**4.40 ± 1.07**
SpamAssassin	6.01 ± 1.30	0.48 ± 0.37	0.43 ± 0.36	0.33 ± 0.32	**0.33 ± 0.32**
SMS	18.34 ± 4.17	24.80 ± 4.84	25.14 ± 7.61	44.46 ± 6.46	**23.14 ± 4.75**
FP rate [%]					
Enron	**0.82 ± 0.76**	5.33 ± 2.05	6.62 ± 6.11	5.21 ± 1.98	2.32 ± 0.99
SpamAssassin	21.46 ± 6.09	2.30 ± 2.02	2.20 ± 2.53	6.15 ± 2.94	**1.48 ± 1.69**
SMS	4.57 ± 0.86	1.55 ± 0.58	1.63 ± 1.76	**0.79 ± 0.38**	0.90 ± 0.48

* significantly higher at $p = 0.05$.

The results of classification are summarized in Table 1. Here, the best classification accuracy and related performance measures are presented over 200, 1000 and 2000 words. We employed Student's paired t-test at $p = 0.05$ to test the differences in classification accuracy.

The results show that the RANN-ReL achieved the highest classification accuracy on Enron and SMS datasets. However, the SVM performed best on SpamAssassin and statistically similar with only slightly worse accuracies on the remaining datasets. For the SpamAssassin datasets, the MLP and k-NN also performed well. Although the NB classifier performed significantly worse in terms of accuracy, it achieved the lowest FP rate for the Enron dataset. For the remaining two datasets, the RANN-ReL achieved the best results. Since the SpamAssassin and SMS datasets were strongly imbalanced in favour of legitimate messages, the classification performance in terms of FN and FP rate was particularly important. It is therefore favourable that the RANN-ReL performed reasonably well in terms of both measures.

Detailed results on spam filtering are presented in Tables 2, 3 and 4, separately for 200, 1000 and 2000 words, respectively. Considering the number of features, the best

Table 3. Classification results on Enron, SpamAssassin and SMS datasets for 1000 words

Dataset	NB	SVM	MLP	k-NN	RANN-ReL
Acc [%]					
Enron	91.32 ± 1.18	96.53 ± 0.78*	95.90 ± 2.95	83.36 ± 2.02	**98.33 ± 0.57***
SpamAssassin	92.38 ± 1.33	99.66 ± 0.31*	99.52 ± 0.39*	96.60 ± 0.62	**99.75 ± 0.27***
SMS	94.93 ± 0.84	97.58 ± 0.68*	95.77 ± 3.78	92.43 ± 0.86	**98.41 ± 0.52***
FN rate [%]					
Enron	12.05 ± 1.64	2.82 ± 0.79	3.21 ± 2.88	21.93 ± 2.99	**1.72 ± 0.74**
SpamAssassin	7.70 ± 1.41	0.24 ± 0.27	0.32 ± 0.35	0.75 ± 0.45	**0.19 ± 0.26**
SMS	13.56 ± 4.16	12.89 ± 3.98	28.98 ± 29.55	55.49 ± 5.81	**10.35 ± 2.75**
FP rate [%]					
Enron	**0.45 ± 0.57**	5.07 ± 1.67	6.28 ± 11.18	1.95 ± 1.47	1.63 ± 0.99
SpamAssassin	7.19 ± 3.58	0.94 ± 1.51	1.34 ± 1.75	4.95 ± 3.15	**0.56 ± 1.06**
SMS	3.84 ± 0.98	1.46 ± 0.59	**0.40 ± 0.45**	0.62 ± 0.51	0.50 ± 0.30

* significantly higher at $p = 0.05$.

Table 4. Classification results on Enron, SpamAssassin and SMS datasets for 2000 words

Dataset	NB	SVM	MLP	k-NN	RANN-ReL
Acc [%]					
Enron	91.35 ± 1.18	96.55 ± 0.78*	94.23 ± 5.25	81.74 ± 1.79	**98.31 ± 0.54***
SpamAssassin	92.49 ± 1.38	**99.85 ± 0.22***	97.74 ± 4.66*	98.50 ± 0.58*	99.80 ± 0.22*
SMS	95.07 ± 0.77	97.39 ± 0.60*	86.60 ± 0.08	92.25 ± 0.71	**98.54 ± 0.43***
FN rate [%]					
Enron	12.04 ± 1.65	2.93 ± 0.87	5.54 ± 6.85	25.11 ± 3.13	**1.69 ± 1.70**
SpamAssassin	7.94 ± 1.51	**0.09 ± 0.17**	0.31 ± 0.36	0.63 ± 0.44	0.17 ± 0.22
SMS	12.44 ± 3.63	11.68 ± 3.68	28.30 ± 1.30	57.13 ± 5.42	**8.67 ± 3.19**
FP rate [%]					
Enron	**0.34 ± 0.51**	4.73 ± 1.64	6.33 ± 14.22	1.48 ± 1.16	1.71 ± 0.93
SpamAssassin	5.17 ± 2.99	0.50 ± 1.11	12.97 ± 31.79	6.29 ± 2.98	**0.40 ± 0.90**
SMS	3.77 ± 0.79	1.20 ± 0.48	0.30 ± 0.25	**0.11 ± 0.15**	0.34 ± 0.25

* significantly higher at $p = 0.05$.

Table 5. Average elapsed training times on Enron, SpamAssassin and SMS datasets

Dataset	NB	SVM[a]	MLP[b]	k-NN	RANN-ReL[b]
Enron (200)	0.3	27.0	2087.1	0.0	40.3
Enron (1000)	3.3	12.1	8846.8	0.0	668.8
Enron (2000)	8.4	12.7	12018.1	0.0	2054.4
SpamAssassin (200)	0.5	0.7	578.4	41.2	1353.9
SpamAssassin (1000)	1.2	1.6	1250.9	0.0	2070.0
SpamAssassin (2000)	2.1	0.7	5154.1	0.0	4079.3
SMS (200)	0.2	16.9	1261.8	24.8	12.7
SMS (1000)	0.7	12.2	3541.5	8.5	187.2
SMS (2000)	1.5	6.2	8500.8	0.0	1536.9

[a] $C = 2^4$, [b] the number of hidden units = 100.

results were obtained, as expected, for 2000 words in case of the RANN-ReL consistently over all the three datasets. However, we observed a statistically similar performance for 1000 words and a significant drop in accuracy occurred only when 200 words were used.

6 Conclusion

The results showed that the RANN-ReL outperformed the remaining methods in terms of classification accuracy, except for the SpamAssassing dataset. More importantly, it classified well both major (legitimate) and minor (spam) classes. SVM achieved statistically similar accuracies on all the three benchmark spam datasets. By contrast, the remaining algorithms (NB, MLP and k-NN) performed relatively poorly.

However, the RANN-ReL was significantly more computationally intensive than the remaining algorithms. On average, the elapsed training time of the RANN-ReL was substantially higher than that of the SVM (Table 5), particularly for higher dimensions. This fact limits the application of the RANN-ReL as an online spam filter. On the other hand, the results suggest that it can be effectively used for static training datasets. Returning to the issues specific for spam filters, it is now possible to state that the RANN-ReL classifier may effectively address imbalanced class distributions, uncertain misclassification costs, as well as complex text patterns. Another minor point of the present study is that it uses traditional *tf.idf* features (and BoW approach) rather than, for example, exploiting word vectors. Moreover, the high computational expenses make it difficult to tackle the problem of concept drift. Further investigation and experimentation into the concept drift issue is therefore strongly recommended. It would be also interesting to assess the effects of various feature selection methods on the classification accuracy.

Acknowledgments. This work was supported by the grant No. SGS_2016_023 of the Student Grant Competition.

References

1. Cormack, G.V.: Email spam filtering: a systematic review. Found. Trends Inf. Retrieval **1**(4), 335–455 (2006)
2. Delany, S.J., Buckley, M., Greene, D.: SMS spam filtering: methods and data. Expert Syst. Appl. **39**(10), 9899–9908 (2012)
3. Hoanca, B.: How good are our weapons in the spam wars? IEEE Technol. Soc. Mag. **25**(1), 22–30 (2006)
4. Laorden, C., Ugarte-Pedrero, X., Santos, I., Sanz, B., Nieves, J., Bringas, P.G.: Study on the effectiveness of anomaly detection for spam filtering. Inf. Sci. **277**, 421–444 (2014)
5. Shen, H., Li, Z.: Leveraging social networks for effective spam filtering. IEEE Trans. Comput. **63**(11), 2743–2759 (2014)

6. Androutsopoulos, I., Koutsias, J., Chandrinos, K.V., Spyropoulos, C.D.: An experimental comparison of naive bayesian and keyword-based anti-spam filtering with personal E-mail messages. In: Proceedings of the 23rd Annual International ACM SIGIR Conference on Research and Development in Information Retrieval, pp. 160–167. ACM (2000)

7. Metsis, V., Androutsopoulos, I., Paliouras, G.: Spam filtering with naive bayes - which naive bayes? In: Third Conference on Email and AntiSpam (CEAS), pp. 27–28 (2006)

8. Carreras, X., Marquez, L.: Boosting trees for anti-spam email filtering. In: Proceedings of RANLP 2001, Bulgaria, pp. 58–64 (2001)

9. Drucker, H., Wu, D., Vapnik, V.: Support vector machines for spam categorization. IEEE Trans. Neural Netw. **10**(5), 1048–1054 (1999)

10. Jiang, S., Pang, G., Wu, M., Kuang, L.: An Improved K-nearest-neighbor algorithm for text categorization. Expert Syst. Appl. **39**(1), 1503–1509 (2012)

11. Clark, J., Koprinska, I., Poon, J.: A neural network based approach to automated e-mail classification. In: Proceedings of the IEEE/WIC International Conference on Web Intelligence (WI 2003), pp. 702–705. IEEE Computer Society (2003)

12. Zhou, B., Yao, Y., Luo, J.: Cost-sensitive three-way email spam filtering. J. Intell. Inf. Syst. **42**(1), 19–45 (2014)

13. Guzella, T., Caminhas, W.: A review of machine learning approaches to spam filtering. Expert Syst. Appl. **36**(7), 10206–10222 (2009)

14. Caruana, G., Li, M.: A survey of emerging approaches to spam filtering. ACM Comput. Surv. **44**(2), 1–27 (2012)

15. Nam, J., Kim, J., Mencía, E.L., Gurevych, I., Fürnkranz, J.: Large-scale multi-label text classification - revisiting neural networks. In: Calders, T., Esposito, F., Hüllermeier, E., Melo, R. (eds.) Machine Learning and Knowledge Discovery in Databases, pp. 437–452. Springer, Berlin Heidelberg (2014)

16. Hinton, G., Srivastava, N., Krizhevsky, A., Sutskever, I., Salakhutdinov, R.: Improving neural networks by preventing co-adaptation of feature detectors. arXiv:1207.0580 (2012)

17. Khan, A., Baharudin, B., Lee, L.: A review of machine learning algorithms for text-documents classification. J. Adv. Inf. Technol. **1**(1), 4–20 (2010)

18. Carpinter, J., Hunt, R.: Tightening the net: a review of current and next generation spam filtering tools. Comput. Secur. **25**(8), 566–578 (2006)

19. Talbot, D.: Where Spam is born. MIT Technol. Rev. **111**(3), 28 (2008)

20. Fawcett, T.: In vivo spam filtering: a challenge problem for KDD. ACM SIGKDD Explor. Newsl. **5**(2), 140–148 (2003)

21. Zhang, Y., Wang, S., Phillips, P., Ji, G.: Binary PSO with mutation operator for feature selection using decision tree applied to spam detection. Knowl.-Based Syst. **64**, 22–31 (2014)

22. Sahami, M., Dumais, S., Heckerman, D., Horvitz, E.: A bayesian approach to filtering junk E-Mail. In: Papers from the 1998 Workshop Learning for Text Categorization, vol. 62, pp. 98–105 (1998)

23. Zhang, L., Zhu, J., Yao, T.: An evaluation of statistical spam filtering techniques. ACM Trans. Asian Lang. Inf. Process. **3**(4), 243–269 (2004)

24. Koprinska, I., Poon, J., Clark, J., Chan, J.: Learning to classify E-mail. Inf. Sci. **177**(10), 2167–2187 (2007)

25. Lai, C.: An empirical study of three machine learning methods for spam filtering. Knowl.-Based Syst. **20**(3), 249–254 (2007)

26. Vyas, T., Prajapati, P., Gadhwal, S.: A survey and evaluation of supervised machine learning techniques for spam E-mail filtering. In: IEEE International Conference on Electrical, Computer and Communication Technologies (ICECCT), pp. 1–7. IEEE (2015)

27. Almeida, T.A., Hidalgo, J.M.G., Yamakami, A.: Contributions to the study of SMS spam filtering: new collection and results. In: Proceedings of the 11th ACM Symposium on Document Engineering, pp. 259–262. ACM (2011)
28. Maas, A.L., Hannun, A.Y., Ng, A.Y.: Rectifier nonlinearities improve neural network acoustic models. In: Proceedings of the 30th International Conference on Machine Learning, vol. 30, pp. 1–6 (2013)
29. Jaitly, N., Hinton, G.: Learning a better representation of speech soundwaves using restricted boltzmann machines. In: IEEE International Conference on Acoustics, Speech and Signal Processing (ICASSP), pp. 5884–5887. IEEE (2011)
30. Hajek, P., Bohacova, J.: Predicting abnormal bank stock returns using textual analysis of annual reports - a neural network approach. In: Jayne, C., Iliadis, L. (eds.) Engineering Applications of Neural Networks (EANN), pp. 67–78. Springer, New York (2016)

User Mood Tracking for Opinion Analysis on Twitter

Giuseppe Castellucci[1(✉)], Danilo Croce[2], Diego De Cao[1], and Roberto Basili[2]

[1] Reveal s.r.l., Rome, Italy
{castellucci,decao}@reavealsrl.it
[2] Department of Enterprise Engineering,
University of Roma Tor Vergata, Rome, Italy
{croce,basili}@info.uniroma2.it

Abstract. The huge variability of trends, community interests and jargon is a crucial challenge for the application of language technologies to Social Media analysis. Models, such as grammars and lexicons, are exposed to rapid obsolescence, due to the speed at which topics as well as slogans change during time. In Sentiment Analysis, several works dynamically acquire the so-called opinionated lexicons. These are dictionaries where information regarding subjectivity aspects of individual words are described. This paper proposes an architecture for dynamic sentiment analysis over Twitter, combining structured learning and lexicon acquisition. Evidence about the beneficial effects of a dynamic architecture is reported through large scale tests over Twitter streams in Italian.

Keywords: Social media analytics · Sentiment analysis · Opinion mining · Polarity lexicons

1 Introduction

One of the most complex challenges in social media analytics is the huge variability characterizing Social Web data. Traditional knowledge models, such as grammars or lexicons, are exposed to rapid obsolescence. Sentiment recognition has been recently proposed as a dynamic process that analyzes the opinion flows across linguistic structures and texts [1,2]: given a sentence and a discrete structure (e.g., a dependency graph reflecting its syntactic information) these approaches assign contextual polarity to words [1] or concepts [2], and then make the polarity flows through dependency arcs. The polarity class assigned to each sentence is given by the flows of such polarities through the adopted linguistic structure. These sentiment flows allow to model the contextual role of words (as well as their underlying polarity) in sentences and seem to support effective dynamic polarity inference.

Most work on Sentiment Analysis (SA) in Twitter focused on labeling individual tweets. It is more recent the idea that inferring user-level sentiments is important. Studies about both tweet and user-level sentiments often rely on user

© Springer International Publishing AG 2016
G. Adorni et al. (Eds.): AI*IA 2016, LNAI 10037, pp. 76–88, 2016.
DOI: 10.1007/978-3-319-49130-1_7

sentiment "profiles" obtained from his/her posts. An important challenge is the modeling of the temporal evolution of such profiles, as users tend to change their mind, through the interactions with other users. In [3], a two-level graph clustering is applied to a tripartite graph, including three mutually related bipartite graphs: a tweet-feature, a user-feature and a user-tweet graph. Co-clustering is used to capture joint constraints between features, users and tweets. On-line versions of the co-clustering algorithms can account for the dynamics of vocabularies as well as user profiles. In general, assumptions in these works consider that the vocabulary distributions change over time, but not their polarities. One often neglected issue is the changes in lexical polarities and their impact on the dynamics of the underlying target SA system. While a variety of works on Natural Language Processing (NLP) and SA have proposed automatic or semi-automatic processes for sentiment lexicons acquisition ([4–7]), most of them regarded the lexicon as a static: the emotional aspects of lexical entries are considered as given *a priori* and stable across time. While this is true on a relatively short time basis, emotional nuances are also dynamic lexical properties, and SA models that are sensitive to their changes should be investigated. The portability issue about opinionated lexicons is of significant interest for a large-scale SA within Social Web sources, and it refers not only to the *domain* changes but mostly concerns the opinion dynamics characterizing lexical entries. Words denoting specific events or people tend to be polarized according to the social consensus. For example, words denoting the places of the recent terrorist attacks in Paris (e.g., *Bataclan*) can be biased towards negative opinions, while before these events they could be considered mostly neutral.

In this paper, we promote an adaptive Web-based SA architecture, that accounts for the acquisition and automatic adaptation of the different involved knowledge resources: the semantic lexicon, the adopted opinionated lexicon as well as the sentiment recognition model. In the proposed SA architecture, the adaptation characterizes different steps: the stage of acquisition of a general (i.e., domain independent) semantic lexicon that can be driven by corpus analysis; the automatic acquisition of a sentiment lexicon, where polarity features for the major sentiment categories (e.g., positive, neutral and negative) are derived for large sets of the vocabulary; finally, the learning of an opinion model, through traditional supervised statistical algorithms. Here, we propose a first investigation on how general-purpose lexicons and sentiment oriented lexicons can be made more precise over data observed within a given time period. We compare the adoption of different lexicons across datasets referring to different time periods. Experiments show that a dynamic approach to SA (based on incremental opinionated lexicon acquisition) is more robust to model the dynamics of opinions in social contexts, w.r.t. an approach based on static opinionated lexicons.

In Sect. 2, we introduce related works. The proposed dynamic SA architecture is introduced in Sect. 3, while evaluation is discussed in Sect. 4.

2   Related Work

The tracking of evolving opinions is a recent research topic, where most works focused on domain adaptation for Sentiment Analysis, in analogy with tasks such as the automatic tracking of topical changes. For example, [8] investigates domain adaptation for sentiment classifiers, focusing on online reviews for different types of products. They extend the Structural Correspondence Learning algorithm with a specific selection of pivot features for linking the source and target domains. In [9] statistical approaches for the modeling of social media dynamics are proposed to predict collective sentiment dynamics. In particular, they build a statistical model to predict the sentiment change in the social media by considering three factors: the tweet history about a specific product/brand; how long it takes for the sentiment to change given the occurrence of a specific event; the duration this change lasts. These factors are synthesized in features that are adopted within statistical learning algorithms to predict future moods about products/brands. [10] proposes to model the opinion dynamics based on social media data in order to understand and explain the dynamic changes of public attitudes toward certain social events or hot topics. In particular, they propose to model and detect opinion dynamics in security-related social media data through a feature based approach. In [11] the authors focus on the definition of a semi-supervised topic-adaptive sentiment classification model, which starts with a classifier built on common features and mixed labeled data from various topics. The work in [3] doesn't focus on the domain adaptation issue. They build a graph clustering algorithm applied over a tripartite graph of users, tweet and words to detect the general sentiment of specific users, proposing a model that considers the temporal evolution of the user-level sentiment.

In this paper, we propose to model the evolution of opinion changes by adequately capturing the language dynamics at different time points. In order to capture and represent lexical information from messages written in a Social Network, we will make use of distributional models of lexical semantics ([12,13]). Moreover, we model the sentiment changes of single words within a dynamic notion of Polarity Lexicons. We will not manually annotate the sentiment of words, as it has been done in many works (see for example [14,15] or [16]). We will rely on an automatic process for polarity lexicon acquisition [7], i.e., a corpus-based approach inspired by the same distributional approaches used to represent the meaning of words. Corpus-based approaches acquire statistics of the words usage while considering the sentiment orientation. For example, in [6] a minimally-supervised approach based on Social Media data is proposed by exploiting hashtags or emoticons related to positivity and negativity, e.g., #happy, #sad, :) or :(. They compute a score, reflecting the polarity of each word, through a PMI based measure between a word and an emotion. Polarity Lexicon acquisition methods have been also proposed according to graph-based algorithms, as in [4] or [5]. These methods adopt word graphs, seed them with polarity information regarding some specific words and exploit sentiment-related information flows through words.

3   Tracking Opinion Across Time

The variability of topics and community interests in different time periods brings people to adapt their language in the Web, and in particular in the Social Media.

This phenomenon makes the language continuously evolving, by the introduction of new words that quickly become part of the dictionary people can use. Such evolution makes the linguistic systems obsolete, shortly. We promote the need of addressing this issue with adaptive architectures, that rely on systems whose update and deployment can be carried out with minimal effort. Under this perspective, an appropriate support is given for the retraining of an Opinion Mining architecture, in terms of the revision of its involved knowledge resources: data, semantic and opinionated lexicons, for promptly facing language changes. In order to achieve such goal, we strongly rely on statistical learning methods. In this way, by minimizing the need of manual annotation, we adopt semi-supervised and unsupervised methods by automatically gathering noisy data.

In Sentiment Analysis in Social Media, the language evolution phenomenon is prevalent, as for example in Twitter: users are encouraged to make creative uses of language. Tracking opinion with a *static* Opinion Mining (OM) system can be thus problematic. A possible dynamic OM workflow for Social Media analysis is outlined in Fig. 1. It shows a weakly supervised architecture integrating three modules based on learning algorithms, i.e., that can be adapted along time if exposed to *fresh* and *timely* data. Automatic semantic lexicon acquisition is supported by the adoption of low-dimensional lexical vector representations, i.e., Word Embeddings (WE) as in [17], in order to enable significant levels of lexical generalization for supervised learning algorithms. Moreover, corpus analysis is also adopted for the acquisition of sentiment oriented lexical knowledge, in the so-called Distributional Polarity Lexicon (DPL) [7]: a DPL characterizes lexical items with polarity information through a weakly supervised

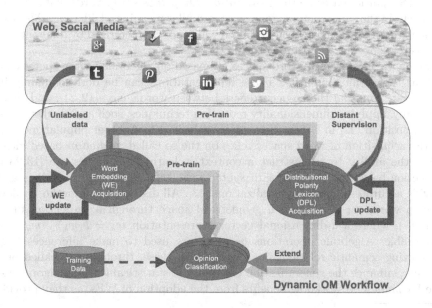

Fig. 1. Dynamic OM architecture.

acquisition process[1]. Finally, supervised Opinion Classification (OC) is applied alone, given the small set of manually annotated data, or in combination with the previous two knowledge sources. While technical details of the above modules will be provided later, it is useful to emphasize here how the language dynamics can be captured by this architecture. In a first scenario, the workflow can adapt to changes in the semantic lexicon by changing the underlying WEs: this can be achieved by renewing the lexicon with more recent texts. The WEs acquisition can be refreshed in different time periods, better reflecting language changes. Consequently, the OC model, pre-trained through different WEs, can be refreshed as well. In a second scenario, also the DPL can be updated. This can be done in two ways: by *a change of space*, i.e., expressing the opinionated lexicon over a new WEs, or *by retraining it* possibly over the same WE. It will be clear in the following Sections that only a change of space (i.e., a new WE) strictly requires the acquisition (by retraining) of a novel OC model. In the second case, retraining a DPL can be carried out without retraining also the OC stage. In this case, the new DPL *provides novel input vectors* for the OC classifier, i.e., more timely sentiment oriented lexical representations. In the rest of this section, technical details on the modules will be provided, and the ways different upgrades affect the performances of the Dynamic OM workflow over Twitter will be discussed.

Semantic Lexicon Acquisition through Word Embeddings. Word Embeddings (WEs) are broadly adopted in NLP to represent lexical items in compact geometrical representations, where vectors *embed* information about lexical semantics properties and relationships. WEs are Distributional Models (DMs) of Lexical Semantics based on the *Distributional Hypothesis* [19]: words that are used and occur in the same contexts tend to have similar meanings. Different DMs, also named Word Spaces, have been proposed which are characterized by different lexical vector representations for words.

Semantic inferences in word spaces are based on algebraic vector operations, such as the cosine similarity. These representations can be derived mainly in two ways[2]: *counting* the co-occurrences between words, e.g., [12,21], and then, optionally, applying dimensionality reduction techniques, such as Singular Value Decomposition [22] in Latent Semantic Analysis [12]. Another popular method for the acquisition of word spaces relies on the so-called *prediction* based view, where the ability to reconstruct a context is learned (examples are [13,23]). Prediction-based methods capture *syntagmatic* aspects of word meanings and provide accurate semantic generalizations [13]. All distributional methods allow to map words $w_k \in \mathbb{W}$ into a geometrical space through a projection function $\Phi(\cdot)$, where a d-dimensional vector representation $\boldsymbol{w_k} = \Phi(w_k) \ \forall w_k \in \mathbb{W}$ is available. Algebraic operations over $\boldsymbol{w_k}$ are used to make inferences (e.g. recognizing semantic relationships) over words: these, through a so-called pre-training, enhance the generalization capability of statistical learning algorithms. The Dynamic OM workflow benefits from the adoption of WEs, as they provide

---

[1] It is based on a Distant Supervision [18] selection of training instances.

[2] For an in-depth comparison between the two methods, refer to [20].

rich lexical information to learning algorithms. In this work, we will experiment with different WEs obtained from data gathered in different times, through the application of the *prediction-based* method, called Skip-gram model [13].

Polarity Lexicon Acquisition. The semantic similarity (closeness) functions established by WE models do not correspond well to emotional similarity, as word clusters in these spaces may correspond to opposite polarities (see the first and second column of Table 1). In Sentiment Analysis, Polarity Lexicons (PLs) are widely used to inject sentiment-oriented information into statistical learning models. A PL describes associations of individual words to some polarity information. Many lexicons exist for English, acquired through fully manual, or semi-automated, processes. In the dynamic OM workflow perspective pursued in this paper, we are interested in an automatic methodology, as proposed in [7]. There, a polarity classification system is trained from annotated sentences, to *transfer* polarity information to individual words. This method exploits the characteristics of WEs to represent both words and sentences in a common space, e.g., [12,17]. The transfer of polarity information from known subjective sentences to words is carried out through supervised classification. The resulting lexicons are called Distributional Polarity Lexicons (DPLs). The process proceeds through three steps. First, sentences labeled into some polarity classes are gathered and projected in the embedding space through linear combination of word vectors. Then, a SVM based polarity classifier is trained over these sentences in order to capture dimensions (i.e., sub-spaces) more expressive for polarity classes. Finally, a lexicon (DPL) is generated by classifying each word (i.e., a lexical vector w) with respect to each targeted polarity class, C: confidence levels of the classification are used to estimate polarity probability scores $p(C|w)$. For example, in the DPL used in the experiments the distribution for the word *sadly* is 0.91 negative, 0.07 positive and 0.02 neutral (as in Table 1). More details on the DPL acquisition stage are in [7]. These polarity scores can be arranged into a 3-dimensional vector that can be then adopted within statistical learning algorithms.

The above method has two merits: first, it allows deriving a polarity signature for each word in the embedding; second, the sentiment information depends strongly on the word usage as it is observed in labeled sentences. Notice how sentence polarity is a much clearer notion than *a priori* word polarity, as sentiment is a social and contextual, not a purely lexical, phenomena. The resulting DPLs allow us to represent words both in terms of their *semantics*, through the WE embedding, as well as through its *sentiment orientation*, through the DPL distributions. We can thus measure similarity along both dimensions by employing both vectors. In Table 1, a comparison of the most similar words is reported for some polarity carriers (in the first column). In the second column, measures not taking into account the DPL are reported. When the DPL distributions are also considered than the closest words are those in the third column. Notice how DPL positively affects the sentiment information. For example, the word *bad* is no longer the 3-most similar word to *good*. Polarity information in the DPL forces closeness in the vector space to strongly depend on polarity.

Table 1. Similar words in the embedding without ($2^{nd}column$) and with ($3^{rd}column$) DPL, whose scores (*positivity, negativity, neutrality*) are reported in parenthesis in the first column.

Term	w/o DPL	w/ DPL
joy (0.62,0.08,0.30)	happiness	happiness
	sorrow	positivity
	laughter	enjoyment
love (0.52,0.11,0.37)	adore	adore
	luv	loves
	hate	loove
worse (0.13,0.80,0.07)	better	worser
	worser	sadder
	funnier	shittier
sadly (0.07,0.91,0.02)	unfortunately	unfortunately
	alas	alas
	thankfully	nope

The DPL acquisition requires the availability of annotated sentences used to train an SVM classifier. In order to make the training a dynamic process we can use a general methodology, portable across different domains, time periods and even languages: *Distant Supervision* [18], whereas heuristic rules (e.g., the presence of specific emoticons) are used to gather labeled sentences from Twitter. We select Twitter messages whose last token is either a positive smile, e.g., :) (or :D), or a negative one, e.g., :((or :-(). Neutral messages are detected by selecting those messages ending with a url: this characterizes tweets from newspapers (e.g., @nytimes), mostly non subjective launches. In order to improve accuracy, those messages containing elements of more than one class (e.g., ending with a positive smile but including also a negative smile, or a url) are discarded. This selection strategy makes the DPL acquisition an unsupervised process: it allows to update DPLs once new data is available, making it suitable for the Dynamic OM workflow in Fig. 1.

Supervised Opinion Classification. The Opinion Classification (OC) stage of Fig. 1 aims at enabling the automatic labeling of messages within the Dynamic OM workflow. We will adopt a kernel-based approach within a Support Vector Machine framework [24]. Kernel methods allow us to integrate different representations for the training instances. In particular, for Sentiment Analysis it allows to combine lexical information (i.e., the words appearing in the message as in a Bag-Of-Words (BOW) representation), semantic generalization as expressed in WE-based representations as well as polarity information, as described into a DPL. Each message is represented by a discrete BOW representation, the linear combination of WE vectors of its words, and the aggregate polarity summed up across the DPL of its words. Linear combinations of the DPL vectors for words occurring in a message correspond to a DPL-based sentence representation.

The kernel approach allows us to integrate (i.e. linearly combine) different kernels, one for each representation. For example, pure lexical overlap between two messages and WEs can be captured by a linear kernel (K_{BOW}) and the cosine similarity between WE sentence vectors (K_{WE}) in a combined kernel $K = K_{BOW} + K_{WE}$. When sentiment vectors are employed (through cosine similarity) we will denote as $K = K_{BOW} + K_{DPL}$ the resulting kernel. The overall framework gives us enough flexibility to test the contribution of individual dimensions to the OM task.

The OC model must be retrained according to the kernel K, whenever a new embedding WE is available. On the contrary when only the DPL is changed, by retraining against novel distant supervision material, no strict need exists to retrain the OM model against the training labeled data. The OM model is trained over manually annotated data (the *Training Data* in Fig. 1). In this paper, we will study a scenario where no update is applied to training data, but we will only change to the representations of messages according to lexical data acquired in different time periods.

4   Measuring Language Variation Effects on SA

In this Section, we report on experiments over the proposed adaptive architecture, aiming at measuring the impact of lexical changes and independent training steps at different time stamps onto the opinion tracking task. The experimental evaluation has been carried out over the Evalita 2014 Sentipolc dataset [25]. This dataset includes data from Twitter in Italian, about various topics, mostly in the political domain. Each message is labeled with respect to subjectivity information, in particular *positivity*, *negativity* and *irony*. The targeted task is the *polarity classification*, that is the recognition of the *positive*, *negative* and *neutral* polarity in individual input tweet. In the following evaluations we filtered out messages expressing *ironic* content, resulting in a training and a testing set made of 2, 566 and 1, 175 messages, respectively.

In order to conduct time dependent experiments, we collected tweets in the same period of the Evalita 2014 dataset. The resulting dataset (d2014) is made of about 2, 000, 000 messages. We downloaded the same amount of messages in a different time (dataset d2016), with the more recent tweets referring to the year 2016. This latter set has been crawled with a Streaming Twitter API with the keyword *Roma*, from January to June 2016. In order to manage phenomena like emoticons, urls, hashtags and user mentions, each tweet has been pre-processed with a customized version of the Chaos parser [26].

The d2014 and d2016 datasets have been employed as input corpus to acquire two WEs with the Skip-gram model [13]. Here, the *word2vec* tool has generated WEs filtering out those words appearing less than 20 times in the d2014 corpus and less than 5 times for the d2016 one. In this way, both embeddings has been lead to seemingly cover the test set material: both WEs can cover about

Table 2. Sentiment Analysis performances over the Evalita 2014 test set.

System	F1Pn	F1Pnn
BOW	63.19 %	58.77 %
BOW+DPL2014	63.94 %	59.10 %
BOW+DPL2016	63.19 %	58.40 %
BOW+WE2014	67.78 %	63.02 %
BOW+WE2016	67.11 %	61.65 %

70 % and 80 % of the training and testing data, respectively. The WE obtained from the 2014 and 2016 are denoted by WE2014 and WE2016, respectively. Two DPLs have been acquired for 2014 and the 2016 data. In particular, we selected about 12, 000 labeled messages, 4, 000 message for each polarity class. with the distant supervision heuristic. The same setting as in [7] has been applied to generate the DPL2014 and the DPL2016 lexicons. They cover exactly the same sets of words of the WE2014 and WE2016 embeddings. Experiments have been carried out within a kernel-based Support Vector Machine implemented in the KeLP[3] framework [27]. We adopted linear kernel functions over the following representations: Bag-Of-Words (BOW), i.e., the set of words appearing in a message; Word Embedding (WE), i.e., the linear combination of WE vectors of nouns, verbs, adjectives and adverbs in a message; Distributional Polarity Lexicons (DPL), i.e., the linear combination of DPL vectors of nouns, verbs, adjectives and adverbs in a message. The sum of these individual linear kernels is adopted as the final K kernel function.

The first experiments are reported in Table 2, where a comparison in the adoption of resources in different time periods is carried out. *In-time* data and resources refers to lexicons derived from texts and data contemporary of the test dataset. The reported performances are the F1Pn and the F1Pnn: the former is the arithmetic mean between the F1-Measure of the *positive* and *negative* classes; the latter is the arithmetic mean between all the involved classes, i.e., *positive, negative* and *neutral.* When adopting the 2014 resources, i.e., DPL2014 and WE2014 the performance is higher than the baseline provided by the *BOW* measure (58.77 % in F1Pnn). Moreover, the adoption of *in-time* data is beneficial with respect to the adoption of the polarity lexicon of different (i.e., 2016) time periods, see the 59.10 % F1Pnn of the *BOW+DPL2014* with respect to the 58.40 % of the similar configuration *BOW+DPL2016*. Similar trends are observable with the WEs: the *BOW+WE2014* outperforms the *BOW+WE2016* of about 2 points in F1Pnn (63.02 % vs. 61.65 %). For example, in-time resources can be more effective in predicting the polarity of political tweets of the 2014 period. The tweet *"Il governo Monti richiede ai cittadini comuni grandi sacrifici ma ha avuto uno scarso coraggio nell'aggredire i privilegi consolidati"* is correctly classified as *negative* by the *BOW+DPL2014* classifier, while it is classified as

---

[3] http://sag.art.uniroma2.it/demo-software/kelp/.

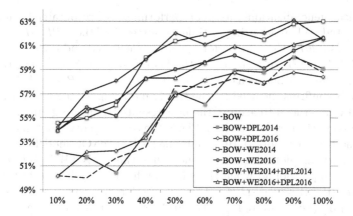

Fig. 2. Learning curves (F1Pnn) of the SA task (full update).

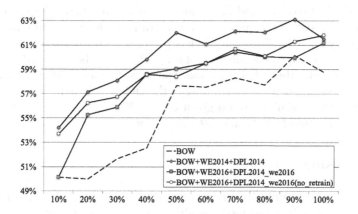

Fig. 3. Learning curves (F1Pnn) of the SA task (DPL/OC update).

neutral by the *BOW+DPL2016* classifier. We speculate that in the adopted lexicalized model *BOW+DPL2016*, the negative bias provided by the words *sacrifici*
and *scarso* is balanced with the positive bias provided by *coraggio* and *privilegi*.
The different result obtained from the *BOW+DPL2014* is mainly due to the
significant negative bias induced by the token *Monti*[4].

Given the cost of annotating training material, a robust OM system is expected
to achieve good results even with a reduced amount of training examples. We
thus estimated the learning curve of the OC (reported in Figs. 2 and 3) in terms
of achievable F1Pnn at incremental size of the training set. In-time resources
also improve learning rates, as reported in Fig. 2. Embeddings seem crucial for

---

[4] In the DPL2014 and DPL2016 the vectors associated to the word *monti* are
(0.15,0.53,0.32) and (0.09,0.13,0.78), respectively.

proper lexical generalizations, as poor performances are obtained when WEs are not adopted. When adopting in-time resources, i.e., the *DPL2014* combined with *BOW+WE2014*, the best performances are achieved and the corresponding *BOW+WE2014+DPL2014* curve has the fastest growth. Notice that there is a considerable difference between the *BOW+WE2014* and the *BOW+WE2016* configurations. It confirms the need for the timely modeling of language phenomena. *In-time* data are crucial to acquire the best resources capturing language statistics correlated to the targeted task. In Fig. 3, the learning curves obtained without updating the WE are reported. Here, only the DPL or the OC model, are updated. The *BOW* curve is plotted as a baseline, while the *BOW+WE2014+DPL2014* represents an upper bound when the whole architecture relies on in-time data. The curve *BOW+WE2016+DPL2014_we2016* refers to a configuration when the DPL is in-time (i.e., acquired on the 2014 data) but making use of a different embedding, i.e., *WE2016*: in this configuration the OC model is re-trained. The *BOW+WE2016+DPL2014_we2016(no_retrain)* refers to the same configuration when the OC model is not re-trained. From these results we can infer that refreshing the full model could be avoided. This is particularly important if we have not enough data to acquire a new WE. In fact, the Skip-gram model, and more generally the Distributional Models perform better when exposed to millions of data. Instead, the acquisition of a new DPL can be obtained by collecting smaller datasets, about 4,000 messages per class through Distant Supervision. Moreover, the effects of a full re-training of the OC model are limited, as demonstrated by the very similar curves given by *BOW+WE2016+DPL2014_we2016* and *BOW+WE2016+DPL2014_we2016(no_retrain)*.

5   Conclusion

In this paper, the complexity underlying the dynamics of trends, community interests and jargon in social media analytics, has been investigated. We showed that adaptive SA architectures can be designed sensitive to language changes through different resources, i.e., a semantic and sentiment lexicons and a statistical opinion recognition model. A first investigation on how these factors impact the quality of the SA on data from different time periods has been carried out. Experiments demonstrate that the adaptive architecture, trained with up-to-dated data, outperforms other configurations. In future, we need to consider the right order of time by annotating more recent material. In fact, we inverted the time line in our evaluations, i.e., considering older data as the test set. This is due to the current availability of manually annotated material in Italian. Moreover, we need to assess the architecture across longer and more finely sampled time periods. A comparison with Online Learning settings [28] could be also adopted to test the ability of *tracking* concept and sentiment shifts across time.

References

1. Socher, R., Perelygin, A., Wu, J., Chuang, J., Manning, C.D., Ng, A.Y., Potts, C.: Recursive deep models for semantic compositionality over a sentiment treebank. In: Proceedings of EMNLP, pp. 1631–1642. ACL (2013)
2. Poria, S., Cambria, E., Gelbukh, A., Bisio, F., Hussain, A.: Sentiment data flow analysis by means of dynamic linguistic patterns. IEEE CIM **10**(4), 26–36 (2015)
3. Zhu, L., Galstyan, A., Cheng, J., Lerman, K.: Tripartite graph clustering for dynamic sentiment analysis on social media. In: Proceedings of the 2014 ICMD, pp. 1531–1542. ACM (2014)
4. Esuli, A., Sebastiani, F.: SentiWordNet: a publicly available lexical resource for opinion mining. In: Proceedings of 5th LREC, pp. 417–422 (2006)
5. Rao, D., Ravichandran, D.: Semi-supervised polarity lexicon induction. In: Proceedings of the EACL, pp. 675–682. ACL (2009)
6. Kiritchenko, S., Zhu, X., Mohammad, S.M.: Sentiment analysis of short informal texts. JAIR **50**, 723–762 (2014)
7. Castellucci, G., Croce, D., Basili, R.: Acquiring a large scale polarity lexicon through unsupervised distributional methods. In: Biemann, C., Handschuh, S., Freitas, A., Meziane, F., Métais, E. (eds.) NLDB 2015. LNCS, vol. 9103, pp. 73–86. Springer, Heidelberg (2015). doi:10.1007/978-3-319-19581-0_6
8. Blitzer, J., Dredze, M., Pereira, F.: Biographies, bollywood, boom-boxes and blenders: domain adaptation for sentiment classification. In: ACL (2007)
9. Nguyen, L.T., Wu, P., Chan, W., Peng, W., Zhang, Y.: Predicting collective sentiment dynamics from time-series social media. In: Proceedings of the 1st Workshop on Issues of Sentiment Discovery and Opinion Mining, pp. 6:1–6:8. ACM (2012)
10. Zhang, Y., Mao, W., Zeng, D., Zhao, N., Bao, X.: Exploring opinion dynamics in security-related microblog data. In: JISIC, pp. 284–287 (2014)
11. Liu, S., Cheng, X., Li, F., Li, F.: TASC: topic-adaptive sentiment classification on dynamic tweets. IEEE Trans. KDE **27**(6), 1696–1709 (2015)
12. Landauer, T., Dumais, S.: A solution to Plato's problem: the latent semantic analysis theory of acquisition, induction and representation of knowledge. Psychol. Rev. **104**, 211–240 (1997)
13. Mikolov, T., Chen, K., Corrado, G., Dean, J.: Efficient estimation of word representations in vector space. CoRR abs/1301.3781 (2013)
14. Stone, P.J., Dunphy, D.C., Smith, M.S., Ogilvie, D.M.: The General Inquirer: A Computer Approach to Content Analysis. MIT Press, Cambridge (1966)
15. Wilson, T., Wiebe, J., Hoffmann, P.: Recognizing contextual polarity in phrase-level sentiment analysis. In: Proceedings of EMNLP. ACL (2005)
16. Mohammad, S.M., Turney, P.D.: Emotions evoked by common words and phrases: using mechanical turk to create an emotion lexicon. In: Proceedings of CAAGET Workshop (2010)
17. Mikolov, T., Yih, W.t., Zweig, G.: Linguistic regularities in continuous space word representations. In: Proceedings of NAACL, pp. 746–751 (2013)
18. Go, A., Bhayani, R., Huang, L.: Twitter sentiment classification using distant supervision. Technical report, Stanford (2009)
19. Harris, Z.: Distributional structure. In: Katz, J.J., Fodor, J.A. (eds.) The Philosophy of Linguistics. Oxford University Press, Oxford (1964)
20. Baroni, M., Dinu, G., Kruszewski, G.: Don't count, predict! A systematic comparison of context-counting vs. context-predicting semantic vectors. In: Proceedings of ACL, pp. 238–247 (2014)

21. Sahlgren, M.: The word-space model. Ph.D. thesis, Stockholm University (2006)
22. Golub, G., Kahan, W.: Calculating the singular values and pseudo-inverse of a matrix. J. Soc. Ind. Appl. Math. Ser. B Numer. Anal. **2**(2), 205–224 (1965)
23. Bengio, Y., Ducharme, R., Vincent, P., Janvin, C.: A neural probabilistic language model. J. Mach. Learn. Res. **3**, 1137–1155 (2003)
24. Shawe-Taylor, J., Cristianini, N.: Kernel Methods for Pattern Analysis. Cambridge University Press, New York (2004)
25. Basile, V., Bolioli, A., Nissim, M., Patti, V., Rosso, P.: Overview of the Evalita 2014 sentiment polarity classification task. In: Proceedings of the 4th EVALITA (2014)
26. Basili, R., Zanzotto, F.M.: Parsing engineering and empirical robustness. Nat. Lang. Eng. **8**(3), 97–120 (2002)
27. Filice, S., Castellucci, G., Croce, D., Basili, R.: KeLP: a kernel-based learning platform for natural language processing. In: Proceedings of ACL-IJCNLP, pp. 19–24 (2015)
28. Wang, Z., Vucetic, S.: Online passive-aggressive algorithms on a budget. JMLR **9**, 908–915 (2010)

Using Random Forests for the Estimation of Multiple Users' Visual Focus of Attention from Head Pose

Silvia Rossi[1]([⊠]), Enrico Leone[1], and Mariacarla Staffa[2]

[1] Department of Electrical Engineering and Information Technology,
University of Naples Federico II, Naples, Italy
`silvia.rossi@unina.it`
[2] Department of Engineering, University of Naples Parthenope, Naples, Italy
`mariacarla.staffa@uniparthenope.it`

Abstract. When interacting with a group of people, a robot requires the ability to compute people's visual focus of attention in order to regulate the turn-taking, to determine attended objects, as well as to estimate the degree of users' engagement. This work aims at evaluating the possibility of computing real-time multiple users' focus of attention by combining a random forest approach for head pose estimation with the user's head joint tracking. The system has been tested both on single users and on couples of users interacting with a simple scenario designed to guide the user attention towards a specific space region. The aim is to highlight the possible requirements and problems arising when dealing with the presence of multiple users. Results show that while the approach is promising, datasets that are different from the ones available in the literature are required in order to improve performance.

Keywords: Visual focus of attention · Human-robot interaction · Random forests

1 Introduction

When interacting with a human being, a robot should be able to dynamically detect and track users and objects [1,2] in order to determine the person's visual focus of attention (VFOA) [3]. This task is inherently difficult, especially when applied to real world conditions, where real time responses and effective mechanisms for controlling the robot sensors and effectors with respect to the available computational resources and the external environment are required [4]. Typically, VFOA estimation mainly relies on the eye-gaze analysis on high-resolution images, while personal robots are typically equipped with commercial cameras. However, in addition to eye-gaze, people use the head orientation to convey information during a conversation. For example, a person can direct her head towards an object to observe. Head movements are also essential for analyzing complex meaningful gestures such as pointing [5,6], as a nonverbal cue [7], and as a form

© Springer International Publishing AG 2016
G. Adorni et al. (Eds.): AI*IA 2016, LNAI 10037, pp. 89–102, 2016.
DOI: 10.1007/978-3-319-49130-1_8

of gestural communication themselves; in fact, through a simple nod or shake of the head, it is possible to express disagreement, confusion, and so on.

When interacting with more than one person, additional observations can be made by establishing their VFOA (through a head pose estimation). For example, if two persons focus their attention on one another (e.g., mutual gaze) or towards the same object, this is often a sign that the individuals are engaged in a conversation. The speaker and the hearers typically, indeed, attend to each others' states of attention by alternating gaze between each other and a target object [8]. Head gaze can hence be interpreted as an explicit social cue communicating attention towards a subject or an object, but also, it is used to regulate the turn-taking [9], to decide whether or not to initiate an interaction [10], as well as to estimate the degree of user's engagement during a multi-party interaction [11]. Such information can be of extreme importance for a robot that has to interact with a group of people or mediate a multi-party interaction [12].

This work aims at evaluating the possibility of computing real-time multiple users' focus of attention, by combining the skeleton tracking of an RGB-D camera with the Random Regression Forest approach (DRRF) developed by [13] for the case of head pose estimation. Specifically, in the latter, such calculation is performed on low-quality static depth images. Head poses are described by six parameters, namely the 3D coordinates of the user's nose tip (and, thus, the position of the head), and the Euler angles (yaw, pitch, and roll) that describe the head rotation. The use of random forest makes it possible to achieve real-time performance on standard processors, with the possibility of balancing between accuracy and computational cost. Moreover, they are very powerful in learning complex mappings from large datasets without over-fitting of data. In this work, we will extend the approach proposed in [13] in order to evaluate performance when different heads are in the robot field of view.

While different VFOA applications have been developed in the literature, none of these aims at evaluating the scalability of the approach while changing the number of users. This research takes an initial step towards VFOA classification when interacting in multi-party settings, using a simple and efficient method, in order to analyze its limits as well as its potential. We will evaluate the performance of the system by considering a simple scenario where one or two human users are requested to gaze with head movements images appearing on a vertical screen.

2   Related Works

In human-robot interaction (HRI) applications, it is common that VFOA is analyzed post-study and often annotated by hand [3,14]. However, in the more general human-machine interaction area, many approaches dealt with the problem of VFOA automatic estimation, especially in common highly constrained domains, as the ones defined in HRI research [3]. However, in these cases, most works use either invasive eye-tracker devices [15], multiple cameras, or high definition images.

In order to determine the VFOA of a human user, we only considered the head pose estimation. In the literature, there are several methods used to address the problem with respect to the available data that is 2D images or depth data. Within the first category, appearance-based methods consider the entire face region. A typical approach is to discretize the head pose, and learn a separate detector for each pose, or to focus on mapping from high-dimension spaces of facial images in spaces with fewer dimensions [16]. Feature-based methods, on the contrary, are based on the estimation of the location of specific facial points. The use of these methods requires that the same facial features are visible in all poses [5].

In general, methods that solely rely on 2D images are sensitive to light changes, the lack of visible characteristics and partial occlusions. Fortunately, the additional information provided by multiple cameras [17] or the depth data to detect the head pose of a single user, can help in dealing with some of the previously described limitations; therefore, many recent works are based on the availability of depth data (eventually, in addition to the standard 2D images). For example, in [18], a real-time system is developed to recognize a large number of variations of the poses, partial occlusions (until the nose is visible) and facial expressions, from a set of images. The method uses geometric features to generate possible candidates for the nose by exploiting the potentiality of the parallel calculation of the GPU to compare all hypothesized poses with a generic model of the face.

An alternative can be the use of the Random Forests (RF) [19], which thanks to the ability to handle huge training dataset, the high power of generalization, the speed of calculation and the ease of implementation, are becoming very popular in the field of computer vision. In [13], a variant of RF, known as Discriminative Regression Random Forest (DRRF), has been proposed, which calculates the head poses starting from a depth images dataset. The term discriminative refers to the fact that the algorithm firstly discriminates the image parts belonging to the head (eventually more than one) and then uses only those to produce votes to determine the pose. This method could be very useful in contexts where multiple users are considered, who can be found in different parts of the image.

A very few works have been presented dealing with the problem of simultaneously tracking of multiple users. Multiparty interaction and VFOA estimation were mainly investigated in the context of meetings [20]. In these cases, different cameras were used to estimate the meeting participants head poses and, consequently, their VFOAs. A single panoramic camera was used in [21], where the authors presented a Neural Network approach to estimate gaze direction of multiple persons. In [22], the authors estimated the VFOA in well-separated areas of a multi-party interaction with two persons and a robot by using the Kinect sensor. However, each user was monitored by a different Kinect. A similar study was conducted in [23], with two homemade datasets, and the distribution of head poses estimated with a K-component Gaussian mixture model (where K is the number of existing targets). Finally, in [14], the authors combine the RGB-D data used for head pose estimation and a manual estimation, obtaining that head gaze data from the RGB-D camera is not consistent with the manual one. However, authors do not specify how such RGB-D head pose is computed. In all these settings, VFOA is identified starting from a few predetermined areas, as in our case. However, up to our knowledge, this is the first attempt to use DRRF to multiple users real-time gaze tracking.

3 Materials and Methods

In the following sections, we recall some basic information on the use of Discriminative Random Forest for the head pose estimation problem. Please refer to the work of Fanelli et al. [13,24] for further information.

3.1 Discriminative Random Regression Forests

A Discriminative Random Regression Forest (DRRF) is a set of trees that are randomly trained in order to reduce the over-fitting in comparison to trees trained on the whole dataset. Specifically, a random forest is a classifier consisting of a collection of L decision trees $T_k = h(x, \Theta_k)$, with $k = 1, ..., L$, used as classifiers, where $\{\Theta_k\}$ is a randomly generated vector (i.e., a subset of training examples) used to build the k-th tree. With respect to an input x, each tree produces a single vote for the class of the input. The forest determines the final classification considering all the trees in it; the result corresponds to an average or a weighted average of all reached terminal nodes. Starting from the root, a binary test ϕ (i.e., the considered attributes and thresholds for the node in the case of continuous values) is chosen, for each non-leaf node, from a random set. Among all the tests, the test $\phi*$, the one that maximizes a specific function optimization, is selected. In this case, the one that maximizes the information gain $IG(\phi)$ of the split. The purpose of the split is to have a subset of input instances in the left and right children of the node as pure as possible, i.e., containing homogeneous instances. Every tree is grown up to the maximum set depth.

3.2 DRRF Training

In general, it is assumed that a human head can be modeled as a rigid disembodied body. Under this assumption, the movements of a human head are limited to three degrees of freedom, which can be described by the egocentric rotation angles yaw, pitch, and roll. The used training set is the Biwi Kinect Head Pose Database. In the dataset, the training samples have been acquired, frame-by-frame, via a Kinect sensor. A training sample is a depth image containing a single head that is annotated with the 3D locations of the nose tip, and the Euler angles of the head orientation. Fix-sized patches are extracted from a training image. Each patch is annotated with two real-valued vectors $\theta = (\theta^v, \theta^a)$, where, $\theta^v = \{\theta_y, \theta_x, \theta_z\}$ is the offset computed between the 3D point falling at the patch center and the nose tip, while the head orientation is encoded as Euler angles $\theta^a = \{\theta_{pi}, \theta_{ya}, \theta_{ro}\}$. The class of each patch is determined with the aid of the vector θ^v; if its length is below a certain value (that is set to 10 mm), then the corresponding patch assumes the '1' label. In this way, the patches that assume label '0' are not only those extracted from parts of the body such as the torso or arms, but also those extracted from regions that contain the hairs.

3.3   Head Pose Estimation

From each image, a set of patches is extracted (i.e., regions of rectangular pixels of a fixed size). The probability $p(c = k \mid P)$, stored in a leaf during the training phase, evaluates the level of information of the patch with respect to the class k; in this case, the classification is binary (class '1' and class '0'). Whenever a patch is identified in class '1', two Gaussian distributions, related to the pose of the head, are provided in the leaf node. The distributions are expressed in the following way: $p(\theta^v) = \mathcal{N}(\theta^v, \overline{\theta}^v, \Sigma^v)$, and $p(\theta^a) = \mathcal{N}(\theta^a, \overline{\theta}^a, \Sigma^a)$, where θ^v and θ^a are, respectively, the distance vector and the vector that contains the rotation angles associated with each patch; $\overline{\theta}^v$ and $\overline{\theta}^a$ represent, respectively, the average of the distance vectors and of the rotation angles; Σ^v and Σ^a are, respectively, the covariance matrix of the distance vectors and the covariance matrix of the rotation angles.

After that, in each tree, a patch has reached a leaf, the algorithm chooses the leaves that are useful for the pose estimation, that is, those that contain $p(c = k|P) = 1$ as probability value (i.e., the patch containing pixels that belong to the head) and that have a variance value less than a certain threshold ($Trace(\Sigma^v) \leq max_v$). Finally, such approach lets every image region to vote for the head pose and, therefore, it is not constrained to a certain area of the face to be visible. Hence, this allows handling partial occlusions, even of the nose.

4   Online Estimation of the VFOA

In Sect. 3, the process of building and training the random forest was presented. However, the presented method is designed to work on single images extracted by a Kinect sensor and the dataset training samples are collected frame-by-frame; this means that each image, representing a particular head pose, is individually acquired and is not necessarily linked to the others within a sequence. Indeed, the random forest calculates head pose parameters regardless of what happens with the head tracking.

Here, we adopted an approach that combines OpenNi[1] standard functions for user tracking based on skeleton data gathered by the Kinect sensor, and a random forest method trained on the set of depth images. Hence, let us assume to have a visual OpenNI-compliant device, such as the Kinect, enabled to acquire a stream of images, and let us consider one or more users located at a distance of 1.3 m from the sensor; with the support of the OpenNI library, for each

---

[1] The OpenNI framework provides a set of open source APIs for writing 3D sensing applications that facilitates communication with low-level devices including vision and audio sensors.

Fig. 1. Users identified by the skeleton tracking and the DRRF (yellow dots). (Color figure online)

image acquired by the RGB-D sensor, a data structure is maintained, where each element is of $XnUserID$ type (i.e., a numeric identifier i for each user). By considering these IDs, it is possible to access the skeleton joint positions of each user. Since the main objective is to determine the VFOA, for each user, the XN_SKEL_HEAD joint, more precisely, the (x, y) coordinates of the user's head, that is $(head[i].x, head[i].y)$, are stored. In Fig. 1, two users are identified and labeled by using a Kinect sensor.

As soon as a user is identified by the skeleton tracking, the DepthImage of the current frame is passed as input to the trained DRRF. In addition to DepthImage, it receives in input the *stride* parameter, which determines the density with which pixel values, within each patch, are controlled. For example, if $stride = 1$, the value of each single pixel of a patch is checked. Greater values of stride imply the jumping of some pixels, so this allows to increase the speed of the tests. From the DepthImage a set of patches, of the same dimension as the ones used in the training phase, is extracted and sent to the forest. The work of [13] allows identifying different heads in the same image. In fact, once that the votes from each patch are obtained, a clustering step is performed to detect the presence of one or more heads within the image. Therefore, all the votes that are within a certain distance from each other (in the implementation a *larger_radius_ratio* parameter is considered, which corresponds to the average diameter of a head) are collected within a cluster. A cluster is identified as a head only if it contains a certain number of votes. Since the number of votes is directly proportional to the number of trees, and since the number of selected patches is inversely proportional to the square of the *stride* parameter, the threshold of votes in order to detect a head is defined as: $\beta * (\#trees/stride^2)$, where β is a constant, and #*trees* is the number of trees in the DRRF. Subsequently, for each head cluster, all the votes located at a distance greater than a sixth of the average head diameter are eliminated. Finally, for each cluster, the average of the votes is an estimation (in a continuous space) of the position of the tip of the nose and of its orientation.

In [13], the pose estimation is obtained frame-by-frame, regardless of any tracking process; Hence, once it is applied for multiple users' pose identification, could cause that the number of users (and then head poses to trace) is different (smaller or greater) than the number of heads identified by the Kinect sensor. For this reason, in this work, a mapping between the coordinates of the user's head joints, as identified by the RGB-D camera, is combined with the results of the Random Forest. Such mapping associates, for each head joint, the nearest head pose (if any, or the head pose obtained in a previous frame). This also means that if the DRRF identifies further user heads, they are not taken into account.

4.1 VFOA Calculation

The pose parameters θ_i (3D coordinates of the tip of the nose and rotation angles), as returned by DRRF for each i detected head, are stored within a matrix $means[][]$, in which the element $means[i][j]$ is the j-th parameter of the i-th head. In particular: $means[i][0]$, $means[i][1]$, and $means[i][2]$ are, respectively, the x, y and z coordinates of the nose tip; $means[i][3]$ is the pitch rotation angle; $means[i][4]$ is the yaw rotation angle and $means[i][5]$ is the roll rotation angle. Thanks to $\theta_k = (\theta_k^v, \theta_k^a)$ parameters, it is now possible to determine the users' focus of attention, with respect to a vertical plane in front of the users. For a

Fig. 2. VFOA visualization for two users (red and green dots). (Color figure online)

user i, which has been correctly identified, the final coordinates (after yaw and pitch rotations), which are the real focus of attention, are computed in this way:

$$Fx_i = f * \frac{means[i][0]}{means[i][2]} + \frac{n\_cols}{2} - (means[i][2] * tan(means[i][4] * \frac{\pi}{180.0})) \quad (1)$$

$$Fy_i = f * \frac{means[i][1]}{means[i][2]} + \frac{n\_rows}{2} + (means[i][2] * tan(means[i][3] * \frac{\pi}{180.0})) \quad (2)$$

where f is a constant equal to the camera focal length divided by the pixel size, n_cols is the width and n_rows is the height of the camera image, $means[i][2]$ represents the distance between the user and the screen, and $means[i][3]*\pi/180.0$ and $means[i][4] * \pi/180.0$ are the rotation computed in degree with respect to the final coordinates.

In Fig. 2, the calculated output after the head tracking phase and pose estimation of a user is shown; the red and the green dots represent VFOA.

5   Experimental Results

The system has been tested on both single users and couple of users interacting within the same scenario. The goal of this experiment was to observe the change in performance while having more that one user at the time. The scenario was designed in order to guide the user attention towards a specific region of the space, and to compare the VFOA values, as computed by the random forest, with respect to the expected ones. In order to achieve this goal, a large TV was used to get the user attention on specific areas without distractions. The TV area was divided into 6 quadrants (i.e., areas of interest).

Single User Testing. 18 users have been considered for the individual tests. Namely, 14 males and 4 females. In Fig. 3, we show the scenario setting organized as follows:

– The user is positioned in front of a 47 in. smart TV with a resolution of 1920 × 1080 px.
– A Microsoft Kinect camera is located on top of the monitor.

Each user was asked to perform the two following tasks:

1. *Head tracking activation*: which consists in a calibration procedure needed at the beginning of the interaction. During this phase, the monitor displays a black screen;
2. *Testing Session*: during this phase, the monitor shows a sequence of 10 different images located in one of the six different screen regions, randomly selected. Each image is shown for 4 s. The user is requested to look at the currently displayed image (See Fig. 3).

Users' head movements were completely natural and solely guided by the change in position of the images within the quadrants.

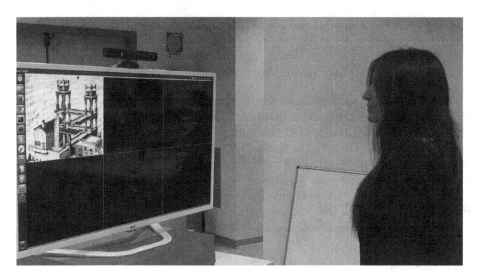

Fig. 3. Single-user testing session.

Multi-user Testing. As for the multi-user setting, the same users of the previous cases were then randomly coupled, so obtaining 14 couples. The testing procedure has been the same of that of the single user setting, apart from the calibration phase that must be performed sequentially by the users. Thus, both users of the selected pair have been asked to look at each of the ten images randomly displayed on the screen. In the two-user case, the VFOA of each user is computed independently from the other user according to Eqs. 1 and 2 since, in a different, more general, interaction setting, users may have different VFOA.

5.1 Evaluation Measures

For each user, the VFOA coordinates, as computed by the system, were compared to the position of the image displayed on the screen and observed by the user. A correct estimation of visual attention is considered when the user watches an image displayed in a specific area of the screen, whose area contains the focus of attention coordinates. Note that here the main focus is not to evaluate the DRRF performance in terms, for example, of angular error or root mean square error that are typical for regression models, but simply as in [5,20,22] to discriminate the capability of the system to correctly individuate a wider area containing the VFOA. Since both for the single and multi-user case, each image was displayed, within a given quadrant, for 4 s, during which the system processed a certain number of frames ($30\,fps$ in the average). Therefore, to correctly determine the values described above, it is necessary to adopt a method for measuring, during the 4 s period, the presence of the focus of attention inside the currently observed region. For this purpose, for each displayed image, we considered these two measures:

- HIT_COUNT: counts the number of times the coordinates of the focus calculated are recognized within the quadrant that contains the showed image;
- MISS_COUNT: counts the number of times the coordinates of the focus computed by the system are localized outside such quadrant.

These counters are initialized every time a new image is displayed; if at the end of each 4 s period the value of HIT_COUNT is strictly greater than MISS_COUNT, then the system counts the obtained result as a true positive (consequently, the TP value is increased, with five more true negatives TN); otherwise, it is a false positive FP; in such a case, the related variables, FP and FN, are increased.

In order to evaluate the performance of our system, we considered the Positive Predictive Values (i.e., the *Precision* $P = TP/(TP + FP)$). In our case, since $FP = FN$, this performance value identifies also the *Sensitivity* of the system.

5.2   Results

The evaluation tests have been executed by considering the following features: $L = 10$; $stride = 5$; Patch dimension $= 80x80$; $max_v = 800$.

Table 1. Single-user test results.

User_id	TP	FN	FP	TN	Precision
1	10	0	0	50	1
2	9	1	1	49	0.90
3	10	0	0	50	1
4	10	0	0	50	1
5	10	0	0	50	1
6	6	4	4	46	0.60
7	10	0	0	50	1
8	9	1	1	49	0.90
9	3	7	7	43	0.30
10	8	2	2	48	0.80
11	8	2	2	48	0.80
12	10	0	0	50	1
13	9	1	1	49	0.90
14	2	8	8	42	0.20
15	7	3	3	47	0.70
16	8	2	2	48	0.80
17	10	0	0	50	1
18	8	2	2	48	0.80
Average					0.82
Baseline					0.20

Single User Test Results. The results of the single user tests are reported in Table 1. For these tests, the achieved average Precision is 0.82. The Precision values are satisfying for almost all the users, except for users 9 and 14, whose precision values are near the random baseline value. The reported random baseline value of 0.20 is obtained by computing the same measures on a set of tests where six different images were displayed to the user, but a different one, every four seconds, was randomly selected to compute the TPs. This means that in the case of users 9 and 14, the system was not able to identify at all the VFOA, probably due to a wrong calibration or distance from the sensor. Moreover, since TP are evaluated with respect to the HIT_COUNT number in 4 s, users with slow reactions towards a change of image could lead to FNs. Since the performance of the DRRF was already evaluated in [13,24] in terms of the errors for the 3D head localization, here the precision values of the single user case are evaluated in order to represent a benchmark for the multi-user case.

Multi-user Test Results. In the case of multi-user tests, the Precision values were firstly calculated by separately considering the VFOA estimation of each member of the couples. Hence, in Table 2, $Precision_i$ values are the average of each couple and 0.62 is the result of averaging on 28 users. With respect to the single user case, the performance, while tracking two users simultaneously, is degraded. One-way ANOVA test on Precision values of single users and multiple users test shows a significant difference in the two cases (with $p = 0.0157$). However, Precision values are still comparable with values in the state of the art [23]. Unfortunately, if we consider, to compute Precision, only the simultaneous TPs of the two users of a couple (i.e., the times that both users' VFOA have

Table 2. Multi-user test results.

Couple_id	TP_{i1}	FP_{i1}	TN_{i1}	TP_{i2}	FP_{i2}	TN_{i2}	$Precision_i$	TP_t	FP_t	TN_t	$Precision_t$
1	7	3	47	9	1	49	0.80	6	4	46	0.60
2	2	8	42	9	1	49	0.55	2	8	42	0.20
3	2	8	42	4	6	44	0.30	0	10	40	0
4	7	3	47	7	3	47	0.70	4	6	44	0.40
5	7	3	47	6	4	46	0.65	3	7	43	0.30
6	6	4	46	5	5	45	0.55	4	6	44	0.40
7	10	0	50	8	2	48	0.90	8	2	48	0.80
8	4	6	44	4	6	44	0.40	2	8	42	0.20
9	3	7	43	9	1	49	0.60	3	7	43	0.30
10	2	8	42	7	3	47	0.45	1	9	41	0.10
11	10	0	50	8	2	48	0.90	8	2	48	0.80
12	7	3	47	5	5	45	0.60	3	7	43	0.30
13	4	6	44	2	8	42	0.30	0	10	40	0
14	10	0	50	10	0	50	1	10	0	50	1
Average							0.62				0.39

been simultaneously correctly recognized), this value drops to 0.39. This means that, while two users' VFOA can be tracked with a loss in performance, joint attention on a specific area is unattainable. In our opinion, multi-user results might be affected by a change of perspective with respect to the considered dataset since users are not located in the center of the camera anymore.

6   Conclusions

In this work, we evaluated a system that allows determining the VFOA of one or more users identified by a suitable RGB-D sensor. Our approach is an extension of the approach proposed in [13] that allows learning a mapping between simple depth features (extracted from the depth map generated by the visual sensor) and a probabilistic estimation of the nose tip and the head Euler rotation angles. In addition to the regression, the Random Forests method also includes a classification mechanism that allows identifying different heads in the image. The adopted approach for the user's head pose estimation is based on a combination of standard OpenNi skeleton tracking function, and the DRRF approach.

The system has been tested in a simple scenario, where one or more users are placed in front of a smart TV endowed with a Kinect camera at a distance of 1.3 meters. After being traced by the sensor, each user must follow with the gaze the images as they are presented on the screen within the six possible areas. We observed that the use of random forests makes it possible to achieve real-time performance on standard processors ($30fps$) while tracking two user poses, and the desired trade-off between accuracy and computation cost are kept for the single use case achieving a good Precision performance that is degraded when the number of users grows.

One of the main issues emerged from the experimental tests is the constraint imposed by the distance allowed for a correct interaction with the Kinect sensor, which affects the performance in case of multiple users, and that depends on the used dataset. In this setting, in fact, the available space is reduced producing a negative impact on the system global performance. It is expected that this problem will become even more apparent with a greater number of users and this reflects the need to improve and expand the constraint on the distance from the sensor. This also endorses the outcome that the choice of the training set is a crucial issue affecting the overall performance of the system and suggests to enrich the training set by considering a large amount of data during the construction of the random forest. Random forests show their power when using large datasets, on which they can be trained efficiently. However, because the accuracy of a regression mechanism depends on the amount of annotated training data, the acquisition and the labeling of a training set are key issues. Hence, in our opinion, collecting data dealing with people who take different poses of the head, at different distances from the sensor, and at different position with respect to the center of the image could greatly improve the accuracy of the pose estimation and, consequently, of the visual attention in the presence of many more users.

Acknowledgment. The research leading to these results has been supported by the RoDyMan project, which has received funding from the European Research Council FP7 Ideas under Advanced Grant agreement number 320992 and supported by the Italian National Project Security for Smart Cities PON-FSE Campania 2014-20. The authors are solely responsible for the content of this manuscript. The Authors thank Silvano Sorrentino for his contribution in code development.

References

1. Sidobre, D., Broqure, X., Mainprice, J., Burattini, E., Finzi, A., Rossi, S., Staffa, M.: Humanrobot interaction. In: Advanced Bimanual Manipulation. Volume 80 of Springer Tracts in Advanced Robotics. Springer Berlin Heidelberg (2012) 123–172
2. Staffa, M., De Gregorio, M., Giordano, M., Rossi, S.: Can you follow that guy? In: 22th European Symposium on Artificial Neural Networks, ESANN 2014, 23–25 April 2014, Bruges, Belgium, pp. 511–516 (2014)
3. Clabaugh, C., Ram, T., Matarić, M.J.: Estimating visual focus of attention in dyadic human-robot interaction for planar tasks. In: International Conference on Social Robotics, October 2015
4. Burattini, E., Finzi, A., Rossi, S., Staffa, M.: Monitoring strategies for adaptive periodic control in behavior-based robotic systems. In: Advanced Technologies for Enhanced Quality of Life, AT-EQUAL 2009, pp. 130–135, July 2009
5. Vatahska, T., Bennewitz, M., Behnke, S.: Feature-based head pose estimation from images. In: 7th IEEE-RAS International Conference on Humanoid Robots, pp. 330–335 (2007)
6. Burattini, E., Finzi, A., Rossi, S., Staffa, M.: Attentional human-robot interaction in simple manipulation tasks. In: Proceedings of the Seventh Annual ACM/IEEE International Conference on Human-Robot Interaction. HRI 2012, pp. 129–130. ACM, New York (2012)
7. Vinciarelli, A., Salamin, H., Polychroniou, A., Mohammadi, G., Origlia, A.: From nonverbal cues to perception: personality and social attractiveness. In: Esposito, A., Esposito, A.M., Vinciarelli, A., Hoffmann, R., Müller, V.C. (eds.) Cognitive Behavioural Systems. LNCS, vol. 7403, pp. 60–72. Springer, Heidelberg (2012). doi:10.1007/978-3-642-34584-5_5
8. Brinck, I.: Joint attention, triangulation and radical interpretation: a problem and its solution. Dialectica **58**(2), 179–205 (2004)
9. Broz, F., Kose-Bagci, H., Nehaniv, C.L., Dautenhahn, K.: Towards automated human-robot mutual gaze. In: Proceedings of International Conference on Advances in Computer-Human Interactions (ACHI) (2011)
10. Das, D., Rashed, M.G., Kobayashi, Y., Kuno, Y.: Supporting human-robot interaction based on the level of visual focus of attention. IEEE Trans. Hum. Mach. Syst. **45**(6), 664–675 (2015)
11. Nakano, Y.I., Ishii, R.: Estimating user's engagement from eye-gaze behaviors in human-agent conversations. In: Proceedings of the 15th International Conference on Intelligent User Interfaces, pp. 139–148. ACM (2010)
12. Short, E., Matarić, M.J.: Towards robot moderators: understanding goal-directed multi-party interactions. In: AAAI Fall Symposium on Artificial Intelligence and Human-Robot Interaction, November 2015
13. Fanelli, G., Weise, T., Gall, J., Gool, L.: Real time head pose estimation from consumer depth cameras. In: Mester, R., Felsberg, M. (eds.) DAGM 2011. LNCS, vol. 6835, pp. 101–110. Springer, Heidelberg (2011). doi:10.1007/978-3-642-23123-0_11

14. Kennedy, J., Baxter, P., Belpaeme, T.: Head pose estimation is an inadequate replacement for eye gaze in child-robot interaction. In: Proceedings of the Tenth Annual ACM/IEEE International Conference on Human-Robot Interaction Extended Abstracts, HRI 2015, pp. 35–36. Extended Abstracts (2015)
15. Babcock, J.S., Pelz, J.B.: Building a lightweight eyetracking headgear. In: Proceedings of the 2004 Symposium on Eye Tracking Research & Applications, pp. 109–114. ACM (2004)
16. Balasubramanian, V., Ye, J., Panchanathan, S.: Biased manifold embedding: a framework for person-independent head pose estimation. In: IEEE Conference on Computer Vision and Pattern Recognition, CVPR, pp. 1–7 (2007)
17. Muoz-Salinas, R., Yeguas-Bolivar, E., Saffiotti, A., Medina-Carnicer, R.: Multi-camera head pose estimation. Mach. Vis. Appl. **23**(3), 479–490 (2012)
18. Breitenstein, M., Kuettel, D., Weise, T., Van Gool, L., Pfister, H.: Real-time face pose estimation from single range images. In: IEEE Conference on Computer Vision and Pattern Recognition, pp. 1–8 (2008)
19. Breiman, L.: Random forests. Mach. Learn. **45**(1), 5–32 (2001)
20. Voit, M., Stiefelhagen, R.: Deducing the visual focus of attention from head pose estimation in dynamic multi-view meeting scenarios. In: Proceedings of the 10th International Conference on Multimodal Interfaces, ICMI 2008, pp. 173–180. ACM (2008)
21. Stiefelhagen, R., Yang, J., Waibel, A.: Simultaneous tracking of head poses in a panoramic view. In: Proceedings of 15th International Conference on Pattern Recognition, vol. 3, pp. 722–725 (2000)
22. Johansson, M., Skantze, G., Gustafson, J.: Head pose patterns in multiparty human-robot team-building interactions. In: Herrmann, G., Pearson, M.J., Lenz, A., Bremner, P., Spiers, A., Leonards, U. (eds.) ICSR 2013. LNCS (LNAI), vol. 8239, pp. 351–360. Springer, Heidelberg (2013). doi:10.1007/978-3-319-02675-6_35
23. Sheikhi, S., Odobez, J.-M.: Recognizing the visual focus of attention for human robot interaction. In: Salah, A.A., Ruiz-del-Solar, J., Meriçli, Ç., Oudeyer, P.-Y. (eds.) HBU 2012. LNCS, vol. 7559, pp. 99–112. Springer, Heidelberg (2012). doi:10.1007/978-3-642-34014-7_9
24. Fanelli, G., Dantone, M., Gall, J., Fossati, A., Van Gool, L.: Random forests for real time 3D face analysis. Int. J. Comput. Vis. **101**(3), 437–458 (2013)

Multi-agent Systems

An Analytic Study of Opinion Dynamics in Multi-agent Systems with Additive Random Noise

Stefania Monica$^{(\boxtimes)}$ and Federico Bergenti

Dipartimento di Matematica e Informatica, Università degli Studi di Parma,
Parco Area Delle Scienze 53/A, 43124 Parma, Italy
{stefania.monica,federico.bergenti}@unipr.it

Abstract. In this paper we analytically derive macroscopic properties of the temporal evolution of opinion distribution in multi-agent systems under the influence of additive random noise. Proper rules which describe how the opinion of agents are updated after their interactions are given. Such rules involve a deterministic part, related to compromise, and a stochastic part, in terms of an additive random noise. Starting from the microscopic interaction rules among agents, macroscopic properties of the system are derived using an approach based on kinetic theory of gases. In particular, the stationary profiles of opinion distribution are derived analytically. In the last part of the paper, some illustrative examples of stationary profiles are presented.

Keywords: Opinion dynamics · Kinetic theory · Multi-agent systems

1 Introduction

In recent years, a new discipline called *sociophysics* was introduced in [1]. The main idea of sociophysics is that social interactions in multi-agent systems can be described using models inspired by kinetic theory of gases, namely a branch of physics which aims at describing properties of gases. The approach of kinetic theory is called *mesoscopic*, since it collocates between the microscopic and the macroscopic approaches. As a matter of fact, kinetic theory aims at deriving macroscopic properties of gases, such as pressure and temperature, starting from the microscopic details of interactions among molecules. This is done by using a proper integro-differential equation, typically known as the Boltzmann equation [2].

Using a parallelism between the molecules of gases and agents, macroscopic features of multi-agent systems can be analytically found on the basis of the proper description of the effects of microscopic interactions. While molecules of gases are typically associated with their velocities and positions, agents can be associated with attributes which represent their characteristics. According to this idea, collisions among molecules in gases can be reinterpreted as interactions among agents. Opinion dynamics models, such as that in [3], can be studied using

© Springer International Publishing AG 2016
G. Adorni et al. (Eds.): AI*IA 2016, LNAI 10037, pp. 105–117, 2016.
DOI: 10.1007/978-3-319-49130-1_9

the kinetic framework and we envisage that cooperation models, such as that in [4], and agent-based large scale systems, such as those in [5], can also be studied with the presented framework.

In the literature, various models have been used to describe the effects of interactions among agents, such as those based on graph theory (see, e.g., [6]), on Brownian motion (see, e.g., [7]), and on cellular automata (see, e.g., [8]). In this paper we focus on opinion dynamics models based on sociophysics and we aim at modeling the temporal evolution of social characteristics of the system as agents interact with each other. In order to model interactions among agents, we make some assumptions inspired by those of kinetic theory. In particular, we assume that interactions are binary, namely that they involve two agents at a time. Moreover, each agent can interact with any other agent in the system just like molecules can freely interact with each other in gases. According to the proposed approach, each agent is associated with a scalar attribute that represents its opinion. From a microscopic viewpoint, the opinion of an agent can change through interactions with other agents. A stochastic rule to update the opinions of two agents after an interaction is considered. More precisely, such a rule is meant to model two processes, namely *compromise*, which corresponds to the idea that the opinions of two agents get closer after their interaction, and *diffusion*, which is modeled as additive random noise. These two processes are commonly studied when dealing with opinion dynamics [9]. Based on the microscopic rules for opinion updates, we derive macroscopic properties of the considered multi-agent system, such as the average opinion and the stationary profile of the opinion. In order to do so, we rely on a generalization of the Boltzmann equation, which has been specifically adapted for the considered microscopic model and it is first introduced in this paper.

The paper is organized as follows. Section 2 describes the considered kinetic model from an analytical viewpoint. Section 3 derives explicit formulas for the stationary profiles in a specific case. Section 4 shows results for different values of the parameters of the model. Finally, Sect. 5 concludes the paper.

2   Kinetic Model of Opinion Formation

In order to describe opinion formation in multi-agent systems using a kinetic approach, we first need to introduce some notation and to define the microscopic equations which model the effects of interactions among agents. Then, relying on such rules, the explicit expression of the Boltzmann equation can be derived and used to obtain macroscopic properties of the system, concerning, for instance, the average opinion and the stationary profile of the opinion. Throughout the paper, we assume that the opinion of agents can be represented as a continuous variable, denoted as $v \in \mathbb{R}$, so that values of v close to 0 correspond to moderate opinions while if v has a large absolute value it represents an extremal opinion.

2.1   The Microscopic Model

According to the considered model, the opinions of agents can be updated as a consequence of (binary) interactions among them. Therefore, proper rules for opinion updates after interactions need to be introduced. More precisely, in our assumptions the post-interaction opinions of two agents are obtained by adding two terms to their respective pre-interaction opinions. The first term is related to compromise through a deterministic parameter, while the second term is stochastic and it is modeled as a random additive noise. More precisely, denoting as v' and w' the post-interaction opinions of two agents whose pre-interaction opinions were v and w, the following formula is considered

$$\begin{cases} v' = v + \gamma(w - v) + \eta_1 \\ w' = w + \gamma(v - w) + \eta_2. \end{cases} \tag{1}$$

In (1), γ is a deterministic parameter and η_1 and η_2 are two random variables with the same support $\mathbb{B}$ and the same distribution $\Theta(\cdot)$. The second terms on the right hand side of the two equations in (1) are proportional to the difference between the pre-interaction opinions according to the parameter γ. Throughout the paper, we assume that γ varies in $(0,1)$ in order to model the idea of compromise, according to which the absolute value of the difference between the post-interaction opinions of two agents has to be smaller than the absolute value of the difference between the pre-interaction opinions. As a matter of fact, if the last terms in (1) are neglected, the following inequality holds

$$|v' - w'| = |(1 - 2\gamma)(v - w)| < |v - w|$$

where we used the fact that $\gamma \in (0,1)$ is equivalent to $|(1 - 2\gamma)| < 1$. The third terms on the right hand side of the two equations in (1) are two random variables related to diffusion, which model the idea that agents can change their opinions due to external events which are independent from their own opinion and from that of the interacting agents. Diffusion is hence modeled as additive random noise. In the following, we assume that the average value of η_1 and η_1 is 0 and that their variance is σ^2. We remark that the rules to update the opinions of two interacting agents, defined in (1), are different from those considered in other papers, such as [10–12].

2.2   The Boltzmann Equation for Opinion Dynamics

Given the microscopic rules that describe the effects of microscopic interactions, we now need to define a framework to derive macroscopic properties of the considered multi-agent system. To this purpose, as it happens in kinetic theory, we rely on a so called density function $f(v, t)$ which represents the number of agents with opinion v at time t and which is defined for each opinion $v \in \mathbb{R}$ and for each time $t \geq 0$. Notably, this function generalises the distribution function commonly used in kinetic theory and related models. Of course, due to its definition, the function $f(v,t)$ is non-negative, namely it satisfies $f(v,t) \geq 0$. Moreover, defining as $N(t)$ the number of agents at time t, the following equality needs to hold

$$\int_{\mathbb{R}} f(v,t)\mathrm{d}v = N(t). \tag{2}$$

Relying on the density function $f(v,t)$, the average opinion $u(t)$ of the multi-agent system at time t can also be introduced. It is defined as

$$u(t) = \frac{1}{N(t)} \int_{\mathbb{R}} v f(v,t)\mathrm{d}v. \tag{3}$$

As in kinetic theory, we assume that function $f(v,t)$ evolves on the basis of the Boltzmann equation, whose homogeneous form is

$$\frac{\partial f}{\partial t} = \mathcal{Q}(f,f)(v,t). \tag{4}$$

The left-hand side of (4) represents the temporal evolution of the density function $f(v,t)$ and the right-hand side $\mathcal{Q}$ is an integral operator which takes into account the effects of interactions. For this reason, $\mathcal{Q}$ is known as *collisional operator* and it is strictly related to the definition of the post-interaction opinions in (1).

Instead of considering (4), we focus on the *weak form* of the Boltzmann equation. In functional analysis, the weak form of a differential equation in the unknown function $f(v)$ is obtained by multiplying both sides of the considered equation by a test function $\phi(v)$, namely a smooth function with compact support, and then integrating the obtained equation with respect to v. The weak form of the Boltzmann equation in (4) can then be written as

$$\frac{\mathrm{d}}{\mathrm{d}t} \int_{I} f(v,t)\phi(v)\mathrm{d}v = \int_{\mathbb{R}} \mathcal{Q}(f,f)\phi(v)\mathrm{d}v \tag{5}$$

where on the left-hand side we used the fact that, under proper regularity conditions, the integral and the derivative commute. In order to clarify the importance of the weak form of the Boltzmann equation, let us observe that setting $\phi(v) = 1$ and recalling (2), the left-hand side of (5) represents the temporal derivative of the number of agents $N(t)$. Similarly, setting $\phi(v) = v$ and recalling (3), the left-hand side of (5) represents the temporal derivative of the average opinion $u(t)$. It can then be concluded that the weak form of the Boltzmann equation is useful, at least, to study properties related to the number of agents and to the average opinion.

To proceed, the explicit expression of the weak form of the collisional operator which appears at the right-hand side of (5) needs to be found. In order to take into account the effects of the stochastic part in (1), the following *transition rate* [2] is defined as

$$W(v,w,v',w') = \Theta(\eta_1)\Theta(\eta_2). \tag{6}$$

The explicit expression of the weak form of the collisional operator can then be written as

$$\int_{I} \mathcal{Q}(f,f)\phi(v)\mathrm{d}v = \int_{\mathbb{B}^2} \int_{\mathbb{R}^2} \left[{}'W \frac{1}{J} f({}'v)f({}'w) - W f(v)f(w) \right] \phi(v)\mathrm{d}v\mathrm{d}w\mathrm{d}\eta_1\mathrm{d}\eta_2$$

where $'v$ and $'w$ are the pre-interaction variables which lead to v and w, respectively, $'W$ is the transition rate relative to the 4−uple $('v,'w,v,w)$ and J is the Jacobian of the transformation of $('v,'w)$ in (v,w) [2]. Using a proper change of variable in the collisional operator, the weak form of the Boltzmann Eq. (5) can be finally written as

$$\frac{d}{dt}\int_I f(v,t)\phi(v)dv = \int_{\mathbb{B}^2}\int_{\mathbb{R}^2} Wf(v)f(w)(\phi(v') - \phi(v))dvdwd\eta_1 d\eta_2. \qquad (7)$$

2.3   The Macroscopic Properties

Setting $\phi(v) = 1$ in (7) leads to

$$\frac{d}{dt}\int_{\mathbb{R}} f(v,t)dv = 0 \qquad (8)$$

where the left-hand side represents the temporal derivative of the number of agents and the right-hand side is 0 because the difference $(\phi(v') - \phi(v))$ inside the integral in (7) is 0 for any constant function $\phi(\cdot)$. According to (8), it can then be concluded that the number of agents is constant and, for this reason, from now on we omit the dependence of N on time t. This property is analogous to mass conservation of the molecules in gases.

Considering instead $\phi(v) = v$ as a test function in (7) and recalling (1) we obtain

$$\frac{d}{dt}\int_{\mathbb{R}} f(v,t)vdv = \gamma\int_{\mathbb{B}^2}\int_{\mathbb{R}^2} Wf(v)f(w)(w - v)dvdwd\eta_1 d\eta_2$$
$$+ \int_{\mathbb{B}^2}\int_{\mathbb{R}^2} Wf(v)f(w)\eta_1 dvdwd\eta_1 d\eta_2. \qquad (9)$$

Observe that, according to (3), the left-hand side of (9) is proportional through the constant N to the temporal derivative of the average opinion $u(t)$. The first integral in the right-hand side of (9) can be written as

$$\gamma\left(\int_{\mathbb{R}} f(v)dv \int_{\mathbb{R}} wf(w)dw - \int_{\mathbb{R}} f(v)vdv \int_{\mathbb{R}} f(w)dw\right) \qquad (10)$$

where we used the fact that $\Theta(\cdot)$ is a probability density function and, therefore, its integral equals 1. Observe that, using the definitions of N and $u(t)$ in (2), (3) and (10) can be written as $\gamma(N^2u(t) - N^2u(t)) = 0$. The second integral in (9) can be written as

$$\int_{\mathbb{B}} \eta_1\Theta(\eta_1)d\eta_1 \int_{\mathbb{R}} f(v)dv \int_{\mathbb{R}} f(w)dw \qquad (11)$$

and it equals 0 because the average value of the random variable η_1 is 0. It can be concluded that (9) can be written as

$$N\dot{u}(t) = 0 \qquad (12)$$

and, since N is constant, the average opinion is conserved, namely $u(t) = u(0)$. From now on, we denote the average opinion as u, thus neglecting the dependence on time t.

3   Analytic Derivation of Stationary Profiles

In order to derive the stationary profile, let us define a new temporal value $\tau = \gamma t$, where γ is the constant coefficient related to compromise in (1). A new function can then be defined as

$$g(v, \tau) = f(v, t). \tag{13}$$

If $\gamma \simeq 0$, then the function $g(v, \tau)$ describes the asymptotic behaviour of $f(v, t)$. Observe that the hypothesis $\gamma \simeq 0$ corresponds to assuming that each interaction causes small opinion exchange, as it happens in realistic cases. By substituting $f(v, t)$ with $g(v, \tau)$ in (7) and using a Taylor series expansion of $\phi(v)$ around v, one obtains [2]

$$\frac{\mathrm{d}}{\mathrm{d}\tau} g(v, \tau) = \frac{\lambda}{2} \frac{\partial^2}{\partial v^2} (g(v, \tau)) + \frac{\partial}{\partial v} ((v - u)g(v, \tau)) \tag{14}$$

where

$$\lambda = \sigma^2 / \gamma. \tag{15}$$

Observe that, according to (15), the parameter λ is defined as the ratio between the variance σ^2 of the random variables η_1 and η_2, which is related to the additive random noise, and the parameter γ, which is related to compromise. Therefore, it can be concluded that λ quantifies the impact of compromise with respect to that of the random additive noise in the rules for opinion updates.

The stationary profiles, denoted as $g_\infty(v)$ in the rest of this paper, can be found as solutions of the equation

$$\frac{\mathrm{d}}{\mathrm{d}\tau} g(v, \tau) = 0 \tag{16}$$

which, according to (14), corresponds to

$$\frac{\lambda}{2} \frac{\partial}{\partial v} g(v, \tau) + (v - u)g(v, \tau) = C \tag{17}$$

where u is the (constant) average opinion, λ is defined in (15), and C is a proper constant. Observe that the constant C must be 0. As a matter of fact, considering the integral of (17) over the interval $[-v_1, v_2]$, one obtains

$$\frac{\lambda}{2} \int_{-v_1}^{v_2} \frac{\partial}{\partial v} g(v, \tau) \mathrm{d}v + \int_{-v_1}^{v_2} (v - u)g(v, \tau) \mathrm{d}v = C(v_2 + v_1). \tag{18}$$

The result in (18) holds for every v_1 and v_2 in $\mathbb{R}$. The first integral in (18) can be written as the difference between $g(v_2, \tau)$ and $g(-v_1, \tau)$. Since, according to its definition and to that of $f(v, t)$, the integral of $g(v, \tau)$ over $\mathbb{R}$ is finite, as it equals N, it can be concluded that, considering the limits for $v_1, v_2 \to +\infty$, both $g(v_2, \tau)$ and $g(-v_1, \tau)$ tend to 0. Concerning the second integral in (18), it is easy to show that

$$\int_{\mathbb{R}} vg(v, \tau) \mathrm{d}v - u \int_{\mathbb{R}} g(v, \tau) \mathrm{d}v = Nu - Nu = 0.$$

Since as $v_1, v_2 \to +\infty$ both integrals on the left-hand side of (18) tend to 0, it is necessary that that $C = 0$ to guarantee that (18) holds.

Using standard analysis on (17) and omitting the dependence of g on τ for the sake of readability, one obtains

$$\frac{g'(v)}{g(v)} = -\frac{2}{\lambda}(v - u). \tag{19}$$

The left-hand side of the previous equation is the derivative of $\log g(v)$ and, hence, integrating both sides of (19) leads to

$$\log g(v) = -\frac{2}{\lambda} \int (v - u)dv = -\frac{1}{\lambda}(v - u)^2 \tag{20}$$

Finally, we can write the explicit expression of the stationary profile found as solution of (20) is

$$g_\infty(v) = c_{\lambda,N} \exp\left(-\frac{1}{\lambda}(v - u)^2\right) \tag{21}$$

where $c_{\lambda,N}$ is a constant depending on the value of λ and, as shown hereafter, on the number of agents N. Such a constant must be set in order to guarantee that

$$\int_{\mathbb{R}} g_\infty(v)dv = N. \tag{22}$$

Since

$$\int_{\mathbb{R}} \exp\left(-\frac{1}{\lambda}(v - u)^2\right) dv = \sqrt{\lambda\pi},$$

algebraic manipulations of (22) lead to the following expression of the normalization constant $c_{\lambda,N}$

$$c_{\lambda,N} = \frac{N}{\sqrt{\pi\lambda}}. \tag{23}$$

Finally, observe that, according to (21), $g_\infty(v)$ is an even function if $u = 0$.

According to (2), the integral over $\mathbb{R}$ of the stationary profile normalized by the number of agents equals 1, namely

$$\int_{\mathbb{R}} \frac{g_\infty(v)}{N} dv = 1. \tag{24}$$

Defining $\hat{g}_\infty(v) = g_\infty(v)/N$, it can be concluded that $\hat{g}_\infty(v)$ represents a probability distribution function of the opinion in the considered multi-agent system. Using (21) and (23), the explicit expression of $\hat{g}_\infty(v)$ is

$$\hat{g}_\infty(v) = \frac{1}{\sqrt{\pi\lambda}} \exp\left(-\frac{(v - u)^2}{\lambda}\right) \tag{25}$$

which represents a Gaussian distribution with average value u and variance $\sigma_g^2 = \lambda/2$. Due to the linear relationship between the variance σ_g^2 and the parameter λ, it can be concluded that the variance σ_g^2 increases as λ increases. According to (15), λ is directly proportional to the variance σ^2 of the two random variables η_1 and η_2 and inversely proportional to the constant γ, which is related to compromise. Therefore, the growth of λ corresponds to an increase of the impact of random additive noise with respect to that of compromise. Hence, it is expected that the variance σ_g^2 of $\hat{g}_\infty(v)$ increases as the impact of additive random noise increases with respect to that of compromise.

In order to see if the stationary profile is characterized by maxima and/or minima, we now aim at studying the derivative of g_∞. From (19) the derivative of g_∞ can be written as

$$g_\infty'(v) = -g_\infty(v)\frac{2}{\lambda}(v - u) \tag{26}$$

Since $g_\infty(v)$ is an exponential function, it is always positive. Therefore, from (26) it can be concluded that the unique stationary point of $g_\infty(v)$ is $v = u$, corresponding to the average opinion.

4   Verification of Analytic Results

In this section, various stationary profiles for different values of the average opinion u and of the parameter λ are shown. Besides plotting the functions $\hat{g}_\infty(v)$ defined in (25) for various values of u and λ, we also show some simulation results. All these results are obtained by using 10^3 agents and by simulating pairwise interactions between them. The opinions of the considered multi-agent system are randomly initialized. In order to reproduce the behaviors of the stationary profiles, 10^5 iterations are simulated. Each iteration corresponds to an interaction between two agents. More precisely, at each iteration, two agents in the considered multi-agent system are randomly chosen and their opinions are updated according to (1). The profiles obtained after 10^5 iterations are denoted as $\tilde{g}(v)$ in the following. We remark that simulation results are obtained by only considering the microscopic rules defined in (1) to update the opinions of two interacting agents. All the kinetic-based framework which has been used throughout the paper to derive macroscopic properties of the system is ignored when performing simulations.

We start by considering a multi-agent system where the average opinion is $u = 0$. Under this assumption, according to (21), the stationary profile $g_\infty(v)$ is represented by the following even function

$$g_\infty(v) = \frac{N}{\sqrt{\pi\lambda}}\exp\left(-\frac{v^2}{\lambda}\right). \tag{27}$$

According to (25), in this case the function $\hat{g}_\infty(v)$ represents a Gaussian distribution with average value 0 and variance $\sigma_g^2 = \lambda/2$.

Various values of the parameter λ are considered. First, we consider $\lambda = 1$. According to (15), this assumption corresponds to setting $\sigma^2 = \gamma$, so that the variance of the random variables η_1 and η_2 equals the value of the parameter γ,

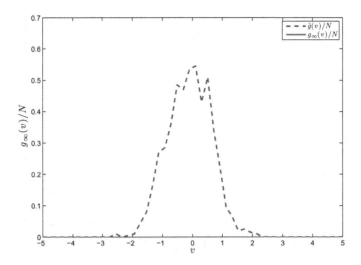

Fig. 1. The stationary profile g_∞, normalized by the number of agents N, relative to the average opinion $u = 0$ is shown for $\lambda = 1$ (red line). The values of $\tilde{g}_\infty$ obtained by simulations also normalized by the number of agents N are shown (dashed blue line). (Color figure online)

which is related to compromise. Figure 1 shows the stationary profile normalized by the number of agents N (red line), namely $\hat{g}_\infty(v)$, which, according to (24), represents the probability distribution function of the opinion. In this case, the variance σ_g^2 of the probability distribution function $\hat{g}_\infty(v)$ is $1/2$. In Fig. 1 the function $\hat{g}_\infty(v)$ is shown for values of $v \in [-5, 5]$, since outside this interval the values of $\hat{g}_\infty(v)$ are very close to 0. Moreover, according to (27), $\hat{g}_\infty(v)$ tends to 0 as v tends to $\pm\infty$. Figure 1 also shows the profile of the distribution of opinion obtained according to simulations. To obtain this result, we consider a multi-agent system composed of 10^3 agents, as stated at the beginning of this section. The opinions of all agents are randomly initialized in order to match the condition on the average opinion, namely $u = 0$. The profile $\tilde{g}(v)$ obtained after the simulation of 10^5 pairwise interactions between agents is derived and Fig. 1 shows the values of $\tilde{g}(v)/N$ (dashed blue line). A comparison between the result analytically derived and that obtained by simulation shows that the two distributions of the opinion are very similar, thus assessing the validity of the considered analytic model.

We now consider $\lambda = 3$, under the assumption that $u = 0$, as in the previous case. According to (15), this corresponds to assuming that the variance of the random variables η_1 and η_2 is $\sigma^2 = 3\gamma$. It can then be observed that the contribution of additive random noise with respect to that of compromise is more important than in the case with $\lambda = 1$. Figure 2 shows the function $\hat{g}_\infty(v)$ relative to $\lambda = 3$, which represents the probability distribution function of the opinion. In this case, according to (25), the variance σ_g^2 of the probability distribution function $\hat{g}_\infty(v)$ is $3/2$ and it is larger than in the previous case. This result

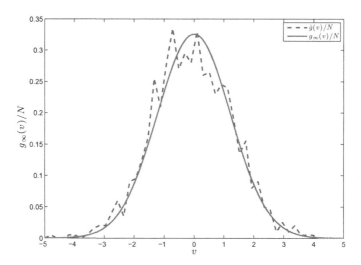

Fig. 2. The stationary profile g_∞, normalized by the number of agents N, relative to the average opinion $u = 0$ is shown for $\lambda = 1$ (red line). The values of $\tilde{g}_\infty$ obtained by simulations also normalized by the number of agents N are shown (dashed blue line). (Color figure online)

can also be observed by comparing Figs. 1 and 2. The function $\hat{g}_\infty(v)$ is shown in Fig. 2 for values of $v \in [-5, 5]$. Outside this interval, the values of $\hat{g}_\infty(v)$ are close to 0, and, according to (27), they tend to 0 as v tends to $\pm\infty$. Figure 2 also shows the profile of the distribution of opinion obtained from simulation. The opinions of all agents are randomly initialized at the beginning such that the average opinion u equals 0 and 10^5 pairwise interactions among agents are simulated. After each interaction, the opinions of the two involved agents are updated according to (1). In Fig. 2, the values of $\tilde{g}(v)/N$ are shown (dashed blue line) and, as in the case with $\lambda = 1$, they are similar to the values of $\hat{g}_\infty(v)$ analytically obtained.

Finally, we reduce the value of λ to $1/3$, considering once again $u = 0$ as average opinion. According to (15), this corresponds to $\gamma = 3\sigma^2$ and, hence, with respect to previous cases the impact of compromise is more important than that of additive random noise. Figure 3 shows the probability distribution function of the opinion, namely $\hat{g}_\infty(v)$ (red line). According to (25), the variance σ_g^2 of the probability distribution function $\hat{g}_\infty(v)$ is $1/6$. The value of σ_g^2 is smaller than those relative to the previous cases as it can also be observed from a comparison among the three figures. Figure 2 is restricted to values of $v \in [-5, 5]$ since outside this interval the values of $\hat{g}_\infty(v)$ rapidly tend 0. Figure 3 also shows the profile of the distribution of opinion obtained by simulating a multi-agent system composed of 10^3 agents. The opinions of all agents are randomly initialized in order to guarantee that $u = 0$ and pairwise interactions between agents are simulated. The values of $\tilde{g}(v)/N$ obtained after the simulation of 10^5 interactions are shown in Fig. 3 (dashed blue line). Once again, the analytic result

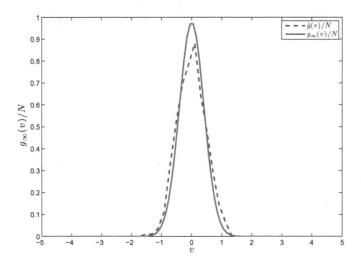

Fig. 3. The stationary profile g_∞, normalized by the number of agents N, relative to the average opinion $u = 0$ is shown for $\lambda = 1$ (red line). The values of $\tilde{g}_\infty$ obtained by simulations also normalized by the number of agents N are shown (dashed blue line). (Color figure online)

is similar to that obtained by simulation, confirming the validity of the considered analytic model.

As observed at the end of Sect. 3, by comparing the graphs obtained for different values of λ it can be observed that any increase of λ also increases the variance of the stationary profiles.

As a final example, we now consider a multi-agent system where the average opinion is $u = 50$. According to (25), in this case the function $\hat{g}_\infty(v)$ can be written as

$$\hat{g}_\infty(v) = \frac{1}{\sqrt{\pi\lambda}} \exp\left(-\frac{(v - 50)^2}{\lambda}\right) \tag{28}$$

and it represents a Gaussian distribution with average value 50 and variance $\sigma_g^2 = \lambda/2$. Figure 4 shows the function $\hat{g}_\infty(v)$ (red line), obtained with $\lambda = 1$. This graph is very similar to that in Fig. 1, due to the fact that both functions represent Gaussian distribution with standard deviation $\sigma_g^2 = 1/2$. The only difference between Figs. 1 and 4 is that the latter is relative to $u = 50$ and, therefore, it is centered in $v = 50$, while the former is centered in $v = 0$ since it corresponds to a multi-agent system with average opinion $u = 0$. The values of $\tilde{g}(v)/N$ obtained by simulation are also shown in Fig. 3 (dashed blue line). This result has been obtained by considering a multi-agent system composed by 10^3 agents whose opinions are initialized in order to guarantee that the average opinion is $u = 50$. Also in this last case, the result obtained by simulation matches the analytic one.

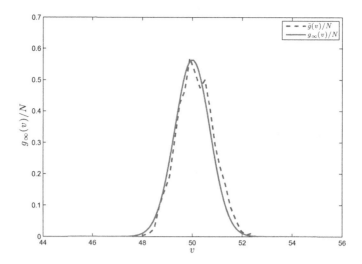

Fig. 4. The stationary profile g_∞, normalized by the number of agents N, relative to the average opinion $u = 50$ is shown for $\lambda = 1$ (red line). The values of $\tilde{g}_\infty$ obtained by simulations also normalized by the number of agents N are shown (dashed blue line). (Color figure online)

5   Conclusions

This paper presents a kinetic model to analytically study the evolution of opinion in multi-agent systems. We start from the description of a stochastic model which describes the effects of interactions among agents and which reproduces two phenomena, namely compromise and diffusion. The latter is modeled as an additive random noise. Then, we propose an analytic framework to study macroscopic properties of the system and to derive the asymptotic behaviour of the opinion distribution. More precisely, we show that the average opinion of the system is conserved and that the probability distribution function of the opinion tends to a Gaussian function, whose variance is related to the parameters of the model. According to the obtained analytic results, the larger is the impact of additive random noise with respect to that of compromise, the larger is the variance of the stationary profile. Analytic results are confirmed by simulations.

As a future work, it is possible to model different sociological phenomena, such as *negative influence, homophily,* and *striving for uniqueness* [13]. Moreover, taking inspiration from kinetic theory of gas mixtures, which studies gases composed of different types of molecules, it is also possible to analyse heterogeneous multi-agent systems, namely those where classes of agents, characterized by different features, coexist. A preliminary investigation of such an extension is discussed in [14], where two classes of agents are considered. The characteristics associated to each class of agents are the number of agents in each class, the initial distribution of opinion, and the propensity of agents in each class to change their opinions.

References

1. Chakraborti, B.K., Chakrabarti, A., Chatterjee, A.: Econophysics and Socio-physics: Trends and Perspectives. Wiley, Berlin (2006)
2. Toscani, G.: Kinetic models of opinion formation. Commun. Math. Sci. **4**, 481–496 (2006)
3. Monica, S., Bergenti, F.: A Study of Consensus Formation using Kinetic Theory. In: Omatu, S., Semalat, A., Bocewicz, G., Sitek, P., Nielsen, I.E., García, J.A.G., Bajo, J. (eds.) DCAI, 13th International Conference. AISC, vol. 474, pp. 213–221. Springer, Heidelberg (2016)
4. Bergenti, F., Poggi, A., Somacher, M.: A collaborative platform for fixed and mobile networks. Commun. ACM **45**(11), 39–44 (2002)
5. Bergenti, F., Caire, G., Gotta, D.: Large-scale network and service management with WANTS. In: Industrial Agents: Emerging Applications of Software Agents in Industry, pp. 231–246. Elsevier (2015)
6. Tsang, A., Larson, K.: Opinion dynamics of skeptical agents. In: Proceedings of 13th International Conference on Autonomous Agents and Multiagent Systems (AAMAS 2014), Paris, France, May 2014
7. Schweitzer, F., Holyst, J.: Modelling collective opinion formation by means of active brownian particles. Eur. Phys. J. B **15**, 723–732 (2000)
8. Monica, S., Bergenti, F.: A stochastic model of self-stabilizing cellular automata for consensus formation. In: Proceedings of 15th Workshop "Dagli Oggetti agli Agenti" (WOA 2014), Catania, Italy, September 2014
9. Hegselmann, R., Krause, U.: Opinion dynamics and bounded confidence models, analysis, and simulation. J. Artif. Soc. Soc. Simul. **5**(3), 1–33 (2002)
10. Monica, S., Bergenti, F.: Simulations of opinion formation in multi-agent systems using kinetic theory. In: Proceedings of 16th Workshop "Dagli Oggetti agli Agenti" (WOA 2015), Napoli, Italy, June 2015
11. Monica, S., Bergenti, F.: A kinetic study of opinion dynamics in multi-agent systems. In: Gavanelli, M., Lamma, E., Riguzzi, F. (eds.) AI*IA 2015. LNCS (LNAI), vol. 9336, pp. 116–127. Springer, Heidelberg (2015). doi:10.1007/978-3-319-24309-2_9
12. Monica, S., Bergenti, F.: Kinetic description of opinion evolution in multi-agent systems: analytic model and simulations. In: Chen, Q., Torroni, P., Villata, S., Hsu, J., Omicini, A. (eds.) PRIMA 2015. LNCS (LNAI), vol. 9387, pp. 483–491. Springer, Heidelberg (2015). doi:10.1007/978-3-319-25524-8_30
13. Mark, N.P.: Culture and competition: Homophily and distancing explainiations for cultural niches. Amm. Sociol. Rev. **68**, 319–345 (2003)
14. Bergenti, F., Monica, S.: Analytic study of opinion dynamics in multi-agent systems with two classes of agents. In: Proceedings of 17th Workshop "Dagli Oggetti agli Agenti" (WOA 2016), Catania, Italy, July 2016

Combining Avoidance and Imitation to Improve Multi-agent Pedestrian Simulation

Luca Crociani[1], Giuseppe Vizzari[1(✉)], and Stefania Bandini[1,2]

[1] Complex Systems and Artificial Intelligence Research Centre,
University of Milano-Bicocca, Viale Sarca 336/14, 20126 Milan, Italy
{Luca.Crociani,Giuseppe.Vizzari,Stefania.Bandini}@disco.unimib.it
[2] Researche Center for Advanced Science and Technology,
The University of Tokyo, Tokyo, Japan

Abstract. Simulation of pedestrian and crowd dynamics is a consolidated application of agent-based models but it still presents room for improvement. Wayfinding, for instance, is a fundamental task for the application of such models on complex environments, but it still requires both empirical evidences as well as models better reflecting them. In this paper, a novel model for the simulation of pedestrian wayfinding is discussed: it is aimed at providing general mechanisms that can be calibrated to reproduce specific empirical evidences like a proxemic tendency to avoid congestion, but also an imitation mechanism to stimulate the exploitation of longer but less congested paths explored by emerging leaders. A demonstration of the simulated dynamics on a large scale scenario will be illustrated in the paper and the achieved results will show the achieved improvements compared to a basic floor field Cellular Automata model.

Keywords: Agent-based modeling and simulation · Pedestrian simulation · Wayfinding

1 Introduction

Simulating crowds dynamics in the built environment is an established but still lively application area with contributions from different disciplines: the automated analysis and the synthetic generation of pedestrians' and crowd behavior (and attempts to integrate these activities [1] still present challenges and potential developments, for instance in a smart environment perspective [2]. Although available commercial tools are used on a day-to-day basis by designers and planners, according to a report commissioned by the Cabinet Office [3] there is still room for innovations in models, to improve their effectiveness, expressiveness (i.e. simplifying the modeling activity or introducing the possibility of representing phenomena that were still not considered by existing approaches) and efficiency.

Even if we only consider choices and actions related to walking, modeling human decision making activities and actions is a complicated task: different

© Springer International Publishing AG 2016
G. Adorni et al. (Eds.): AI*IA 2016, LNAI 10037, pp. 118–132, 2016.
DOI: 10.1007/978-3-319-49130-1_10

types of decisions are taken at different levels of abstraction, from path planning to the regulation of distance from other pedestrians and obstacles present in the environment. Moreover, the measure of success and validity of a model is definitely not the *optimality* with respect to some cost function, as (for instance) in robotics, but the *plausibility*, the adherence of the simulation results to data that can be acquired by means of observations or experiments.

In a recent work, we started to investigate the actual need of modeling wayfinding decisions in situations characterized by the fact that pedestrians have different options (i.e. paths leading them from a region to a desired target, passing through intermediate gateways and regions) potentially in presence of congestion [4]. This work has highlighted the fact that not considering these kind of decisions would lead to a potential overestimation of travel times, since baseline models such as the floor field CA approach [5] is based on a "least effort" principle, essentially choosing the shortest path with corrections to the low level trajectory considering obstacles and basic proxemic but just on local neighboring cells. This means, that agents representing pedestrians do not anticipate what could generally be perceived, that is, the fact that the passage on the shortest trajectory towards their goal is actually heavily congested. In the present work, we extend the previous research by investigating the interplay between avoidance and imitation, at the level of wayfinding decisions: an avoidance tendency leads an agent to choose a suboptimal path whenever the optimal one has become less attractive due to congestion, and this kind of decision can be imitated by other nearby agents, also reconsidering their choices. These conflicting tendencies can be calibrated according to empirical evidences. After a discussion of relevant related works, an analysis of different alternatives for modeling and simulating this kind of scenario will be illustrated in Sect. 3. Results of the application of the proposed model in a real world scenario, initially described in [6], will then be described, with reference to their plausibility. Conclusions and future works will end the paper.

2 Related Works

The inclusion in simulation models of decisions related to trade off scenarios, such as the one between overall trajectory length and presumed travel time (considering congestion in perceived alternative gateways), represent an issue in current modeling approaches.

Commercial instruments, for instance, mostly provide basic tools to the modelers, that are enabled and required to specify how the population of pedestrians will behave: this implies that the operator constructing the simulation model needs to specify how the pedestrians will generally choose their trajectory (generally selecting among different alternatives defined by means of annotation of the actual spatial structure of the simulated environment through landmarks representing intermediate or final destinations [7]), as well the conditions generating exceptions to the so called "least effort principle", suggesting that pedestrians generally try to follow the (spatially) shortest path toward their destination.

Space, in fact, represents just one of the relevant aspects in this kind of choice: since most pedestrians will generally try to follow these "best paths" congestion can arise and pedestrians can be pushed to make choices that would be sub-optimal, from the perspective of traveled distance.

Recent works in the area of pedestrian and crowd simulation started to investigate this aspect. In particular, [8] proposed the modification of the floor field Cellular Automata [5] approach for considering pedestrian choices not based on the shortest distance criterion but considering the impact of congestion on travel time. [6] explored the implications of four different strategies for the management of route choice operations, through the combination of applying the shortest or quickest path, with a local (i.e., minimize time to vacate the room) or global (i.e., minimize overall travel time) strategies.

Iterative approaches, borrowing models and even tools from vehicular transportation simulation, propose to adopt a more coarse grained representation of the environment, i.e. a graph in which nodes are associated to intersections among road sections, but the process can be also adopted in buildings [7]. In this kind of scenario, pedestrians can start by adopting shortest paths on a first round of simulation: as suggested before, the fact that all pedestrians take the best path generally leads to congestion and sub-optimal travel times. Some selected pedestrians, especially those whose actual travel time differs significantly from the planned one, will change their planned path and a new simulation round will take place. The iteration of this process will lead to an equilibrium or even to system optimum, according to the adopted travel cost function [9]. This iterative scheme has also been employed in multi-scale modeling approaches [10,11].

The above approach naturally leads to consider that this kind of problem has been paid considerable attention in the field of Artificial Intelligence, in particular by the planning community. Hierarchical planning [12] approaches, in particular, provide an elegant and effective framework in which high level abstract tasks can be decomposed into low level activities. Despite the fact that the formulation of the approach date to the seventies, it is still widely considered and employed in the close area of computer graphics [13], in which actions of virtual pedestrians are planned with the aim of being visually plausible and decided within real-time constraints. Within this framework, also issues related to the reconsideration of choices and plans were analyzed, mostly within the robotics area [14]. In the pedestrian simulation context, one could consider that microscopic decisions on the steps to be taken can follow a high-level definition of a sequence of intermediate destinations to be reached by the pedestrian. This kind of approach, which we experimentally investigated in [4], also allows exploiting already existing models dealing with low level aspects of pedestrian actions and perceptions.

The main issues in transferring AI planning results within this context of application, and more generally producing generally applicable contributions to the field, are partly due to the above suggested fundamental difference between the measures of success between *simulation* and *control* applications. Whereas the latter are targeted at *optimal* solutions, the former have to deal with the notions of *plausibility* and *validity*. Moreover, we are specifically dealing with a *complex system*, in which different and conflicting mechanisms are active at the same time (e.g. proxemics [15] and imitative behaviours [16]). Finally, whereas recent extensive observations and analyses (see, e.g., [17]) produced extensive data that can be used to validate simulations within relatively simple scenarios (in which decisions are limited to basic choices on the regulation of mutual distances among other pedestrian while following largely common and predefined paths like corridors with unidirectional or bidirectional flows, corners, bottlenecks), we still lack comprehensive data on way-finding decisions.

3   A Model to Encompass the Pedestrian Movement and Route Choice

This Section will propose a multi-agent model designed for the simulation of pedestrian movement and route choice behavior. The model of agent is composed of two elements, respectively dedicated to the low level reproduction of the movement towards a target (i.e. the operational level, considering a three level model described in [18]) and to the decision making activities related to the next destination to be pursued (i.e. the route choice at the tactical level). The component dedicated to the operational level behavior of the agent is not extensively described since, for this purpose, the model described in [19] has been applied (this choice allows inheriting the validation of operational level properties of the above model). For a proper understanding of the approaches and mechanisms that will be defined at the tactical level, a brief description on the representation of the environment, with different levels of abstractions, is provided in this Section. More attention will then be dedicated to the introduction and discussion of the model for the management of the route choice, which represents the main contribution of this paper.

3.1   The Representation of the Environment and the Knowledge of Agents

The adopted agent environment [20] is discrete and modeled with a rectangular grid of 40 cm sided square cells. The size is chosen considering the average area occupied by a pedestrian [21], and also respecting the maximum densities usually observed in real scenarios. The cells have a state that informs the agents about the possibilities for movement: each one can be vacant or occupied by obstacles or pedestrians (at most two, so as to be able to manage locally high density situations).

To allow the configuration of a pedestrian simulation scenario, several *markers* are defined with different purposes. This set of objects has been introduced to allow the movement at the operational level and the reasoning at the tactical level, identifying intermediate and final targets:

– *start areas* ▭, places were pedestrians are generated: they contain information for pedestrian generation both related to the type of pedestrians (e.g. the distribution of their destinations), and to the frequency of generation;

- *openings* ▭, sets of cells that divide, together with the obstacles, the environment into regions. These objects constitutes the decision elements, intermediate destinations, for the route choice activities;
- *regions* ▬, markers that describe the type of the region where they are located: with them it is possible to design particular classes of regions (e.g. stairs, ramps) and other areas that imply a particular behavior of pedestrians;
- *final destinations* ▬, the ultimate targets of pedestrians;
- *obstacles* ▬, non-walkable cells defining obstacles and non-accessible areas.

An example of environment annotated with this set of markers is proposed in Fig. 1(b). This model uses the *floor fields* approach [5], using the agents' environment as a container of information for the management of the interactions between entities. In this particular model, discrete potentials are spread from cells of obstacles and destinations, informing about distances to these objects. The two types of floor fields are denoted as *path field*, spread from openings and final destinations (one per destination object), and *obstacle field*, a unique field spread from all the cells marked as obstacle. In addition, a *dynamic* floor field that has been denoted as *proxemic field* is used to reproduce a proxemic behavior [15] in a repulsive sense, letting the agents to maintain distances with other agents. This approach generates a plausible navigation of the environment as well as an anthropologically founded means of regulating interpersonal distances among pedestrians.

This framework, on one hand, enables the agents to have a position in the discrete environment and to perform movement towards a user configured final destination. On the other hand, the presence of intermediate targets allows choices at the tactical level of the agent, with the computation of a graph-like, topological, representation of the walkable space, based on the concept of *cognitive map* [22]. The method for the computation of this environment abstraction has been defined in [23] and it uses the information of the scenario configuration, together with the floor fields associated to openings and final destinations. In this way a data structure for a complete knowledge of the environment is precomputed. Recent approaches explores also the modeling of partial knowledge of the environment by agents (e.g. [24]), but this aspect goes beyond the scope of the current work. The cognitive map identifies *regions* (e.g. a room) as nodes of the labeled graph and *openings* as edges. An example of the data structure associated to the sample scenario is illustrated in Fig. 1(c). Overall the cognitive map allows the agents to identify their topological position in the environment and it constitutes a basis for the generation of an additional knowledge base, which will enable the reasoning for the route calculation.

This additional data structure has been called *Paths Tree* and it contains the information about *plausible* paths towards a final destination, starting from each region of the environment. The concept of plausibility of a path is encoded in the algorithm for the computation of the tree, which is discussed in [4] and only briefly described here. The procedure starts by defining the destination as the root of the tree and it recursively adds child nodes, each of them mapped to an intermediate destination reachable in the region. Nodes are added if the

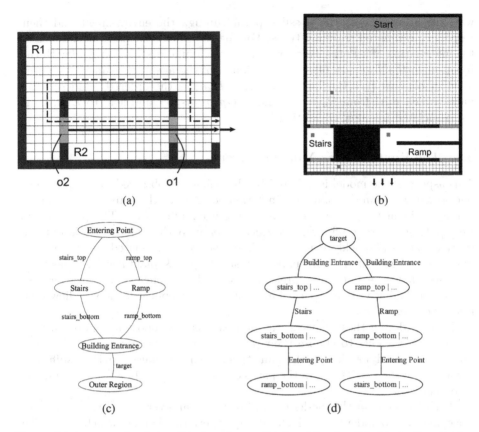

Fig. 1. (a) An example of plausible (continuous line) and implausible (dashed) paths in a simple environment. (b) A simulation scenario with the considered annotation tools and its respective cognitive map (c) and the shortest path tree (d).

constraints describing the plausibility of a path are satisfied: in this way, trajectories that imply cycles or a not reasonable usage of the space (e.g. passing inside a room to reach the exit of a corridor, as illustrated in Fig. 1(a)) are simply avoided. The results of the computation is a tree whose nodes are mapped to targets in the environment and each edge refers to a particular path between two targets. The root of the tree is mapped to a final destination, while the underlying nodes are only mapped to openings. Hence, each branch from the root to an arbitrary node describes a *minimal* (i.e. plausible) path towards the final destination. To complete the information, each node n is labeled with the free flow travel time[1] associated to the path starting from the center of the opening associated to n and passing through the center of all openings mapped by the parent nodes of n, until the final destination. In this

---

[1] The travel time that the agent can employ without encountering any congestion in the path, thus moving at its free flow speed.

way, the agents knows the possible paths through the environment and their respective estimated traveling times. For the choice of their path, agents access the information of a Paths Tree generated from a final destination End with the function $Paths(R, End)$. Given the region R of the agent, the function returns a set of couples $\{(P_i, tt_i)\}$. $P_i = \{\Omega_k, \ldots, End\}$ is the ordered set describing paths which start from Ω_k, belonging to $Openings(R)$, and lead to End. tt_i is the associated free flow travel time.

3.2 The Route Choice Model of Agents

This aspect of the model is inspired by the behaviors observed in a experiment performed with human volunteers in November 2015 at the University of Tokyo, aiming at identifying basic behavior at the wayfinding level. The participants were put into a trade off scenario, since different paths were available but the shortest one was quickly congested. Empirical analysis related to this experiment are not presented in this paper for lack of space. Qualitatively, it has been observed that several persons preferred to employ a longer trajectory for preserving his higher walking speed, but many times it appear to do so following a first *emerging leader*.

By considering these aspects, the objective is to propose an approach that would enable agents to choose their path considering distances as well as the evolution of the dynamics. At the same time, the model must provide a sufficient variability of the results (i.e. of the paths choices) and a calibration over possible empirical data.

The discussion of the model must starts with an overview of the agent life-cycle, in order to understand which activity is performed and in which order. The work-flow of the agent, encompassing the activities at operational and tactical level of behavior at each time-step, is illustrated in Fig. 2.

First of all, the agent performs a perception of his situation considering his knowledge of the environment, aimed at understanding its position in the environment and the markers perceivable from its region (e.g. intermediate targets). At the very beginning of its life, the agent does not have any information about its location, thus the first assignment to execute is the *localization*. This task analyses the values of floor fields in its physical position and infers the location in the Cognitive Map. Once the agent knows the region where it is situated, it loads the Paths Tree and evaluates the possible paths towards its final destination.

The evaluation has been designed with the concept of *path utility*, assigned to each path to successively compute a probability to be chosen by the agent. The probabilistic choice of the path outputs a new intermediate target of the agent, used to update the reference to the floor field followed at the operational layer with the local movement.

The utility-based approach fits well with the needs to easily calibrate the model and to achieve a sufficient variability of the results.

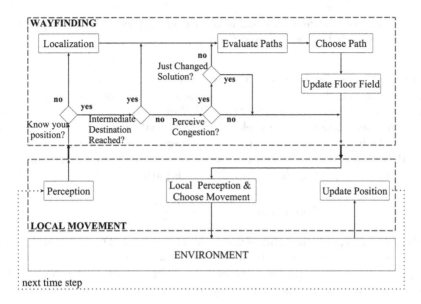

Fig. 2. The life-cycle of the agent, emphasizing the two components of the model.

Utility and Choice of Paths. The function that computes the probability of choosing a path is exponential with respect to the utility value associated to it. This is completely analogous to the choice of movement at the operational layer:

$$Prob(P) = N \cdot e^{U(P)} \tag{1}$$

The usage of the exponential function for the computation of the probability of choosing a path P is a good solution to emphasize the differences in the perceived utility values of paths, limiting the choice of relatively bad solutions (that in this case would lead the agent to employ relatively long paths). $U(P)$ comprises the three observed components influencing the route choice decision, which are aggregated with a weighted sum:

$$U(P) = \kappa_{tt} Eval_{tt}(P) - \kappa_q Eval_q(P) + \kappa_f Eval_f(P) \tag{2}$$

where the first element evaluates the expected travel times; the second considers the *queueing* (crowding) conditions through the considered path and the last one introduces a positive influence of perceived choices of nearby agents to pursue the associated path P (i.e. imitation of emerging leaders). All the three functions provide values normalized within the range $[0, 1]$, thus the value of $U(P)$ is included in the range $[-\kappa_q, \kappa_{tt} + \kappa_f]$.

The only way to evaluate the reliability of this model is through a validation procedure considering empirical evidences gathered through experiments or observations. Hence, these three mechanisms have been designed with the main objective of allowing their calibration over empirical datasets, preferring the usage of simple functions where possible.

Traveling Times Evaluation. The evaluation of traveling times is a crucial element of the model. First of all, the information about the travel time tt_i of a path P_i is derived from the Paths Tree with $Paths(R, End)$ (where End is the agent's final destination, used to select the appropriate Paths Tree, and R is the region in which the agent is situated and it is used to select the relevant path P_i in the Paths Tree structure) and it is integrated with the free flow travel time to reach the first opening Ω_k described by each path:

$$TravelTime(P_i) = tt_i + \frac{PF_{\Omega_k}(x, y)}{Speed_d} \tag{3}$$

where $PF_{\Omega_k}(x, y)$ is the value of the path field associated to Ω_k in the position (x, y) of the agent and $Speed_d$ is the *desired velocity* of the agent, that can be an arbitrary value (see [19] for more details of this aspect of the model). The value of the traveling time is then evaluated by means of the following function:

$$Eval_{tt}(P) = N_{tt} \cdot \frac{\min\limits_{P_i \in Paths(r)} (TravelTime(P_i))}{TravelTime(P)} \tag{4}$$

where N_{tt} is the normalization factor, i.e., 1 over the sum of $TravelTime(P)$ for all paths. By using the minimum value of the list of possible paths leading the agent towards its own destination from the current region, the range of the function is set to (0,1], being 1 for the path with minimum travel time and decreasing as the difference with the other paths increases. This modeling choice, makes this function describe the *utility* of the route in terms of travel times, instead of its *cost*.

This design is motivated by the stability of its values with the consideration of relatively long path, which might be represented in the simulation scenario. By using a cost function, in fact, the presence of very high values of $TravelTime(P)$ in the list would flatten the differences among cost values of other choices after the normalization: in particular, in situations in which most relevant paths have relatively similar costs, excluding a few outliers (even just one), the normalized cost function would provide very similar values for most sensible paths, and it would not have a sufficient discriminating power among them.

Evaluation of Congestion. The behavior modeled in the agent in this model considers congestion as a negative element for the evaluation of the path. This does not completely reflect the reality, since there could be people who could be attracted by congested paths as well, showing a mere *following* behavior. On the other hand, by acting on the calibration of the parameter κ_q it is possible to define different classes of agents with customized behaviors, also considering attraction to congested paths with the configuration of a negative value.

For the evaluation of this component of the route decision making activity associated to a path P, a function is first introduced for denoting agents a' that precede the evaluating agent a in the route towards the opening Ω of a path P:

$$Forward(\Omega, a) = |\{a' \in Ag \setminus \{a\} : Dest(a') = \Omega \wedge PF_\Omega(Pos(a')) < PF_\Omega(Pos(a))\}| \tag{5}$$

where Pos and $Dest$ indicates respectively the position and current destination of the agent; the fact that $PF_\Omega(Pos(a')) < PF_\Omega(Pos(a))$ assures that a' is closer to Ω than a, due to the nature of floor fields. Each agent is therefore able to perceive the main direction of the others (its current destination). This kind of perception is plausible considering that only preceding agents are counted, but we want to restrict its application when agents are sufficiently close to the next passage (i.e. they perceive as important the choice of continuing to pursue that path or change it). To introduce a way to calibrate this perception, the following function and an additional parameter γ is introduced:

$$PerceiveForward(\Omega, a) = \begin{cases} Forward(\Omega, a), & \text{if } PF_\Omega(Pos(a)) < \gamma \\ 0, & \text{otherwise} \end{cases} \tag{6}$$

The function $Eval_q$ is finally defined with the normalization of $PerceiveForward$ values for all the openings connecting the region of the agent:

$$Eval_q(P) = N \cdot \frac{PerceiveForward(FirstEl(P), myself)}{width(FirstEl(P))} \tag{7}$$

where $FirstEl$ returns the first opening of a path, $myself$ denotes the evaluating agent and $width$ scales the evaluation over the width of the door (larger doors sustain higher flows).

Imitative Behavior. This component of the decision making model aims at representing the effect of an additional stimulus perceived by the agents associated to sudden decision changes of other persons that might have an influence. An additional grid has been introduced to model this kind of event, whose functioning is similar to the one of a dynamic floor field. The grid, called *ChoiceField*, is used to spread a gradient from the positions of agents that, at a given time-step, change their plan due to the perception of congestion.

The functioning of this field is described by two parameters ρ_c and τ_c, which defines the diffusion radius and the time needed by the values to *decay*. The diffusion of values from an agent a, choosing a new target Ω', is performed in the cells c of the grid with $Dist(Pos(a), c) \leq \rho_c$ with the following function:

$$Diffuse(c, a) = \begin{cases} 1/Dist(Pos(a), c) & \text{if } Pos(a) \neq c \\ 1 & \text{otherwise} \end{cases} \tag{8}$$

The diffused values persist in the *ChoiceField* grid for τ_c simulation steps, then they are simply discarded. The index of the target Ω' is stored together with the diffusion values, thus the grid contains in each cell a vector of couples $\{(\Omega_m, \mathit{diff}_{\Omega_m}), \ldots, (\Omega_n, \mathit{diff}_{\Omega_n})\}$, describing the values of influence associated to each opening of the region where the cell is situated. While multiple neighbor agents changes their choices towards the opening Ω', the values of the diffusion are summed up in the respective $\mathit{diff}_{\Omega'}$. In addition, after having changed its decision, an agent spreads the gradient in the grid for a configurable amount of time steps represented by an additional parameter τ_a. In this way it influences the choices of its neighbors for a certain amount of time.

The existence of values $\mathit{diff}_{\Omega_k} > 0$ for some opening Ω_k implies that the agent is influenced in the evaluation phase by one of these openings, but the probability for which this influence is effective is, after all, regulated by the utility weight κ_f. In case of having multiple $\mathit{diff}_{\Omega_k} > 0$ in the same cell, a individual influence is chosen with a simple probability function based on the normalized weights diff associated to the cell. Hence, for an evaluation performed by an agent a at time-step t, the utility component $Eval_f$ can be equal to 1 only for one path $\overline{P}$, between the paths having $\mathit{diff}_{\Omega_k} > 0$ in the position of a.

4 Experimentation in a Large Scale Scenario

The evaluation of the model is here discussed with a simulation of an egress from a large scenario, with the aim of verifying the behavior of the model in a real-world environment and to perform a qualitative comparison of the results with a baseline model, not considering an explicit form of wayfinding decisions. In particular, we employed a previously implemented model described in [25] essentially extending the already cited floor field CA approach [5] for better managing high density situations. The rationale of the experimentation is to evaluate if the expected advantage brought by the adaptiveness to congestions in bottlenecks is reflected by simulation results, both in terms of a higher measured walking speed and a better distribution of pedestrians in available exits.

All the presented results have been achieved with the calibration weights of the utility function configured as $\Omega_{tt} = 100, \Omega_q = 27; \Omega_f = 5$, while the parameters related to the *ChoiceField* are set to $\rho_c = 1.2 \, \text{m}, \tau_c = 2$ time-steps $= 0.44 \, \text{s}$ and $\tau_d = 4$ time-steps $= 1 \, \text{s}$. The desired speed of agents have been configured with a normal distribution centered in $1.4 \, \text{m/s}$ and with standard deviation of $0.2 \, \text{m/s}$, in accordance with the pedestrians speeds usually observed in the real world (e.g. [26]). The distribution is discretized in classes of $0.1 \, \text{m/s}$, and cut by configuring a minimum velocity of $1.0 \, \text{m/s}$ and a maximum one of $1.8 \, \text{m/s}$ (see the blue boxes in Fig. 3(a)). To allow a maximum speed of $1.8 \, \text{m/s}$ the time-step duration is assumed to $\overline{\tau} = 0.22 \, \text{s}$.

The simulation scenario describes the outflow from a portion of the Düsseldorf Arena, as described in [6]. The environment used for the simulation with the discussed model is essentially composed of 4 starting areas (representing the bleachers of the stadium) that generate the agents in the simulation; the goal of the agents is to vacate the area through an intermediate passage leading to a common L-shaped area connecting all the starting ares to the stadium exits. 400 agents are generated at the beginning of the simulation from each start area, producing a total of 1600 pedestrians.

The heat map shown in Fig. 3(c) provides information about the usage of the space during the simulation, by describing the average local densities perceived by the agents (so-called *cumulative mean density*). The major congested areas are located in front of the exit doors, given their relatively small width of 1.2 m. The corridors connecting each bleacher to the atrium are affected as well by high densities (around 2.5–3 persons/m^2) but their widths guarantee a sensibly higher flow, causing lower congestion inside the starting regions. The achieved results can be qualitatively compared to the analogous spatial utilization analysis of results achieved by means of the baseline floor field model in the same scenario,

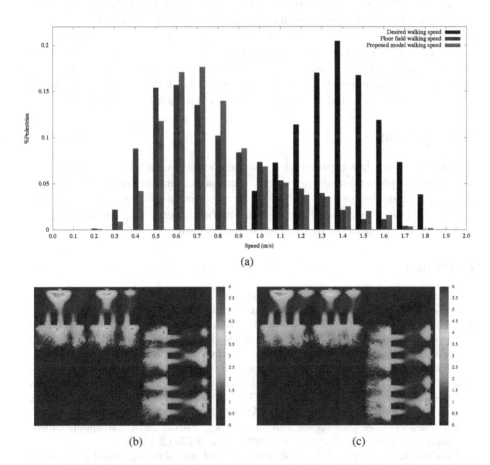

Fig. 3. (a) Walking speed distributions respectively desired (black), achieved by the floor field (blue) and proposed model (red). (b) and (c) show respectively the cumulative mean density maps for the floor field and proposed model. (Color figure online)

shown in Fig. 3(b). The proposed model is apparently better at exploiting the available exits, distributing pedestrians more evenly among them and therefore reducing the congestion level: for instance, the second exit from the top-left corner is almost not used by floor field agents, due to the fact that it is positioned a little bit on the right with respect to an intermediate passage from the first start area. The proposed approach makes this exit instead significantly more used by pedestrians from the first start area, that are able to avoid congestion and achieve a shorter travel time.

This is also reflected by the histogram depicted in Fig. 3(a) that shows the walking speed distributions in the simulation scenario, respectively desired (black), achieved by the floor field baseline model (blue), and achieved by the proposed model (red). The congestion arisen in the exit doors of the atrium sensibly affected the walking speed of the agents, for both modeling approaches: even though a small portion of the simulated population succeeded in maintaining its desired speed (i.e. the agents generated in positions closer to the exits), most of them experienced a significant delay during their way, caused by frequent blocks. The difference among the distribution of achieved walking speed of the floor field and proposed model is quite noticeable: by choosing less congested exits, agents follow longer trajectories but they are able to incur less frequently in block situations, leading to a higher mode in the walking speed distribution.

5   Conclusions

The present paper has introduced a general model for decision making activities related to pedestrian route choices. The model encompasses three aspects influencing these choices, as observed in an experimental observation: expected travel time, perceived level of congestion on the chosen path, and decisions of other preceding pedestrian to pursue a different path.

References

1. Vizzari, G., Bandini, S.: Studying pedestrian and crowd dynamics through integrated analysis and synthesis. IEEE Intell. Syst. **28**(5), 56–60 (2013)
2. Sassi, A., Borean, C., Giannantonio, R., Mamei, M., Mana, D., Zambonelli, F.: Crowd steering in public spaces: approaches and strategies. In: Wu, Y., Min, G., Georgalas, N., Hu, J., Atzori, L., Jin, X., Jarvis, S.A., Liu, L.C., Calvo, R.A. (eds.) 15th IEEE International Conference on Computer and Information Technology, CIT 2015, 14th IEEE International Conference on Ubiquitous Computing and Communications, IUCC 2015, 13th IEEE International Conference on Dependable, Autonomic and Secure Computing, DASC 2015, 13th IEEE International Conference on Pervasive Intelligence and Computing, PICom 2015, Liverpool, United Kingdom, October 26–28, 2015, pp. 2098–2105. IEEE (2015)
3. Challenger, R., Clegg, C.W., Robinson, M.A.: Understanding crowd behaviours: supporting evidence. Technical report, University of Leeds (2009)
4. Crociani, L., Piazzoni, A., Vizzari, G., Bandini, S.: When reactive agents are not enough: tactical level decisions in pedestrian simulation. Intell. Artif. **9**(2), 163–177 (2015)

5. Burstedde, C., Klauck, K., Schadschneider, A., Zittartz, J.: Simulation of pedestrian dynamics using a two-dimensional cellular automaton. Phys. A Stat. Mech. Appl. **295**(3–4), 507–525 (2001)
6. Wagoum, A.U.K., Seyfried, A., Holl, S.: Modelling dynamic route choice of pedestrians to assess the criticality of building evacuation. Adv. Complex Syst. **15**(7), 15 (2012)
7. Kretz, T., Lehmann, K., Hofsäß, I., Leonhardt, A.: Dynamic assignment in microsimulations of pedestrians. Annual Meeting of the Transportation Research Board, 14-0941(2014)
8. Guo, R.Y., Huang, H.J.: Route choice in pedestrian evacuation: formulated using a potential field. J. Stat. Mech. Theory Exp. **2011**(4), P04018 (2011)
9. Lämmel, G., Klüpfel, H., Nagel, K.: The MATSim network flow model for traffic simulation adapted to large-scale emergency egress and an application to the evacuation of the Indonesian City of Padang in case of a tsunami warning, pedestrian behavior. In: Pedestrian Behavior: Models, Data Collection and Applications, pp. 245–265 (2009)
10. Lämmel, G., Seyfried, A., Steffen, B.: Large-scale and microscopic: a fast simulation approach for urban areas. In: Transportation Research Board 93rd Annual Meeting. Number 14-3890 (2014)
11. Crociani, L., Lämmel, G., Vizzari, G.: Multi-scale simulation for crowd management: a case study in an urban scenario. In: Proceedings of the 1st Workshop on Agent Based Modelling of Urban Systems (ABMUS 2016) (2016)
12. Sacerdoti, E.D.: Planning in a hierarchy of abstraction spaces. Artif. Intell. **5**(2), 115–135 (1974)
13. Kapadia, M., Beacco, A., Garcia, F.M., Reddy, V., Pelechano, N., Badler, N.I.: Multi-domain real-time planning in dynamic environments. In: Chai, J., Yu, Y., Kim, T., Sumner, R.W. (eds.) The ACM SIGGRAPH/Eurographics Symposium on Computer Animation, SCA 2013, Anaheim, CA, USA, July 19–21, 2013, pp. 115–124. ACM (2013)
14. Levihn, M., Kaelbling, L.P., Lozano-Pérez, T., Stilman, M.: Foresight and reconsideration in hierarchical planning and execution. In: 2013 IEEE/RSJ International Conference on Intelligent Robots and Systems, Tokyo, Japan, November 3–7, 2013, pp. 224–231. IEEE (2013)
15. Hall, E.T.: The Hidden Dimension. Doubleday, New York (1966)
16. Helbing, D., Schweitzer, F., Keltsch, J., Molnár, P.: Active walker model for the formation of human and animal trail systems. Phys. Rev. E **56**(3), 2527–2539 (1997)
17. Boltes, M., Seyfried, A.: Collecting pedestrian trajectories. Neurocomputing **100**, 127–133 (2013)
18. Michon, J.A.: A critical view of driver behavior models: What do we know, What should we do? In: Evans, L., Schwing, R.C. (eds.) Human Behavior and Traffic Safety, pp. 485–524. Springer, New York (1985)
19. Bandini, S., Crociani, L., Vizzari, G.: An approach for managing heterogeneous speed profiles in cellular automata pedestrian models. Journal of Cellular Automata (in press)
20. Weyns, D., Omicini, A., Odell, J.: Environment as a first class abstraction in multiagent systems. Auton. Agents Multi-Agent Syst. **14**(1), 5–30 (2007)
21. Weidmann, U.: Transporttechnik der Fussgänger - Transporttechnische Eigenschaftendes Fussgängerverkehrs (Literaturstudie). Literature Research 90, Institut füer Verkehrsplanung, Transporttechnik, Strassen- und Eisenbahnbau IVT an der ETH Zürich (1993)

22. Tolman, E.C.: Cognitive maps in rats and men. Psychol. Rev. **55**(4), 189–208 (1948)
23. Crociani, L., Invernizzi, A., Vizzari, G.: A hybrid agent architecture for enabling tactical level decisions in floor field approaches. Transp. Res. Procedia **2**, 618–623 (2014)
24. Andresen, E., Haensel, D., Chraibi, M., Seyfried, A.: Wayfinding and cognitive maps for pedestrian models. In: Proceedings of Traffic and Granular Flow 2015 (TGF2015), Springer (in press)
25. Bandini, S., Mondini, M., Vizzari, G.: Modelling negative interactions among pedestrians in high density situations. Transp. Res. Part C Emerg. Technol. **40**, 251–270 (2014)
26. Willis, A., Gjersoe, N., Havard, C., Kerridge, J., Kukla, R.: Human movement behaviour in urban spaces: implications for the design and modelling of effective pedestrian environments. Environ. Plan. B Plan. Design **31**(6), 805–828 (2004)

Knowledge Representation
and Reasoning

Educational Concept Maps for Personalized Learning Path Generation

Giovanni Adorni[✉] and Frosina Koceva[✉]

Department of Informatics, Bioengineering, Robotics and Systems Engineering,
University of Genoa, Genoa, Italy
`adorni@unige.it`, `frosina.koceva@edu.unige.it`

Abstract. The paper focuses on the delivery process of learning materials of a course for students by means of Educational Concept Maps (ECM), while in previous works, we presented the ECM model and its implementation, ENCODE system, as a tool to assist the teacher during the process of instructional design of a course. An ECM is composed of concepts and educational relationships, where a concept represents a learning argument, its prerequisites, learning outcomes and associated learning materials. We propose the learning materials generation founded on the ECM with suggested learning path for accessing educational resources personalized on the base of the student's knowledge. The personalized document creation is based on a self-evaluation process of his/her knowledge and learning objectives, by pruning concepts on the original ECM and verifying for propaedeutic inconsistency. An algorithm that linearize the map generates the suggested learning path for the student.

Keywords: Knowledge representation · Learning path · e-learning

1 Introduction

E–learning has become an important component of our educational life. Different web based Learning Content Management Systems (LCMS) have been developed to support teachers for constructing and updating the course materials, as well as learners in the learning process. Does Artificial Intelligence (AI) play any role in this process?

Maybe the question is ill posed; the right question should be: What are the methods and AI applications that are of interest for e-learning?

The list, which does not pretend to be comprehensive, certainly includes: (i) systems able to learn from the interaction with the user and adapt themselves to the user by providing personalized suggestions and recommendations; (ii) systems able to make intelligent web searches; (iii) systems are able to personalize the content on the basis of a user profile, designing and reusing learning objects; (iv) intelligent agents for user assistance (see, for example, [1–4]).

In this picture, even if partial, of methods and AI applications for e-learning, emerges an emblematic aspect: AI helps people find information as needed and intelligent systems allow people to rely less on learning and memory to solve problems. The learner

G. Adorni et al. (Eds.): AI*IA 2016, LNAI 10037, pp. 135–148, 2016.
DOI: 10.1007/978-3-319-49130-1_11

must perceive it being at the center of the learning process, and when AI takes this responsibility away, it inhibits learning [5].

AI has, however, already in some way influenced the web based learning through contributions on various aspects of cognitive science, as is clear not only from the scientific works available in the conference proceedings and journals, but also through discussions of the online community of professionals [see, for example, elearningindustry.com]. This can be a shared vision or not, but it is certainly true that the potential of methodologies and applications of AI to web based learning is high.

Scanning professional literature are highlighted certain concepts on which it is necessary to make some considerations.

The first of these concepts is that of Reusable Learning Object (LO) [6]. LOs and the Instructional Design theory [7] are daughters of Rapid Application Development methodologies, typical of software engineering, and have established themselves in recent years as a reference point in the design and implementation of online courses of the latest generation. However, it is necessary to point out that the LCMS still seem weak in several key functions, such as:

- Semantic retrieval using LO metadata for both authoring and for their use;
- Sharing of LO repository on the basis of common and shared ontologies;
- Intelligent management of annotations and revision of concepts in collaborative learning.

In parallel, but with incredible synergy, is worth noting that semantic Web has revived and strengthened some classical AI fields such as knowledge representation, the formalities for the construction of ontologies, intelligent agents for the Web, with one big goal: building materials for the Web that can be manipulated significantly from the semantic point of view by programs and software agents, and not only designed to be readable by humans [8].

In this scenario in which on one side converge:

- the needs of e-learning researchers to have international standards effective for the indexing, the packaging and distribution of learning objects [9, 10];
- the declared intent for governments, organizations, companies, universities to invest in the design and creation for large repository of digitized structured knowledge;
- the joint efforts of technologists, documentary makers, librarians, educators in standardization of the metadata required to make available for the latest generation of search engines vocabularies "adequate" and semantically valid (take for example the problems of polysemy, synonyms, and linguistic-terminological ambiguities in textual documents, not to mention the semantic interpretation of multimedia objects);

and on the other are emerging:

- technological solutions and formalisms for the construction of suitable ontologies to the semantic Web;
- working environments (possibly open-source) for the implementation of management systems for cooperative and collaborative processes by means of agents (web agents for user profile management, web services for brokering information and knowledge management, ..);

- applications for soft-computing and neural networks for data mining and classification for concepts;

it becomes more and more interesting to see what points of contact emerge and what opportunities you can glimpse in the near future in the implementation of technologies and in the testing methodologies for "intelligent design" for e-learning.

In this paper we want to make a contribution to some of previous discussed topics; more precisely, we would like to focus on personalization of a lesson or, in more general terms, of a subject matter.

Nowadays, learning a new subject does not always correspond to following a real course, but rather to finding information independently and in a disorganized way, which leads to disorientation during the learning process. From the other hand, the design of a good lesson plan requires diverse perspicacity and particular attention should be given to the student target, learner's abilities and therefore to the specific content and difficulty level associated with the concepts. Another important factor is the course length that could affect the course efficiency and the learner overload. Thus, one of the main issues in e-learning systems is providing a personalized user experience and in specific personalized learning paths that reflect the learner needs and as consequence improve and optimize learner performance. A learning path is a sequence of arguments/activities ordered in a way that satisfy the prerequisite constraint between the arguments, leading the student to achieve his/her learning goals. Personalized learning leans on various techniques as data mining, machine learning, and learner preference classification.

In order to provide a learning path in an automatic or semi-automatic way one, among various approaches, is to rely on existing reference paths and/or concept extraction and correlation satisfying the prerequisite requirement to create a concept graph, from which than a learning path is created [11]. This approach is demanding if the course structure is not known, thus ontology analysis or statistical algorithms usually support it. On the contrary, of the aforementioned solution, the prior knowledge of the course structure in our case is known since the course is built as a result of the creation of a subject structure based on the Educational Concept Map (ECM) model [12]. The subject structure is represented as an ECM where nodes are the main arguments of the subject and are characterized by resources, prerequisites and learning outcomes, while the map is built upon relating the nodes in respect of the predefined relations as, for instance, the prerequisite relation. Modeling the knowledge states and/or course content and/or knowledge units as a graph in the e-learning system is not a novelty (see, for example, [13–18]). In our model, the subject structure is a graph that contains all possible learning paths. The teacher customizes the ECM in a more target student oriented map, called Course Concept Map (CCM). The latter represents the specific vision the teacher has for its course. From the CCM and the results of the student profiling, an induced subgraph is created which is linearized in a personalized learning path that aligns with the student self-stated knowledge, needs and learning skills. The personalized learning path is delivered by downloading the learning document consisting of a set of web pages following a guided navigation.

The ECM and CCM are founded on the ECM model [12] that will be discussed more in-depth in the next section following by its implementation and architecture of the ENCODE (ENvironment for COntent Design and Editing) tool [19]. In Sect. 3 we describe the learner profiling on which is based the learning path personalization. Finally, conclusions and a possible future roadmap are presented in Sect. 4.

2   ECM Model and Its Implementation

An ECM is a formal representation of the domain of knowledge in the context of learning environments. It is a logical and abstract annotation model created with the aim of guaranteeing the reusability of the knowledge structure, as well as of the teaching materials associated to each node of the structure. ECM model has been designed taking into account the pedagogical requirements defined by Educational Modelling Language research group [20], namely: pedagogical flexibility, learner centrality, centrality of learning objects, personalization, domain-independence, reusability, interoperability, medium neutrality, compatibility and formalization. The ECM model has been developed by means of an ontological structure characterized by the integration of hierarchical and associative relationships. Firstly, it asks teachers and instructional designers to focus their attention on learner's' profile (in particular educational background, learning and cognitive styles) and objectives. Taking into account these elements, the model suggests how to identify, within the discipline's subject matter, the key concepts and their relationships to identify effective strategies of contents presentation and to support the activation of meaningful learning processes. According to the ECM model (see Fig. 1), a profiled learner has a Lesson Plan with a goal identified by an objective (or a composition of objectives) that is (are) achieved by a *Unit of Learning* (UoL), or by a composition of UoLs. A UoL is characterized by an *Effort*, i.e. an evaluation of the needed commitment that the learner requires in dealing with the learning materials. A UoL is composed by key concepts and their relationships, where the key concepts can be of two types:

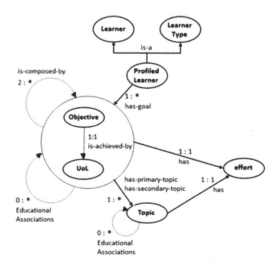

Fig. 1.   Educational concept map model

- Primary Notion, the starting point of a UoL that identifies the "prerequisites", i.e., the concepts (Primary Topics) that a student must know before attending a given UoL;
- Secondary Notion, identifies the concepts (Secondary Topics) that will be explained in the present UoL (this kind of concepts go with learning materials).

The semantics of the knowledge map is given by connecting the concepts with predefined hierarchical and associative relationships. Thus, four educational relations are defined (see [19] for a detailed description of the relations):

- is-requirement-of: denoted as is_req(x,y), a propaedeutic relation, e.g. it may be used with the aim of specifying the logical order of contents, i.e. used for creating learning linearized paths;
- is-item-of: denoted as is_item(x,y), representing a hierarchical structure of generalization (aggregation or membership type);
- is-related-to: denoted as is_rel(x,y), represents a relation between closely related concepts;
- is-suggested-link-of: denoted as is_sug(x,y), relates a main concept with its in-depth examination; e.g., this relationship type may be used in order to suggest in-depth resources.

These relation types have been selected with the aim of allowing teachers to create different learning paths (with or without precedence constraints among Topics). The Topics represent the concepts of the domain: any subjects a teacher may want to talk about. Moreover, the units of learning are connected to the Topics through two relationships: (i) has-primary-topic: where a primary Topic identifies the "prerequisites", in other words the concept that a student must know before attending a given unit of learning; (ii) has-secondary-topic: where secondary Topic identifies the concepts that will be explained in the present unit of learning (this kind of Topics will have specific learning materials associated).

Suppose the general Subject Matter is taken to be the content of an undergraduate first course in statistics. Then, the content of such a course might differ when given under the contexts associated with a group of English majors in contrast to a group of electrical engineering students. In many instances, the differences in content result from using the terminology peculiar to a discipline in presenting examples and exercises. Other less trivial differences arise by including special techniques useful in particular disciplines. Specifying context is thus seen to partially define subsets of the general Subject Matter represented by the ECM. In a general, the boundaries and content of these subsets begin to be identified and we use the CCM to represent them. Thus a CCM is the structure of the subject matter, e.g., a specific teacher vision of the course domain tailored for a specific class of students. As to reusability, the ECMs are designed to maintain the concept layer separate from the resources, making it possible to provide courses with the same CCM from the ECM but with different resources, as in the case of a course for beginners and a course on the same argument for advanced students.

The ECM model development is not intended to define the domain ontologies, in other words, is not a model of representation of the proper knowledge, rather it should support the structuring of the contents of a subject matter. This from one side results in a loss in terms of semantic expressiveness of the model, but from another results in a gain in terms of applicability of the same.

Based on the ECM model and on the Topic Map standard [21], the ENCODE - ENvironment for COntent Design and Editing system, has been implemented [19] in a pre-α release. For the Topic Map core engine, the open source Wandora framework has been used [22].

2.1 ENCODE and Learning Path Generation

The ECM model and the ENCODE implementation are aiming to provide course realization conformed to a pedagogical model starting from the knowledge of a subject matter. Through ENCODE the domain knowledge is encoded in the ECM, where the concepts describing the domain are the Topics of that map, and the relationships are those defined by the ECM model.

ENCODE uses this subject-based classification flexible model of Topic Maps (TM) with close ECM lightweight ontology. By adding the topic/occurrence axis to the topic/ association model (see Fig. 2), TMs provide a means of "bridging the gap", between knowledge representation (knowledge layer) and the field of information management (information layer) [23]. The aforementioned characteristics of the TM model satisfy the requirement of reusability imposed by the ECM model guaranteeing reusability of the knowledge layer (ECM/CCM) and of the information layer (learning resources). Figure 2 illustrates the basic elements of a TM, known as the TAO [23], i.e., Topics, Association and Occurrence. In order to use the ECM classification the ECM ontology is used as a fixed ontology for ENCODE, thus, as it is shown in the example in Fig. 2, it is possible to define the concepts "ICT", "Abstraction" and "Generalization" as

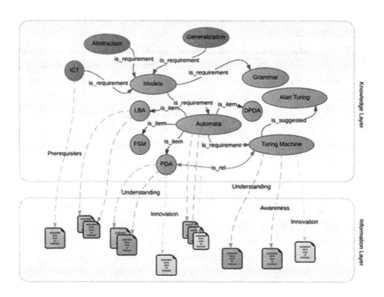

Fig. 2. E ENCODE TAO example

primary notions and all others as secondary notions; to define the structure of educational relationships between them independently of the learning materials documenting the relative concepts.

The solution of the problem of creating a lesson plan resides in the generation of topological sort [25] sequence of the topics of CCM resolved in a graph, since this linearization preserves the precedence of the prerequisites defined by the is_req relationship. For an example, if topic t_1 representing the concept "Automata" is a propaedeutic requirement of the topic t_2 representing the concept of the "Turing Machine" (see Fig. 2), thus in the map (graph) an association of type is_req exists between the two topics (nodes), instead, in the linearized sequence this means that t_1 precede t_2.

If we analyze the CCM graph, it is not of immediate identification what is the best path to be taken in order to satisfy all intermediate knowledge for fulfilling the learning objectives. Thus, ENCODE implements an algorithm whose task is to explore the acyclic graph G(T, E) formed by the nodes T and the edges E and produces learning paths that satisfy the prerequisite constraint. Each node represents a Topic where a topic is a container of educational resources and materials, and the edges are the is_req association.

The algorithm for learning path generation firstly classifies the Topics by increasing path distance level from the primary notion, where the maximum path distance is taken into consideration. With reference to the map of n Topics, we indicate the maximum distance of the Topic t_x from the Primary Topic t_p with D, where $0 \leq D \leq n-1$, i.e. function level(u), then we order the topics in a sequence by increasing distance. For instance in Fig. 3 for the graph representing a simplified CCM with Primary Topic A and Learning Outcome U, the maximum path distance D is 6 and the topics classified per distance are shown in the left table. A tree structure is created where the root is the Primary Topic; the children at Level 1 are all the permutations of the list containing the topics with distance equal to 1 from the Primary Topic; ... the children at Level D are all the permutations of the list containing the topics with distance equal to D from the Primary Topic. The tree traversals from the root to the leaves (see Fig. 3) represent the

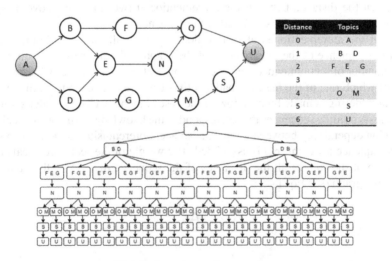

Fig. 3. Learning path generation

reduced solution space of the possible topological orders according to the heuristic in respect of all possible orders.

```
/*Classify Topics by increasing path distance level*/
Input DAG G=(T,E)
topo_sort(G)
for every u in T from topo_sort
   function level(u){
      if(u=PN) level(u)=0
      else level(u)=max(v,u)∈ E{(level(v))+1}
   }
   levels.put(level,u)
end for
/*Tree creation with all children permutations of topics
with same distance*/
tree.root(PN)
for every l in levels and l>0
   permutations = permute_topics(levels.get(l))
   for every leaf in leafs
      for every p in permutations
         leaf.addChildren(p)
      end for
   end for
end for
paths=traverse(tree)
```

One of the consequences of the linearization of the graph is that two Topics that are adjacent in the graph could not necessarily be adjacent in the sequence. Thus, in order to solve the problem of concept continuity in the learning path a Topic Aider is introduced when the distance between the aforementioned two nodes is above a certain threshold the distanced topic is recalled in the sequence [12].

Even if the aforementioned algorithm does not give all the possible orders, still another consequence of the linearization of the graph is the not unique solution, thus limited number of recommended sequence are presented to the teacher where the final choice of which one of the learning paths is the best solution depends on the teacher. Hence, an ordering is made by minimizing the cases in which a Topic Aider should be added, thus providing sequences that correspond to the knowledge structure and enhance the learning dependency between topics, i.e. adjacent prerequisite. In order to do so, to the k-th sequence a weight (σ_k) is associated. The weight of the sequence is calculated on the base of the sum of all the distances of the sequence divided by the number of edges in the graph (see Eq. 1):

$$\sigma_k = \frac{\sum_{\forall (t_i, t_j) \in E} \Delta s_k^{(t_i, t_j)}}{|E|} \tag{1}$$

The minimum value of σk is zero; this happens only if all the pairs of adjacent nodes in the graph are also adjacent in the sequence.

At last, the sequence is completed by adding in alphabetic order the adjacencies Topics from the graph related to the Topics involved in the sequence with is_item, is_rel and is_sug.

2.2 Association Rules

The ontology of the Topic Map is defined based on the ECM model, respecting from one side the requirements of knowledge and information that should be represented and from the other side the application for which it is designed, e.g. ENCODE. The conditional educational associations are the key to intelligent enhanced ECM/CCM and learning path generation. Some of the constraints on the association are defining the relation property, for instance: is_req is asymmetric and transitive association; is_sug is asymmetric association that has only incoming arcs; is_item is asymmetric association; is_rel is symmetric association; two topics x and y can be related only by one type of association [12].

For every argument(topic) in the map prerequisite and learning outcome are defined as occurrences (see Fig. 4), where if two topics t_i and t_j are related with an is_req association and t_i is prerequisite of t_j, the learning outcome of the t_i become prerequisites of t_j. This prerequisite propagation is done only on explicit save of the working project.

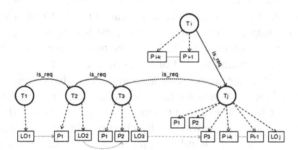

Fig. 4. Prerequisite propagation

3 Course Delivery

3.1 Profiling

The fundamental outcome from the Kolb's learning style theory is that there is no common and unique way for efficient learning; on the contrary there are different ways depending on a learner's preferred way of how she/he memorizes, comprehends and processes the information, her/his culture, personal characteristics, educational background, etc. [26, 27]. The Gardner's multiple intelligence theory, as the name emphasizes, takes into consideration different types of intelligence like for instance: visual, body, musical, interpersonal, intrapersonal, naturalistic, existential, linguistic and logical intelligence [28]. McKenzie additionally found a correspondence between learning styles and types of intelligences and regrouped them into three domains: thinking critically (logical, musical, naturalistic), thinking outward (linguistic, body, interpersonal) and thinking within (visual, intrapersonal, existential) [29]. The learner's style and intelligences revealed by a learner's profile analysis can provide concrete incomes that then should be used for the creation of effective learning resources and furthermore for providing personalized or individualized learning. As highlighted in the final report of the LEADLAB Grundtvig LL European project, we can find different meanings of the aforementioned learning types [30]. Thus, individualized learning refers to individual learning activities for a student who works in an individually, i.e., alone, on the individualized learning resources according to her/his learning style and attitude. In spite of this, personalized learning refers to didactic actions aimed at adding value to the individual in relation to the group.

Therefore, a learner profile explicates the learner educational background, knowledge, learning and cognitive styles, interest and learner objectives, and it's used in order to tailor the learning materials and learning path to each learner aiming to efficient learning. In order to adapt the learning path to the student's needs and knowledge a simplified cooperative overlay student model is implemented in ENCODE. The aforementioned is created by explicitly collecting student information with a self-evaluation form consisting of a list of objectives and test for identification of the learning style and individuation of learners' intelligence.

The first part of the form (see Fig. 5), called know-what, is build in a tree structure from the learning outcomes associated to every argument in the CCM. The tree structure is obtained by traversing the graph in a Breadth First Search in respect of the is_req association. The learning outcomes of the Topics are on the same level in the tree if they are part of the traversal, instead the LO of the Topics related to the visited topic with one of the is_item or is_sug association are children of the visited Topic learning outcomes. The is_rel association is not taken into consideration during the creation of the tree representing the self-evaluation form. The student auto-certificates her/his knowledge state (background knowledge) by choosing the learning outcomes corresponding to the topics known by him/her, in this manner a induced subgraph pertinent to the student is created.

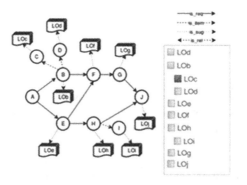

Fig. 5. Know-what test generation

The second part of the form individuates the cognitive attitude of the learner, her/his learning style and type of intelligences; to this purpose the McKenzie test could be used [31] or Felder and Soloman learning preference questionnaire [32]. This evaluation can be done only once, for example at learner registration to the system. By defining appropriate scope of the occurrence, e.g. McKenzie domain [31], a filtering of the learning resources is made in respect of the learner test results. The scope, as defined by the TMDM [24], represents the context within which a statement is valid and relevant.

3.2 Personalized Path Generation

The ENCODE tool assists the teacher during the design of a course, i.e., creation of ECM, CCM and the Learning Paths. Its output consists in reusable ECM, CCM or Learning Path in XTM format or simple html pages. As to the delivery, the Learning Path can be imported in a LCMS, specifically we used Moodle [33], or it can be personalized in respect of the learner knowledge and preference. In the first case, by importing the Learning Path the course structure is automatically created in Moodle. In the later case, the output is a set of simple web pages representing the arguments (topics) of the course incorporating the learning materials and providing navigational links as defined by the educational association in the CCM enhanced by suggested navigation indicating the learning path. Thus, the full workflow process involving both teacher and learner consists in:

1. Teacher: Creation or importation of an ECM representing the structure of the subject domain knowledge and the relative learning resources classified by difficulty and cognitive abilities. Starting from the ECM a know-what test is generated. The map subsequently is shared with the learners of the relative subject.
2. Student: Self-verification of her/his knowledge that drives the pruning process of the ECM map from the known topics, and filters the learning resources in respect of the learner cognitive attitude.
3. The output of the pruning process involving additional consistency checks is a personalized CCM.
4. Learning path is generated from the CCM, by student specifying the initial point (primary topic) of the learning path and the target point (learning outcome).
5. The course is delivered by providing a compressed file of html pages representing the not known arguments of the course with suggested navigational links between the topics representing the suggested propaedeutic order of delivery.

There are potential pitfalls to the cooperative overlay student model including: the student subjective and/or biased opinion or inability to accurately evaluate her/his knowledge state [34]; or the fact that the student knowledge is subset of the subject matter structure, thus the diverse and/or incorrect student knowledge and misconception is not addressed [35]. These problems will be taken into consideration for the future work, likewise the fact that the student model is not responding dynamically on student progress during the fruition of the course.

4 Conclusions

Through this work, we presented how a model, designed to assist the teacher during the process of instructional design of a lesson or an entire course, can also be an interesting tool for students assisting them during the delivery process of learning materials via browsing personalized learning pathway. The idea originates from the analysis of the open issues in instructional authoring system, and from the lack of a well-defined process able to merge pedagogical strategies with systems for the knowledge organization of the domain. In particular, we introduced the ECM model. By means of ECMs, is possible to design lessons and/or learning paths from an ontological structure characterized by the integration of hierarchical and associative relationships among the educational objectives. Within this context, we addressed also the problem to find a "suitable" learning path through an ECM, i.e., a sequence of concept characterizing the subject matter (a lesson or an entire course) under definition, and how the maps can be implemented by means of ISO/IEC 13250 TM - Topic Maps standard [24]. A Topic Map possesses interchangeable features that allow it to describe the relativeness between concepts and even link it to online information resources easily.

A learning path can be personalized taking into account the learner's style and intelligences of a student [29]. The learner's style and intelligences revealed by a learner's profile analysis can provide concrete incomes that then should be used for the creation of effective learning resources and furthermore for providing personalized or individualized learning.

Currently, ENCODE is being testing with a community of teachers and students for its learning path model and learning path delivery, its roadmap envisage user-friendliness improvements and recommendation of learning resources during the course design.

References

1. Jacko, J.A. (ed.): Human Computer Interaction Handbook: Fundamentals, Evolving Technologies and Emerging Applications, 3rd edn. CRC Press, Boca Raton (2012)
2. Lingras, P., Akerkar, R.: Building an Intelligent Web: Theory and Practice. J&B Publishers Inc., Boston (2008)
3. Kanoje, S., Girase, S., Mukhopadhyay, D.: User profiling trends, techniques and applications. Int. J. Adv. Found. Res. Comput. 1(11), 119–125 (2014)
4. Uskov, V.L., Howlett, R.J., Jain, L.C. (eds.): Smart Education and Smart e-Learning. Springer International Publishing AG, New York (2015)
5. Poddiakov, A.: Development and inhibition of learning abilities in agents and intelligent systems. In: Proceedings of IADIS International Conference "Intelligent Systems and Agents", Lisbon, Portugal, pp. 235–238, 3–8 July 2007
6. Anane, R.: The learning object triangle. In: IEEE 14th ICALT International Conference on Advanced Learning Technologies, Athens, Greece, pp. 719–721, 7–10 July 2014
7. Wiley, D.A. et al.: The instructional use of learning objects (2001). http://www.reusability.org/read/. Accessed 28 Aug 2016
8. Shadbolt, N., Hall, W., Berners-Lee, T.: The semantic web revisited. IEEE Intell. Syst. J. 21, 96–101 (2006)

9. ADL, SCORM 2004 (4th Edn.) (2009). https://www.adlnet.gov/adl-research/scorm/scorm-2004-4th-edition/. Accessed 28 Aug 2016

10. IEEE Learning Technology Standard Committee - WG12, 1484.12.1-2002 - IEEE Standard for Learning Object Metadata (2009). http://standards.ieee.org/findstds/standard/1484.12.1-2002.html. Accessed 28 Aug 2016

11. Fung, S., Tam, V., Lam, E.Y.: Enhancing learning paths with concept clustering and rule-based optimization. In: 2011 IEEE 11th ICALT International Conference on Advanced Learning Technologies, 6–8 July 2011, Athens, Greece, pp. 249–253 (2011)

12. Adorni, G., Koceva, F.: Designing a knowledge representation tool for subject matter structuring. In: Croitoru, M., Marquis, P., Rudolph, S., Stapleton, G. (eds.) GKR 2015. LNCS (LNAI), vol. 9501, pp. 1–14. Springer, Heidelberg (2015). doi:10.1007/978-3-319-28702-7_1

13. Doignon, J.-P., Falmagne, J.-C.: Knowledge Spaces-Applications in Education. Springer, New York (2013)

14. Marwah, A., Riad, J.: A shortest adaptive learning path in eLearning systems: mathematical view. J. Am. Sci. **5**(6), 32–42 (2009)

15. Sterbini, A., Temperini, M.: Adaptive construction and delivery of web-based learning paths. In: Proceedings - Frontiers in Education Conference, FIE, pp. 1–6 (2009)

16. Anh, N.V., Ha, N.V., Dam, H.S.: Constructing a Bayesian belief network to generate learning path in adaptive hypermedia system. J. Comput. Sci. Cybern. **24**(1), 12–19 (2008)

17. Pirrone, R., Pilato, G., Rizzo, R., Russo, G.: Learning path generation by domain ontology transformation. In: Bandini, S., Manzoni, S. (eds.) AI*IA 2005. LNCS (LNAI), vol. 3673, pp. 359–369. Springer, Heidelberg (2005). doi:10.1007/11558590_37

18. Latha, C.B.C., Kirubakaran, E.: Personalized learning path delivery in web based educational systems using a graph theory based approach. J. Am. Sci. **9**(12s), 981–992 (2013)

19. Koceva, F.: ENCODE - ENvironment for COntent Design and Editing, Ph.D. thesis, University of Genoa (2016)

20. Koper, R., Manderveld, J.: Educational modelling language: modelling reusable, interoperable, rich and personalised units of learning. Br. J. Educ. Technol. **35**(5), 537–551 (2004)

21. Garshol, L.M., Moore, G. (eds.) Topic Maps - Data Model. ISO/IEC JTC1/SC34 Information Technology - Document Description and Processing Languages (2008). http://www.isotopicmaps.org/sam/sam-model/. Accessed 28 Aug 2016

22. WANDORA Project, Documentation Sit. http://wandora.org/. Accessed 28 Aug 2016

23. Pepper, S., Graham, M.: XML topic maps (XTM) 1.0. TopicMaps. Org Specification xtm1-20010806 (2001)

24. ISO/IEC 13250-2:2006 Topic Maps Data Model. http://www.iso.org/iso/home/store/catalogue_tc/catalogue_detail.htm?csnumber=40017. Accessed 28 Aug 2016

25. Kahn, A.B.: Topological sorting of large networks. Commun. ACM **5**(11), 558–562 (1962)

26. Kolb, A.Y., Kolb, D.A.: The Kolb Learning Style Inventory—Version 3.1 (2005)

27. Yamazaki, Y.: Learning styles and typologies of cultural differences: a theoretical and empirical comparison. Working paper. Department of Organizational Behavior, Case Western Reserve University (2002)

28. Gardner, H.: Frames of Mind. Basic Book Inc., New York (1983)

29. McKenzie, W.: Intelligenze Multiple e Tecnologie per la Didattica. Erikson, Trento (2006)

30. LEADLAB Project, European Model of Personalization for Adult Learners, Final Report 502057-LLP-1-2009-1-IT-GRUNDTVIG-GMP (2009). http://leadlab.euproject.org/. Accessed 28 Aug 2016

31. McKenzie, N., Knipe, S.: Research dilemmas: paradigms, methods and methodology. Issues Educ. Res. **16**, 193–205 (2006)

32. NC State University. Index of Learning Styles Questionnaire (2016). http://www.engr.ncsu.edu/learningstyles/ilsweb.html
33. Moodle, Community. Moodle - Open Source Learning Platform (2015). https://moodle.org/
34. Beck, J., Stem, M., Woolf, B.P.: Cooperative student models. In: Artificial Intelligence in Education, 1997: Knowledge and Media in Learning Systems, Proceedings of AI-ED 97, World Conference on Artificial Intelligence in Education, Kobe, Japan, vol. 39. IOS Press (1997)
35. Carr, B., Goldstein, I.P.: Overlays: a theory of modelling for computer aided instruction. No. AI-M-406, MIT, Cambridge, AI Lab (1977)

Answer Set Enumeration via Assumption Literals

Mario Alviano and Carmine Dodaro$^{(\boxtimes)}$

Department of Mathematics and Computer Science,
University of Calabria, 87036 Rende (cs), Italy
{alviano,dodaro}@mat.unical.it

Abstract. Modern, efficient Answer Set Programming solvers implement answer set search via non-chronological backtracking algorithms. The extension of these algorithms to answer set enumeration is nontrivial. In fact, adding blocking constraints to discard already computed answer sets is inadequate because the introduced constraints may not fit in memory or deteriorate the efficiency of the solver. On the other hand, the algorithm implemented by CLASP, which can run in polynomial space, requires invasive modifications of the answer set search procedure. The algorithm is revised in this paper so as to make it almost independent from the underlying answer set search procedure, provided that the procedure accepts as input a logic program and a list of assumption literals, and returns either an answer set (and associated branching literals) or an unsatisfiable core. The revised algorithm is implemented in WASP, and compared empirically to the state of the art solver CLASP.

Keywords: Answer Set Programming · Enumeration · Assumption literals

1 Introduction

Answer Set Programming (ASP) is a declarative formalism for knowledge representation and reasoning based on stable model semantics [1]. In ASP, logic programs are associated with classical models satisfying a stability condition: only necessary information is included in a model of the input program under the assumptions provided by the model itself for the *unknown knowledge* in the program, where unknown knowledge is encoded by means of *default negation*. Models satisfying such a stability condition are called answer sets, and describe plausible scenarios for the knowledge represented in the input program.

A rational agent reasoning in presence of unknown knowledge may want to consider more than one plausible scenario before taking any decision on her next action. It is for this reason that ASP solvers usually implement at least two reasoning tasks, namely *answer set search* and *answer set enumeration*. Answer set search amounts to compute one answer set of a given program, if it exists. Answer set enumeration, instead, amounts to compute all answer sets of a given

© Springer International Publishing AG 2016
G. Adorni et al. (Eds.): AI*IA 2016, LNAI 10037, pp. 149–163, 2016.
DOI: 10.1007/978-3-319-49130-1_12

program, and usually can be stopped at *anytime* [2,3] if a sufficient number of answer sets has been provided to the user.

Different algorithms have been considered for answer set search. Early approaches implement a chronological backtracking algorithm extending DPLL [4] to the richer syntax of ASP [5–8]. Without going into much detail, these algorithms propagate deterministic inferences that can be efficiently detected (e.g., unit resolution), and then nondeterministically select a branching literal according to some heuristic. The branching literal is assigned true, and flagged as *to be flipped*. The process is reiterated until either an answer set is found, or a conflict arises. In case of conflict, the latest assigned literals are retracted until a flagged branching literal is found; the branching literal is then *flipped*, i.e., it is assigned false and its flag is removed, so that the process can continue to search for an answer set. The main drawback of chronological backtracking is that the only information gained from a conflict is that the latest flagged branching literal has to be flipped. This is a naive observation, and it is often the case that the conflict is independent from the latest flagged branching literal, meaning that the same conflict will be discovered again and again.

Non-chronological backtracking, instead, aims at gathering more information from any conflict that may arise during the search. In particular, Conflict-Driven Clause Learning (CDCL) algorithms [9] analyze conflicts so to materialize propositional formulas (specifically, clauses) that are entailed from the input program, and whose knowledge would had avoided the arising of the conflicts. Usually, conflict analysis is achieved by performing backward resolution on the reasons (represented by clauses) that lead to assign the latest literals, until the learned clause contains exactly one unassigned literal in the current partial assignment. Such a literal is called *first unique implication point* [10], and its inference allows the process to continue to search for an answer set. CDCL algorithms often also take advantages of *restarts* [11], which essentially retract all assigned literals so that the heuristic may select different branching literals taking into account the previously arose conflicts.

Answer set enumeration clearly includes answer set search as a subtask: once an answer set is found, the process has to continue and search for the next, until no other answer set can be discovered. However, the way answer set search is implemented may impact considerably in the definition of an algorithm for answer set enumeration. For example, if answer set search is obtained by chronological backtracking, its extension to answer set enumeration is straightforward [6,12]: after returning the answer set to the user, a fake conflict is arisen so that the process can continue to search for the next answer set. The extension of non-chronological backtracking is less obvious. Simple arguments such as adding *blocking constraints*, or blocking clauses, that discard already computed answer sets may introduce crucial inefficiencies. In fact, in order to block a single answer set, a constraint has to include a set of literals that uniquely identify the answer set, which is usually the set of branching literals used in the computation of the answer set. The space required to store such constraints is in general exponential with respect to the size of the input program, as exponentially many answer sets may be admitted. Practically, blocking constraints may not even fit in memory, or anyhow they often deteriorate the performance of answer set enumeration.

A less obvious solution was proposed by Gebser et al. [12], and successfully implemented in CLASP [13]. It combines chronological and non-chronological backtracking: a first answer set is searched by means of non-chronological backtracking, and after one is found, the used branching literals are flagged as in a chronological backtracking; the process then continues with the non-chronological backtracking for the new branching literals, and with the chronological backtracking for the branching literals that lead to the latest discovered answer set. According to this scheme, the chronological backtracking guarantees that answer sets are enumerated requiring only polynomial space, still taking advantage from the efficiency of the non-chronological backtracking. The downside of the algorithm proposed by Gebser et al. is that it requires invasive modifications of the answer set search procedure.

The aim of this paper is to redesign the algorithm proposed by Gebser et al. in terms of *assumption literals*, i.e., a list of literals that the answer set search procedure has to use as the first branching literals. In a nutshell, after the first answer set is found, branching literals are stored in the list of assumptions and flagged as in a chronological backtracking. The process then continues by flipping the latest flagged assumption literal: if an answer set is found, the list of assumptions is extended with the new branching literals; if an incoherence is detected, literals are removed from the list of assumptions. According to this redesigned scheme, the answer set search procedure is used as a black box: whether it implements a chronological or non-chronological backtracking is irrelevant for the enumeration procedure. The only requirements on the answer set search procedure are that the input comprises a Boolean theory (e.g., a program) and a list of assumption literals, and the output is either a solution (e.g., an answer set) or some explanation of the lack of solutions (e.g., an unsatisfiable core; see Sect. 2 for a definition). These requirements are satisfied by almost all modern ASP solvers, but also by solvers for other Boolean theories, among them SAT solvers.

2   Preliminaries

This section recalls syntax and semantics of propositional ASP programs. A quick overview of the main steps of answer set search is also reported.

Syntax. Let $\mathscr{A}$ be a countable set of propositional atoms comprising $\bot$. A *literal* is either an atom (a positive literal), or an atom preceded by the *default negation* symbol $\sim$ (a negative literal). The complement of a literal ℓ is denoted $\overline{\ell}$, i.e., $\overline{p} = \sim p$ and $\overline{\sim p} = p$ for an atom p. This notation extends to sets of literals, i.e., $\overline{L} := \{\overline{\ell} \mid \ell \in L\}$ for a set of literals L. A *program* is a finite set of rules of the following form:

$$p_1 \vee \ldots \vee p_n \leftarrow q_1, \ldots, q_j, \sim q_{j+1}, \ldots, \sim q_m \qquad (1)$$

where $p_1, \ldots, p_n, q_1, \ldots, q_m$ are atoms and $n \geq 0$, $m \geq j \geq 0$. The disjunction $p_1 \vee \ldots \vee p_n$ is called *head*, and the conjunction $q_1, \ldots, q_j, {\sim}q_{j+1}, \ldots, {\sim}q_m$ is referred to as *body*. For a rule r of the form (1), the following notation is also used: $H(r)$ denotes the set of head atoms; $B(r)$ denotes the set of body literals. A rule r is said to be a *constraint* if $H(r) = \{\bot\}$.

Example 1. Let Π_1 be the program comprising the following rules:

$$a \vee b \leftarrow \qquad c \vee d \leftarrow \qquad \bot \leftarrow {\sim}a, {\sim}c.$$

Program Π_1 is used as a running example in the remainder of this paper. ∎

Semantics. An *interpretation* I is a set of literals containing ${\sim}\bot$ (for the sake of simplicity, literal ${\sim}\bot$ is possibly omitted when reporting interpretations in the remainder of this paper). An interpretation I is *total* if for each $p \in \mathscr{A}$ either $p \in I$ or ${\sim}p \in I$, and I is *inconsistent* if there is $p \in \mathscr{A}$ such that $\{p, {\sim}p\} \subseteq I$. Relation $\models$ is inductively defined as follows: for $\ell \in \mathscr{A} \cup \overline{\mathscr{A}}$, $I \models \ell$ if $\ell \in I$; for a rule r, $I \models B(r)$ if $B(r) \subseteq I$, and $I \models r$ if $I \cap H(r) \neq \emptyset$ whenever $I \models B(r)$; for a program Π, $I \models \Pi$ if $I \models r$ for all $r \in \Pi$. Note that $I \not\models \bot$, and $I \models {\sim}\bot$, for any interpretation I. A *model* of a program Π is a consistent, total interpretation I such that $I \models \Pi$. The *reduct* Π^I of a program Π with respect to an interpretation I is obtained from Π as follows: (i) any rule r such that $I \not\models B(r)$ is removed; (ii) all negative literals are removed from the remaining rules. A model I is an *answer set* (or stable model) of a program Π if there is no interpretation J such that both $J \models \Pi^I$, and $J^+ \subset I^+$. Let $AS(\Pi)$ denote the set of answer sets of Π. A program Π is *coherent* if $AS(\Pi) \neq \emptyset$; otherwise, Π is *incoherent*.

Example 2. Consider program Π_1 from Example 1. Its answer sets are the following: $I_1 = \{{\sim}a, b, c, {\sim}d\}$, $I_2 = \{a, {\sim}b, {\sim}c, d\}$, and $I_3 = \{a, {\sim}b, c, {\sim}d\}$. ∎

Answer Set Search. Answer set search is implemented in modern ASP solvers by extending CDCL to ASP. The input comprises a propositional program Π, and a list A of literals, called *assumption literals* (or simply assumptions). Its output is an answer set I of Π such that $A \subseteq I$, and a set $B \subseteq I$ of *branching literals*, if such an I does exist; in this case, $B \supseteq A$ holds, and I is the only answer set in $AS(\Pi)$ such that $B \subseteq I$. Otherwise, if there is no $I \in AS(\Pi)$ such that $A \subseteq I$, the algorithm returns as output a set $C \subseteq A$ such that $\Pi \cup \{\bot \leftarrow \overline{\ell} \mid \ell \in C\}$ is incoherent, which is called *unsatisfiable core*. In the remainder of this paper, $search(\Pi, A)$ is used to denote a call to the answer set search algorithm.

Example 3. Consider again program Π_1 from Example 1. If $[{\sim}a, c]$ is the list of assumptions, the answer set returned by function *search* is necessarily $I_1 = \{{\sim}a, b, c, {\sim}d\}$. On the other hand, if $[{\sim}a, b, {\sim}c, d]$ is the list of assumptions, the returned unsatisfiable core is $\{{\sim}a, {\sim}c\}$, or one of its supersets. ∎

In more detail, I is initially set to $\{{\sim}\bot\} \cup A$, and B to A. Subsequently, a *propagation* step extends I by inferring deterministic consequences of the literals in

I that are efficiently detectable. For each inferred literal, a subset of I that is responsible of the inference is also identified; it is referred to as *reasons*. Three cases are possible after a propagation step is completed:

1. *I is consistent, but not total.* In this case, a branching literal ℓ is chosen according to some heuristic criterion, and added to both I and B. After that, the algorithm continues with a propagation step.
2. *I is consistent, and total.* In this case, I is an answer set, and the algorithm terminates returning I and B. (For some programs, an additional check is required to ensure stability of I; see [14–16].)
3. *I is inconsistent.* In this case, the consistency of I is restored by retracting the latest assigned literals. While literals are removed from I, their reasons are used to learn a new constraint via backward resolution, until the learned constraint is such that the complement of a retracted literal can be inferred. After that, the algorithm continues with a propagation step. When the consistency cannot be restored, i.e., when the literal inferred by the learned constraint is in $\{\top\} \cup \overline{A}$, the algorithm terminates returning an unsatisfiable core.

Example 4. Consider again program Π_1 from Example 1, and an empty list of assumptions. The answer set search algorithm starts by setting I to $\{\sim\!\bot\}$, and B to $\emptyset$. The propagation step does not infer any new literal, and therefore a branching literal is chosen according to some heuristic. Let us say that $\sim\!a$ is chosen, and added to both I and B. The propagation step can now infer b from rule $a \vee b \leftarrow$ (because $\sim\!a$ is in I). At this point, I is $\{\sim\!\bot, \sim\!a, b\}$, B is $\{\sim\!a\}$, and a new branching literal, say c, is chosen. The propagation step infers $\sim\!d$ from rule $c \vee d \leftarrow$ (because c is in I and d does not occur in any other rule head), and the algorithm terminates returning $I = \{\sim\!\bot, \sim\!a, b, c, \sim\!d\}$, a consistent and total interpretation. This is answer set I_1 from Example 2.

Consider now the case in which the list of assumptions is $[\sim\!a, c]$. The algorithm initially sets I to $\{\sim\!\bot, \sim\!a, c\}$, and B to $\{\sim\!a, c\}$. The propagation step extends I to I_1 (for the same reasons already given in the previous paragraph), which is then returned.

Let us now consider the case in which the list of assumptions is $[\sim\!a, b, \sim\!c, d]$. After initializing I to $\{\sim\!\bot, \sim\!a, b, \sim\!c, d\}$, an inconsistency is detected. Specifically, $\bot \leftarrow \sim\!a, \sim\!c$ raises a conflict in the assignment of $\sim\!a$ and $\sim\!c$. These two literals are possibly returned as an unsatisfiable core, since they are eventually responsible of the lack of answer sets for the given list of assumptions. However, the algorithm may also return a superset of $\{\sim\!a, \sim\!c\}$, as any superset of an unsatisfiable core is in turn an unsatisfiable core. ∎

3   Answer Set Enumeration

The computational problem analyzed in this paper is referred to as *answer set enumeration*: Given a program Π, compute all answer sets I of Π. Currently, there are two different algorithms that are implemented in modern ASP solvers. The first algorithm, which is the simplest, is based on the introduction of *blocking constraints*, that is, constraints that block already printed answer sets. This algorithm is recalled in Sect. 3.1. The second algorithm, which was introduced in CLASP, is an in-depth modification of the CDCL algorithm, and essentially combines the non-chronological backtracking of CDCL with a more classical chronological backtracking on the set of branching literals that lead to the latest printed answer set. This algorithm is revised in Sect. 3.2 in terms of assumption literals, so as to make it almost independent from the underlying answer set search procedure.

3.1   Enumeration via Blocking Constraints

An immediate algorithm for enumerating answer sets is based on *blocking constraints*: an answer set is searched, and a (blocking) constraint is added to the program forbidding the repetition of the computed answer set, until an incoherence arises. This naive approach is reported as Algorithm 1. In particular, the algorithm iteratively searches for the answer sets of Π. When an answer set is found, the set of branching literals B is returned by function *search*. Subsequently, a constraint containing all branching literals in B is added to Π. This constraint guarantees that the same answer set will not be returned by future calls to function *search*.

Example 5. Consider again the program Π_1 of Example 2 and suppose that the function *search* $(\Pi_1, \emptyset)$ returns ($"COHERENT", I_1, \{b, c\}, \emptyset$). Algorithm 1 first prints the answer set I_1 and then extends Π_1 with the constraint $\bot \leftarrow b, c$. Let us assume that the subsequent call to *search* $(\Pi_1, \emptyset)$ returns ($"COHERENT", I_2, \{d\}, \emptyset$). Thus, I_2 is printed and Π_1 is extended with the constraint $\bot \leftarrow d$. The subsequent call to *search* $(\Pi_1, \emptyset)$ returns ($"COHERENT", I_3, \emptyset, \emptyset$), thus I_3 is printed and Π_1 is extended with $\bot \leftarrow$. After that, *search* $(\Pi_1, \emptyset)$ returns ($"INCOHERENT", \emptyset, \emptyset, \emptyset$) and the algorithm terminates. ∎

The main drawback of this algorithm is that it adds a blocking constraint for each printed answer set. The algorithm is therefore not practical for programs with a large number of answer sets, and in general the number of answer sets may be exponential in the number of distinct atoms occurring in the input program.

Algorithm 1. Answer Set Enumeration of a Program Π via Blocking Constraints

1 $(res, I, B, C) := search(\Pi, \emptyset)$;   // search $I \in AS(\Pi)$
2 **if** $res = "COHERENT"$ **then**   // found I using branching literals B
3   **print** I;   // print answer set
4     $\Pi := \Pi \cup \{\bot \leftarrow B\}$;   // add blocking constraint to inhibit I
5   **goto** 1;   // search for another answer set

Algorithm 2. Answer Set Enumeration of a Program Π via Assumption Literals

```
1  A := [∼⊥];     F := ∅;             // initialize assumptions and flipped literals
2  while A ≠ [⊥] do                    // there are still assumptions to be flipped
3  |   (res, I, B, C) := search(Π, A);          // search I ∈ AS(Π) such that A ⊆ I
4  |   if res = "COHERENT" then          // found I using branching literals B
5  |   |   print I;                                        // print answer set
6  |   |   for ℓ ∈ B \ A do  push(A, ℓ);     // extend A with new branching literals
7  |   else                                     // found unsatisfiable core C ⊆ A
8  |   └   while A ≠ [∼⊥] and top(A) ∉ C do  F := F \ {pop(A)};    // backjumping
9  |   while top(A) ∈ F do  F := F \ {pop(A)};       // remove flipped assumptions
10 └   push(A, pop(A));    F := F ∪ top(A);              // flip top assumption
```

3.2   Enumeration via Assumption Literals

The idea underlying the algorithm presented in this section is the following: answer sets differ from each other by at least one literal, thus whenever an answer set is found, at least one of the branching literals must be flipped; the process is repeated until all meaningful combinations of branching literals have been explored. The algorithm is reported as Algorithm 2, where a list A of assumption literals and a set F of (flipped) literals are used to track branching literals to consider and already flipped, respectively. The following stack operations are applied to the list A: *push*, to append an element to A; *top*, to return the last element of A; *pop*, to return and remove the last element of A.

Function $search(\Pi, A)$ is iteratively called until there are still assumptions to be flipped. Two cases are possible:

1. Π *admits an answer set* $I \supseteq A$. In this case, I is printed (line 5), and A is extended with all new branching literals (line 6), i.e., branching literals not already contained in A.
2. Π *is incoherent under the assumptions* A. In this case an unsatisfiable core $C \subseteq A$ is returned. Since $\Pi \cup \{\bot \leftarrow \bar{\ell} \mid \ell \in C\}$ is incoherent by definition of unsatisfiable core, the algorithm has to flip at least one of the assumption literals in A (line 8).

In both cases, the algorithm continues by flipping the latest added assumption that has not been already flipped (lines 9–10).

Example 6. Consider again program Π_1 from Example 1. The main steps of Algorithm 2 are the following:

– Initially, A is $[\sim\bot]$, and F is $\emptyset$.
– The result of $search(\Pi_1, A)$ is, say, ($"COHERENT"$, I_1, $\{b, c\}$, $\emptyset$). After printing I_1 (line 5), A is extended to $[\sim\bot, b, c]$ (line 6). Later on, the latest element of A, i.e., c, is flipped (line 10); at this point, A is $[\sim\bot, b, \sim c]$, and F is $\{\sim c\}$.

- The result of $search(\Pi_1, A)$ is ("*INCOHERENT*", $\emptyset, \emptyset, \{b, \sim c\}$). Thus, $\sim c$ is removed from F (line 9), the latest element of A, i.e., b, is flipped (line 10), and $\sim b$ is added to F (line 10). At this point, A is $[\sim\perp, \sim b]$, and F is $\{\sim b\}$.
- The result of $search(\Pi_1, A)$ is, say, ("*COHERENT*", $I_2, \{\sim b, d\}, \emptyset$). Thus, I_2 is printed (line 5), and d is added to A (line 6). After that, A is modified by flipping the latest element, i.e., d, and $\sim d$ is added to F. At this point, A is $[\sim\perp, \sim b, \sim d]$, and F is $\{\sim b, \sim d\}$.
- The result of $search(\Pi_1, A)$ is ("*COHERENT*", $I_3, \{\sim b, \sim d\}, \emptyset$). Thus, I_3 is printed, and no new branching literal is added to A. At this point, $\sim d$ and $\sim b$ are removed from F (line 9), and the latest element of A, i.e., $\sim\perp$, is flipped. Hence, A is $[\perp]$, and F is still $\emptyset$.

Since A is $[\perp]$, the algorithm terminates.   ∎

3.3 Correctness

Correctness of Algorithm 2 is stated in the following claim:

Theorem 1. *Given a program Π, Algorithm 2 prints all and only answer sets of Π, with no repetitions.*

In order to formalize the proof of the above theorem, let us first consider a slight variant of our algorithm, called Algorithm 2*, obtained by replacing line 6 as follows:

---
6 **for** $\ell \in I \setminus A$ **do** $push(A, \ell)$;   // extend A with new true literals
---

and by removing lines 7–8. After establishing the correctness of Algorithm 2*, we will extend the proof to Theorem 1. To this aim, let us introduce three lemmas concerning interpretations printed by Algorithm 2*.

Lemma 1. *Given a program Π, if Algorithm 2* prints an interpretation I, then I is an answer set of Π.*

Proof. By definition, $search(\Pi, A)$ returns an answer set I in $AS(\Pi)$ such that $A \subseteq I$, if it exists. Hence, if I is printed at line 5, I is an answer set of Π.   □

Lemma 2. *Given a program Π, if I is an answer set of Π, then I is printed by Algorithm 2* at some point.*

Proof. Let us assume that Π is coherent, otherwise the claim is trivial. Thus, the algorithm finds a first answer set $I := \{\sim\perp, \ell_1, \ldots, \ell_n\}$ $(n \geq 1)$ of Π. At this point, the algorithm implements a backtracking procedure by flipping assumption literals, from ℓ_n to ℓ_1. Hence, the answer sets printed by the algorithm are those in set S_0, where S_i is defined as follows for $i \in [0..n]$:

$$S_i := \{I\} \cup \bigcup_{j=i+1}^{n} \{I' \in AS(\Pi) \mid \{\sim\perp, \ell_1, \ldots, \ell_{j-1}, \overline{\ell_j}\} \subseteq I'\}. \tag{2}$$

Now consider the following set S_i', for $i \in [0..n]$:

$$S_i' := \{I' \in AS(\Pi) \mid \{\sim\bot, \ell_1, \ldots, \ell_i\} \subseteq I'\}. \tag{3}$$

For all $i \in [0..n]$, $S_i = S_i'$. In fact, $S_i \subseteq S_i'$ is immediate. As for $S_i' \subseteq S_i$, for any $I' \in S_i' \setminus \{I\}$ there is $j \in [i+1..n]$ such that $I' = \{\sim\bot, \ell_1, \ldots, \ell_{j-1}, \overline{\ell_j}, \ell'_{j+1}, \ldots, \ell'_n\}$, where $\ell'_k \in \{\ell_k, \overline{\ell_k}\}$, for each $k \in [j+1..n]$. Hence, $I' \in S_i$ holds, and our proof is complete because $S_0' = \{I' \in AS(\Pi) \mid \{\sim\bot\} \subseteq I'\} = AS(\Pi)$.   □

Lemma 3. *Given a program Π, if Algorithm 2* prints two interpretations I and J, then $I \neq J$.*

Proof. Let us assume that I is printed before J. After printing I, list A contains all literals in I, which are then iteratively flipped in subsequent calls to $search(\Pi, A)$. Since J has to contain all assumptions in A, among them those flipped, J is necessarily different from I.   □

Correctness of Algorithm 2* follows from the three lemmas above.

Theorem 2. *Given a program Π, Algorithm 2* prints all and only answer sets of Π, with no repetitions.*

We can eventually complete this section by extending the proof of correctness to Algorithm 2.

Proof (Theorem 1). First of all, removing non-branching literals from A is correct. In fact, according to the definition of $search$ given in Sect. 2, if $search(\Pi, A)$ returns ("$COHERENT$", I, B, C), then I is the unique answer set of Π such that $B \subseteq I$. Hence, flipping any non-branching literal necessarily results into an incoherence.

The second observation concerns lines 7–8, which are executed when $search(\Pi, A)$ returns ("$INCOHERENT$", I, B, C). In this case, C is an unsatisfiable core, and by definition $\Pi \cup \{\bot \leftarrow \overline{\ell} \mid \ell \in C\}$ is incoherent. Hence, for all A' such that $C \subseteq A' \subseteq A$, function $search(\Pi, A')$ would result into an incoherence. Lines 7–8 allow to skip such useless calls to function $search$, so that the search can continue by flipping at least one literal in the unsatisfiable core.   □

4 Experiments

The two algorithms for answer set enumeration given in Sect. 3 have been compared empirically on WASP [17,18], an ASP solver implementing non-chronological backtracking, handling assumption literals, and returning unsatisfiable cores in case of incoherences. Actually, answer set enumeration via blocking constraints was already supported by the solver, and we had only to implement answer set enumeration via assumption literals. For our comparison, all coherent instances from the fourth ASP Competition were considered [19–23]. The experiments were run on an Intel Xeon 2.4 GHz with 16 GB of memory, and time and memory were limited to 10 min and 15 GB, respectively.

Table 1. Aggregated results on the average number of enumerated answer sets within 600 s by using blocking constraints (BC), assumption literals (AL), and backtracking on branching literals (BT). The percentage gain obtained by WASP and CLASP when BC is replaced by AL or BT is also reported at the bottom of the table.

Problems	CLASP-BC	CLASP-BT	WASP-BC	WASP-AL
BottleFilling	888 332	2 114 259	41 200	804 688
GracefulGraphs	56	66	21	36
GraphColoring	1 016 672	40 460 090	116 846	16 139 214
HanoiTower	46	44	46	46
Labyrinth	78 320	184 680	4 573	76 965
NoMistery	432 287	679 618	13 095	235 918
PPM	232 469	3 024 614	88 575	2 173 629
QSR	311 638	430 011	8 880	83 683
RicochetRobot	965	1 056	579	362
Sokoban	478 293	2 194 185	45 936	702 147
Solitaire	294 936	1 234 253	51 103	1 029 140
StableMarriage	277 024	1 796 476	60 513	1 073 491
VisitAll	582 113	3 133 040	72 641	2 635 478
Weighted-Sequence	35	111	45	47

Aggregated results are shown in Table 1, where the average number of answer sets computed is reported for each benchmark. As a first comment, answer set enumeration via assumptions literals (WASP-AL) is on average more efficient than answer set enumeration via blocking constraints (WASP-BC) on the considered benchmarks. The performance improvement obtained by switching to the new algorithm is quite evident by observing the percentage gain of WASP-AL over WASP-BC: it is on average 2 187 % with a peak of 13 712 % on GraphColoring. The improvements is also shown in the cactus plot of Fig. 1. In the cactus plot a line is reported for each compared method. For each considered algorithm, instances are sorted (in ascending order) according to the number of enumerated answer sets. A point (x, y) in the line represents that the method enumerated y answer sets of the xth instance. The graph highlights that WASP-AL enumerates far more answer sets than WASP-BC. On average, for each answer set found by WASP-BC, 463 answer sets are found by WASP-AL.

WASP-AL and WASP-BC exhibit a similar performance only on four benchmarks, namely GracefulGraphs, HanoiTower, RicochetRobot and Weighted-Sequence. This behavior can be explained by observing the number of answer sets admitted by these instances. In fact, the number of answer sets observed in those benchmarks is on average 134. As a consequence, the number of blocking constraints introduced by WASP-BC is low, and the performance of the solver is not deteriorated in this benchmark.

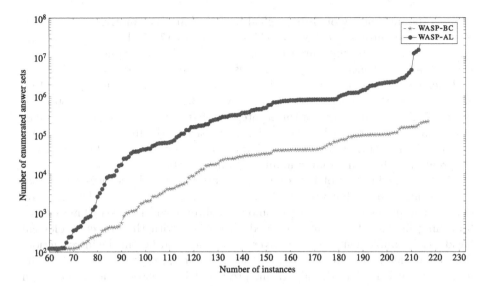

Fig. 1. Cactus plot of enumerated answer sets by WASP within 600 s using blocking constraints (BC) and assumption literals (AL).

For the sake of completeness, WASP was also compared with the state of the art ASP solver CLASP [24]. Answer set enumeration in CLASP is achieved either via blocking constraints (CLASP-BC), or by means of a backtracking algorithm on branching literals (CLASP-BT) [12] which actually inspired Algorithm 2 presented in Sect. 3.2. As expected, CLASP is in general faster than WASP to complete answer set searches: it can be observed by comparing WASP-BC and CLASP-BC, that is, the two solvers enumerating answer sets via blocking constraints. In fact, for instances of Crossing Minimization and Stable Marriage, the number of answer sets produced by CLASP-BC is on average 9 times higher than those printed by WASP-AL. However, the advantage of CLASP is considerably reduced when CLASP-BT and WASP-AL are compared. In fact the number of answer sets enumerated by CLASP-BT is on average 2 times higher than those printed by WASP-AL.

5   Related Work

Answer set enumeration is an important computational task of ASP solving, and was supported already in early proposed ASP solvers such as DLV [7,25,26] and SMODELS [8]. These two solvers are based on chronological backtracking, and therefore their answer set search procedures are easily extended to implement answer sets enumeration: once an answer set is found and printed, the search continues by unrolling the latest assigned branching literal. However, chronological backtracking proved to be quite inefficient in addressing answer set search, and as a consequence it is usually inefficient also for answer set enumeration.

It is the inefficiency of chronological backtracking that motivated the introduction of non-chronological backtracking in ASP [13,17,27]. However, extending an answer set search procedure to obtain answer set enumeration is less obvious when non-chronological backtracking is used. In fact, the easy solution is based on blocking constraints and was briefly recalled in Sect. 3.1. It is supported by all modern ASP solvers; among them, CLASP [13,24], CMODELS [6,28–30], and WASP [17,27]. The downside of this strategy is that in the worst case the number of introduced constraints is exponential in the size of the input program. Practically, this means that the enumeration of answer sets may not fit in memory for programs admitting a large number of answer sets.

Such an inefficiency of blocking constraints motivated the development of the innovative algorithm presented by Gebser et al. [12] and implemented by CLASP. The underlying idea is to combine a chronological backtracking on the branching literals of the latest printed answer set with the non-chronological backtracking implemented in the solver. As already clarified in the introduction, this genial idea is also at the basis of the algorithm presented in Sect. 3.2. The main difference is that the algorithm presented by Gebser et al. requires an invasive modification of the answer set search procedure, while the algorithm given in this paper is almost independent from the underlying solver.

The strength of the proposed solution is precisely its few requirements: the algorithm can be used to enumerate solutions of any solver accepting as input a Boolean theory (e.g., a program) and a list of assumption literals, and proving as output either a solution (e.g., an answer set) or some explanation of the lack of solutions (e.g., an unsatisfiable core). According to these requirements, the algorithm presented in Sect. 3.2 is suitable to enumerate classical models of propositional logic theories: modern SAT solvers accepts as input a set of clauses and a list of assumption literals, and provide as output either a classical model or an unsatisfiable core. Prominent examples are GLUCOSE [31] and LINGELING [32].

Abstract argumentation and abstract dialectical framework [33–35] are other prominent examples of knowledge representation formalisms in which the enumeration of solutions is a relevant computational task, as witnessed by specific tracks in the First International Competition on Computational Models of Argumentation (ICCMA'15) [36,37]. Several semantics are available in these frameworks, nevertheless the algorithm proposed in this paper can be used for all of them thanks to its generality.

ASP programs may also include other constructs that ease the representation of complex knowledge, such as aggregates [8,38–41] and weak constraints [42]. The enumeration algorithm presented in this paper can be used in presence of these constructs, even if in presence of weak constraints some optimization of the solver has to be disabled (this is the case, for example, of the *hardening* procedure [43] in unsatisfiable core-based algorithms such as ONE [44]).

6   Conclusion

Answer set enumeration can be implemented in modern ASP solvers without any invasive modification of the answer set search procedure: if assumption literals can be provided as input, a list of linear size with respect to the input program is sufficient to inhibit the computation of already discovered answer sets. The algorithm for answer set enumeration presented in this paper allows to combine the strength of non-chronological backtracking for answer set search with the compactness of chronological backtracking for discarding previous answer sets. However, since the answer set search procedure is seen as a black box, the presented algorithm is not forced to use a specific strategy to complete answer set searches: whether chronological or non-chronological backtracking are used is irrelevant for the presented enumeration algorithm; for the same reason, the presented algorithm is also open to any future answer set search procedure. Another advantage of the generality of the presented algorithm is that it can be used for other formalisms. For example, we implemented enumeration of classical models of propositional theories in few lines of code, using GLUCOSE as a SAT oracle for completing model searches. The source code is available online at the following URL: https://github.com/dodaro/ModelsEnumeration.

Concerning future work, we plan to consider the application of the enumeration algorithm based on literal assumptions for implementing query answering over graphs with preferences [45] and complex reasoning on combinatorial auctions [46].

Acknowledgement. This work was partially supported by the National Group for Scientific Computation (GNCS-INDAM), by the Italian Ministry of Economic Development under project "PIUCultura (Paradigmi Innovativi per l'Utilizzo della Cultura)" n. F/020016/01–02/X27, and by the Italian Ministry of University and Research under PON project "Ba2Know (Business Analytics to Know) Service Innovation - LAB", No. PON03PE_00001_1.

References

1. Gelfond, M., Lifschitz, V.: Classical negation in logic programs and disjunctive databases. New Gener. Comput. **9**(3/4), 365–386 (1991)
2. Alviano, M., Dodaro, C., Ricca, F.: Anytime computation of cautious consequences in answer set programming. TPLP **14**(4–5), 755–770 (2014)
3. Bliem, B., Kaufmann, B., Schaub, T., Woltran, S.: ASP for anytime dynamic programming on tree decompositions. In: IJCAI. AAAI Press (2016)
4. Davis, M., Logemann, G., Loveland, D.W.: A machine program for theorem-proving. Commun. ACM **5**(7), 394–397 (1962)
5. Brochenin, R., Lierler, Y., Maratea, M.: Abstract disjunctive answer set solvers. In: ECAI, vol. 263. Frontiers in Artificial Intelligence and Applications, pp. 165–170. IOS Press (2014)
6. Giunchiglia, E., Maratea, M.: On the relation between answer set and SAT procedures (or, Between CMODELS and SMODELS). In: Gabbrielli, M., Gupta, G. (eds.) ICLP 2005. LNCS, vol. 3668, pp. 37–51. Springer, Heidelberg (2005). doi:10.1007/11562931_6
7. Leone, N., Pfeifer, G., Faber, W., Eiter, T., Gottlob, G., Perri, S., Scarcello, F.: The DLV system for knowledge representation and reasoning. ACM Trans. Comput. Log. **7**(3), 499–562 (2006)
8. Simons, P., Niemelä, I., Soininen, T.: Extending and implementing the stable model semantics. Artif. Intell. **138**(1–2), 181–234 (2002)

9. Silva, J.P.M., Sakallah, K.A.: GRASP: a search algorithm for propositional satisfiability. IEEE Trans. Comput. **48**(5), 506–521 (1999)

10. Zhang, L., Madigan, C.F., Moskewicz, M.W., Malik, S.: Efficient conflict driven learning in boolean satisfiability solver. In: ICCAD, pp. 279–285. IEEE Computer Society (2001)

11. Audemard, G., Simon, L.: Refining Restarts Strategies for SAT and UNSAT. In: Milano, M. (ed.) CP 2012. LNCS, vol. 7514, pp. 118–126. Springer, Heidelberg (2012)

12. Gebser, M., Kaufmann, B., Neumann, A., Schaub, T.: Conflict-driven answer set enumeration. In: Baral, C., Brewka, G., Schlipf, J. (eds.) LPNMR 2007. LNCS (LNAI), vol. 4483, pp. 136–148. Springer, Heidelberg (2007). doi:10.1007/978-3-540-72200-7_13

13. Gebser, M., Kaufmann, B., Schaub, T.: Conflict-driven answer set solving: from theory to practice. Artif. Intell. **187**, 52–89 (2012)

14. Alviano, M., Dodaro, C., Ricca, F.: Reduct-based stability check using literal assumptions. In: ASPOCP (2015)

15. Gebser, M., Kaufmann, B., Schaub, T.: Advanced conflict-driven disjunctive answer set solving. In: IJCAI, IJCAI/AAAI (2013)

16. Koch, C., Leone, N., Pfeifer, G.: Enhancing disjunctive logic programming systems by SAT checkers. Artif. Intell. **151**(1–2), 177–212 (2003)

17. Alviano, M., Dodaro, C., Leone, N., Ricca, F.: Advances in WASP. In: Calimeri, F., Ianni, G., Truszczynski, M. (eds.) LPNMR 2015. LNCS (LNAI), vol. 9345, pp. 40–54. Springer, Heidelberg (2015). doi:10.1007/978-3-319-23264-5_5

18. Dodaro, C., Alviano, M., Faber, W., Leone, N., Ricca, F., Sirianni, M.: The birth of a WASP: preliminary report on a new ASP solver. In: Fioravanti, F. (ed.) CILC 2011, vol. 810. CEUR Workshop Proceedings, pp. 99–113 (2011)

19. Alviano, M., et al.: The fourth answer set programming competition: preliminary report. In: Cabalar, P., Son, T.C. (eds.) LPNMR 2013. LNCS (LNAI), vol. 8148, pp. 42–53. Springer, Heidelberg (2013). doi:10.1007/978-3-642-40564-8_5

20. Calimeri, F., Gebser, M., Maratea, M., Ricca, F.: Design and results of the fifth answer set programming competition. Artif. Intell. **231**, 151–181 (2016)

21. Calimeri, F., Ianni, G., Ricca, F.: The third open answer set programming competition. TPLP **14**(1), 117–135 (2014)

22. Gebser, M., Maratea, M., Ricca, F.: The design of the sixth answer set programming competition. In: Calimeri, F., Ianni, G., Truszczynski, M. (eds.) LPNMR 2015. LNCS (LNAI), vol. 9345, pp. 531–544. Springer, Heidelberg (2015). doi:10.1007/978-3-319-23264-5_44

23. Gebser, M., Maratea, M., Ricca, F.: What's hot in the answer set programming competition. In: AAAI, pp. 4327–4329. AAAI Press (2016)

24. Gebser, M., Kaminski, R., Kaufmann, B., Romero, J., Schaub, T.: Progress in *clasp* series 3. In: Calimeri, F., Ianni, G., Truszczynski, M. (eds.) LPNMR 2015. LNCS (LNAI), vol. 9345, pp. 368–383. Springer, Heidelberg (2015). doi:10.1007/978-3-319-23264-5_31

25. Alviano, M., Faber, W., Leone, N., Perri, S., Pfeifer, G., Terracina, G.: The disjunctive datalog system DLV. In: Moor, O., Gottlob, G., Furche, T., Sellers, A. (eds.) Datalog 2.0 2010. LNCS, vol. 6702, pp. 282–301. Springer, Heidelberg (2011). doi:10.1007/978-3-642-24206-9_17

26. Maratea, M., Ricca, F., Faber, W., Leone, N.: Look-back techniques and heuristics in DLV: implementation, evaluation, and comparison to QBF solvers. J. Algorithms **63**(1–3), 70–89 (2008)

27. Alviano, M., Dodaro, C., Faber, W., Leone, N., Ricca, F.: WASP: a native ASP solver based on constraint learning. In: Cabalar, P., Son, T.C. (eds.) LPNMR 2013. LNCS (LNAI), vol. 8148, pp. 54–66. Springer, Heidelberg (2013). doi:10. 1007/978-3-642-40564-8_6
28. Giunchiglia, E., Leone, N., Maratea, M.: On the relation among answer set solvers. Ann. Math. Artif. Intell. **53**(1–4), 169–204 (2008)
29. Giunchiglia, E., Lierler, Y., Maratea, M.: Answer set programming based on propositional satisfiability. J. Autom. Reasoning **36**(4), 345–377 (2006)
30. Lierler, Y., Maratea, M.: Cmodels-2: SAT-based answer set solver enhanced to non-tight programs. In: Lifschitz, V., Niemelä, I. (eds.) LPNMR 2004. LNCS (LNAI), vol. 2923, pp. 346–350. Springer, Heidelberg (2003). doi:10.1007/ 978-3-540-24609-1_32
31. Audemard, G., Simon, L.: Predicting learnt clauses quality in modern SAT solvers. In: IJCAI, pp. 399–404 (2009)
32. Biere, A.: Lingeling essentials, a tutorial on design and implementation aspects of the the SAT solver lingeling. In: POS, vol. 27. EPiC Series, p. 88. EasyChair (2014)
33. Alviano, M., Faber, W.: Stable model semantics of abstract dialectical frameworks revisited: a logic programming perspective. In: IJCAI, pp. 2684–2690. AAAI Press (2015)
34. Brewka, G., Strass, H., Ellmauthaler, S., Wallner, J.P., Woltran, S.: Abstract dialectical frameworks revisited. In: IJCAI, IJCAI/AAAI (2013)
35. Brewka, G., Woltran, S.: Abstract dialectical frameworks. In: KR, AAAI Press (2010)
36. Bistarelli, S., Rossi, F., Santini, F.: A comparative test on the enumeration of extensions in abstract argumentation. Fundam. Inform. **140**(3–4), 263–278 (2015)
37. Thimm, M., Villata, S., Cerutti, F., Oren, N., Strass, H., Vallati, M.: Summary report of the first international competition on computational models of argumentation. AI Mag. **37**(1), 102 (2016)
38. Alviano, M., Faber, W.: The complexity boundary of answer set programming with generalized atoms under the FLP semantics. In: Cabalar, P., Son, T.C. (eds.) LPNMR 2013. LNCS (LNAI), vol. 8148, pp. 67–72. Springer, Heidelberg (2013). doi:10.1007/978-3-642-40564-8_7
39. Alviano, M., Faber, W., Gebser, M.: Rewriting recursive aggregates in answer set programming: back to monotonicity. TPLP **15**(4–5), 559–573 (2015)
40. Alviano, M., Leone, N.: Complexity and compilation of gz-aggregates in answer set programming. TPLP **15**(4–5), 574–587 (2015)
41. Faber, W., Pfeifer, G., Leone, N.: Semantics and complexity of recursive aggregates in answer set programming. Artif. Intell. **175**(1), 278–298 (2011)
42. Alviano, M., Dodaro, C.: Anytime answer set optimization via unsatisfiable core shrinking. TPLP (2016, in press)
43. Alviano, M., Dodaro, C., Marques-Silva, J., Ricca, F.: Optimum stable model search: algorithms and implementation. J. Logic Comput. (2015)
44. Alviano, M., Dodaro, C., Ricca, F.: A MaxSAT algorithm using cardinality constraints of bounded size. In: IJCAI 2015, pp. 2677–2683. AAAI Press (2015)
45. Fionda, V., Pirrò, G.: Querying graphs with preferences. In: CIKM, pp. 929–938. ACM (2013)
46. Fionda, V., Greco, G.: The complexity of mixed multi-unit combinatorial auctions: tractability under structural and qualitative restrictions. Artif. Intell. **196**, 1–25 (2013)

On the Application of Answer Set Programming to the Conference Paper Assignment Problem

Giovanni Amendola, Carmine Dodaro[(✉)], Nicola Leone, and Francesco Ricca

Department of Mathematics and Computer Science,
University of Calabria, Rende, Italy
{amendola,dodaro,leone,ricca}@mat.unical.it

Abstract. Among the tasks to be carried out by conference organizers is the one of assigning reviewers to papers. That problem is known in the literature as the Conference Paper Assignment Problem (CPAP). In this paper we approach the solution of a reasonably rich variant of the CPAP by means of Answer Set Programming (ASP). ASP is an established logic-based programming paradigm which has been successfully applied for solving complex problems arising in Artificial Intelligence. We show how the CPAP can be elegantly encoded by means of an ASP program, and we analyze the results of an experiment, conducted on real-world data, that outlines the viability of our solution.

Keywords: Answer Set Programming · Conference Paper Assignment Problem · Applications

1 Introduction

Among the tasks to be carried out by conference organizers is the one of assigning reviewers to papers. That problem is known in the literature as the Conference Paper Assignment Problem (CPAP). The CPAP has quickly attracted the interest of researchers, and several formulations of the problem as well as a range of different solutions have been proposed [1,2]. Actually, there is no recognized canonical form of the CPAP, and there is debate around the optimality criterion to be used for computing "fair" or "desiderable" assignments of papers to reviewers [1,2]. In this paper we focus on a reasonably rich formulation of the problem where: (i) each paper has to be assigned to a given number of reviewers, (ii) each reviewer receives at most a given number of papers, and (iii) assignments are not done in case of (declared) conflict of interest. Moreover additional preference criteria have to be satisfied. In particular, the reviewer preferences (expressed by means of a numeric score) are maximized and the number of papers assigned to each reviewer is balanced. Note that this formulation of the CPAP complies (in terms of input data and parameters) with the information usually available to conference organizers in well-known conference paper management system such as Easychair (http://www.easychair.org). Moreover, it contemplates a set of requirements that are common to the majority of CPAP formulations in the

G. Adorni et al. (Eds.): AI*IA 2016, LNAI 10037, pp. 164–178, 2016.
DOI: 10.1007/978-3-319-49130-1_13

literature [2]. It is worth noting that the CPAP variant we consider in this paper is a computationally hard problem, indeed it can be proved to be NP-hard [3].

Complex combinatorial optimization problems, such as the CPAP, are usually the target for the application of formalisms developed in the area of Artificial Intelligence. Among these, Answer Set Programming (ASP) [4], a well-known declarative programming paradigm which has been proposed in the area of logic programming and non-monotonic reasoning, is an ideal candidate. Indeed, ASP combines a comparatively high knowledge-modeling power [4] with a robust solving technology [5–12]. For these reasons ASP has become an established logic-based programming paradigm with successful applications to complex problems in Artificial Intelligence [13,14], Bioinformatics [15–17], Databases [18,19], Game Theory [20]; more recently ASP has been applied to solve industrial applications [21,22].

Despite ASP can be used –in principle– for solving the CPAP, no specific investigation has been done [1,2] (to the best of our knowledge) about the suitability of the ASP framework for solving real-world instances of the CPAP. The goal of this paper is to provide an assessment of the applicability of ASP to the CPAP. To this end, we consider a variant of the CPAP including constraints and optimization criteria commonly considered in the literature (see Sect. 3), and we show that it can be compactly encoded by means of an ASP program (see Sect. 4). Moreover, we analyze and discuss on the results of an experiment, conducted on real-world data, that outline the viability of an ASP-based solution (see Sect. 5). This work paves the way for the development of a more comprehensive ASP-based system for the CPAP.

2   Answer Set Programming

Answer Set Programming (ASP) [4] is a programming paradigm developed in the field of nonmonotonic reasoning and logic programming. In this section we overview the language of ASP, and we recall a methodology for solving complex problems with ASP. The reader is referred to [23] for a more detailed introduction.

Syntax. The syntax of ASP is similar to the one of Prolog. Variables are strings starting with uppercase letter and constants are non-negative integers or strings starting with lowercase letters. A *term* is either a variable or a constant. A *standard atom* is an expression $p(t_1, \ldots, t_n)$, where p is a *predicate* of arity n and $t_1, \ldots, t_n$ are terms. An atom $p(t_1, \ldots, t_n)$ is ground if $t_1, \ldots, t_n$ are constants. A *ground set* is a set of pairs of the form $\langle consts : conj \rangle$, where $consts$ is a list of constants and $conj$ is a conjunction of ground standard atoms. A *symbolic set* is a set specified syntactically as $\{Terms_1 : Conj_1; \cdots ; Terms_t : Conj_t\}$, where $t > 0$, and for all $i \in [1, t]$, each $Terms_i$ is a list of terms such that $|Terms_i| = k > 0$, and each $Conj_i$ is a conjunction of standard atoms. A *set term* is either a symbolic set or a ground set. Intuitively, a set term $\{X : a(X, c), p(X); Y : b(Y, m)\}$ stands for the union of two sets: The first one contains the X-values making the conjunction $a(X, c), p(X)$ true, and the second one contains the Y-values making the conjunction $b(Y, m)$ true. An *aggregate function* [24] is of the form $f(S)$, where S is a set term, and f is an *aggregate function symbol*. Basically, aggregate functions map multisets of constants to a constant. The most common functions implemented in ASP systems are the following: #min, minimal term, undefined for the empty set; #max, maximal term, undefined for the empty set; #count, number of terms; #sum, sum of integers. An *aggregate atom* is of the form $f(S) \prec T$, where $f(S)$ is an aggregate function, $\prec \in \{<, \leq, >, \geq, =, \neq\}$ is a comparison operator, and T is a term called guard. An aggregate atom $f(S) \prec T$ is ground if T is a constant and S is a ground set. An *atom* is either a standard atom or an aggregate atom. A *rule r* is of the form:

$$a_1 \mid \ldots \mid a_n \leftarrow b_1, \ldots, b_k, \text{ not } b_{k+1}, \ldots, \text{ not } b_m.$$

where $a_1, \ldots, a_n$ are standard atoms, $b_1, \ldots, b_k$ are atoms, $b_{k+1}, \ldots, b_m$ are standard atoms, and $n, k, m \geq 0$. A literal is either a standard atom a or its negation not a. The disjunction $a_1 | \ldots | a_n$ is the *head* of r, while the conjunction $b_1, \ldots, b_k, \text{not } b_{k+1}, \ldots, \text{not } b_m$ is its *body*. A rule is a *fact* if its body is empty ($\leftarrow$ is omitted), whereas it is a *constraint* if its head is empty. A variable appearing uniquely in set terms of a rule r is said to be *local* in r, otherwise it is *global* in r. An ASP program is a set of *safe* rules. A rule r is *safe* if both the following conditions hold: *(i)* for each global variable X of r there is a positive standard atom ℓ in the body of r such that X appears in ℓ; *(ii)* each local variable of r appearing in a symbolic set $\{Terms : Conj\}$ also appears in $Conj$.

A *weak constraint* [25] ω is of the form:

$$\leftsquigarrow b_1, \ldots, b_k, \text{ not } b_{k+1}, \ldots, \text{ not } b_m. \ [w@l]$$

where w and l are the weight and level of ω. (Intuitively, $[w@l]$ is read "as weight w at level l", where weight is the "cost" of violating the condition in the body of w, whereas levels can be specified for defining a priority among preference criteria). An ASP program with weak constraints is $\Pi = \langle P, W \rangle$, where P is a program and W is a set of weak constraints. A standard atom, a literal, a rule, a program or a weak constraint is *ground* if no variable appears in it.

Semantics. Let P be an ASP program. The *Herbrand universe* U_P and the *Herbrand base* B_P of P are defined as usual [23]. The ground program G_P is the set of all the ground instances of rules of P obtained by substituting variables with constants from U_P.

An *interpretation I* for P is a subset I of B_P. A ground atom a is true w.r.t. I if $a \in I$, and false otherwise. Literal not a is true in I if a is false in I, and true otherwise. An aggregate atom is true w.r.t. I if the evaluation of its aggregate function (i.e., the result of the application of f on the multiset S) w.r.t. I satisfies the guard; otherwise, it is false. A ground rule r is *satisfied* by I if at least one atom in the head is true w.r.t. I whenever all conjuncts of the body of r are true w.r.t. I. A model is an interpretation that satisfies all the rules of a program. Given a ground program G_P and an interpretation I, the *reduct* [26] of G_P w.r.t.

I is the subset G_P^I of G_P obtained by deleting from G_P the rules in which a body literal is false w.r.t. I. An interpretation I for P is an *answer set* (or stable model [27]) for P if I is a minimal model (under subset inclusion) of G_P^I (i.e., I is a minimal model for G_P^I) [26]. A program having an answer set is called coherent, otherwise it is incoherent [28]. Given a program with weak constraints $\Pi = \langle P, W \rangle$, the semantics of Π extends from the basic case defined above. Thus, let $G_\Pi = \langle G_P, G_W \rangle$ be the instantiation of Π; a constraint $\omega \in G_W$ is violated by I if all the literals in ω are true w.r.t. I. An *optimum answer set* O for Π is an answer set of G_P that minimizes the sum of the weights of the violated weak constraints in a prioritized way.

Problem Solving in ASP. ASP can be used to encode problems in a declarative way usually employing a *Guess&Check&Optimize* programming methodology [29]. This method requires that a database of facts is used to specify an instance of the problem; a set of rules, called "guessing part", is used to define the search space; admissible solutions are then identified by other rules, called the "checking part", which impose some admissibility constraints; finally weak constraints are used to single out solutions that are optimal with respect to some criteria, the "optimize part". As an example, consider the Traveling Salesman Problem (TSP). Given a weighted graph $G = \langle N, A \rangle$, where N is the set of nodes and A is the set of arcs with integer labels, the problem is to find a path of minimum length containing all the nodes of G. TSP can be encoded as follows:

$$
\begin{aligned}
r_1: \quad & node(n). \qquad\qquad \forall\, \mathbf{n} \in N \\
r_2: \quad & arc(i, j, w). \qquad\quad \forall\, (\mathbf{i}, \mathbf{j}, \mathbf{w}) \in A \\
r_3: \quad & inPath(X, Y) \mid outPath(X, Y) \leftarrow arc(X, Y, W). \\
r_4: \quad & \leftarrow node(X),\ \#count\{I : inPath(I, X)\} \neq 1. \\
r_5: \quad & \leftarrow node(X),\ \#count\{O : inPath(X, O)\} \neq 1. \\
r_6: \quad & \leftarrow node(X),\ \mathbf{not}\ reached(X). \\
r_7: \quad & reached(X) \leftarrow inPath(M, X),\ \#min\{N : node(N)\} = M. \\
r_8: \quad & reached(X) \leftarrow reached(Y),\ inPath(Y, X). \\
r_9: \quad & \rightsquigarrow\ inPath(X, Y),\ arc(X, Y, W). \quad [W@1]
\end{aligned}
$$

The first two rules introduce suitable facts, representing the input graph G. Then, rule r_3, which can be read as "each arc may or may not be part of the path", guesses a solution (a set of *inPath* atoms). Rules $r_4 - r_6$ select admissible paths. In particular, rule r_4 (r_5) is satisfied if each node has exactly one incoming (resp. outgoing) arc in the solution. Moreover, rule r_6 ensures that the path traverses (say, reaches) all the nodes of G. Actually, this condition is obtained by checking that there exists a path reaching all the nodes of G and starting from the first node of N, say M. In particular, a node X is reached either if there is an arc connecting M to X (rule r_7), or if there is an arc connecting a reached node Y to X (rule r_8). Finally, solutions of minimal weight are selected by minimizing the cost W of arcs in the solution (rule r_9).

3   The Conference Paper Assignment Problem

Let $P = \{p_1, ..., p_s\}$ be a set of s papers and let $R = \{r_1, ..., r_t\}$ be a set of t reviewers. Each paper must be revised by ρ reviewers ($\rho \leq t$), and each reviewer must revise at most π papers ($\pi \leq s$). Moreover, to identify qualified reviewers, it is required that a reviewer r cannot review a paper p if there is a *conflict of interest* with some author of p. To formalize this property, it is introduced a *conflict function*, $\chi : R \times P \to \{0, 1\}$, which assigns to each pair (r, p) the value 1 in case of conflict of interest, and 0, otherwise. Let $\chi(R, p) = \{r \in R | \chi(r, p) = 1\}$ be the set of all reviewers with a conflict of interest with p. A tuple $\langle P, R, \rho, \pi, \chi \rangle$ is called a *Paper Revision System* (PRS).

Definition 1 (Allocation solution). *An allocation solution for a PRS* $\Sigma = \langle P, R, \rho, \pi, \chi \rangle$ *is a function* $\psi : P \to 2^R$ *such that,*

$$|\psi(p)| = \rho, \text{ for each } p \in P; \tag{1}$$

$$\bigcap_{j \in M} \psi(p_j) = \emptyset, \text{ for each } M \subset \{1, \ldots, s\}, \text{ with } |M| = \pi + 1; \tag{2}$$

$$\psi(p) \cap \chi(R, p) = \emptyset, \text{ for each } p \in P; \tag{3}$$

$$R = \bigcup_{p \in P} \psi(p). \tag{4}$$

A PRS admitting an allocation solution is called *consistent*. Intuitively, first condition claims that each paper is assigned to exactly ρ reviewers. Second one states that it is not possible that a reviewer $r \in R$ revises more than π papers. Indeed, more formally, in such a case there would exist at least $\pi + 1$ papers $p^1, ..., p^{\pi+1} \in P$, such that $r \in \psi(p^j)$, for each $j = 1, ..., \pi + 1$. Hence, $r \in \bigcap_{j \in \{1, ..., \pi+1\}} \psi(p^j)$, and so $\bigcap_{j \in \{1, ..., \pi+1\}} \psi(p^j) \neq \emptyset$. Note that the number of papers assigned to a reviewer $r \in R$ is given by

$$\nu(r) = |\{p | r \in \psi(p)\}|.$$

In particular, we proved the following result.

Proposition 1. *Let* ψ *be an allocation solution for a consistent PRS* $\Sigma = \langle P, R, \rho, \pi, \chi \rangle$. *Then* $\nu(r) \leq \pi$, *for each* $r \in R$.

Third condition claims that an allocation solution cannot admit conflictual assignments. In particular, if $\chi = 0$ is the zero constant function ($\chi(r, p) = 0$, for each $(r, p) \in R \times P$), then condition (3) is always satisfied, because $\chi(R, p) = \emptyset$, for each $p \in P$. A PRS $\Sigma = \langle P, R, \rho, \pi, \chi \rangle$, where $\chi = 0$, is called a *non-conflictual* PRS, and we denote it by $\Sigma_0 = \langle P, R, \rho, \pi \rangle$. Finally, fourth condition states that to each reviewer r at least a paper p is assigned, i.e., $r \in \psi(p)$, for some $p \in P$.

It is important to establish sufficient or necessary conditions to have a consistent PRS, avoiding useless computations.

Proposition 2. *If* $\Sigma = \langle P, R, \rho, \pi, \chi \rangle$ *is a consistent PRS, then* $|R \setminus \chi(R, p)| \geq \rho$, $\forall p \in P$.

We give the following characterization for consistent non-conflictual PRS.

Proposition 3. *A non-conflictual PRS $\Sigma_0 = \langle P, R, \rho, \pi \rangle$ is consistent iff $|P| \cdot \rho \leq |R| \cdot \pi$.*

Example 1. Consider a non-conflictual PRS $\Sigma_0 = \langle P, R, \rho, \pi \rangle$ such that $P = \{p_1, p_2, p_3\}$ is a set of 3 papers and $R = \{r_1, r_2, r_3, r_4, r_5\}$ is a set of 5 reviewer. Each paper must be revised by $\rho = 3$ reviewers, and a reviewer must revise at most $\pi = 2$ papers. Note that Σ is consistent, since $3 \cdot \rho = 9 < 5 \cdot \pi = 10$. An allocation solution is given by $\psi(p_1) = \{r_1, r_2, r_3\}$, $\psi(p_2) = \{r_1, r_4, r_5\}$, $\psi(p_3) = \{r_2, r_3, r_4\}$.

In general, in a conference paper assignment, it is preferable that each reviewer has "more or less" the same number of papers of each other reviewer. Now, we introduce a notion of distance from a desiderata number of papers to formalize this request.

Definition 2 (Distance). *Given a consistent PRS $\Sigma = \langle P, R, \rho, \pi, \chi \rangle$, an allocation solution ψ, a reviewer $r \in R$, and a desiderata number of papers D, we define the distance of r from D as $\delta_D(r) = |D - \nu(r)|$, and the distance of R from D as $\delta_D(R) = \sum_{r \in R} \delta_D(r)$.*

Example 2. Consider the PRS Σ_0 and the allocation solution ψ of Example 1, and a desiderata number of papers $D = 1$. Therefore $\delta_1(r_i) = |1 - \nu(r_i)| = 1$, for $i = 1, 2, 3, 4$, and $\delta_1(r_5) = |1 - \nu(r_5)| = 0$. Hence, the distance of R from D is $\delta_1(R) = 4$.

Definition 3 (Minimal Allocation Solution). *Let $\Sigma = \langle P, R, \rho, \pi, \chi \rangle$ be a PRS, and let D be a desiderata number of papers for each reviewer. An allocation solution ψ for Σ is called minimal, if the distance of R from D is minimized.*

Another main feature of conference paper assignment is the possibility given to each reviewer of bidding some papers from the most desirable to the least desired. To this end, a *preference function* ϕ_r from P to a finite set $N = \{0, 1, ..., n\}$, assigning a preference value to each paper, is associated to each reviewer $r \in R$.

Definition 4 (Satisfaction degree). *Given an allocation solution ψ for a consistent PRS $\Sigma = \langle P, R, \rho, \pi, \chi \rangle$, and a preference function ϕ_r, for each $r \in R$, we define the satisfaction degree of ψ for Σ as the number*

$$d(\psi, \Sigma) = \sum_{p \in P} \sum_{r \in \psi(p)} \phi_r(p).$$

Example 3. Consider again the PRS Σ_0 and the allocation solution ψ of Example 1. Let $N = \{0, 1\}$ be a boolean set of preferences. Hence, a reviewer can just specify if a paper is desired (value 1) or not (value 0). Suppose that reviewers r_1 and r_2 desire paper p_3; reviewer r_3 desires papers p_1 and p_2; reviewer r_4 desires paper p_2; and reviewer r_5 desires paper p_1. Therefore, we have the following preference functions $\phi_{r_1}(p_3) = 1$, $\phi_{r_2}(p_3) = 1$, $\phi_{r_3}(p_1) = \phi_{r_3}(p_2) = 1$,

$\phi_{r_4}(p_2) = 1$, $\phi_{r_5}(p_1) = 1$, and in all other cases the value is zero. The satisfaction degree of ψ for Σ is $d(\psi, \Sigma) = \phi_{r_1}(p_1) + \phi_{r_2}(p_1) + \phi_{r_3}(p_1) + \phi_{r_1}(p_2) + \phi_{r_4}(p_2) + \phi_{r_5}(p_2) + \phi_{r_2}(p_3) + \phi_{r_3}(p_3) + \phi_{r_4}(p_3) = 0+0+1+0+1+0+1+0+0 = 3$. Note that there exist others allocation solutions whose satisfaction degree is greater that this. Moreover, for this PRS, it is even possible to obtain the maximum satisfaction degree, that is 6, considering, for instance, $\psi'(p_1) = \{r_1, r_3, r_5\}$, $\psi'(p_2) = \{r_2, r_3, r_4\}$, $\psi'(p_3) = \{r_1, r_2, r_5\}$.

Definition 5 (Maximal Satisfying Allocation Solution). *Let* $\Sigma = \langle P, R, \rho, \pi, \chi \rangle$ *be a PRS, and let* ϕ_r *be a preference function, for each* $r \in R$. *An allocation solution* ψ *for* Σ *is called maximal satisfying, if the satisfaction degree of* ψ *for* Σ *is maximized.*

In the next, we consider the following formulations of CPAP:

1. Given a PRS Σ, a desiderata number of papers for each reviewer, and a preference function for each reviewer, finding among the minimal allocation solutions for Σ the one that is maximal satisfying.
2. Given a PRS Σ, a desiderata number of papers for each reviewer, and a preference function for each reviewer, finding among the maximal satisfying allocation solutions for Σ the one that is minimal.

4   Encoding CPAP in ASP

This section illustrates the ASP program which solves the Conference Paper Assignment problem specified in the previous section. First, the input data is described (Sect. 4.1), then the ASP encoding is presented (Sect. 4.2).

4.1   Data Model

The input is specified by means of factual instances of the following predicates:

- Instances of the predicate *paper(id)* represent the information about papers, where *id* represents a numerical identifier of a specific paper.
- Instances of the predicate *reviewer(id)* represent the information about reviewers, where *id* represents a numerical identifier of a specific reviewer.
- Instances of the predicate *score(id_reviewer, id_paper, score)* represent the information about the preference of reviewers for papers, where *id_reviewer* is the identifier of the reviewer, *id_paper* is the identifier of the paper, and *score* represents a numerical preference ($0 \le score \le 4$) assigned by the reviewer to the paper, where a lower score is associated with a higher confidence.
- Instances of the predicate *conflict(id_reviewer, id_paper)* represent a conflict of the reviewer with the paper.
- The only instance of the predicate *reviewersToPaper(ρ)* represents the number of reviewers that must be assigned to each paper.
- The only instance of the predicate *maxPaperPerReviewer(π)* represents the maximum number of papers that can be assigned to each reviewer.
- The only instance of the predicate *desiderata(d)* represents the number of papers that organizers are willing to assign to each reviewer.

4.2 ASP Encoding

In this section we describe the ASP rules used for solving the conference paper assignment problem. We follow the *Guess&Check&Optimize* programming methodology [29]. In particular, the following rule guesses the reviewers to assign to each paper:

$$assign(R,P)| \; nassign(R,P) \leftarrow \; paper(P), reviewer(R), \text{not } conflict(R,P). \tag{5}$$

The guess is limited to the reviewers that are not in conflict with the specific paper.

Each paper must be assigned to exactly N reviewers, thus all assignments violating this requirement are filtered out by the following constraint:

$$\leftarrow \; paper(P), \; \#count\{R : assign(R,P)\} \neq N. \tag{6}$$

Then, assignments exceeding the maximum number of paper assigned to each reviewer are filtered out by the following constraint:

$$\leftarrow \; reviewer(R), \; maxPaperPerReviewer(M), \; \#count\{P : assign(R,P)\} > M. \tag{7}$$

Moreover, each reviewer must be assigned at least to one paper:

$$workload(R,N) \leftarrow \; \#count\{P : assign(R,P)\} = N, reviewer(R),$$
$$maxPaperPerReviewer(M), N \leq M.$$
$$\leftarrow reviewer(R), \; workload(R,N), \; N < 1. \tag{8}$$

The predicate *workload(reviewer, number)* stores the association between a reviewer and the number of papers assigned to him/her.

Theoretical Improvements. In the following some constraints exploiting the theoretical results obtained in Sect. 3 are given. Since each paper must be assigned to exactly ρ reviewers, if the number of reviewers with no conflicts for a paper is less than ρ then a solution cannot exist (see Proposition 2). This is modeled by the following constraint:

$$\leftarrow paper(P_1), \; reviewersToPaper(N), \; \#count\{R_1 : reviewer(R_1)\} = R,$$
$$\#count\{R_1 : conflict(R_1,P_1)\} = C, \; R - C < N. \tag{9}$$

Moreover, the results presented in Proposition 3 are exploited by adding:

$$\leftarrow reviewersToPaper(N), maxPaperPerReviewer(M),$$
$$\#count\{R_1 : reviewer(R_1)\} = R, \#count\{P_1 : paper(P_1)\} = P, \tag{10}$$
$$P * N > R * M.$$

Intuitively, if P (number of papers) times N (number of reviewers per paper) exceeds R (number of reviewers) times M (maximum number of papers assigned to a reviewer) a solution cannot exist.

Optimization Requirements. The *satisfaction degree* and the *minimal allocation* requirements are obtained in our encoding by means of two weak constraints, where the numerical values ℓ_p and ℓ_w represent their levels; an order on the preferences can be later on specified by properly assigning a value to those levels.

Concerning the satisfaction degree of reviewers, the assignment of a reviewer to a paper is associated with a cost depending on the preference assigned from the reviewer to the paper. The maximum preference for a paper, i.e. a score equal to zero, is associated with no cost. Thus, the minimization of the cost (i.e. the maximization of satisfaction degree) is obtained by means of the following weak constraint:

$$\rightsquigarrow reviewer(R),\; assign(R, P),\; score(R, P, S).\; [S@\ell_p] \qquad (11)$$

Finally, the minimization of the distance between the desiderata number of papers to be assigned to each reviewer and the number of papers assigned by the solution is obtained by means of the following weak constraint:

$$\rightsquigarrow reviewer(R),\; workload(R, N),\; desiderata(D),\; V = |D - N|.\; [V@\ell_w] \qquad (12)$$

Intuitively, for each reviewer the distance is computed as the difference between the number of assigned papers to him/her and the desiderata number of papers. Then, a greater distance corresponds to a greater cost associated to the solution.

5   Experiments

In our experiments we considered a set of four real events held in the recent years, whose names are omitted for protecting our sources. For each event, we considered $\pi = 4$ and the desiderata number of papers to be assigned to each reviewer equal to 4. *Event 1* was composed of 31 papers, 46 reviewers, and $\rho = 4$; *event 2* was composed of 59 papers, 55 reviewers, and $\rho = 3$; *event 3* was composed of 16 papers, 31 reviewers, and $\rho = 4$; *event 4* was composed of 15 papers, 30 reviewers, $\rho = 4$. Concerning the preferences we considered two settings for the levels ℓ_p and ℓ_w of (11) and (12), i.e. $\ell_w > \ell_p > 0$ and $\ell_p > \ell_w > 0$ corresponding to formulations 1. and 2. of CPAP, respectively.

We executed the ASP solvers CLASP [8] and WASP [30]. The former has been configured with the model-guided algorithm called *bb* [8], which basically searches for an answer set so to initialize an upper bound of the optimum cost, and new answer sets of improved cost are iteratively searched until the optimum cost is found. WASP has been configured with the core-guided algorithm called *one* [5], which searches for an answer set satisfying all weak constraints. If there is no answer set of this kind, an unsatisfiable core is identified, i.e. a subset of the weak constraints that cannot be jointly satisfied, representing a lower bound of the optimum cost. In addition, WASP is able to produce upper bounds of the optimum cost during the search of an unsatisfiable core. The experiments were run on an Intel Xeon 2.4 GHz with 16 GB of RAM, and time and memory were limited to 60 min and 15 GB, respectively.

Table 1. Workload of reviewers computed by CLASP when $\ell_w > \ell_p > 0$.

	Time (s)	Rev. with 1 paper	Rev. with 2 papers	Rev. with 3 papers	Rev. with 4 papers
event1	≤ 1	7	4	31	4
	≤ 2	6	5	32	3
	≤ 10	3	8	35	0
	≤ 60	3	8	35	0
	≤ 3600	3	8	35	0
event2	≤ 1	8	1	17	29
	≤ 2	4	3	25	23
	≤ 10	0	0	43	12
	≤ 60	0	0	43	12
	≤ 3600	0	0	43	12
event3	≤ 1	13	3	15	0
	≤ 2	13	3	15	0
	≤ 10	9	11	11	0
	≤ 60	9	11	11	0
	≤ 3600	9	11	11	0
event4	≤ 1	11	8	11	0
	≤ 2	12	6	12	0
	≤ 10	12	6	12	0
	≤ 60	11	8	11	0
	≤ 3600	13	4	13	0

Formulation 1 ($\ell_w > \ell_p > 0$). An overview of the obtained results is given in Table 1. For each event the number of reviewers receiving one, two, three or four papers within different time limits is reported. Concerning the first event the 76 % of reviewers received exactly the desiderata number of papers, i.e. 3. The remaining 17 % and 7 % received 2 and 1 paper, respectively. According to this solution no reviewer has to review more than 3 papers. Even better results are obtained for the second event, where 78 % of reviewers received 3 papers, and the remaining 22 % received 4 papers. None of the reviewers received less than 3 papers. Concerning events 3 and 4, solutions found by CLASP assign 3 papers only to few reviewers. Similar results are found by WASP where 29 out of 31 and 30 out of 30 reviewers are associated to exactly 2 papers, respectively. This might be explained by the few number of papers w.r.t. the number of reviewers, which makes it difficult to assign the desiderata number of papers to each reviewer.

Formulation 2 ($\ell_p > \ell_w > 0$). An overview of the obtained results is given in Table 2, where for each event lower and upper bounds found by CLASP and WASP are reported. The analysis of lower and upper bounds allows us to estimate the *error* of the best found solution, reported in the last column of the table and computed as follows:

$$\epsilon(ub, lb) := \begin{cases} \frac{ub - lb}{lb} & \text{if } ub \neq \infty \text{ and } lb \neq 0; \\ \infty & \text{if } ub = \infty, \text{ or both } ub \neq 0 \text{ and } lb = 0; \\ 0 & \text{if } ub = lb = 0. \end{cases}$$

Table 2. Lower and upper bounds computed by CLASP and WASP when $\ell_p > \ell_w > 0$.

Time (s)		CLASP (ub)		WASP (ub)		WASP (lb)		ε		Optimum	
		ℓ_p	ℓ_w	ℓ_p	ℓ_w	ℓ_p	ℓ_w	ℓ_p	ℓ_w	ℓ_p	ℓ_w
event1	≤ 1	56	28	∞	∞	18	0	2.11	∞	-	-
	≤ 2	46	26	**25**	18	**25**	10	**0**	0.80	25	-
	≤ 10	**25**	16	**25**	18	**25**	10	**0**	0.60	25	-
	≤ 60	**25**	14	**25**	18	**25**	10	**0**	0.40	25	-
	≤ 3600	**25**	14	**25**	18	**25**	10	**0**	0.40	25	-
event2	≤ 1	54	48	∞	∞	8	0	5.75	∞	-	-
	≤ 2	52	46	∞	∞	8	0	5.50	∞	-	-
	≤ 10	45	30	∞	∞	8	0	4.62	∞	-	-
	≤ 60	28	24	37	56	14	0	1	∞	-	-
	≤ 3600	17	28	37	56	14	0	0.21	∞	-	-
event3	≤ 1	25	31	**18**	29	**18**	18	**0**	0.61	18	-
	≤ 2	20	29	**18**	29	**18**	20	**0**	0.45	18	-
	≤ 10	**18**	29	**18**	29	**18**	20	**0**	0.45	18	-
	≤ 60	**18**	29	**18**	29	**18**	20	**0**	0.45	18	-
	≤ 3600	**18**	29	**18**	29	**18**	20	**0**	0.45	18	-
event4	≤ 1	24	36	**18**	30	**18**	26	**0**	0.15	18	-
	≤ 2	20	34	**18**	30	**18**	26	**0**	0.15	18	-
	≤ 10	**18**	32	**18**	30	**18**	26	**0**	0.15	18	-
	≤ 60	**18**	30	**18**	30	**18**	26	**0**	0.15	18	-
	≤ 3600	**18**	30	**18**	30	**18**	27	**0**	0.11	18	-

As first observation, WASP is able to find the solution maximizing the satisfaction of reviewers within 2 s for all the events but *event2*. Concerning *event2*, the best result is obtained by CLASP that is able to provide a solution with an error equal to 1 within 60 s. Results are far better if we look at the solution found within 3600 s, where CLASP provides a solution with an error equal to 0.21. For the sake of completeness, we also mention that intermediate solutions were found within 600 and 900 s with errors equal to 0.36 and 0.22, respectively. Concerning the minimization of the distance, the solution produced by CLASP within 60 s has an error less than 0.5 for all the events but *event2*.

6   Related Work

The problem of conference paper assignment has been attracting the interest of researchers in the last two decades [1,2]. Researchers from different areas have focused on different aspects of CPAP [31–35]. Data mining techniques have been applied for inferring preferences and desiderata of reviewers; operational research tools have been used to compute assignments; in Economy the CPAP has been related to the allocation of indivisible goods to a set of agents. The solving methods in the literature range from dedicated algorithms, to genetic algorithms, integer programming-based methods, and approximation algorithms. All these are different from our approach in terms of modeling language, whereas our formulation of the problem shares often the constraints on assignments and in some cases the optimization criteria with some of these works. Since an exhaustive description of the state of the art can be found on existing survey papers [1,2], in the following we locate our contribution by comparing some of the recent papers on CPAP.

In [36] a fuzzy mathematical model for the assignment of experts to project proposal is investigated, which is a problem similar to CPAP. The method imposes assignment constraints that are similar to the ones considered in this paper, but it does not consider conflicts and focuses on a matching criteria that is defined using linguistic variables denoting the expertise of experts with respect to proposals. The resultant fuzzy model was solved with the selected fuzzy ranking methods. The approach of [36] cannot be directly compared to ours, since the modeling itself would not fit the standard ASP framework that works on problems formulations where all the information is crisp.

In [37] authors considered the problem of determining a prediction of reviewer's preference, and provided some empirical evidence that an accurate identification of preferences can improve satisfaction of reviewers. Thus, the goal of [37] is to improve the modeling of reviewer preferences. The problem of determining reviewer relevance by automatically identifying reviewer profiles was also the subject of research [38]. These studies could be employed for designing better models of reviewer's preferences that could be used as input of a method that computes the optimal assignment as the one considered is in this paper.

A formulation of reviewer preferences based on a combination of information about topics and direct preference of reviewers is considered in [39]. Here a matching criteria based on a matching degree function is proposed. The criterion of [39] can be modeled in ASP, and can be considered as one possible extension of our current model. Topic coverage of the paper reviewer assignment is the main optimization employed in [40], where also an algorithm for computing an approximation of the optimal solution is proposed in presence of conflicts of interest.

An alternative formulation of the CPAP has been proposed in [41], where a group-to-group reviewer assignment problem is defined. The idea is that manuscripts and reviewers are divided into groups, with groups of reviewers assigned to groups of manuscripts. A two-phase stochastic-biased greedy algorithm is then proposed to solve the problem. This variant of the problem is less similar to the traditional CPAP formulation that we consider in this paper, so a direct comparison is not feasible.

The approach that is most related ours is [42] where ASP has been also employed, but the CPAP is studied as an example of *reconfiguration* problem. Thus, the focus is on *updating a given solution* rather than in the computation of a new assignment.

7   Conclusion

The main goal of this paper was to provide an assessment of the applicability of ASP to the CPAP. We first provided a formal description of the problem, which combines the most common constraints and optimization criteria considered in the literature, and we outlined some theoretical properties of the problem. Then, we provided a disjunctive logic program modeling the CPAP. The ASP encoding is natural and intuitive, in the sense that it was obtained by applying the standard modeling methodology, and it is easy to understand. We also modeled in a natural way the theoretical conditions ensuring the absence of solutions, so that ASP solvers can easily recognize unsatisfiable instances. Finally, the performance of our ASP-based approach was studied in an experiment.

We conclude that ASP is suitable from the perspective of modeling, since we obtained a natural ASP encoding of the problem. Moreover, the results of an experiment outline that an ASP-based approach can perform well on real-world data.

Future work will focus on extending the framework with additional information (e.g., topics, coauthor information, etc.) and with additional preference criteria (e.g., different models of fairness, coauthors distance, etc.). Indeed, the flexibility of ASP as a modeling language should allow us to enrich current model or encode some of its variants. We also planned to extend the experimental analysis by considering more data and more computation methods. Moreover, we are investigating the application of ASP to other hard problems [43,44].

Acknowledgments. This work was partially supported by MIUR under PON project "Ba2Know (Business Analytics to Know) Service Innovation - LAB", N. PON03PE_00 001_1, and by MISE under project "PIUCultura (Paradigmi Innovativi per l'Utilizzo della Cultura)", N. F/020016/01-02/X27.

References

1. Goldsmith, J., Sloan, R.H.: The AI conference paper assignment problem. In: Proceedings of thr AAAI Workshop on Preference Handling for Artificial Intelligence, pp. 53–57 (2007)
2. Wang, F., Chen, B., Miao, Z.: A Survey on Reviewer Assignment Problem. In: Nguyen, N.T., Borzemski, L., Grzech, A., Ali, M. (eds.) IEA/AIE 2008. LNCS (LNAI), vol. 5027, pp. 718–727. Springer, Heidelberg (2008). doi:10.1007/978-3-540-69052-8_75
3. Manlove, D., Irving, R.W., Iwama, K., Miyazaki, S., Morita, Y.: Hard variants of stable marriage. Theor. Comput. Sci. **276**(1–2), 261–279 (2002)
4. Brewka, G., Eiter, T., Truszczynski, M.: Answer set programming at a glance. Commun. ACM **54**(12), 92–103 (2011)
5. Alviano, M., Dodaro, C., Ricca, F.: A MaxSAT algorithm using cardinality constraints of bounded size. In: IJCAI, pp. 2677–2683. AAAI Press (2015)
6. Alviano, M., Faber, W., Leone, N., Perri, S., Pfeifer, G., Terracina, G.: The disjunctive datalog system DLV. In: Moor, O., Gottlob, G., Furche, T., Sellers, A. (eds.) Datalog 2.0 2010. LNCS, vol. 6702, pp. 282–301. Springer, Heidelberg (2011). doi:10.1007/978-3-642-24206-9_17

7. Calimeri, F., Gebser, M., Maratea, M., Ricca, F.: Design and results of the fifth answer set programming competition. Artif. Intell. **231**, 151–181 (2016)
8. Gebser, M., Kaminski, R., Kaufmann, B., Romero, J., Schaub, T.: Progress in *clasp* series 3. In: Calimeri, F., Ianni, G., Truszczynski, M. (eds.) LPNMR 2015. LNCS (LNAI), vol. 9345, pp. 368–383. Springer, Heidelberg (2015). doi:10.1007/978-3-319-23264-5_31
9. Giunchiglia, E., Leone, N., Maratea, M.: On the relation among answer set solvers. Ann. Math. Artif. Intell. **53**(1–4), 169–204 (2008)
10. Giunchiglia, E., Maratea, M.: On the relation between answer set and SAT procedures (or, Between CMODELS and SMODELS). In: Gabbrielli, M., Gupta, G. (eds.) ICLP 2005. LNCS, vol. 3668, pp. 37–51. Springer, Heidelberg (2005). doi:10.1007/11562931_6
11. Maratea, M., Pulina, L., Ricca, F.: A multi-engine approach to answer-set programming. TPLP **14**(6), 841–868 (2014)
12. Maratea, M., Ricca, F., Faber, W., Leone, N.: Look-back techniques and heuristics in DLV: implementation, evaluation, and comparison to QBF solvers. J. Algorithms **63**(1–3), 70–89 (2008)
13. Balduccini, M., Gelfond, M., Watson, R., Nogueira, M.: The USA-advisor: a case study in answer set planning. In: Eiter, T., Faber, W., Truszczyński, M. (eds.) LPNMR 2001. LNCS (LNAI), vol. 2173, pp. 439–442. Springer, Heidelberg (2001). doi:10.1007/3-540-45402-0_39
14. Gaggl, S.A., Manthey, N., Ronca, A., Wallner, J.P., Woltran, S.: Improved answer-set programming encodings for abstract argumentation. TPLP **15**(4–5), 434–448 (2015)
15. Campeotto, F., Dovier, A., Pontelli, E.: A declarative concurrent system for protein structure prediction on GPU. J. Exp. Theor. Artif. Intell. **27**(5), 503–541 (2015)
16. Erdem, E., Öztok, U.: Generating explanations for biomedical queries. TPLP **15**(1), 35–78 (2015)
17. Fionda, V., Palopoli, L., Panni, S., Rombo, S.E.: A technique to search for functional similarities in protein-protein interaction networks. IJDMB **3**(4), 431–453 (2009)
18. Manna, M., Ricca, F., Terracina, G.: Taming primary key violations to query large inconsistent data via ASP. TPLP **15**(4–5), 696–710 (2015)
19. Marileo, M.C., Bertossi, L.E.: The consistency extractor system: Answer set programs for consistent query answering in databases. Data Knowl. Eng. **69**(6), 545–572 (2010)
20. Amendola, G., Greco, G., Leone, N., Veltri, P.: Modeling and reasoning about NTU games via answer set programming. In: IJCAI, pp. 38–45. IJCAI/AAAI Press (2016)
21. Grasso, G., Leone, N., Manna, M., Ricca, F.: ASP at work: spin-off and applications of the DLV system. In: Balduccini, M., Son, T.C. (eds.) Logic Programming, Knowledge Representation, and Nonmonotonic Reasoning. LNCS (LNAI), vol. 6565, pp. 432–451. Springer, Heidelberg (2011). doi:10.1007/978-3-642-20832-4_27
22. Dodaro, C., Gasteiger, P., Leone, N., Musitsch, B., Ricca, F., Shchekotykhin, K.: Combining answer set programming and domain heuristics for solving hard industrial problems. TPLP **16**(5-6) (2016, to appear)
23. Baral, C.: Knowledge Representation, Reasoning and Declarative Problem Solving. Cambridge University Press, Cambridge (2003)
24. Alviano, M., Faber, W., Gebser, M.: Rewriting recursive aggregates in answer set programming: back to monotonicity. TPLP **15**(4–5), 559–573 (2015)

25. Buccafurri, F., Leone, N., Rullo, P.: Enhancing disjunctive datalog by constraints. IEEE Trans. Knowl. Data Eng. **12**(5), 845–860 (2000)
26. Faber, W., Pfeifer, G., Leone, N.: Semantics and complexity of recursive aggregates in answer set programming. Artif. Intell. **175**(1), 278–298 (2011)
27. Gelfond, M., Lifschitz, V.: Classical negation in logic programs and disjunctive databases. New Gener. Comput. **9**(3/4), 365–386 (1991)
28. Amendola, G., Eiter, T., Fink, M., Leone, N., Moura, J.: Semi-equilibrium models for paracoherent answer set programs. Artif. Intell. **234**, 219–271 (2016)
29. Leone, N., Pfeifer, G., Faber, W., Eiter, T., Gottlob, G., Perri, S., Scarcello, F.: The DLV system for knowledge representation and reasoning. ACM Trans. Comput. Log. **7**(3), 499–562 (2006)
30. Alviano, M., Dodaro, C., Leone, N., Ricca, F.: Advances in WASP. In: Calimeri, F., Ianni, G., Truszczynski, M. (eds.) LPNMR 2015. LNCS (LNAI), vol. 9345, pp. 40–54. Springer, Heidelberg (2015). doi:10.1007/978-3-319-23264-5_5
31. Bogomolnaia, A., Moulin, H.: A new solution to the random assignment problem. J. Econ. Theory **100**(2), 295–328 (2001)
32. Demange, G., Alkan, A., Gale, D.: Fair allocation of indivisible goods and money and criteria of justice. Econometrica **59**(4), 1023–1039 (1991)
33. Hartvigsen, D., Wei, J.C., Czuchlewski, R.: The conference paper-reviewer assignment problem. Decis. Sci. **30**(3), 865–876 (1999)
34. Janak, S.L., Taylor, M.S., Floudas, C.A., Burka, M., Mountziaris, T.J.: Novel and effective integer optimization approach for the NSF panel-assignment problem: a multiresource and preference-constrained generalized assignment problem. Ind. Eng. Chem. Res. **45**(1), 258–265 (2006)
35. Svensson, L.G.: Strategy-proof allocation of indivisible goods. Soc. Choice Welf. **16**(4), 557–567 (1999)
36. Das, G.S., Göçken, T.: A fuzzy approach for the reviewer assignment problem. Comput. Ind. Eng. **72**, 50–57 (2014)
37. Conry, D., Koren, Y., Ramakrishnan, N.: Recommender systems for the conference paper assignment problem. In: RecSys, pp. 357–360. ACM (2009)
38. Mimno, D.M., McCallum, A.: Expertise modeling for matching papers with reviewers. In: ACM SIGKDD, pp. 500–509. ACM (2007)
39. Li, X., Watanabe, T.: Automatic Paper-to-reviewer Assignment, based on the Matching Degree of the Reviewers. In: KES, vol. 22. Procedia Computer Science, pp. 633–642. Elsevier (2013)
40. Long, C., Wong, R.C., Peng, Y., Ye, L.: On good and fair paper-reviewer assignment. In: ICDM, pp. 1145–1150. IEEE Computer Society (2013)
41. Wang, F., Zhou, S., Shi, N.: Group-to-group reviewer assignment problem. Comput. OR **40**(5), 1351–1362 (2013)
42. Ryabokon, A., Polleres, A., Friedrich, G., Falkner, A.A., Haselböck, A., Schreiner, H.: (Re)Configuration using web data: a case study on the reviewer assignment problem. In: Krötzsch, M., Straccia, U. (eds.) RR 2012. LNCS, vol. 7497, pp. 258–261. Springer, Heidelberg (2012). doi:10.1007/978-3-642-33203-6_28
43. Fionda, V., Greco, G.: The complexity of mixed multi-unit combinatorial auctions: tractability under structural and qualitative restrictions. Artif. Intell. **196**, 1–25 (2013)
44. Fionda, V., Pirrò, G.: Querying graphs with preferences. In: CIKM, pp. 929–938. ACM (2013)

A Subdivision Approach to the Solution of Polynomial Constraints over Finite Domains Using the Modified Bernstein Form

Federico Bergenti[✉], Stefania Monica, and Gianfranco Rossi

Dipartimento di Matematica e Informatica, Università degli Studi di Parma,
Parco Area delle Scienze 53/A, 43124 Parma, Italy
{federico.bergenti,stefania.monica,gianfranco.rossi}@unipr.it

Abstract. This paper discusses an algorithm to solve polynomial constraints over finite domains, namely constraints which are expressed in terms of equalities, inequalities and disequalities of polynomials with integer coefficients whose variables are associated with finite domains. The proposed algorithm starts with a preliminary step intended to rewrite all constraints to a canonical form. Then, the modified Bernstein form of obtained polynomials is used to recursively restrict the domains of variables, which are assumed to be initially approximated by a bounding box. The proposed algorithm proceeds by subdivisions, and it ensures that each variable is eventually associated with the inclusion-maximal finite domain in which the set of constraints is satisfiable. If arbitrary precision integer arithmetic is available, no approximation is introduced in the solving process because the coefficients of the modified Bernstein form are integer numbers.

Keywords: Bernstein form · Finite domain constraints · Constraint logic programming

1 Introduction and Motivation

This paper proposes an algorithm to solve systems of polynomial constraints, namely constraints expressed as equalities, inequalities, and disequalities of polynomials. We assume that the coefficients of all polynomials are integer numbers, and we assume that each variable is associated with a finite domain, i.e., a finite set of integer numbers where the variable takes values. We call such constraints *polynomial constraints over finite domains* [1], and we acknowledge their importance because they are often used in many practical applications. A common problem related to polynomial constraints over finite domains is to reduce the domains of variables to make them coincident with inclusion-maximal sets containing only values for which constraints are satisfiable. *Finite Domain* (*FD*) techniques to effectively treat polynomial constraints already exist (see, e.g., [2]), but they often trade off efficiency with accuracy at reducing the domains of variables (see, e.g., [3]). Commonly used FD techniques opt for efficiency and

© Springer International Publishing AG 2016
G. Adorni et al. (Eds.): AI*IA 2016, LNAI 10037, pp. 179–191, 2016.
DOI: 10.1007/978-3-319-49130-1_14

they do not always succeed at completely reducing the domains of variables, especially when treating constraints involving non-linear polynomials [1]. As an example, let us consider the following constraint expressed using the syntax of SWI-Prolog [3,4]

X in -20..20, (X-2)^2 #>= 3.

This constraint involves only one variable which takes values in the finite domain $I = [-20..20]$. Simple algebraic manipulations show that the constraint is satisfied only for values in $I_X = [-20..0] \cup [4..20]$. However, when trying to solve the above constraint using the FD solver of SWI-Prolog (version 6.6.6), the obtained result is

X in -20..20, Y+2 #=X, Y in -22..-1\/1..18, Y^2 #= Z, Z in 5..sup

where Y and Z are two variables added by the solving process. As an example with more than one variable, let us consider the following non-linear constraint

X in -20..20, Y in -20..20, X+X*Y #>= 100

where both variables take values in the finite domain I. Such a constraint is satisfied for $I_X = [-20.. - 6] \cup [5..20]$ and $I_Y = [-20.. - 6] \cup [4..20]$. When trying to solve the constraint using SWI-Prolog (version 6.6.6), the obtained result is

X+W #= Z, X*Y#= W, Z in 100..sup

where W and Z are two variables added by the solving process.

In this paper we describe an algorithm to solve polynomial constraints over finite domains that guarantees to reduce the domain of each variable to the inclusion-maximal set containing only values for which constraints are satisfiable. In a preliminary step, all constraints are rewritten to a canonical form according to which the right-hand sides of all the constraints equal 0 and only one inequality symbol ($\geq$) is used. Then, a method based on the *Bernstein Form* (*BF*) (see, e.g., [5–7]) of polynomials is used to recursively reduce the domains of all variables. To be precise, we consider a variant of the BF of polynomials, namely the *Modified Bernstein Form* (*MBF*) [1,8]. This choice is due to the fact that the coefficients of the BF of a polynomial may not be integer numbers even if the considered polynomial has integer coefficients. At the opposite, the MBF of a polynomial with integer coefficients has integer coefficients and, therefore, the use of the MBF of polynomials allows avoiding the use of rational or real numbers. The MBF of a polynomial is useful to obtain information about its sign, which can be used to process polynomial constraints over finite domains, and to reduce the domains of the variables, as detailed in the rest of this paper. The implementation of the proposed algorithm is available for download (cmt.dmi.unipr.it/software/clppolyfd.zip).

The paper is organized as follows. In Sect. 2 the constraint language is introduced and the canonical form of all types of polynomial constraints is derived. In Sect. 3 the constraint solving algorithm based on the MBF of polynomials is presented and discussed. In Sect. 4 the details of the proposed algorithm are shown by means of illustrative examples. Finally, in Sect. 5 conclusions are drawn and ongoing and future research on the topic is briefly outlined.

2   The Proposed Approach

The interesting features of the proposed constraint solving algorithm are closely related to the specific form of considered constraints, namely polynomial constraints over finite domains. Such constraints are expressed as equalities, inequalities, and disequalities of polynomials with integer coefficients whose variables take values in finite sets of integer numbers. In this section, we describe the constraint language and we show how to derive the proposed canonical form of constraints.

2.1   The Constraint Language

The syntax of the considered language for polynomial constraints is defined by a signature Σ, namely a triple $\Sigma = < \mathcal{V}, \mathcal{F}, \Pi >$ where $\mathcal{V}$ is a denumerable set of variable symbols, $\mathcal{F}$ is a set which contains constant and function symbols, and Π is the finite set of constraint predicate symbols. The set $\mathcal{F}$ is defined as $\mathcal{F} = O \cup Z$, where $O = \{+, *\}$ is a set containing two function symbols used to represent binary operations over integer numbers, and $Z = \{0, 1, -1, 2, -2, \ldots\}$ is the denumerable set of constants representing integer numbers. The set Π of constraint predicate symbols is composed of six binary predicate symbols, namely $\Pi = \{=, \neq, <, \leq, >, \geq\}$. The constraints that we consider can be *primitive* or *non-primitive*. Primitive constraints are atomic predicates built using the symbols defined in signature Σ, while non-primitive constraints, normally simply called constraints, are defined as conjunctions of primitive constraints. The semantics of the language defined over signature Σ is trivial and it allows expressing systems of polynomial constraints with integer coefficients in terms of equalities, inequalities, and disequalities of polynomials.

Observe that parenthesized expressions and usual operator precedences can be added to the language defined by signature Σ. Moreover, syntactic abbreviations can be adopted to support other common function symbols and predicate symbols. More precisely, the following syntactic abbreviations can be used

$$v \text{ in } a..b \rightarrow (v + (-1) * a) * (v + (-1) * b) \leq 0 \tag{1}$$

$$v \text{ nin } a..b \rightarrow (v + (-1) * a) * (v + (-1) * b) > 0 \tag{2}$$

$$-t \rightarrow (-1) * t \tag{3}$$

$$t_1 - t_2 \rightarrow t_1 + (-1) * t_2 \tag{4}$$

$$t^0 \rightarrow 1 \tag{5}$$

$$t^n \rightarrow \underbrace{t * t * \cdots * t}_{n > 0}. \tag{6}$$

where v is a variable symbol, a and b are two integer constant symbols, and t, t_1, and t_2 are terms. While syntactic rules in (3)–(6) are straightforward, those expressed in (1) and (2) deserve a short explanation. Observe that the expression $(v + (-1) * a) * (v + (-1) * b)$ represents an upward parabola which annihilates in a and b. It is then easy to understand that values in the interval $[a..b]$ correspond to non-positive values of the parabola, thus leading to the constraint at the right-hand side of (1). Analogous considerations can be made to show that values not in the interval $[a..b]$ correspond to positive values of the parabola, as stated at the right-hand side of (2). We remark that the set of constraints expressible with the extended language introduced by parenthesized expressions, usual operator precedences, and syntactic abbreviations in (1)–(6) equals the set of constraints expressible with the language with signature Σ.

2.2 Constraints in Canonical Form

Any primitive constraint can be rewritten to a canonical form under the assumption that a total order on variable symbols is fixed. In order to define the canonical form of constraints, let us introduce the canonical form of terms. A term t is in the proposed canonical form if it is structured as a sum of terms $\{t_i\}_{i=1}^h$ (i.e., $t = t_1 + \ldots + t_h$) with the following structure

$$t_i = k * \underbrace{x_1 * x_1 * \cdots * x_1}_{n_1} * \underbrace{x_2 * x_2 * \cdots * x_2}_{n_2} * \cdots * \underbrace{x_m * x_m * \cdots * x_m}_{n_m} \qquad i \in [1..h]$$

where k is a constant symbol and $\{x_i\}_{i=1}^m$ are distinct variable symbols. The fixed total order of variable symbols $\{x_i\}_{i=1}^m$ induces a lexicographic order of product terms, which can be used to arrange terms $\{t_i\}_{i=1}^h$ in t. It is then possible to define a function on product terms, denoted as $deg(v, t)$, which returns n_v if variable symbol v is repeated n_v times in term t. Such a function is useful to write product terms in a compact form as

$$t_i = k * \underbrace{x_1 * x_1 * \cdots * x_1}_{deg(x_1, t_i)} * \underbrace{x_2 * x_2 * \cdots * x_2}_{deg(x_2, t_i)} * \cdots * \underbrace{x_m * x_m * \cdots * x_m}_{deg(x_m, t_i)}$$
$$= k * x_1 \hat{\ } deg(x_1, t_i) * x_2 \hat{\ } deg(x_2, t_i) * \cdots * x_m \hat{\ } deg(x_m, t_i) \qquad i \in [1..h].$$

The proposed canonical form of constraints relies on the total order of variable symbols and it allows any primitive constraint to be expressed as $t \geq 0$, where t is a term in canonical form. To illustrate how to rewrite a given primitive constraint to its canonical form, let us first note that if $f(\mathbf{x})$ and $g(\mathbf{x})$ are two polynomials with integer coefficients, where $\mathbf{x}$ is the vector of their variables and $\odot \in \{=, \neq, <, \leq, >, \geq\}$, then

$$f(\mathbf{x}) \odot g(\mathbf{x}) \tag{7}$$

can be written as

$$p(\mathbf{x}) \odot 0 \tag{8}$$

where $p(\mathbf{x}) = f(\mathbf{x}) - g(\mathbf{x})$, regardless of $\odot$. Since $p(\mathbf{x})$ is defined as the difference between two polynomials with integer coefficients, it is a polynomial with integer coefficients too. Observe that the canonical form of a constraint requires that the right-hand side equals 0, and also that the constraint predicate symbol is $\geq$. Hence, (8) may not be in canonical form since the constraint predicate symbol $\odot$ may differ from $\geq$. For this reason, if $\odot$ differs from $\geq$, a second step is necessary to write (8) in canonical form. To this purpose, the following lemma provides explicit rules that allow deriving the canonical form of (8) for all constraint predicate symbols.

Lemma 1. *The following equivalences hold for every polynomial $p(\mathbf{x})$ with integer coefficients and variables* $\mathbf{x}$

$$p(\mathbf{x}) \leq 0 \iff -p(\mathbf{x}) \geq 0 \tag{9}$$

$$p(\mathbf{x}) > 0 \iff p(\mathbf{x}) \geq 1 \iff p(\mathbf{x}) - 1 \geq 0 \tag{10}$$

$$p(\mathbf{x}) < 0 \iff p(\mathbf{x}) \leq -1 \iff -p(\mathbf{x}) - 1 \geq 0 \tag{11}$$

$$p(\mathbf{x}) = 0 \iff p^2(\mathbf{x}) \leq 0 \iff -p^2(\mathbf{x}) \geq 0 \tag{12}$$

$$p(\mathbf{x}) \neq 0 \iff p^2(\mathbf{x}) > 0 \iff p^2(\mathbf{x}) - 1 \geq 0 \tag{13}$$

Proof. The proof of all equivalences is trivial and it requires only ordinary algebraic manipulations.   □

Since the reformulation of constraints to their canonical forms can be seen as a preprocessing step, from now on we assume that all constraints are expressed in canonical form. This allows the constraint solving process to be simplified, as it can be reduced to the study of the sign of polynomials. In next section, a detailed description of the proposed constraint solving algorithm is presented.

3   The Constraint Solving Algorithm

As discussed in Sect. 1, we assume that the user provides not only a set of constraints but also a set of initial domains for each one of the $m \geq 1$ variables. More precisely, it is assumed that each variable is initially associated with an interval, namely

$$x_i \in [\underline{x}_i .. \overline{x}_i] \qquad i \in [1..m] \tag{14}$$

where $\{x_i\}_{i=1}^m$ are the variables involved in constraints, $\{\underline{x}_i\}_{i=1}^m$ are their lower bounds, and $\{\overline{x}_i\}_{i=1}^m$ are their upper bounds. Such bounds are given before starting the solving process and they identify a box in $\mathbb{Z}^m$ that can be used as an initial approximation of the domains of variables. The constraint solving algorithm considers all primitive constraints, which are assumed to be in canonical form, and it verifies if each constraint is consistent in the current box. If neither consistency nor inconsistency can be proved, then a variable in the constraint is chosen and its current domain is split, thus leading to two disjoint boxes, which are recursively processed. The recursive process continues until the set of constraints is either consistent or inconsistent in each box. Termination of the considered algorithm is always guaranteed because, in the worst case, the consistency or the inconsistency of a constraint can always be verified in a box formed by intervals that contain only one element. The performance of the algorithm depends significantly on the initial bounds of each variable, and on the details of the subdivision into disjoint boxes. A significant number of alternatives has been studied in the literature (see, e.g., [9,10]) to address this last problem. In the rest of this paper, we assume that variables are selected using a fixed total order and that once a variable x_i is selected, its domain $[\underline{x}_i..\overline{x}_i]$ is split at

$$s_i = \left\lfloor \frac{(\underline{x}_i + \overline{x}_i)}{2} \right\rfloor \tag{15}$$

into two disjoint intervals $[\underline{x}_i..s_i]$ and $[(s_i+1)..\overline{x}_i]$. This is the simplest choice and further research on this topic is ongoing to possibly improve the performance of the algorithm.

In order to test the satisfiability of a constraint in a box, for each primitive constraint expressed in canonical form $p(\mathbf{x}) \geq 0$, we assume that we can compute suitable lower bound l and upper bound u, with $l \leq u$, for the polynomial $p(\mathbf{x})$ in the box. Such values can be used to verify if the currently analysed constraint is consistent in the box as follows

1. If $u < 0$ the constraint is not consistent in the box;
2. If $l \geq 0$ the constraint is consistent in the box;
3. Otherwise the box should be split and analysed recursively.

It is worth noting that, since constraints are in canonical form, we are not interested in the actual values of l and u, but only on their signs. Any algorithm that allows the effective computation of such signs can be used to support the proposed algorithm. In the following, we present an algorithm to effectively compute the signs of suitable l and u.

3.1   The Modified Bernstein Form of Polynomials

In this section we introduce the *Modified Bernstein Form* (MBF) [1] of polynomials and we explain how to use it to solve polynomial constraints over finite domains. The MBF is a variant of the *Bernstein Form* (*BF*) of polynomials (see, e.g., [11–13]) that, if the coefficients of polynomials are all integer numbers, can be computed with integer arithmetic only.

Let us denote as $\{x_j\}_{j=1}^k$ the considered variables where $k \geq 1$ and j is an index that we associate with each variable. Then, it is possible to define a *multi-index*

$$I = (i_1, \ldots, i_k) \quad i_j \geq 1 \quad 1 \leq j \leq k. \tag{16}$$

Two multi-indices can be summed component-wise and they can be compared using the lexicographic order. For a multi-index I and a vector of variables

$$\mathbf{x} = (x_1, \ldots, x_k) \tag{17}$$

we allow the notation $\mathbf{x}^I = x_1^{i_1} \cdot \ldots \cdot x_k^{i_k}$.

We now consider a single multivariate polynomial $p(\mathbf{x})$ with real coefficients $\{a_I\}_{I \leq N}$ and $k \geq 1$ variables

$$p(\mathbf{x}) = \sum_{I \leq N} a_I \mathbf{x}^I \tag{18}$$

and we assume that each variable $\{x_j\}_{j=1}^k$ is defined over a finite interval $[\underline{x}_j, \overline{x}_j]$, so that the box $\mathbf{I}$ where the polynomial $p(\mathbf{x})$ is defined can be denoted as $\mathbf{I} = [\underline{\mathbf{x}}, \overline{\mathbf{x}}]$ where $\underline{\mathbf{x}} = (\underline{x}_1, \ldots, \underline{x}_k)$ and $\overline{\mathbf{x}} = (\overline{x}_1, \ldots, \overline{x}_k)$. To derive the explicit expression of the polynomial $p(\mathbf{x})$ defined in (18) in the *Bernstein basis* (see, e.g., [13]), it is necessary to consider an affine change of variable between $\mathbf{x}$ and a new variable $\mathbf{t}$ defined over the unit box $[0, 1]^k$. Using the affine transformation

$$x_i = \underline{x}_i + (\overline{x}_i - \underline{x}_i)t_i \tag{19}$$

it can be shown that

$$\sum_{I \leq N} a_I \mathbf{x}^I = \sum_{I \leq N} c_I \mathbf{t}^I \tag{20}$$

where the coefficients c_I are related to a_I as follows

$$c_I = \sum_{L=I}^N a_I \binom{L}{I} \underline{\mathbf{x}}^{L-I} (\overline{\mathbf{x}} - \underline{\mathbf{x}})^I \tag{21}$$

and the binomial coefficients for multi-indices is defined as

$$\binom{L}{I} = \prod_{j=1}^k \binom{l_j}{i_j}. \tag{22}$$

The coefficients $\{c_I\}_{I \leq N}$ can be used to derive the Bernstein form of the polynomial $p(\mathbf{x})$, whose definition relies on the polynomials $\{B_I^N(\mathbf{x})\}_{I \leq N}$ which compose the Bernstein basis of $p(\mathbf{x})$. Such polynomials are defined as

$$B_I^N(\mathbf{x}) = \prod_{j=1}^k B_{i_j}^{n_j}(x_j) \qquad I \leq N \tag{23}$$

where

$$B_{i_j}^{n_j}(x_j) = \binom{n_j}{i_j} \frac{(x_j - \underline{x}_j)^{i_j} (\overline{x}_j - x_j)^{n_j - i_j}}{(\overline{x}_j - \underline{x}_j)^{n_j}} \qquad 0 \leq i_j \leq n_j \quad 1 \leq j \leq k.$$

Given the Bernstein basis $\{B_I^N(\mathbf{x})\}_{I \leq N}$, the polynomial $p(\mathbf{x})$ in (20) can be written as

$$p(\mathbf{x}) = \sum_{I \leq N} b_I B_I^N(\mathbf{x}) \tag{24}$$

where the coefficients $\{b_I\}_{I \leq N}$ are defined as

$$b_I = \sum_{J \leq I} \frac{\binom{I}{J}}{\binom{N}{J}} c_J \qquad I \leq N. \tag{25}$$

Observe that the coefficients $\{b_I\}_{I \leq N}$ are defined in terms of $\{c_I\}_{I \leq N}$ and, hence, in terms of $\{a_I\}_{I \leq N}$.

A well-known property of the coefficients of the Bernstein form of polynomials is that the range of a polynomial $p(\mathbf{x})$ over $\mathbf{I}$ satisfies

$$\text{range}(p) \subseteq [\min_{I \leq N}\{b_I\}, \max_{I \leq N}\{b_I\}]. \tag{26}$$

However, according to (25), the coefficients $\{b_I\}_{I \leq N}$ may not be integer numbers, even if $\{c_I\}_{I \leq N}$ are integer numbers. For this reason, we consider the MBF of the polynomial $p(\mathbf{x})$. In order to do so, let us consider the *modified Bernstein basis* given by the following set of polynomials

$$\tilde{B}_I^N(\mathbf{x}) = \prod_{j=1}^k \tilde{B}_{i_1}^{n_1}(x_1) \tag{27}$$

where

$$\tilde{B}_{i_j}^{n_j}(x_j) = \frac{(x_j - \underline{x}_j)^{i_j}(\overline{x}_j - x_j)^{n_j - i_j}}{(\overline{x}_j - \underline{x}_j)^{n_j}} \qquad 0 \leq i_j \leq n_j \quad 1 \leq j \leq k.$$

Using such a basis, the polynomial $p(\mathbf{x})$ can be written as

$$p(\mathbf{x}) = \sum_{i=0}^n \tilde{b}_I \tilde{B}_I^N(\mathbf{x}) \tag{28}$$

where the coefficients $\{\tilde{b}_I\}_{I \leq N}$ are

$$\tilde{b}_I = \sum_{J \leq I} \binom{N}{I} \frac{\binom{I}{J}}{\binom{N}{J}} c_J = \sum_{J \leq I} \binom{N - J}{I - J} c_J \qquad I \leq N. \tag{29}$$

Observe that, if the coefficient $\{a_I\}_{I \leq N}$ are integer numbers, then coefficients $\{c_I\}_{I \leq N}$ are integer numbers, and therefore $\{\tilde{b}_I\}_{I \leq N}$ are integer numbers since all factors involved in (29) are integer numbers. Moreover, observe that the binomial coefficients in (29) can be computed with no divisions using the following recursive formula valid for $n \geq 2$ and $1 \leq k \leq n - 1$

$$\binom{n}{j} = \binom{n-1}{j} + \binom{n-1}{j-1}, \tag{30}$$

which ensures that $\{\tilde{b}_I\}_{I \leq N}$ can be computed using only sums and products among integer numbers. Moreover, this allows evaluating the coefficients $\{\tilde{b}_I\}_{I \leq N}$ without using rational numbers and, hence, without introducing approximations. Unfortunately, condition (26) holds for the coefficients $\{b_I\}_{I \leq N}$ but not for the coefficients $\{\tilde{b}_I\}_{I \leq N}$. However, we can focus on the (integer) coefficients $\{\tilde{b}_I\}_{I \leq N}$ since we are only interested in the signs of $\{b_I\}_{I \leq N}$, and for all $I \leq N$ the sign of b_I equals that of $\tilde{b}_I$.

3.2   Pseudo-code of the Constraint Solving Algorithm

The pseudo-code of the proposed algorithm is shown in function SOLVE($\cdot$, $\cdot$) in Fig. 1, which takes as inputs a set of constraints C and an initial bounding box $B \subset \mathbb{Z}^k$ for the domains of variables. For all constraints, the coefficients of the MBF of related polynomials are evaluated using function BERNSTEIN($\cdot$, $\cdot$), which returns two of such coefficients $\tilde{l}$ and $\tilde{u}$, with $\tilde{l} \leq \tilde{u}$. If $\tilde{u}$ is negative, then the currently analysed constraint is not satisfiable in the current box. At the opposite, if $\tilde{l}$ is non-negative, then the currently analysed constraint is satisfiable in the current box. If both previous conditions are not met, then a variable is chosen and its domain is properly split in two disjoint intervals, which are recursively processed. The notation $B_{j \to [a..b]}$ is used to create a new box from box B where the interval of variable x_j is replaced with interval $[a..b]$.

```
 1: function SOLVE(C, B)
 2:    global D : subset of Z^k              ▷ domains of variables, initially D = ∅
 3:    input C : set of constraints          ▷ current set of constraints
 4:    input B : box in Z^k                  ▷ box currently analysed
 5:    output satisfiable or unsatisfiable   ▷ satisfiability of C in B
 6:        for all c ∈ C with c = (p(x) ≥ 0) do
 7:            < l̃, ũ >←BERNSTEIN(p, B)
 8:            if ũ < 0 then
 9:                return unsatisfiable
10:            else if l̃ ≥ 0 then
11:                C ← C\{c}
12:            else
13:                select x_j with I_j = [x_j, x̄_j]
14:                if x_j = x̄_j then
15:                    q(x) ← p(x) with x_j replaced by x̄_j
16:                    C ← (C\{c}) ∪ {(q(x) ≥ 0)}
17:                else
18:                    select s ∈ I_j
19:                    s_l ← SOLVE(C, B_{j→[x_j..s]} )
20:                    s_r ← SOLVE(C, B_{j→[s+1..x̄_j]} )
21:                    if s_l = unsatisfiable and s_r = unsatisfiable then
22:                        return unsatisfiable
23:                    end if
24:                end if
25:            end if
26:        end for
27:        if C = ∅ then
28:            D ← D ∪ B
29:        end if
30:        return satisfiable
31: end function
```

Fig. 1. The pseudo-code of the proposed algorithm to determine if a set of constraints C is satisfiable in a given box $B \subset \mathbb{Z}^k$, also updating the domains of variables D.

```
1: function BERNSTEIN(p, B)
2: input p : polynomial              ▷ a multivariate polynomial with k variables
3: input B : box in ℤ^k      ▷ the box used to compute the coefficients of the MBF
4: output < l̃, ũ >                          ▷ two coefficients of the MBF of p
5:     l̃ ← +∞
6:     ũ ← −∞
7:     for all I ≤ N do
8:         b̃_I ← MODIFIEDBERNSTEINCOEFFICIENT(p, B, I)
9:         if b̃_I < l̃ then
10:            l̃ ← b̃_I
11:        else if b̃_I > ũ then
12:            ũ ← b̃_I
13:        end if
14:        if l̃ < 0 and ũ ≥ 0 then
15:            return < l̃, ũ >
16:        end if
17:    end for
18:    return < l̃, ũ >
19: end function
```

Fig. 2. The pseudo-code of function BERNSTEIN($\cdot, \cdot$), which requires a polynomial p and a box $B \subset \mathbb{Z}^k$, and it returns $< \tilde{l}, \tilde{u} >$, two suitable coefficients of the MBF of p.

Function BERNSTEIN($\cdot, \cdot$), shown in Fig. 2, takes as input a polynomial p with $k \geq 1$ variables and a box $B \subset \mathbb{Z}^k$, and it computes the coefficients of the MBF of polynomial p through function MODIFIEDBERNSTEINCOEFFICIENT($\cdot, \cdot, \cdot$). As soon as two of such coefficients with different signs are found, their values are returned. If all coefficients have the same sign, then the minimum and maximum coefficients are returned.

It is worth noting that the proposed algorithm is open to several possible optimizations. In particular, the algorithm does not detail five choices, which can be used to accommodate optimizations, as follows:

1. Function SOLVE($\cdot, \cdot$) does not specify the order in which constraints are selected in the outmost loop, see line 6 in Fig. 1;
2. Function SOLVE($\cdot, \cdot$) does not specify the order in which variables are selected, see line 13 in Fig. 1;
3. Function SOLVE($\cdot, \cdot$) does not specify the order in which values in domains are selected, see line 18 in Fig. 1;
4. Function BERNSTEIN($\cdot, \cdot$) does not specify the order in which the coefficients of the MBF are computed, see line 7 in Fig. 2; and
5. Function BERNSTEIN($\cdot, \cdot$) does not specify how the coefficients of the MBF are computed, see line 8 in Fig. 2.

All such choices can be fixed statically or they can be postponed at execution time to make them dynamic. In both cases, they have the potential to impact significantly on the performances of the proposed algorithm in practical cases (see, e.g., [9,10,14,15]).

4   Illustrative Examples

We now show how the proposed algorithm can be successfully applied to the two motivating examples proposed in Sect. 1. The first example is a quadratic constraint involving only one variable, namely

$$\text{X in -20..20, (X-2)\^2 \#>= 3.}$$

Observe that such a constraint can be expressed in canonical form and interpreted as

$$p(x) = x^2 - 4x + 1 \geq 0 \qquad x \in [-20..20]. \tag{31}$$

The trace of the solving process for the constraint is shown in Fig. 3 where different colors are associated to different types of nodes: blue nodes represent non-terminal nodes, green nodes represent terminal nodes with consistent constraints, and red nodes represent terminal nodes with inconsistent constraints. In non-terminal nodes, we call $\tilde{l}$ and $\tilde{u}$ a negative value and a positive value of $\{\tilde{b}_i\}_{i=0}^2$, respectively, as computed by function BERNSTEIN$(\cdot, \cdot)$. In terminal nodes, we explicitly write whether all coefficients $\{\tilde{b}_i\}_{i=0}^2$ are negative or non-negative. Figure 3 shows that after 4 branches the algorithm computes the domain where the constraint is satisfied, namely

$$x \in [-20..0] \cup [4..20]. \tag{32}$$

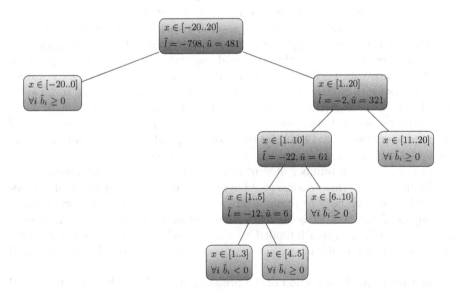

Fig. 3. Description of the reduction of the domain of variable x in (31). (Color figure online)

The second illustrative example shown in Sect. 1 is

$$\texttt{X in -20..20, Y in -20..20, X+X*Y \#>= 100}$$

and it involves two variables. Such a constraint can be expressed in canonical form and interpreted as

$$p(x) = x + xy - 100 \geq 0 \qquad x \in [-20..20] \quad y \in [-20..20]. \tag{33}$$

The trace of the solving process is too complex to be shown because it counts 68 branches. However, after all such branches, the algorithm computes the correct domains for both variables, namely

$$x \in [-20.. - 6] \cup [5..20] \qquad y \in [-20.. - 6] \cup [4..20]. \tag{34}$$

5   Conclusions

In this paper we focused on polynomial constraints over finite domains, namely constraints expressed as equalities, inequalities, and disequalities of polynomials with integer coefficients, and with variables taking values in finite sets of integer numbers. We showed how to write a generic polynomial constraint in canonical form, namely a specific type of inequality which is useful to simplify the constraint solving process. Then, we presented an algorithm based on the modified Bernstein form of polynomials to solve polynomial constraints in canonical form. According to the proposed algorithm, the coefficients of the modified Bernstein form are computed, and the signs of such coefficients are used to reduce the domains of variables by removing values that make constraints inconsistent. Observe that such coefficients are necessarily integer numbers for polynomials with integer coefficients and, therefore, assuming that arbitrary precision integer arithmetic is available, no approximation is introduced in the solving process.

Further developments of the proposed algorithm are currently ongoing in order to improve its performance. First, it is of interest to investigate appropriate heuristics to choose the next variable whose domain is going to be split. Second, the choice regarding the most convenient value to split a domain needs to be further studied. Various static and dynamic heuristics have been proposed in the literature (see, e.g., [9,10,15]) to tackle both such problems, and we believe that use of such heuristics could improve the performance of the proposed algorithm in terms of the number of explored branches. Then, we are investigating how the modified Bernstein form could help in computing the gradient of polynomials, which can be used to dynamically support mentioned heuristics. Finally, an analysis of the type of consistency that the proposed algorithm enforces is currently work-in-progress intended to understand if the proposed framework could host weaker forms of consistency, and how they could impact practical uses.

The current implementation of the proposed algorithm, which has been used for the two examples shown in the last section of this paper, can be downloaded (cmt.dmi.unipr.it/software/clppolyfd.zip) as a reusable Java library that can also be used as a drop-in replacement of the CLP(FD) solver that ships with SWI-Prolog.

References

1. Bergenti, F., Monica, S., Rossi, G.: Polynomial constraint solving over finite domains with the modified Bernstein form. In: Fiorentini, C., Momigliano, A. (eds.) Proceedings of the 31st Italian Conference on Computational Logic, vol. 1645. CEUR Workshop Proceedings, pp. 118–131. RWTH Aachen (2016)
2. Apt, K.: Principles of Constraint Programming. Cambridge University Press, Cambridge (2003)
3. Triska, M.: The finite domain constraint solver of SWI-prolog. In: Schrijvers, T., Thiemann, P. (eds.) FLOPS 2012. LNCS, vol. 7294, pp. 307–316. Springer, Heidelberg (2012). doi:10.1007/978-3-642-29822-6_24
4. Wielemaker, J., Schrijvers, T., Triska, M., Lager, T.: SWI-prolog. Theory Pract. Logic Program. **12**(1–2), 67–96 (2012)
5. Bernstein, S.N.: Démonstration du théorème de Weierstrass fondée sur le calcul des probabilités. Commun. Soc. Math. de Kharkov **2:XIII**(1), 1–2 (1912)
6. Lorentz, G.G.: Bernstein Polynomials. University of Toronto Press, Toronto (1953)
7. Sánchez-Reyes, J.: Algebraic manipulation in the Bernstein form made simple via convolutions. Comput. Aided Des. **35**, 959–967 (2003)
8. Farouki, R.T., Rajan, V.T.: Algorithms for polynomials in Bernstein form. Comput. Aided Geom. Des. **5**(1), 1–26 (1988)
9. Nataraj, P., Arounassalame, M.: A new subdivision algorithm for the Bernstein polynomial approach to global optimization. Int. J. Autom. Comput. **4**(4), 342–352 (2007)
10. Mourrain, B., Pavone, J.: Subdivision methods for solving polynomial equations. J. Symbolic Comput. **44**(3), 292–306 (2009)
11. Garloff, J.: Convergent bounds for the range of multivariate polynomials. In: Nickel, K. (ed.) IMath 1985. LNCS, vol. 212, pp. 37–56. Springer, Heidelberg (1986). doi:10.1007/3-540-16437-5_5
12. Garloff, J.: The Bernstein algorithm. Interval Comput. **2**, 154–168 (1993)
13. Farouki, R.T.: The Bernstein polynomial basis: a centennial retrospective. Comput. Aided Geom. Des. **29**(6), 379–419 (2012)
14. Garloff, J., Smith, A.P.: Solution of systems of polynomial equations by using Bernstein expansion. In: Alefeld, G., Rohn, J., Rump, S., Yamamoto, T. (eds.) Symbolic Algebraic Methods and Verification Methods, pp. 87–97. Springer, Vienna (2001)
15. Ray, S., Nataraj, P.: An efficient algorithm for range computation of polynomials using the Bernstein form. J. Global Optim. **45**, 403–426 (2009)

𝓘-DLV: The New Intelligent Grounder of DLV

Francesco Calimeri[1,2(✉)], Davide Fuscà[1], Simona Perri[1], and Jessica Zangari[1]

[1] Department of Mathematics and Computer Science,
University of Calabria, Rende CS, Italy
{calimeri,fusca,perri,zangari}@mat.unical.it
[2] DLVSystem Srl, Rende CS, Italy
calimeri@dlvsystem.com

Abstract. DLV is a powerful system for knowledge representation and reasoning which supports Answer Set Programming (ASP) – a logic-based programming paradigm for solving problems in a fully declarative way. DLV is widely used in academy, and, importantly, it has been fruitfully employed in many relevant industrial applications. As for the other main-stream ASP systems, in the first phase of the computation DLV eliminates the variables, generating a ground program which is semantically equivalent to the original one but significantly smaller than the Herbrand Instantiation, in general. This phase, called 'grounding', plays a key role for the successful deployment in real-world contexts. In this work we present 𝓘-DLV, a brand new version of the intelligent grounder of DLV. While relying on the solid theoretical foundations of its predecessor, it has been completely redesigned and re-engineered, both in algorithms and data structures; it now features full support to ASP-Core2 standard language, increased flexibility and customizability, significantly improved performances, and an extensible design that eases the incorporation of language updates and optimization techniques. 𝓘-DLV results in a stable and efficient ASP instantiator that turns out to be a full-fledged deductive database system. We describe the main features of 𝓘-DLV and carry out experimental activities for assessing applicability and performances.

Keywords: Answer Set Programming · DLV · Artificial Intelligence · Knowledge Representation and Reasoning

This work was partially supported by the Italian Ministry of University and Research under PON project "Ba2Know (Business Analytics to Know) Service Innovation - LAB", No. PON03PE_00001_1, and by the Italian Ministry of Economic Development under project "PIUCultura (Paradigmi Innovativi per l'Utilizzo della Cultura)" n. F/020016/01-02/X27. Francesco Calimeri has received funding from the European Union's Horizon 2020 research and innovation programme under the Marie Skłodowska-Curie grant agreement No. 690974 for the project: "MIREL: MIning and REasoning with Legal texts".

© Springer International Publishing AG 2016
G. Adorni et al. (Eds.): AI*IA 2016, LNAI 10037, pp. 192–207, 2016.
DOI: 10.1007/978-3-319-49130-1_15

1   Introduction

Answer Set Programming (ASP) is a purely declarative formalism, developed in the field of logic programming and nonmonotonic reasoning [1–6], that become widely used in AI and recognized as a powerful tool for knowledge representation and reasoning (KRR). The language of ASP is based on rules, allowing (in general) for both disjunction in rule heads and nonmonotonic negation in the body; such programs are interpreted according to the answer set semantics [4,7]. Throughout the years a significant amount of work has been carried out for extending the "basic" language and easing the knowledge representation; ASP has been proven to be highly versatile, offering several language constructs and reasoning modes. Moreover, the availability of reliable, high-performance implementations [8,9] made ASP a powerful tool for developing advanced applications in many research areas, ranging from Artificial Intelligence to Databases and Bioinformatics, as well as in industrial contexts [8,10–13].

The "traditional" approach to the evaluation of ASP programs relies on a grounding module (*grounder*), that generates a propositional theory semantically equivalent to the input program, coupled with a subsequent module (*solver*) that applies proper propositional techniques for generating its answer sets. There have been other attempts deviating from this customary approach [14–16]; nonetheless, the majority of the current solutions relies on the canonical "ground & solve" strategy, including DLV [17] and Potassco, the Potsdam Answer Set Solving Collection [18,19]. Among the most widely used ASP systems, DLV has been one of the first solid and reliable; its project dates back a few years after the first definition of answer set semantics [4,7], and encompassed the development and the continuous enhancements of the system. It is widely used in academy, and, importantly, it is still employed in many relevant industrial applications, significantly contributing in spreading the use of ASP in real-world scenarios.

In this work we present 𝓘-DLV, the new intelligent grounder of DLV; starting from the solid theoretical foundations of its predecessor, it has been redesigned and re-engineered aiming at building a modern ASP instantiator marked by improved performances, native support to the ASP-Core-2 standard language [20], high flexibility and customizability, and a lightweight modular design for easing the incorporation of optimization techniques and future updates. 𝓘-DLV is more than an ASP grounder, resulting also in a complete and efficient deductive database system; indeed, it also keeps a set of features and capabilities that made DLV peculiar and, especially in some industrial application domains, very popular and successful. Furthermore, 𝓘-DLV is envisioned as part of a larger project aiming at a tight integration with a state-of-the art ASP solver, a set of mechanisms and tools for interoperability and integration with other systems and formalisms, further improvements on the deductive database side, a solid base for experimenting with techniques and implementations of new approaches and applications of ASP solving, such as stream reasoning. Although the project is at an initial state and the system is still prototypical, it already shows good performances and stability, proving to be competitive both as ASP grounder and deductive database system. Furthermore, the flexible and customizable nature of $\mathcal{I}$-DLV allows to widely experiment with ASP and its applications and to better tailor ASP-based solutions to real-world applications. The newly herein introduced possibility of *annotating ASP code* with external directives to grounder is a bold move in this direction, providing a new way to fine-tune both ASP encodings and systems for any specific scenario at hand.

In the remainder of the paper we introduce $\mathcal{I}$-DLV, its peculiar optimization strategies and new features, including main customization capabilities and annotations; we then discuss a thorough experimental activity that aims at assessing $\mathcal{I}$-DLV capabilities both as ASP grounder and deductive database system.

2   Answer Set Programming

A significant amount of work has been carried out for extending the basic language of ASP, and the community recently agreed on a standard input language for ASP systems: ASP-Core-2 [20], the official language of the ASP Competition series [8,9]. For the sake of simplicity, we focus next on the basic aspects of the language; for a complete reference to the ASP-Core-2 standard, and further details about advanced ASP features, we refer the reader to [20] and the vast literature. In this Section, we briefly recall syntax and semantics, and then we introduce the major approaches to ASP grounding.

Syntax and Semantics. A *term* is either a *simple term* or a *functional term*. A *simple term* is either a constant or a variable. If $t_1 \ldots t_n$ are terms and f is a function symbol of arity n, then $f(t_1, \ldots, t_n)$ is a *functional term*. If $t_1, \ldots, t_k$ are terms and p is a *predicate symbol* of arity k, then $p(t_1, \ldots, t_k)$ is an *atom*. A *literal* l is of the form a or $\mathtt{not}\ a$, where a is an atom; in the former case l is *positive*, otherwise *negative*. A *rule* r is of the form $\alpha_1 | \cdots | \alpha_k :- \beta_1, \ldots, \beta_n, \mathtt{not}\ \beta_{n+1}, \ldots, \mathtt{not}\ \beta_m.$ where $m \geq 0, k \geq 0; \alpha_1, \ldots, \alpha_k$ and $\beta_1, \ldots, \beta_m$ are atoms. We define $H(r) = \{\alpha_1, \ldots, \alpha_k\}$ (the *head* of r) and $B(r) = B^+(r) \cup B^-(r)$ (the *body* of r), where $B^+(r) = \{\beta_1, \ldots, \beta_n\}$ (the *positive body*) and $B^-(r) = \{\mathtt{not}\ \beta_{n+1}, \ldots, \mathtt{not}\ \beta_m\}$ (the *negative body*). If $H(r) = \emptyset$ then r is a *(strong) constraint*; if $B(r) = \emptyset$ and $|H(r)| = 1$ then r is a *fact*. A rule r is safe if each variable of r has an occurrence in $B^+(r)$. An ASP program is a finite set P of safe rules. A program (a rule, a literal) is said to be *ground* if it contains no variables. A predicate is defined by a rule if the predicate occurs in the head of the rule. A predicate defined only by facts is an *EDB* predicate, the remaining predicates are *IDB* predicates. The set of all facts in P is denoted by $Facts(P)$; the set of instances of all *EDB* predicates in P is denoted by $EDB(P)$.

Given a program P, the *Herbrand universe* of P, denoted by U_P, consists of all ground terms that can be built combining constants and function symbols appearing in P. The *Herbrand base* of P, denoted by B_P, is the set of all ground atoms obtainable from the atoms of P by replacing variables with elements from U_P. A *substitution* for a rule $r \in P$ is a mapping from the set of variables of r to the set U_P of ground terms. A *ground instance* of a rule r is obtained applying a substitution to r. The *full instantiation Ground(P)* of P is defined as the set of all ground instances of its rules over U_P. An *interpretation* I for P is a subset of B_P. A positive literal a (resp., a negative literal **not** a) is true w.r.t. I if $a \in I$ (resp., $a \notin I$); it is false otherwise. Given a ground rule r, we say that r is satisfied w.r.t. I if some atom appearing in $H(r)$ is true w.r.t. I or some literal appearing in $B(r)$ is false w.r.t. I. Given a ground program P, we say that I is a *model* of P, iff all rules in $Ground(P)$ are satisfied w.r.t. I. A model M is *minimal* if there is no model N for P such that $N \subset M$. The *Gelfond-Lifschitz reduct* [4] of P, w.r.t. an interpretation I, is the positive ground program P^I obtained from $Ground(P)$ by: (i) deleting all rules having a negative literal false w.r.t. I; (ii) deleting all negative literals from the remaining rules. $I \subseteq B_P$ is an *answer set* for a program P iff I is a minimal model for P^I. The set of all answer sets for P is denoted by $AS(P)$.

Grounding for ASP Computation. All currently competitive ASP systems mimic the definition of the semantics as given above by first creating a program without variables. This phase usually referred to as *instantiation* or *(grounding)* is then followed by the *answer sets search* phase, in which a solver applies propositional algorithm for finding answer sets. The grounding solves a complex problem, which is in general EXPTIME-hard; the produced ground program is potentially of exponential size with respect to the input program [21]. Grounding, hence, may be computationally expensive and have a big impact on the performance of the whole system, as its output is the input for an ASP solver, that, in the worst case, takes exponential time in the size of the input. Thus, given a program P, ASP grounders are geared toward efficiently producing a ground program, that is considerably smaller than $Ground(P)$, but preserves the semantics.

The two first stable grounders have been *lparse* [22] and the DLV instantiator. They accept different classes of programs, and follow different strategies for the computation. The first binds non-global variables by domain predicates, to enforce ω-restrictedness [22], and instantiates a rule r scanning the extensions of the domain predicates occurring in the body of r, generating ground instances accordingly. On the other hand, the only restriction on DLV input is safety. Furthermore, the DLV instantiation strategy is based on semi-naive database techniques [23] for avoiding duplicate work, and domains are built dynamically. Over the years, a new efficient grounder has been released, namely *gringo* [19]. The first versions accepted only domain restricted programs with an extended notion of domain literal in order to support λ-restrictedness [22]; starting form version 3.0, *gringo* removed domain restrictions and instead requires programs to be safe as in DLV and evaluates them relying on semi-naive techniques as well.

3 Towards a New Grounder

In this Section we introduce $\mathcal{I}$-DLV, the new grounder of the DLV system. As both the DLV instantiator and *gringo*, the core strategies rely on a bottom-up evaluation based on a semi-naive approach, which proved to be reliable and efficient over a huge number of scenarios, especially when domain extensions are very large. A detailed description of such basic techniques is out of the scope of this work, thus, we refer the reader to [24] for further insights; we discuss, instead, the general architecture of $\mathcal{I}$-DLV, its typical work flow, peculiar optimization means and new features.

3.1 $\mathcal{I}$-DLV Overview

Figure 1 depicts $\mathcal{I}$-DLV high-level architecture. The PRE-PROCESSOR module parses the input program P and builds the extensional database (*EDB*) from *Facts*(P); then, the Rewriter produces the intensional database (*IDB*) from the rules. The Dependency Analyzer examines *IDB* rules and predicates, identifying program modules and a proper ordering for incrementally evaluating them according to the definitions in [24]. The PROGRAM INSTANTIATOR grounds the program; the process is managed by the Program Modules Instantiator, that, applying a semi-naive schema [23, 24], evaluates one module at a time according to the order provided by the Dependency Analyzer. The core of the computation is performed by the Rule Instantiator: given a (non-ground) rule r and a set of ground atoms S representing predicate extensions, it generates the ground instances of r, finding proper substitutions for the variables. Notably, the set S is dynamically computed: initially, it contains only *Facts*(P) and from then on it is extended by the ground atoms occurring in the head of the newly generated ground rules. Ground rules are also analyzed by the Simplificator, in order to check if some can be simplified or even eliminated, still guaranteeing semantics. Eventually, the output is gathered and properly arranged by the Output Builder. The produced ground program will have the same answer sets of the full theoretical instantiation, yet being possibly smaller [24].

$\mathcal{I}$-DLV easily interoperates with solvers and other ASP systems and tools, thanks to the fully compliance to the ASP-Core-2 standard and the capability of the Output Builder to format the output in different ways, including the numeric format required by state-of-the-art solvers *wasp* [25] and *clasp* [18].

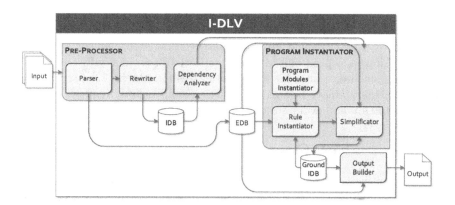

Fig. 1. $\mathcal{I}$-DLV architecture.

3.2 Optimizations

In the following we discuss some of the main optimizations; most of them are related, to different extents, to the rule instantiation task, that constitutes the core of the computation. For the sake of presentation, we will go through a brief overview of the process, discussing optimizations as they intervene. Instantiating a rule basically corresponds to evaluate *relational joins* of the positive body literals: it requires to iterate on the literals, matching them one by one with their instances and binding the free variables accordingly, at each step. When the matching of a predicate fails, backtracking to a previous predicate is needed, trying to match it with another instance.

At the very beginning, rules possibly undergo a rewriting process for removing isolated variables. For instance, consider a non-ground rule r containing a body atom $p(X_1, ..., X_n)$, with a variable X_i that does not appear anywhere else in r. While instantiating r, substitutions for X_i will not affect any failed match for other atoms, nor the instances obtained for the head atoms. Thus, we can safely eliminate X_i by projecting all variables $X_k, k \neq i$ of p to an auxiliary predicate p'. That is, a new (non-ground) rule is added: $p'(X_1, ..., X_{i-1}, X_{i+1}, ..., X_n) :\!- p(X_1, ..., X_n)$. and p is substituted by p' in the body of r. By doing so, the generation of ground instances of r which differ only on the binding of X_i is avoided. In many cases this reduces the size of the ground program, and consequently the instantiation time; nevertheless, an overhead is paid, due to the need for copying the projected instances of p in the extension of p'; such overhead, even if negligible in general, might become significant when the benefits of the projection are limited. For this reason, such projection rewriting can be disabled on request.

Once the rewriting process is over, the order of literals in the rule bodies is analyzed and possibly changed in a way inspired by the database setting, where a crucial task is to find an optimal execution ordering for the join operations. Several ordering strategies have been defined and implemented in *I*-DLV, based on different heuristics; they perform differently, each one featuring some advantages case by case. The one enabled by default is the so-called *Combined*[+] criterion, that we defined by enhancing the *Combined* criterion [26] used by DLV. Basically, given a rule, the *Combined* criterion relies on statistics on involved predicates and variables, such as size of extensions, variable domains and selectivities. However, in many cases this criterion might fail to find a good ordering. As an example, let us consider the rule: $a(X, Y, Z) :\!- b(X), b(Y), b(Z), Z > X, Y < Z$. Here the criterion has no useful information for selecting an order for atoms $b(X)$, $b(Y)$, $b(Z)$: the predicate is the same, and statistics are identical. Nevertheless, their ordering can significantly affect performances, since their variables are involved in the comparisons $Z > X$, and $Y < Z$. The *Combined*[+] criterion, instead, relying on linear interpolations, computes new statistics for variables involved in comparisons, according to different formulae that depend on the type of relational operator at hand, thus providing more information for selecting a good ordering.

The adapted non-ground rules undergo the actual rule instantiation, based on a backtracking search. *I*-DLV employs a non-chronological backtracking strategy, that reduces both the computation time and the size of the instantiation. In particular, while instantiating a rule, when a backtrack step is necessary, by considering the literal that most likely is involved in the *reason* of a failed match, it is possible to jump back many elements, rather than just one as in the chronological algorithm. Such jumps should be designed carefully, in order to avoid that some solution is missed, especially in our case, where we have to compute all (relevant) solutions. Moreover, our algorithm makes use of both semantic and structural information of the rule at hand, for computing only a relevant subset of its ground instances, yet fully preserving the semantics [27].

The process is further optimized by making use of indexing for the retrieval of matching instances from the predicate extensions. $\mathcal{I}$-DLV's indexing schema is very general: any predicate argument can be indexed, allowing both single- and multiple-argument indices. Indices are computed "on-demand", only if needed, while instantiating a rule r, and several choices are available: a predicate p in r can be indexed on a single, a pair or all indexable arguments; the choice might differ from predicate to predicate. The default configuration uses a heuristics that, when different arguments can be indexed in a predicate p, adopts double-argument indices selecting the two indexable arguments that feature the highest number of different values in the extension of p.

The search for an "agreement" between body literals on variable substitutions is further eased: before processing a rule r, for each variable X, we compute the intersection of all sets of possible substitutions for all the occurrences of X in r; this reduces, in general, the number of possible values for X, by skipping those that would not match among distinct variable occurrences. Such technique performs best when the sets of substitutions differ significantly, thus can be enabled on demand.

Another optimization simulates the classic relational algebra operation of pushing selections down the execution tree. Let us consider, for instance, the following rule r: $p(X) :- t(X, Y, Z), q(Z, V, S), V<S$. It is easy to see that, instead of first joining t with q and then selecting what complies with the comparison $V<S$, it is convenient, in general, first selecting, in the extension of q, the instances complying with the comparison, and then joining them with the ones of t. This can be obtained by means of a rewriting process similar to the one already described for projecting-out isolated variables. In order to avoid a possibly significant overhead, we do not actually perform the rewriting, but we simulate it during the instantiation of r: while retrieving the instances of q, only those complying with the comparison are actually taken into account. Such technique can be easily applied to cases in which more than one comparison occurs, and the involved terms appear in more than one atom.

Just like the process of pushing down selections, the projection rewriting described above can be simulated as well during the instantiation, even if only in some cases. In particular, when instantiating a rule r, if a predicate p is *solved* (i.e., it is a fact, or it depends only on facts and other solved predicates), when looking for next matches for it, all instances that differ just because of the substitution of an isolated variable X could be ignored; to this end, the matching of atoms is empowered with a special filter mechanism that suggests only the relevant instances. This filter combined with the backjumping machinery, when the projection rewriting is disabled, can simulate its behaviour without paying the overhead. However, the projection rewriting is more general, since it can safely be applied to all predicates, with no distinction between solved and unsolved; if p is unsolved, instead, all its instances must be considered as relevant in order to preserve semantics of the produced ground program; thus, the use of the described filter mechanism must be prevented.

The output of the rule instantiation process is a set of ground instances of the rule at hand. The size of the output is further reduced by examining the produced ground rules and possibly simplifying, or even eliminating them. Indeed, body literals which are already known to be true can be safely dropped. Moreover, once the rule instance has been created, when some negative literal already known to be false occurs in the body, the rule instance is already (trivially) satisfied: does not contribute to the semantics of the ground program, and it is eliminated. If the input program is non-disjunctive and stratified, the modular evaluation guided by the PROGRAM MODULES INSTANTIATOR along with the simplification allows $\mathcal{I}$-DLV to completely evaluate the program: the output consists of a set of facts, that correspond to the unique answer set of the program.

The described rule grounding process can be applied also to strong constraints, with a few simplifications due to the missing heads. Notably, precisely because of this, their evaluation does not affect the aforementioned dynamic creation of the predicate extensions: thus, they could be safely processed once the instantiation of all program modules is completed. However, just as for ground rules, the body of a ground constraint can be simplified, possibly leading to an empty-body constraint. Such constraints are always violated; thus, the input program is inconsistent, and any pending computation can be aborted. For this reason, $\mathcal{I}$-DLV anticipates constraints evaluation: a constraint is instantiated as soon as the extensions of the body predicates are available.

We conclude this overview with a technique for the effective use of $\mathcal{I}$-DLV in a deductive database setting, where efficient query answering over logic programs is essential; to this end, we integrated an adapted version of the Magic Sets technique [28,29]. It simulates a top-down computation by means of a proper rewriting of the input program for identifying the relevant subset of the instantiation which is sufficient for answering the query. The restriction of the instantiation is obtained by means of additional "magic" predicates, whose extensions represent relevant atoms with respect to the query.

3.3   Flexibility, Customizability and Further Features

As already discussed, one of the main goals of the $\mathcal{I}$-DLV project is to obtain a novel, flexible tool for experimenting with ASP and its applications; to this end, it has been designed in order to allow a fine-grained control over the whole computational process, both via command-line options and inline annotations.

Command-line Customization. In what follows, we describe the most relevant options that the user can set via command-line in order to customize the behaviour of $\mathcal{I}$-DLV. For more options and further details, we refer to the online documentation [30].

$\mathcal{I}$-DLV allows to completely control the indexing strategy, thus providing the user with a mean for handling situations where the default behaviour is not satisfactory. In particular, the indexing module can set per each predicate in the program a single- or multiple-index, on the desired arguments. Since only ground arguments can be indexed, if the user specifies a non indexable argument, the default indexing strategy is adopted.

As for body ordering, the user can currently choose among the following alternatives: (i) a basic ordering that aims at preserving the original literal positions in the rule; (ii) the DLV *Combined* criterion [26]; (iii) the aforementioned *Combined*$^+$ criterion (enabled by default); (iv) an enhanced version of the *Combined* criterion that pushes literals with functional terms down in the body; (v) a strategy that tries to improve the quality of available indices; (vi) a criterion that works in synergy with backjumping in order to facilitate it; (vii) a criterion combining the latter two.

Furthermore, the projection rewriting and the filter mechanism for isolated variables in solved predicates, both enabled by default, can be disabled. Conversely, the technique for aligning variable substitutions can be enabled at will: it is disabled by default, given that its benefits strictly depend from the distribution of the input data.

Choice rules are natively managed by default, possibly undergoing a conservative rewriting/rearrangement that aims at optimizing the efficiency of instantiation; however, the user can ask for a rewriting approach that makes use of disjunction and removes them from the program. Both approaches preserve the semantics.

Insights on the grounding process are available via a number of statistics that can be produced on demand. In particular, the following information is provided for each input rule: total instantiation time, number of produced ground instances, number of iterations required for instantiation (in case the rule is recursive); in addition, size of extension and selectivity of all arguments are reported, for each predicate in the rule.

$\mathcal{I}$-DLV is able to process both ground and non-ground queries. By default, it simply produces the instantiation; upon request, for non-disjunctive and stratified programs, that are completely evaluated by the system, it can directly provide the query answer. The magic-set technique, enabled by default, can be disabled if wanted.

Inline Customization and Optimizations: Annotations. $\mathcal{I}$-DLV introduces a new special feature for facilitating system customization and tuning: *annotations* of ASP code. Annotations and meta-data have been applied in different programming paradigms and languages; Java annotations, for instance, have no direct effect on the code they annotate: a typical usage consists in analyzing them at runtime in order to change the code behaviour. Some examples of annotations have been proposed also for declarative paradigms, although to different extents and purposes with respect to our setting; for space reasons, we refer to the literature (see, e.g., [31–33]). $\mathcal{I}$-DLV annotations allow to give explicit directions on the internal grounding process, at a more fine-grained level with respect to the command-line options: they "annotate" the ASP code in a Java-like fashion while embedded in comments, so that the resulting programs can still be given as input to other ASP systems, without any modification. Syntactically, all annotations start with the prefix "%@" and end with a dot ("."). The current $\mathcal{I}$-DLV release supports annotations for customizing two of the major aspects of the grounding process, body ordering and indexing; additional options are being developed.

A specific body ordering strategy can be explicitly requested for any rule, simply preceding it with the line:

```
%@rule_ordering(@value=Ordering_Type).
```

where *Ordering_Type* is a number representing an ordering strategy (Sect. 3.3). In addition, it is possible to specify a particular partial order among atoms, no matter the employed ordering strategy, by means of **before** and **after** directives. For instance, in the next example $\mathcal{I}$-DLV is forced to always put literals $a(X,Y)$ and $X = \#count\{Z : c(Z)\}\}$ before literal $f(X,Y)$, whatever the order chosen:

```
%@rule_partial_order(@before={a(X,Y),X=#count{Z:c(Z)}},
    @after={f(X,Y)}).
```

As for indexing, directives on a per-atom basis can be given; the next annotation, for instance, requests that, in the subsequent rule, atom $a(X,Y,Z)$ is indexed, if possible, with a double-index on the first and third arguments:

```
%@rule_atom_indexed(@atom=a(X,Y,Z),@arguments={0,2}).
```

Multiple preferences can be expressed via different annotations; in case of conflicts, priority is given to the first. In addition, preferences can also be specified at a "global" scope, by replacing the **rule** directive with the **global** one. While a **rule** annotation must precede the intended rule, **global** annotations can appear at any line in the input program. Both **global** and **rule** annotations can be expressed in the same program; in case of overlap on a particular rule/setting, priority is given to the **rule** ones.

A thorough study of the impact of annotations is out of the scope of this work, even though, intuitively, the way they change the grounding mechanisms can noticeably affect performances on the program at hand. Indeed, it is easy to face such cases; in order to give an idea, as an example, let us consider the rule:

$$reach(X,Y,T) :- reach(XX,YY,T), dneighbor(D,XX,YY,X,Y),$$
$$conn(XX,YY,D,T), conn(X,Y,E,T), inverse(D,E), step(T).$$

taken from the encoding of *Labyrinth* from the 6th ASP Competition suite of benchmarks (see Sect. 4). In this case, by annotating the rule with:

```
%@rule_partial_order(@before={inverse(D,E)},
    @after={conn(X,Y,E,T)}).
```

that asks $\mathcal{I}$-DLV for a specific preceding ordering between the two atoms, yet keeping the default behaviour elsewhere in the program, average grounding time over all instances halves, decreasing from 1.843 to 0.8073 s.

4 Experimental Evaluation

Hereafter we report the results of an experimental activity carried out to assess $\mathcal{I}$-DLV performances as both ASP grounder and deductive database system. In order to obtain trustworthy results, we considered tests that have already been largely used and are publicly available. In particular, we relied on the whole 6th ASP Competition suite [34], the latest edition of a series of events [8,9] assessing ASP systems on challenging benchmarks in order to promote the state of the art, and OpenRuleBench [35], an open set of resources comprising a suite of benchmarks for analyzing performances and scalability of different rule engines. Experiments have been performed on a NUMA machine equipped with two 2.8 GHz AMD Opteron 6320 and 128 GiB of main memory, running Linux Ubuntu 14.04.4 (kernel ver. 3.19.0-25). Binaries were generated with the GNU C++ compiler 4.8.4. As for memory and time limits, we allotted 15 GiB and 600 s for each system per each single run.

ASP Grounding Benchmarks. For this setting we tested $\mathcal{I}$-DLV against the two mainstream grounders *gringo* and the (old) intelligent grounder of DLV, and in particular the latest available versions at the time of writing: 4.5.4 and 2012-12-17, respectively. We first launched the 6th ASP Competition suite, that features 28 problems and 20 different instances per each. Results are reported in Fig. 2(a): first column shows the name of the problem, while the next three report the average times. Dashes in the column of a system means that it has not been tested on the corresponding problem: this was due, for *gringo*, to the presence of queries in three cases, and for DLV to the missing support for relevant ASP-Core-2 constructs. When launched, all systems were able to ground all 20 instances in the allotted time. It is evident, while comparing $\mathcal{I}$-DLV against DLV, that the new grounder systematically outperforms its predecessor, enjoying performance gains up to 90 %. Also the comparison with *gringo* is encouraging: despite the prototypical version of $\mathcal{I}$-DLV, it proves to be competitive. More in detail, excluding the 3 domains solved only by $\mathcal{I}$-DLV, times are substantially aligned (time differences below 20 %) in 9 domains out of 25; as for the remaining domains, each outperforms the other in an almost equal number of domains.

In order to find further comparison settings outside of the ASP Competition series, where *gringo* became the de facto standard grounder for all competing solvers, we took into account the problems appearing in OpenRuleBench. This is also motivated by the fact that the ASP Competition mainly focuses on problems where solving task is more relevant with respect to the grounding one (indeed, as Fig. 2(a) shows, all systems completed all instances), while Open-RuleBench tests demand a more significant work from the grounders. Since such suite consists essentially of a query-based set of problems, that *gringo* would not accept "as-is", we removed the query from the encodings and measured just the grounding times. Obviously, we did the same also for the DLV instantiator and $\mathcal{I}$-DLV: otherwise, these might have took advantage from the magic-set technique, thus leading to an unfair test. Results are reported in Fig. 2(b): after domains names and corresponding number of instances, the next three pairs of

Problem	gringo	DLV	I-DLV
Abstract Dialectical Frameworks	1,97	-	0,11
Combined Configuration	18,02	-	11,94
Complex Optimization	9,38	-	69,41
Connected Still Life	0,10	-	0,10
Consistent Query Answering	-	196,99	76,21
Crossing Minimization	0,10	-	0,10
Graceful Graphs	0,20	-	0,29
Graph Coloring	0,10	-	0,10
Incremental Scheduling	50,10	-	18,20
Knight Tour With Holes	10,09	-	2,35
Labyrinth	0,50	15,73	1,93
Maximal Clique	18,36	-	4,18
MaxSAT	7,41	-	3,79
Minimal Diagnosis	3,03	57,92	4,92
Nomistery	2,10	-	4,07
Partner Units	0,34	-	0,41
Permutation Pattern Matching	119,20	-	127,13
Qualitative Spatial Reasoning	8,78	31,78	5,45
Reachability	-	-	128,26
Ricochet Robots	0,17	-	0,36
Sokoban	1,06	-	1,23
Stable Marriage	112,00	-	117,52
Steiner Tree	32,84	-	29,25
Strategic Companies	-	236,57	0,33
System Synthesis	1,40	-	1,11
Valves Location Problem	5,29	-	2,55
Video Streaming	0,10	-	0,10
Visit-all	0,90	-	1,18

(a)

Problem	# inst.	gringo		DLV		I-DLV	
		solved	time	solved	time	solved	time
Join1 A	3	2	312,31	1	117,57	1	47,71
Join1 B1	3	3	182,86	3	172,92	3	106,18
Join1 B2	3	3	42,04	3	35,34	3	18,97
Join Dupl. A	3	1	229,15	0	-	1	239,05
Join Dupl. B1	3	2	188,77	2	261,19	2	129,79
Join Dupl. B2	3	3	183,50	3	168,43	3	85,32
Join2	1	1	22,71	1	88,32	1	24,06
Mondial	1	1	2,29	1	2,05	1	1,18
DBLP	1	1	54,55	1	27,06	1	15,98
Lubm1	2	2	68,69	2	31,98	2	22,51
Lubm2	2	2	70,70	2	37,49	2	23,54
Lubm9	2	2	70,72	2	37,02	2	24,46
Same Gen. R.	10	10	99,99	9	146,92	10	54,57
Trans. Closure	10	7	211,35	6	232,74	8	180,12
Wordnet	15	15	7,64	15	12,18	15	4,99
Wine	1	1	11,00	1	28,95	1	11,54
Magic Set	5	5	5,64	5	6,29	5	1,79
Win	10	10	8,78	10	8,02	10	5,43
Same Gen. S.N.	5	5	123,23	4	160,77	5	67,45
Indexing	15	15	5,06	15	2,13	15	1,30
Queens	5	5	0,10	5	0,34	5	0,10
Sixteen Puzzle	5	5	0,10	5	0,37	5	0,10
Solved Instances		101 / 108		96 / 108		101 / 108	
Total Running Time		10570		13021		8054	

(b)

Fig. 2. Benchmarks from 6th ASP Competition (a) and OpenRuleBench (b) – grounding times.

columns show the number of solved instances and the running time averaged over solved instances. Last line reports the total running times for each system (600 secs is added for unsolved instances, as systems were stopped if unable to finish before). The unique dash in the table corresponds to a domain for which DLV did not solve any instance within the allotted time. Interestingly, even if DLV is outperformed by both *gringo* and ℐ-DLV, its performances are somehow satisfactory, still confirming its solidity and reliability. As for *gringo* and ℐ-DLV, both solved 101 instances out of 108; however, ℐ-DLV appears to enjoy better performances: it clearly outperforms *gringo* in almost all domains.

Deductive Database Benchmarks. For this setting, the natural choice was the query-based set of problems of the OpenRuleBench initiative. Besides DLV, we tested ℐ-DLV against XSB [36] (the latest available version, 3.6) that was among the clear winners of the OpenRuleBench runs [35] and is currently one of the most widespread Logic Programming and Deductive Database systems. All systems support query answering, thus, differently from above, queries have not been removed. Results are reported in Fig. 3: after domain names and corresponding number of instances, the next three pairs of columns show the number of solved instances and the running time averaged over solved instances. Similarly as above, last line reports the total running times for each system. The unique dash in the table corresponds to a domain for which DLV did not solve any instance within the allotted time. Also in this setting results are very encouraging: not only ℐ-DLV behaves better than DLV, but it is definitely competitive against XSB. Indeed, in spite of a non-negligible variability from a problem to another, ℐ-DLV times are, on the overall, comparable with the ones of XSB; in addition, it was able to solve even more instances within the allotted time

Problem	# inst.	XSB		DLV		I-DLV	
		solved	time	solved	time	solved	time
Join1 A free-free	3	1	19,13	1	122,84	1	47,69
Join1 A bound-free	3	2	23,93	3	37,09	3	18,57
Join1 A free-bound	3	1	8,46	3	178,00	3	94,88
Join1 B1 free-free	3	2	12,84	3	181,62	3	107,10
Join1 B1 bound-free	3	3	2,64	3	3,16	3	1,28
Join1 B1 free-bound	3	2	4,95	3	11,64	3	6,02
Join1 B2 free-free	3	3	8,74	3	38,32	3	19,29
Join1 B2 bound-free	3	3	1,50	3	2,89	3	1,21
Join1 B2 free-bound	3	3	4,34	3	2,90	3	1,20
Join Duplicate A	3	1	101,28	0		1	230,99
Join Duplicate B1	3	2	52,33	2	261,63	2	121,59
Join Duplicate B2	3	3	39,82	3	169,82	3	81,83
Join2	1	1	2,34	1	80,09	1	12,50
DBLP	1	1	98,36	1	23,81	1	13,92
Mondial	1	1	3,49	1	0,77	1	0,51
Same Gen. Recursion free-free	10	10	21,73	9	150,67	10	54,71
Same Gen. Recursion bound-free	10	10	18,26	10	153,06	10	48,47
Same Gen. Recursion free-bound	10	10	26,78	9	150,41	10	56,24
Trans. Closure free-free	10	10	141,52	6	217,35	9	186,89
Trans. Closure bound-free	10	10	96,19	6	207,93	9	180,73
Trans. Closure free-bound	10	10	29,01	10	3,13	10	1,42
Wordnet	15	15	1,71	15	13,99	15	5,26
Wine	1	1	5,85	1	27,76	1	10,63
Indexing	15	10	16,140	15	3,849	15	2,00
Total Solved Instances		115 / 130		114 / 130		123 / 130	
Total Running Time		13120		19345		10779	

Fig. 3. Benchmarks from OpenRuleBench – query answering times.

(123 for $\mathcal{I}$-DLV, 115 for XSB). Interestingly, XSB has been launched with the exact OpenRuleBench settings, where the best configuration was set manually per each problem; $\mathcal{I}$-DLV, instead, features the same default configuration over all domains. As already stated, a thorough study of the impact on performances due to $\mathcal{I}$-DLV fine-tuning capabilities is out of the scope of the present work; an analysis of possible performance improvements due to a specific customization on each single problem in this setting will be definitely of interest.

5 Related Work and Conclusions

We introduced $\mathcal{I}$-DLV, the new grounder of the DLV system. With respect to its predecessor, it features full support to the ASP-Core-2 standard, new optimization techniques, new features and customization capabilities as well as interoperability with current state-of-the-art ASP Solvers and improved performances.

Some connections to our work can be found with other rule-based engines and deductive database systems (an interesting overview can be found in [37]); such systems have common roots, but differ in several respects, especially in supported languages and evaluation mechanisms. For instance, XSB, among the most prominent, is a Prolog system based on a top-down evaluation, while $\mathcal{I}$-DLV is an ASP grounder relying on a bottom-up approach. Connections are stronger with other ASP grounders, mainly with *gringo* [19], that supports same input language of $\mathcal{I}$-DLV and shares the basic evaluation approach. However, $\mathcal{I}$-DLV incorporates different optimization techniques, novel customization properties, such annotations, and specific deductive-database-oriented features, such as Magic Sets. Different grounding approaches are pursued by *lparse* [22], that supports ω-restricted [22] programs, and GIDL [38], a grounder for FO$^+$.

As future work, we plan to significantly extend experiments, putting $\mathcal{I}$-DLV at test both against more systems and over additional domains, study and experiment with a tight integration with a state-of-the art ASP solver, equip it with a set of advanced mechanisms and tools for interoperability and integration with other systems and formalisms, enrich it with additional database-oriented functionalities; moreover, we plan to properly experiment with the customization capabilities, further extending them, for instance by widening the range of aspects over which annotations can intervene. $\mathcal{I}$-DLV is available at the official repository [39].

References

1. Brewka, G., Eiter, T., Truszczynski, M.: Answer set programming at a glance. Commun. ACM **54**(12), 92–103 (2011)
2. Eiter, T., Faber, W., Leone, N., Pfeifer, G.: Declarative problem-solving using the DLV system. In: Minker, J. (ed.) Logic-Based Artificial Intelligence. The Springer International Series in Engineering and Computer Science, vol. 597, pp. 79–103. Springer, New York (2000)
3. Eiter, T., Ianni, G., Krennwallner, T.: Answer set programming: a primer. In: Tessaris, S., Franconi, E., Eiter, T., Gutierrez, C., Handschuh, S., Rousset, M.-C., Schmidt, R.A. (eds.) Reasoning Web 2009. LNCS, vol. 5689, pp. 40–110. Springer, Heidelberg (2009). doi:10.1007/978-3-642-03754-2_2
4. Gelfond, M., Lifschitz, V.: Classical negation in logic programs and disjunctive databases. New Gener. Comput. **9**(3/4), 365–385 (1991)
5. Marek, V.W., Truszczyński, M.: Stable models and an alternative logic programming paradigm. In: Apt, K.R., Marek, V.W., Truszczyński, M., Warren, D.S. (eds.) The Logic Programming Paradigm - A 25-Year Perspective, pp. 375–398. Springer, New York (1999)
6. Niemelä, I.: Logic programming with stable model semantics as constraint programming paradigm. Ann. Math. Artif. Intell. **25**(3–4), 241–273 (1999)
7. Gelfond, M., Lifschitz, V.: The stable model semantics for logic programming. In: Proceedings of the Fifth International Conference and Symposium Logic Programming, 15–19 August 1988, Seattle, WA, 2 Volumes, pp. 1070–1080. MIT Press, Cambridge (1988)
8. Calimeri, F., Gebser, M., Maratea, M., Ricca, F.: Design and results of the fifth answer set programming competition. Artif. Intell. **231**, 151–181 (2016)
9. Gebser, M., Maratea, M., Ricca, F.: What's hot in the answer set programming competition. In: Schuurmans, D., Wellman, M.P. (eds.) Proceedings of the 13th AAAI Conference on Artificial Intelligence, 12–17 February 2016, Phoenix, Arizona, USA, pp. 4327–4329. AAAI Press (2016)
10. Leone, N., Ricca, F.: Answer set programming: a tour from the basics to advanced development tools and industrial applications. In: Faber, W., Paschke, A. (eds.) Reasoning Web 2015. LNCS, vol. 9203, pp. 308–326. Springer, Heidelberg (2015). doi:10.1007/978-3-319-21768-0_10
11. Nogueira, M., Balduccini, M., Gelfond, M., Watson, R., Barry, M.: An a-prolog decision support system for the space shuttle. In: Ramakrishnan, I.V. (ed.) PADL 2001. LNCS, vol. 1990, pp. 169–183. Springer, Heidelberg (2001). doi:10.1007/3-540-45241-9_12
12. Ricca, F., Grasso, G., Alviano, M., Manna, M., Lio, V., Iiritano, S., Leone, N.: Team-building with answer set programming in the Gioia-Tauro seaport. Theory Pract. Logic Program. **12**(3), 361–381 (2012). Cambridge University Press

13. Tiihonen, J., Soininen, T., Niemelä, I., Sulonen, R.: A practical tool for mass-customising configurable products. In: Proceedings of the 14th International Conference on Engineering Design (ICED 2003), pp. 1290–1299 (2003)

14. Dal Palù, A., Dovier, A., Pontelli, E., Rossi, G.: GASP: answer set programming with lazy grounding. Fundamenta Informaticae **96**(3), 297–322 (2009)

15. Lefèvre, C., Nicolas, P.: A first order forward chaining approach for answer set computing. In: Erdem, E., Lin, F., Schaub, T. (eds.) LPNMR 2009. LNCS (LNAI), vol. 5753, pp. 196–208. Springer, Heidelberg (2009). doi:10.1007/978-3-642-04238-6_18

16. Lefèvre, C., Nicolas, P.: The first version of a new ASP solver : In: Erdem, E., Lin, F., Schaub, T. (eds.) LPNMR 2009. LNCS (LNAI), vol. 5753, pp. 522–527. Springer, Heidelberg (2009). doi:10.1007/978-3-642-04238-6_52

17. Leone, N., Pfeifer, G., Faber, W., Eiter, T., Gottlob, G., Perri, S., Scarcello, F.: The DLV system for knowledge representation and reasoning. ACM Trans. Comput. Logic (TOCL) **7**(3), 499–562 (2006)

18. Gebser, M., Kaminski, R., Kaufmann, B., Romero, J., Schaub, T.: Progress in clasp series 3, pp. 368–383 [40]

19. Gebser, M., Kaminski, R., König, A., Schaub, T.: Advances in *gringo* Series 3. In: Delgrande, J.P., Faber, W. (eds.) LPNMR 2011. LNCS (LNAI), vol. 6645, pp. 345–351. Springer, Heidelberg (2011). doi:10.1007/978-3-642-20895-9_39

20. Calimeri, F., Faber, W., Gebser, M., Ianni, G., Kaminski, R., Krennwallner, T., Leone, N., Ricca, F., Schaub, T.: ASP-Core-2: 4th ASP competition official input language format (2013). https://www.mat.unical.it/aspcomp2013/files/ASP-CORE-2.01c.pdf

21. Dantsin, E., Eiter, T., Gottlob, G., Voronkov, A.: Complexity and expressive power of logic programming. ACM Comput. Surv. **33**(3), 374–425 (2001)

22. Syrjänen, T.: Omega-restricted logic programs. In: Eiter, T., Faber, W., Truszczyński, M. (eds.) LPNMR 2001. LNCS (LNAI), vol. 2173, pp. 267–280. Springer, Heidelberg (2001). doi:10.1007/3-540-45402-0_20

23. Ullman, J.D.: Principles of Database and Knowledge-Base Systems, Volume I. Computer Science Press, New York (1988)

24. Faber, W., Leone, N., Perri, S.: The intelligent grounder of DLV. In: Erdem, E., Lee, J., Lierler, Y., Pearce, D. (eds.) Correct Reasoning. LNCS, vol. 7265, pp. 247–264. Springer, Heidelberg (2012). doi:10.1007/978-3-642-30743-0_17

25. Alviano, M., Dodaro, C., Leone, N., Ricca, F.: Advances in WASP, pp. 40–54 [40]

26. Leone, N., Perri, S., Scarcello, F.: Improving ASP instantiators by join-ordering methods. In: Eiter, T., Faber, W., Truszczyński, M. (eds.) LPNMR 2001. LNCS (LNAI), vol. 2173, pp. 280–294. Springer, Heidelberg (2001). doi:10.1007/3-540-45402-0_21

27. Perri, S., Scarcello, F., Catalano, G., Leone, N.: Enhancing DLV instantiator by backjumping techniques. Ann. Math. Artif. Intell. **51**(2–4), 195–228 (2007)

28. Alviano, M., Faber, W., Greco, G., Leone, N.: Magic sets for disjunctive datalog programs. Artif. Intell. **187**, 156–192 (2012)

29. Cumbo, C., Faber, W., Greco, G., Leone, N.: Enhancing the magic-set method for disjunctive datalog programs. In: Demoen, B., Lifschitz, V. (eds.) ICLP 2004. LNCS, vol. 3132, pp. 371–385. Springer, Heidelberg (2004). doi:10.1007/978-3-540-27775-0_26

30. Calimeri, F., Perri, S., Fuscà, D., Zangari, J.: *I*-DLV homepage (2016). https://github.com/DeMaCS-UNICAL/I-DLV/wiki

31. De Vos, M., Kisa, D.G., Oetsch, J., Pührer, J., Tompits, H.: Annotating answer-set programs in LANA. Theory Pract. Logic Program. **12**(4–5), 619–637 (2012)

32. Kulas, M.: Debugging prolog using annotations. Electron. Notes Theoret. Comput. Sci. **30**(4), 235–255 (1999)
33. Reeve, L., Han, H.: Survey of semantic annotation platforms. In: Proceedings of the 2005 ACM Symposium on Applied Computing, SAC 2005, pp. 1634–1638. ACM New York (2005)
34. Gebser, M., Maratea, M., Ricca, F.: The design of the sixth answer set programming competition, pp. 531–544 [40]
35. Liang, S., Fodor, P., Wan, H., Kifer, M.: OpenRuleBench: an analysis of the performance of rule engines. In: Proceedings of the 18th International Conference on World Wide Web, WWW 2009, 20–24 April 2009, Madrid, Spain, pp. 601–610. ACM (2009)
36. Swift, T., Warren, D.S.: XSB: extending prolog with tabled logic programming. Theory Pract. Logic Program. **12**(1–2), 157–187 (2012)
37. Liang, S., Fodor, P., Wan, H., Kifer, M.: OpenRuleBench: detailed report (2009). http://semwebcentral.org/docman/view.php/158/69/report.pdf
38. Wittocx, J., Denecker, M.: GidL: A grounder for FO$^+$. In: Proceedings of the Twelfth International Workshop on NonMonotonic Reasoning, pp. 189–198 (1998)
39. Calimeri, F., Perri, S., Fuscà, D., Zangari, J.: $\mathcal{I}$-DLV repository (2016). https://github.com/DeMaCS-UNICAL/I-DLV
40. Calimeri, F., Ianni, G., Truszczynski, M. (eds.): LPNMR 2015. LNCS, vol. 9345. Springer, Heidelberg (2015)

Abducing Compliance of Incomplete Event Logs

Federico Chesani[1], Riccardo De Masellis[2], Chiara Di Francescomarino[2],
Chiara Ghidini[2(✉)], Paola Mello[1], Marco Montali[3], and Sergio Tessaris[3]

[1] University of Bologna, Bologna, Italy
{federico.chesani,paola.mello}@unibo.it
[2] FBK-IRST, Via Sommarive 18, 38050 Trento, Italy
{r.demasellis,dfmchiara,ghidini}@fbk.eu
[3] Free University of Bozen–Bolzano, Piazza Università, 1, 39100 Bozen-Bolzano, Italy
{montali,tessaris}@inf.unibz.it

Abstract. The capability to store data about business processes execution in so-called Event Logs has brought to the diffusion of tools for the analysis of process executions and for the assessment of the *goodness* of a process model. Nonetheless, these tools are often very rigid in dealing with Event Logs that include incomplete information about the process execution. Thus, while the ability of handling incomplete event data is one of the challenges mentioned in the process mining manifesto, the evaluation of compliance of an execution trace still requires an end-to-end complete trace to be performed. This paper exploits the power of abduction to provide a flexible, yet computationally effective, framework to deal with different forms of incompleteness in an Event Log. Moreover it proposes a refinement of the classical notion of compliance into *strong* and *conditional* compliance to take into account incomplete logs.

Keywords: Abductive logic programming · Formal verification · Compliance in business process · Incompleteness in business processes

1 Introduction

The proliferation of IT systems able to store process executions traces in so-called event logs has originated a quest towards tools that offer the possibility of discovering, checking the conformance and enhancing process models based on actual behaviors [1]. Focusing on conformance, that is, on a scenario where the aim is to assess how a *prescriptive* (or "de jure") process model relates to the execution traces, a fundamental notion is the one of *trace compliance*. Compliance results can be used by business analysts to assess the goodness of a process model and understand how it relates to the actual behaviours exhibited by a company, consequently providing the basis for process re-design, governance and improvement.

The use of event logs to evaluate the goodness of a process model becomes hard and potentially misleading when the event log contains only partial information on the process execution. Thus, while the presence of non-monitorable activities (or errors in the logging procedure) makes the ability of handling incomplete

© Springer International Publishing AG 2016
G. Adorni et al. (Eds.): AI*IA 2016, LNAI 10037, pp. 208–222, 2016.
DOI: 10.1007/978-3-319-49130-1_16

event data one of the main challenges of the BP community, as mentioned in the process mining manifesto [1], still trace compliance of an execution trace requires the presence of a complete end-to-end execution trace to be evaluated. Notable exceptions are [2,3] where trace incompleteness is managed in an algorithmic/heuristic manner using log repair techniques.

In this paper, we take an orthogonal approach and throughly address the problem of log incompleteness from a logic-based point of view, adopting an approach based on *abduction* [4]. Differently from techniques that focus on algorithmic/heuristic repairs of an incomplete trace, we are interested in characterising the notion of incomplete log compliance by means of a sound and complete inference procedure. We rely on abduction to combine the partial knowledge about the real executions of a process as reflected by a (potentially) incomplete event log, with the background knowledge captured in a process model. In particular, abductive reasoning handles different forms of missing information by formulating *hypotheses* that explain how the event log may be "completed" with the missing information, so as to reconcile it with the process model. This leads us to refine the classical notion of *conformance-by-alignment* [5] between an execution trace and a process model into **strong** and **conditional** compliance, to account for incompleteness. In detail, the paper provides: (i) a classification of different forms of incompleteness of an event log based on three dimensions: log incompleteness, trace incompleteness, and event description incompleteness (Sect. 2.1); (ii) a reformulation of the notion of compliance into strong and conditional compliance (Sect. 2.2); and (iii) an encoding of structured process models[1] and of event logs in the SCIFF abductive logic framework [8], and a usage of the SCIFF proof procedure to compute strong, conditional and non-compliance with incomplete event log (Sect. 3). The ideas are illustrated by means of a simple example, and related work is contained in Sect. 4.

2   Dealing with Incomplete Event Logs

We aim at solving the problem of the post mortem identification of compliant traces in the presence of incomplete event logs, given the *prescriptive* knowledge contained in a process model. To do this, we first investigate what incomplete event logs are (Sect. 2.1) and then how we can adapt the notion of compliance to deal with incomplete logs (Sect. 2.2). We perform this investigation with the help of a simple example, described using the BPMN (Business Process Modeling Notation) language[2].

Example 1 (Obtaining a Permit of Stay in Italy). Consider the BPMN process in Fig. 1, hereafter called the Permit-Of-Stay (POS) process, which takes inspiration from the procedure for the granting of a permit of stay in Italy.

Upon her arrival in Italy (AI), the person in need of a permit of stay has three different alternatives: if she is from a EU country and remains in Italy for at

---

[1] We follow previous work in the area of BPM and focus on structured process models and on models with no repeating tasks, in the spirit of [6,7], respectively.

[2] For the sake of clarity we use BPMN, but our framework is language-independent.

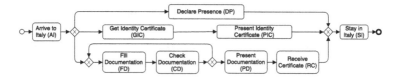

Fig. 1. A process for obtaining a permit of stay (in Italy).

most 30 days, then only indicating her presence in Italy (DP) is needed; if she is from the EU and must remain in Italy for more than 30 days, then she needs to get an identity certificate (GIC) and present it (PIC). In all the remaining cases she needs to fill a documentation (FD) which is then checked (CD). When the documentation is correct, it is presented (PD) and a certificate is received (RC). The procedure concludes with the provision of the permit of stay (SI).

2.1 Classifying Process Execution (In)Completeness

We assume that each execution of the POS process in Fig. 1 is (partially) monitored and logged by an information system. We also assume that activities are atomic, i.e., executing an activity results in an event associated to a single timestamp: event (A, t) indicates that activity A has been executed at time t. A sample trace[3] that logs the execution of a POS instance is:

$$\{(AI, t_1), (FD, t_2), (CD, t_3), (PD, t_4), (RC, t_5), (SI, t_6)\} \tag{1}$$

where $t_i > t_j$ for $i, j \in \{1, \ldots, 6\}$ such that $i > j$. This trace corresponds to the execution of the lower branch of the POC process, where the loop is never executed. A set of execution traces of the same process form an event log.

In many real cases, a number of difficulties may arise when exploiting the data contained in an information system in order to build an event log. Thus, instead of the extremely informative trace reported in (1), we may obtain something like:

$$\{(FD, \_), (\_, t_2), (SI, t_6), (\_, \_)\} \tag{2}$$

This trace does not completely describe an execution of the POS process. For example, the first event logged in the trace is FD. However, by looking at the process description, it is easy to see that the first event of every execution has to be AI. By assuming that the process executors indeed followed the prescriptions of the model, this suggests that the AI-related event has not been logged. Moreover, certain events have been only partially observed. For example, the FD-related event is incomplete, because its exact timestamp is unknown. In this paper, we use "$\_$" to denote a missing information unit.

In accordance with the IEEE standard XES format for representing event logs [9], in general we can describe an event log as a set of execution traces. Each trace,

---

[3] We often present the events in a trace ordered according to their execution time. This is only to enhance readability since the position of an event is fully determined by its timestamp, or unknown if the timestamp is missing.

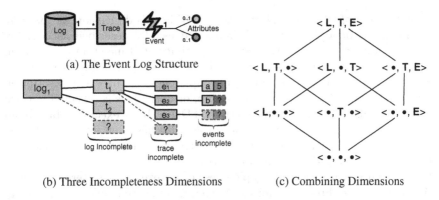

(a) The Event Log Structure

(b) Three Incompleteness Dimensions   (c) Combining Dimensions

Fig. 2. Classifying (in)completeness

in turn, contains events, which are described by means of n-tuples, where each element of the tuple is the value of a given attribute (see Fig. 2a, where we restrict to two attributes as we do in the paper). Consequently, we can classify incompleteness along these three dimensions: incompleteness of the log, incompleteness of the trace, and incompleteness of the event description (see Fig. 2b).

(In)Completeness of the Log. Within this dimension we analyse whether the log contains at least one instance for each possible execution that is allowed in the model. Note that one can account for this form of (in)completeness only by: (a) limiting the analysis to the control flow, without considering complex data objects that may contain values from an unbounded domain; and (b) assuming that there is a maximum length for all traces, thus limiting the overall number of traces that may originate from the unbounded execution of loops. An example of complete log for the POS process is:

$$L_1 = \left\{ \begin{array}{l} \{(\text{AI}, t_{a1}), (\text{DP}, t_{a2}), (\text{SI}, t_{a3})\}, \{(\text{AI}, t_{b1}), (\text{GIC}, t_{b2}), (\text{PIC}, t_{b3})(\text{SI}, t_{b4})\}, \\ \{(\text{AI}, t_{c1}), (\text{FD}, t_{c2}), (\text{CD}, t_{c3}), (\text{PD}, t_{c4}), (\text{RC}, t_{c5}), (\text{SI}, t_{c6})\} \end{array} \right\} \quad (3)$$

where we assume that each trace cannot contain more than 6 event, which intuitively means that the loop is never executed twice.

Assuming this form of completeness is not strictly required to have good process models, and could be unrealistic in practice. In fact, even under the assumption of a maximum trace length, the number of allowed traces could become extremely huge due to (bounded) loops, and the (conditional) interleavings generated by parallel blocks and or choices. Still, analysing the (in)completeness of an event log may be useful to discover parts of the control flow that never occur in practice.

(In)Completeness of the Trace. Within this dimension we focus on a single trace, examining whether it contains a sequence of events that corresponds to an execution foreseen by the process model from start to end. Trace (1) is an example of complete trace. An example of incomplete trace is:

$$\{(\text{AI}, t_1), (\text{PIC}, t_2)(\text{SI}, t_3)\} \quad (4)$$

This trace should also contain an event of the form (GIC, t), s.t. $t_1 < t < t_2$.

(In)Completeness of the Event Description. Within this dimension we focus on the completeness of a single event. Events are usually described as complex objects containing data about the executed activity, its time stamp, and so on [9]. These data can be missing or corrupted. As pointed out before, we consider activity names and timestamps. Thus, incompleteness in the event description may concern the activity name, its timestamp, or both. This is reflected in trace (2): *(i)* event $(\mathsf{FD}, \_)$ indicates that activity FD has been executed, but at an unknown time; *(ii)* $(\_, t_2)$ witnesses that an activity has been executed at time t_2, but we do not know which one; *(iii)* $(\_, \_)$ attests that the trace contains some event, whose activity and time are unknown.

We can characterise the (in)completeness of an event log in terms of (any) combination of these three basic forms. At one extreme, we may encounter a log, such as (3), that is complete along all three dimensions. At the other extreme, we may have a log such as:

$$L_2 = \{\{(\mathsf{AI}, \_), (\_, t_{a2})\}, \{(\mathsf{AI}, t_{b1}), (\_, \_), (\_, t_{b2}), (\mathsf{SI}, t_{b3})\}\} \qquad (5)$$

that is incomplete along all the dimensions. Intermediate situations may arise as well, as graphically depicted in the lattice of Fig. 2c, where $\langle L, T, E \rangle$ indicates the top value (completeness for all three dimensions) and $\langle \bullet, \bullet, \bullet \rangle$ indicated the bottom value (incompleteness of all three dimensions).

2.2   Refining the Notion of Compliance

In our work we consider *prescriptive* process models, that is, models that describe the only acceptable executions. These correspond to the so-called "de jure" models in [5], and consequently call for a definition of *compliance*, so as to characterise the degree to which a given trace conforms/is aligned to the model. The notion of compliance typically requires that the trace represents an end-to-end, valid execution that can be fully replayed on the process model. We call this notion of compliance **strong** compliance. Trace (1) is an example of trace that is fully compliant to the POS process.

Strong compliance is too restrictive when the trace is possibly incomplete. In fact, incompleteness hinders the possibility of replaying it on the process model. However, conformance might be regained by assuming that the trace included additional information on the missing part; in this case we say that the trace is **conditionally** compliant, to reflect that compliance conditionally depends on how the partial information contained in the trace is complemented with the missing one. Consider the partial trace:

$$\{(\mathsf{AI}, t_1), (\mathsf{GIC}, \_), (\mathsf{SI}, t_3)\} \qquad (6)$$

It is easy to see that (6) is compliant with POS, *if* we assume that

$$\mathsf{GIC} \text{ was executed at a time } t_i \text{ s.t. } t_1 < t_i < t_3 \qquad (7)$$

$$\text{an execution of } \mathsf{PIC} \text{ was performed at a time } t_j \text{ s.t. } t_i < t_j < t_3 \qquad (8)$$

Note that the set of assumptions needed to reconstruct full conformance is not necessarily unique. This reflects that alternative strongly compliant real process executions might have led to the recorded partial trace. On the other hand, there are cases in which no assumptions can lead to full conformance. In this case, the partial trace is considered **non**-compliant. For example, the following trace does not comply with POS, since it records that GIC and CD have been both executed, although they belong to mutually exclusive branches in the model:

$$\{(\mathsf{AI}, t_1), (\mathsf{GIC}, \_), (\mathsf{CD}, t_2), (\mathsf{SI}, t_3)\} \tag{9}$$

3 Abduction and Incomplete Logs

Since the aim of this paper is to provide automatic procedures that identify compliant traces in the presence of incomplete event logs, given the *prescriptive* knowledge contained in a process model, we can schematise the input to our problem in three parts: (i) an instance-independent component, the process model, which in this paper is described using BPMN; (ii) an instance-specific component, that is, the (partial) log, and (iii) meta-information attached to the activities in the process model, indicating which are actually *always, never* or *possibly* observable (that is, logged) in the event log. The third component is an extension of a typical business process specification that we propose (following and extending the approach described in [10]) to provide *prescriptive* information about the (non-) observability of activities. Thus, for instance, a business analyst will have the possibility to specify that a certain manual activity is never observable while a certain interaction with a web site is always (or possibly) observable. This information can then be used to compute the compliance of a partial trace. In fact the presence of never observable activities will trigger the need to make hypothesis on their execution (as they will never be logged in the event log), while the presence of always observable activities will trigger the need to find their corresponding event in the execution trace (to retain compliance). Note that this extension is not invasive w.r.t. current approaches to business process modelling, as we can always assume that a model where no information on observability is provided is entirely possibly observable.

Given the input of our problem, in Sect. 3.1 we provide an overview on abduction and on how the SCIFF framework represents *always, never* or *possibly* observable activities; in Sect. 3.2 we show how to use SCIFF to encode a process model and a partial log; in Sect. 3.3 we show how we can formalize the informal different forms of compliance presented in Sect. 2.2; finally, in Sect. 3.4 we illustrate how SCIFF can be used to solve the different forms of incompleteness identified in Sect. 2.1.

3.1 The SCIFF in Short

Abduction is a non-monotonic reasoning process where hypotheses are made to explain observed facts [11]. While deductive reasoning focuses on deciding if a formula ϕ logically follows from a set Γ of logical assertions known to hold, in abductive reasoning it is assumed that ϕ holds (as it corresponds to a set of observed facts) but it cannot be directly inferred by Γ. To make ϕ a consequence of Γ, abduction looks for a further set Δ of hypothesis, taken from a given set of abducibles $\mathcal{A}$, which completes Γ in such a way that ϕ can be inferred (in symbols $\Gamma \cup \Delta \models \phi$). The set Δ is called *abductive explanation* (of ϕ). In addition, Δ must usually satisfy a set of (domain-dependent) integrity constraints $\mathcal{IC}$ (in symbols, $\Gamma \cup \Delta \models \mathcal{IC}$). A typical integrity constraint (IC) is a *denial*, which expresses that two explanations are mutually exclusive.

Abduction has been introduced in Logic Programming in [4]. There, an *Abductive Logic Program (ALP)* is defined as a triple $\langle \Gamma, \mathcal{A}, \mathcal{IC} \rangle$, where: *(i)* Γ is a logic program, *(ii)* $\mathcal{A}$ is a set of abducible predicates, and *(iii)* $\mathcal{IC}$ a set of ICs. Given a goal ϕ, abductive reasoning looks for a set of positive, atomic literals $\Delta \subseteq \mathcal{A}$[4] such that they entail $\phi \cup \mathcal{IC}$.

In this paper we leverage on the SCIFF abductive logic programming framework [8], an extension of the IFF abductive proof procedure [12], and on its efficient implementation in CHR [13]. Beside the general notion of abducible, the SCIFF framework has been enriched with the notions of *happened event*, *expectation*, and *compliance* of an observed execution with a set of expectations. This makes SCIFF suitable for dealing with event log incompleteness. Let a be an event corresponding to the execution of a process activity, and T (possibly with subscripts) its execution time[5]. Abducibles are used here to make hypotheses on events that are not recorded in the examined trace. They are denoted using **ABD**(a, T). Happened events are non-abducible, and account for events that have been logged in the trace. They are denoted with **H**(a, T). Expectations **E**(a, T), instead, model events that should occur (and therefore should be present in a trace). Compliance is described in Sect. 3.3.

ICs in SCIFF are used to relate happened events/abduced predicates with expectations/predicates to be abduced. Specifically, an IC is a rule of the form *body* $\rightarrow$ *head*, where *body* contains a conjunction of happened events, general abducibles, and defined predicates, while *head* contains a disjunction of conjunctions of expectations, general abducibles, and defined predicates. Usually, variables appearing in the body are quantified universally, while variables appearing in the head are quantified existentially.

3.2 Encoding Structured Processes and Their Executions in SCIFF

Let us illustrate how to encode all the different components of an (incomplete) event log and a structured process model one by one.

Event Log. A log is a set of traces, each constituted by a set of observed (atomic) events. Thus trace (4) is represented in SCIFF as $\{\mathbf{H}(\mathsf{Al}, t_1), \mathbf{H}(\mathsf{PIC}, t_2), \mathbf{H}(\mathsf{SI}, t_3)\}$.

---

[4] We slightly abuse the notation of $\subseteq$, meaning that every positive atomic literal in Δ is the instance of a predicate in $\mathcal{A}$.

[5] In the remainder of this paper we will assume that the time domain relies on natural numbers.

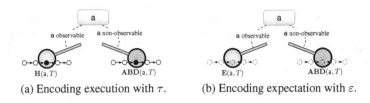

(a) Encoding execution with τ.   (b) Encoding expectation with ε.

Fig. 3. Encoding always/never observable activities.

Always/never Observable Activities. Coherently with the representation of an execution trace, the logging of the execution of an observable activity is represented in SCIFF using an happened event, whereas the hypothesis on the execution of a never observable activity is represented using an abducible **ABD** (see Fig. 3a). Given an event a occurring at T, we use a function τ that represents the execution of a as:

$$\tau(\mathsf{a}, T) = \begin{cases} \mathbf{H}(\mathsf{a}, T) & \text{if a is observable} \\ \mathbf{ABD}(\mathsf{a}, T) & \text{if a is never observable} \end{cases}$$

As for expected occurrences, the encoding again depends on the observability of the activity: if the activity is observable, then its expected occurrence is mapped to a SCIFF expectation; otherwise, it is hypothesised using the aforementioned abducible **ABD** (see Fig. 3b). To this end we use a function ε that maps the expecting of the execution of a at time T as follows:

$$\varepsilon(\mathsf{a}, T) = \begin{cases} \mathbf{E}(\mathsf{a}, T) & \text{if a is observable} \\ \mathbf{ABD}(\mathsf{a}, T) & \text{if a is never observable} \end{cases}$$

Structured Process Model Constructs. A process model is encoded in SCIFF by generating ICs that relate the execution of an activity to the future, expected executions of further activities. In practice, each process model construct is transformed into a corresponding IC. We handle, case-by-case, all the single-entry single-exit block types of structured process models.

Sequence. Two activities a and b are in sequence if, whenever the first is executed, the second is expected to be executed at a later time:

$$\tau(\mathsf{a}, T_a) \to \varepsilon(\mathsf{b}, T_b) \wedge T_b > T_a. \tag{10}$$

Notice that T_a is quantified universally, while T_b is existentially quantified.

And-split activates parallel threads spanning from the same activity. In particular, the fact that activity a triggers two parallel threads, one expecting the execution of b, and the other that of c, is captured using an IC with a conjunctive consequent:

$$\tau(\mathsf{a}, T_a) \to \varepsilon(\mathsf{b}, T_b) \wedge T_b > T_a \wedge \varepsilon(\mathsf{c}, T_c) \wedge T_c > T_a.$$

And-join mirrors the and-split, synchronizing multiple concurrent execution threads and merging them into a single thread. When activities a and b are both executed, then activity c is expected next, is captured using an IC with a conjunctive antecedent:

$$\tau(\mathsf{a}, T_a) \wedge \tau(\mathsf{b}, T_b) \rightarrow \varepsilon(\mathsf{c}, T_c) \wedge T_c > T_a \wedge T_c > T_b.$$

The encoding of **Xor-split/Xor-join** and **Or-split/Or-join** can be found in [14].

Possibly Observable Activities. A possibly observable activity is managed by considering the disjunctive combination of two cases: one in which it is assumed to be observable, and one in which it is assumed to be never observable. This idea is used to refine ICs used to encode the workflow constructs in the case of partial observability. For instance, if a partially observable activity appears in the antecedent of an IC, two distinct ICs are generated, one where the activity is considered to be observable (**H**), and another in which it is not (**ABD**). Thus in the case of a sequence flow from a to b, where a is possibly observable and b is observable, IC (10) generates:

$$\mathbf{H}(\mathsf{a}, T_a) \rightarrow \varepsilon(\mathsf{b}, T_b) \wedge T_b > T_a. \qquad \mathbf{ABD}(\mathsf{a}, T_a) \rightarrow \varepsilon(\mathsf{b}, T_b) \wedge T_b > T_a.$$

If multiple partially observable activities would appear in the antecedent of an IC (as, e.g., in the and-join case), then all combinations have to be considered.

Similarly, if a partially observable activity appears in the consequent of an IC, a disjunction must be inserted in the consequent, accounting for the two possibilities of observable/never observable event. If both the antecedent and consequent of an IC would contain a partially observable activity, a combination of the rules above will be used. For example, in the case of a sequence flow from a to b, where b is possibly observable, IC (10) generates:

$$\mathbf{H}(\mathsf{a}, T_a) \rightarrow \mathbf{E}(\mathsf{b}, T_b) \wedge T_b > T_a \vee \mathbf{ABD}(\mathsf{b}, T_b) \wedge T_b > T_a. \tag{11}$$

With this encoding, the SCIFF proof procedure generates firstly an abductive explanation Δ containing an expectation about the execution of b. If no b is actually observed, Δ is discarded, and a new abductive explanation Δ' is generated containing the hypothesis about b (i.e., $\mathbf{ABD}(\mathsf{b}, T_b) \in \Delta'$). Mutual exclusion between these two possibilities is guaranteed by the SCIFF declarative semantics (cf. Definition 3).

Finally, if both the antecedent and consequent of an IC would contain a possibly observable activity, a combination of the rules above will be used.

Start and End of the Process. We introduce two special activities start and end representing the entry- and exit-point of the process. Two specific ICs are introduced to link these special activities with the process. For example, if the first process activity is a (partially observable), the following IC is added:

$$\mathbf{ABD}(\mathsf{start}, 0) \rightarrow \mathbf{E}(\mathsf{a}, T_a) \wedge T_a > 0 \vee \mathbf{ABD}(\mathsf{a}, T_a) \wedge T_a > 0.$$

To ensure the IC triggering, $\mathbf{ABD}(\mathsf{start}, 0)$ is given as goal to the proof procedure.

3.3 Compliance in SCIFF: Declarative Semantics

We are now ready to provide a formal notion of compliance in its different forms. We do so by extending the SCIFF declarative semantics provided in [8] to incorporate log incompleteness (that is, observability features).

A structured process model corresponds to a SCIFF specification $S = \langle \mathcal{KB}, \mathcal{A}, \mathcal{IC} \rangle$, where: *(i)* $\mathcal{KB}$ is a Logic Program [15] containing the definition of accessory predicates; *(ii)* $\mathcal{A} = \{ \mathbf{ABD}/2, \mathbf{E}/2 \}$, possibly non-ground; *(iii)* $\mathcal{IC}$ is a set of ICs constructed by following the encoding defined in Sect. 3.2. A (execution) trace and an abductive explanation Δ are defined as follows[6]:

Definition 1. *A Trace T is a set of terms of type $\mathbf{H}(e, T_i)$, where e is a term describing the happened event, and $T_i \in \mathbb{N}$ is the time instant at which the event occurred.*

Definition 2. *Given a SCIFF specification S and a trace T, a set $\Delta \subseteq \mathcal{A}$ is an* abductive explanation *for $\langle S, T \rangle$ if and only if $Comp\,(\mathcal{KB} \cup T \cup \Delta) \cup$ CET $\cup T_\mathbb{N} \models \mathcal{IC}$ where* Comp *is the (two-valued) completion of a theory [16], CET stands for Clark Equational Theory [17] and $T_\mathbb{N}$ is the CLP constraint theory [18] on finite domains.*

The following definition fixes the semantics for observable events, and provides the basis for understanding the alignment of a trace with a process model.

Definition 3 (T-Fulfillment). *Given a trace T, an abducible set Δ is T-fulfilled if for every event e and for each time t, $\mathbf{E}(e, t) \in \Delta$ if and only if $\mathbf{H}(e, t) \in T$.*

The "only if" direction defines the semantics of expectation, indicating that an expectation is fulfilled when it finds the corresponding happening event in the trace. The "if" direction captures the prescriptive nature of process models, whose *closed* nature requires that only expected events may happen.

Given an abductive explanation Δ, fulfilment acts as a *compliance classifier*, which separates the legal/correct execution traces with respect to Δ from the wrong ones. Expectations however model the strong flavour of compliance: if T-Fulfillment cannot be achieved because some expectations are not matched by happened events, a **ABD** predicate is abduced as specified in the integrity constraints (see for example the IC 11).

Definition 4 (Strong/Conditional Compliance). *A trace T is compliant with a SCIFF specification S if there exists an abducible set Δ such that: (i) Δ is an abductive explanation for $\langle S, T \rangle$, and (ii) Δ is T-fulfilled. If Δ does not contains any **ABD**(besides the special abducibles for* start *and* end*), then we say that it is* strongly-compliant, *otherwise it is* conditionally-compliant.

If no abductive explanation that is also T-fulfilled can be found, then T is not compliant with the specification of interest. Contrariwise, the abductive explanation witnesses compliance. However, it may contain (possibly non-ground)

---

[6] We do not consider the abductive goal, as it is not needed for our treatment.

ABD predicates, abduced due to the incompleteness of $\mathcal{T}$. In fact, the presence or absence of such predicates determines whether $\mathcal{T}$ is conditionally or strongly compliant. To make an example let us consider traces (6), (1), and (9). In the case of partial trace (6), SCIFF will tell us that it is conditional compliant with the workflow model POS since Δ will contain the formal encoding of the two abducibles (7) and (8) which provide the *abductive explanation* of trace (6). In the case of trace (1), abduction will tell us that it directly follows from Γ without the need of any hypothesis. The case where Δ does not contain any **ABD** coincides in fact, with the classical notion of (deductive) compliance. Finally, in the case of trace (9) SCIFF will tell us that it is not possible to find any set of hypothesis Δ that explains it. This case coincides with the classical notion of (deductive) non-compliance.

We close this section by briefly arguing that our approach is indeed correct. To show correctness, one may proceed in two steps: *(i)* prove the semantic correctness of the encoding w.r.t. semantics of (conditional/strong) compliance; *(ii)* prove the correctness of the proof procedure w.r.t. the SCIFF declarative semantics. Step (i) requires to prove that a trace is (conditionally/strong) compliant (in the original execution semantics of the workflow) with a given workflow if and only if the trace is (conditionally/strong) compliant (according to the SCIFF declarative semantics) with the encoding of the workflow in SCIFF. This can be done in the spirit of [19] (where correctness is proven for declarative, constraint-based processes), by arguing that structured processes can be seen as declarative processes that only employ the "chain-response constraint" [19]. For step (ii), we rely on [8], where soundness and completeness of SCIFF w.r.t. its declarative semantics is proved by addressing the case of closed workflow models (the trace is closed and no more events can happen anymore), as well as that of open workflow models (future events can still happen). Our declarative semantics restricts the notions of fulfilment and compliance to a specific current time t_c, i.e., to open traces: hence soundness and completeness still hold.

3.4   Dealing with Process Execution (In)Completeness in SCIFF

We have already illustrated, by means of the POS example, how Definition 4 can be used to address compliance of a partial trace. In this section we illustrate in detail how SCIFF can be used to solve the three dimensions of incompleteness identified in Sect. 2.1.

Trace and event incompleteness are dealt by with SCIFF in a uniform manner. In fact, the **trace/event incompleteness problem** amounts to check if a given log (possibly equipped with incomplete traces/events), is compliant with a prescriptive process model. We consider as input the process model, together with information about the observability of its activities, a trace, and a maximum length for completed traces. The compliance is determined by executing the SCIFF proof procedure and evaluating possible abductive answers. We proceed as follows:

1. We automatically translate the process model with its observability meta-information into a SCIFF specification. If observability information is missing

for some/all the activities, we can safely assume that some/all activities are possibly observable.

2. The SCIFF proof procedure is applied to the SCIFF specification and to the trace under observation, computing *all* the possible abductive answers Δ_i. The maximum trace length information is used to limit the search, as in the unrestricted case the presence of loop may lead to nontermination.

3. If no abductive answer is generated, the trace is deemed as non-compliant. Otherwise, a set of abductive answers $\{\Delta_1, \ldots, \Delta_n\}$ has been found. If there exists a Δ_i that does not contain any **ABD** predicate, then the trace is strongly compliant. The trace is conditionally compliant otherwise.

Note that, assessing strong/conditional compliance requires the computation of *all* the abductive answers, thus affecting the performances of the SCIFF proof procedure. If only compliance is needed (without classifying it in strong or conditional), it is possible to compute only the first solution.

A different scenario is provided by the **log incompleteness problem**, which instead focuses on an entire event log, and looks if some possible traces allowed by the model are indeed missing in the log. In this case we consider as input the process model, a maximum length for the completed traces, and a log consisting of a number of different traces; we assume each trace is trace- and event-complete. We proceed as follows:

1. We generate the SCIFF specification from the process model, considering all activities as never observable (i.e., their happening must be always hypothesized, so that we can generate all the possible traces).

2. The SCIFF proof procedure is applied to the SCIFF specification. *All* the possible abductive answers Δ_i are computed, with maximum trace length as specified. Each answer corresponds to a different execution instance allowed by the model. Since all the activities are never observable, the generated Δ_i will contain only **ABD**.

3. For each hypothesised trace in the set $\{\Delta_1, \ldots, \Delta_n\}$, a corresponding, distinct trace is looked for in the log. If all the hypothesised traces have a distinct matching observed trace, then the log is deemed as complete.

Notice that, beside the completeness of the log, the proof procedure also generate the missing traces, defined as the Δ_i that do not have a corresponding trace in the log.

An evaluation of the algorithms above, and a study of how different inputs affect their performances is provided in [14] and omitted for lack of space. It shows that the performance of the abductive procedure to evaluate compliance ranges from few seconds when at most a single event description is completely unknown to about 4.5 min when up to 4 event descriptions are missing. A prototype implementation is currently available for download at http://ai.unibo.it/AlpBPM.

4   Related Work

The problem of incomplete traces has been tackled by a number of works in the field of process discovery and conformance. Some of them have addressed the problem of aligning event logs and procedural/declarative process models [2,3]. Such works explore the search space of the set of possible moves to find the best one for aligning the log to the model. Our purpose is not managing generic misalignments between models and logs, but rather focus on a specific type of incompleteness: the model is correct and the log can be incomplete.

We can divide existing works that aim at constructing possible model-compliant "worlds" out of a set of incomplete observations in two groups: quantitative and qualitative approaches. The former rely on the availability of a probabilistic model of execution and knowledge. For example, in [20] the authors exploit stochastic Petri nets and Bayesian Networks to recover missing information. The latter stand on the idea of describing "possible outcomes" regardless of likelihood. For example, in [21] and in [10] the authors exploit Satisfiability Modulo Theory and planning techniques respectively to reconstruct missing information. A different line of work addresses problems of compliance through model checking techniques [22,23]. Here the focus is verifying a broad class of temporal properties rather than specific issues related to incompleteness, which we believe are more naturally represented by abductive techniques.

In this work, the notion of incompleteness has been investigated to take into account its different variants (*log*, *trace* and *event incompleteness*). Similarly, the concept of *observability* has been deeply investigated, by exploring activities *always*, *partially* or *never* observable. This has led to a novel refinement of the notion of compliance.

Abduction and the SCIFF framework have been previously used to model both procedural and declarative processes. In [24], a structured workflow language has been defined, with a formal semantics in SCIFF. In [25], SCIFF has been exploited to formalize and reason about the declarative workflow language Declare.

An interesting work where trace compliance is evaluated through abduction is presented in [26]. Compliance is defined as assessing if actions were executed by users with the right permissions (auditing), and the focus is only on incomplete traces (with complete events). The adopted abductive framework, CIFF [27], only supports ground abducibles, and ICs are limited to denials. The work in [26] explores also the dimension of human confirmation of hypotheses, and proposes a human-based refinement cycle. This is a complementary step with our work, and would be an interesting future direction.

5   Conclusions

We have presented an abductive framework to support business process monitoring (and in particular compliance checking) by attacking the different forms of incompleteness that may be present in an event log. Concerning future development, the SCIFF framework is based on first-order logic, thus paving the way towards (i) the incorporation of data [23], (ii) extensions to further types of workflows (e.g., temporal workflows as in [28]), and (iii) towards the investigation of probabilistic models to deal with incompleteness of knowledge.

References

1. Aalst, W.: Process mining manifesto. In: Daniel, F., Barkaoui, K., Dustdar, S. (eds.) BPM 2011. LNBIP, vol. 99, pp. 169–194. Springer, Heidelberg (2012). doi:10. 1007/978-3-642-28108-2_19
2. Adriansyah, A., van Dongen, B.F., van der Aalst, W.M.P.: Conformance checking using cost-based fitness analysis. In: Proceedings of EDOC. IEEE Computer Society (2011)
3. Leoni, M., Maggi, F.M., Aalst, W.M.P.: Aligning event logs and declarative process models for conformance checking. In: Barros, A., Gal, A., Kindler, E. (eds.) BPM 2012. LNCS, vol. 7481, pp. 82–97. Springer, Heidelberg (2012). doi:10.1007/978-3-642-32885-5_6
4. Kakas, A.C., Kowalski, R.A., Toni, F.: Abductive logic programming. J. Log. Comput. 2(6), 719–770 (1992). http://dblp.uni-trier.de/rec/bibtex/journals/logcom/KakasKT92
5. van der Aalst, W.M.P.: Process Mining - Discovery, Conformance and Enhancement of Business Processes. Springer, Heidelberg (2011)
6. Kiepuszewski, B., ter Hofstede, A.H.M., Bussler, C.J.: On structured workflow modelling. In: Bubenko, J., Krogstie, J., Pastor, O., Pernici, B., Rolland, C., Sølvberg, A. (eds.) 25 Years of CAiSE. Springer, Heidelberg (2013). doi:10.1007/978-3-642-36926-1_19
7. van der Aalst, W.M.P., Weijters, T., Maruster, L.: Workflow mining: discovering process models from event logs. IEEE Trans. Knowl. Data Eng. 16, 1128–1142 (2004)
8. Alberti, M., Chesani, F., Gavanelli, M., Lamma, E., Mello, P., Torroni, P.: Verifiable agent interaction in abductive logic programming: the SCIFF framework. ACM Trans. Comput. Log. 9(4), 29:1–29:43 (2008). http://dl.acm.org/citation.cfm?doid=1380572.1380578
9. On process mining, I.T.F.: XES standard definition (2015). http://www.xes-standard.org/
10. Di Francescomarino, C., Ghidini, C., Tessaris, S., Sandoval, I.V.: Completing workflow traces using action languages. In: Zdravkovic, J., Kirikova, M., Johannesson, J. (eds.) CAiSE 2015. LNCS, vol. 9097, pp. 314–330. Springer, Heidelberg (2015). doi:10.1007/978-3-319-19069-3_20
11. Kakas, A.C., Mancarella, P.: Abduction and abductive logic programming. In: Proceedings of ICLP (1994)
12. Fung, T.H., Kowalski, R.A.: The iff proof procedure for abductive logic programming. J. Log. Program. 33(2), 151–165 (1997). http://dblp.uni-trier.de/rec/bibtex/journals/jlp/FungK97
13. Alberti, M., Gavanelli, M., Lamma, E.: The CHR-based implementation of the SCIFF abductive system. Fundam. Inform. 124, 365–381 (2013)
14. Chesani, F., De Masellis, R., Di Francescomarino, C., Ghidini, C., Mello, P., Montali, M., Tessaris, S.: Abducing compliance of incomplete event logs. Technical report submit/1584687, arXiv (2016)
15. Lloyd, J.W.: Foundations of Logic Programming, 2nd edn. Springer, Heidelberg (1987)
16. Kunen, K.: Negation in logic programming. J. Log. Program. 4(4), 289–308 (1987). http://dblp.uni-trier.de/rec/bibtex/journals/jlp/Kunen87
17. Clark, K.L.: Negation as Failure. In: Proceedings of Logic and Data Bases. Plenum Press (1978)

18. Jaffar, J., Maher, M.J., Marriott, K., Stuckey, P.J.: The semantics of constraint logic programs. J. Log. Program. **37**(1–3), 1–46 (1998). http://dblp.uni-trier.de/rec/bibtex/journals/jlp/JaffarMMS98

19. Montali, M.: Specification and Verification of Declarative Open Interaction Models: A Logic-Based Approach. LNBIP, vol. 56. Springer, Heidelberg (2010)

20. Rogge-Solti, A., Mans, R.S., Aalst, W.M.P., Weske, M.: Improving documentation by repairing event logs. In: Grabis, J., Kirikova, M., Zdravkovic, J., Stirna, J. (eds.) PoEM 2013. LNBIP, vol. 165, pp. 129–144. Springer, Heidelberg (2013). doi:10.1007/978-3-642-41641-5_10

21. Bertoli, P., Francescomarino, C., Dragoni, M., Ghidini, C.: Reasoning-based techniques for dealing with incomplete business process execution traces. In: Baldoni, M., Baroglio, C., Boella, G., Micalizio, R. (eds.) AI*IA 2013. LNCS (LNAI), vol. 8249, pp. 469–480. Springer, Heidelberg (2013). doi:10.1007/978-3-319-03524-6_40

22. Bagheri Hariri, B., Calvanese, D., De Giacomo, G., Deutsch, A., Montali, M.: Verification of relational data-centric dynamic systems with external services, pp. 163–174. ACM Press (2013)

23. De Masellis, R., Maggi, F.M., Montali, M.: Monitoring data-aware business constraints with finite state automata. In: Proceedings of ICSSP. ACM Press (2014)

24. Chesani, F., Mello, P., Montali, M., Storari, S.: Testing careflow process execution conformance by translating a graphical language to computational logic. In: Bellazzi, R., Abu-Hanna, A., Hunter, J. (eds.) AIME 2007. LNCS (LNAI), vol. 4594, pp. 479–488. Springer, Heidelberg (2007). doi:10.1007/978-3-540-73599-1_64

25. Montali, M., Pesic, M., van der Aalst, W.M.P., Chesani, F., Mello, P., Storari, S.: Declarative specification and verification of service choreographiess. TWEB **4**(1), 3:1–3:62 (2010). http://dl.acm.org/citation.cfm?doid=1658373.1658376

26. Mian, U.S., den Hartog, J., Etalle, S., Zannone, N.: Auditing with incomplete logs. In: Proceedings of the 3rd Workshop on Hot Issues in Security Principles and Trust (2015)

27. Mancarella, P., Terreni, G., Sadri, F., Toni, F., Endriss, U.: The CIFF proof procedure for abductive logic programming with constraints: theory, implementation and experiments. TPLP **9**(6), 691 (2009)

28. Kumar, A., Sabbella, S.R., Barton, R.R.: Managing controlled violation of temporal process constraints. In: Motahari-Nezhad, H.R., Recker, J., Weidlich, M. (eds.) BPM 2015. LNCS, vol. 9253, pp. 280–296. Springer, Heidelberg (2015). doi:10.1007/978-3-319-23063-4_20

Boosting the Development of ASP-Based Applications in Mobile and General Scenarios

Francesco Calimeri[1,2]($\boxtimes$), Davide Fuscà[1], Stefano Germano[1], Simona Perri[1], and Jessica Zangari[1]

[1] Department of Mathematics and Computer Science,
University of Calabria, Rende, Italy
{calimeri,fusca,germano,perri,zangari}@mat.unical.it
[2] DLVSystem Srl, Rende, Italy
calimeri@dlvsystem.com

Abstract. Answer Set Programming (ASP) is a well-established declarative programming paradigm in close relationship with other formalisms such as Satisfiability Modulo Theories, Constraint Handling Rules, FO(.) (First-Order logic extensions), Planning Domain Definition Language and many others; it became widely used in AI and recognized as a powerful tool for knowledge representation and reasoning, especially for its high expressiveness and the ability to deal also with incomplete knowledge. In the latest years, the community produced significant theoretical results and a number of robust and efficient implementations; this has been moving the focus from a strict theoretical scope to more practical aspects, and ASP has been increasingly employed in a number of different domains and for the development of industrial-level and enterprise applications. Although different development tools have been released, there is still a lack of proper means for an effective, large-scale applicability of ASP, especially in the mobile setting. In this work we show a general framework for integrating ASP reasoners into external systems and its use for designing and implementing ASP-based applications to different extents. In particular, we illustrate the integration of the ASP system DLV on the Android platform, and a full-native ASP-based mobile app for helping players of a *live* game of checkers.

Keywords: Answer Set Programming · Knowledge representation and reasoning · Logic programs · Education · Software development · Complex systems · Embedded systems · Mobile applications

This work was partially supported by the Italian Ministry of University and Research under PON project "Ba2Know (Business Analytics to Know) Service Innovation - LAB", No. PON03PE_00001_1, and by the Italian Ministry of Economic Development under project "PIUCultura (Paradigmi Innovativi per l'Utilizzo della Cultura)" n. F/020016/01-02/X27. Francesco Calimeri has received funding from the European Union's Horizon 2020 research and innovation programme under the Marie Skłodowska-Curie grant agreement No. 690974 for the project: "MIREL: MIning and REasoning with Legal texts".

© Springer International Publishing AG 2016
G. Adorni et al. (Eds.): AI*IA 2016, LNAI 10037, pp. 223–236, 2016.
DOI: 10.1007/978-3-319-49130-1_17

1   Introduction

Answer Set Programming (ASP) [1–7] is a powerful declarative formalism for knowledge representation and reasoning developed in the field of logic programming and nonmonotonic reasoning. After more than twenty years of scientific research, the theoretical properties of ASP are well understood and the solving technology, as witnessed by the availability of a number of robust and efficient systems [8], is mature for practical applications; indeed, ASP has been increasingly employed in a number of different domains, and used for the development of industrial-level and enterprise applications [9, 10]. Notably, this has been fostered by the release of a variety of proper development tools and interoperability mechanisms for allowing interaction and integration with external systems [11–14]. However, the worldwide commercial, consumer and industrial scenario significantly changed in the latest years; smartphones, or "smart"/wearable devices in general, the "IoT" (Internet Of Things), are constantly gaining popularity as computational power and features increase, in terms of sensors, communication means and applications availability. In this context, that forced the whole ICT industry to radically change, there is still a lack of tools for taking advantage from the knowledge representation and reasoning capabilities of ASP by means of systems that can natively run on mobile devices. This could help ASP developers both at porting already existing ASP-based applications to the mobile world, and at thinking of completely new scenarios for the fruitful application of ASP, both in research and industry.

In this work, we describe how to integrate ASP in external systems, with a special focus on the mobile setting; the integration relies on an abstract framework that we here specialize for the Android[1] platform and the DLV system [15], thus effectively bringing ASP on mobile.

Furthermore, we present *GuessAndCheckers*, a native mobile application that works as a helper for players of a "live" game of the Italian checkers (i.e., with a physical board and pieces). By means of the device camera, a picture of the board is taken, and the information about the current status of the game is translated into facts that, thanks to an ASP-based artificial intelligence module, make the app suggest a move. The app is well-suited to assess applicability of ASP in the mobile context; indeed, while integrating well-established Android technologies, thanks to ASP, it features a fully-declarative approach that eases the development, the improvement of different strategies and also experimenting with combinations thereof.

The herein adopted framework, full documentation and further information about *GuessAndCheckers* are available online [16].

2   Answer Set Programming

We briefly recall here syntax and semantics of Answer Set Programming.

---

[1] http://www.android.com.

2.1   Syntax

A variable or a constant is a *term*. An *atom* is $a(t_1, \ldots, t_n)$, where a is a *predicate* of arity n and $t_1, \ldots, t_n$ are terms. A *literal* is either a *positive literal* p or a *negative literal* not p, where p is an atom. A *disjunctive rule* (*rule*, for short) r is a formula

$$a_1 \mid \cdots \mid a_n :- b_1, \cdots, b_k, \text{ not } b_{k+1}, \cdots, \text{ not } b_m.$$

where $a_1, \cdots, a_n, b_1, \cdots, b_m$ are atoms and $n \geq 0$, $m \geq k \geq 0$. The disjunction $a_1 \mid \cdots \mid a_n$ is the *head* of r, while the conjunction $b_1, \ldots, b_k, \text{ not } b_{k+1}, \ldots, \text{ not } b_m$ is the *body* of r. A rule without head literals (i.e. $n = 0$) is usually referred to as an *integrity constraint*. If the body is empty (i.e. $k = m = 0$), it is called a *fact*.

$H(r)$ denotes the set $\{a_1, \ldots, a_n\}$ of head atoms, and by $B(r)$ the set $\{b_1, \ldots, b_k,$ not $b_{k+1}, \ldots,$ not $b_m\}$ of body literals. $B^+(r)$ (resp., $B^-(r)$) denotes the set of atoms occurring positively (resp., negatively) in $B(r)$. A rule r is *safe* if each variable appearing in r appears also in some positive body literal of r.

An *ASP program* $\mathcal{P}$ is a finite set of safe rules. An atom, a literal, a rule, or a program is *ground* if no variables appear in it. Accordingly with the database terminology, a predicate occurring only in *facts* is referred to as an *EDB* predicate, all others as *IDB* predicates; the set of facts of $\mathcal{P}$ is denoted by $EDB(\mathcal{P})$.

2.2   Semantics

Let $\mathcal{P}$ be a program. The *Herbrand Universe* of $\mathcal{P}$, denoted by $U_\mathcal{P}$, is the set of all constant symbols appearing in $\mathcal{P}$. The *Herbrand Base* of a program $\mathcal{P}$, denoted by $B_\mathcal{P}$, is the set of all literals that can be constructed from the predicate symbols appearing in $\mathcal{P}$ and the constant symbols in $U_\mathcal{P}$.

Given a rule r occurring in $\mathcal{P}$, a *ground instance* of r is a rule obtained from r by replacing every variable X in r by $\sigma(X)$, where σ is a substitution mapping the variables occurring in r to constants in $U_\mathcal{P}$; *ground*$(\mathcal{P})$ denotes the set of all the ground instances of the rules occurring in $\mathcal{P}$.

An *interpretation* for $\mathcal{P}$ is a set of ground atoms, that is, an interpretation is a subset I of $B_\mathcal{P}$. A ground positive literal A is *true* (resp., *false*) w.r.t. I if $A \in I$ (resp., $A \notin I$). A ground negative literal not A is *true* w.r.t. I if A is false w.r.t. I; otherwise not A is false w.r.t. I. Let r be a ground rule in *ground*$(\mathcal{P})$. The head of r is *true* w.r.t. I if $H(r) \cap I \neq \emptyset$. The body of r is *true* w.r.t. I if all body literals of r are true w.r.t. I (i.e., $B^+(r) \subseteq I$ and $B^-(r) \cap I = \emptyset$) and is *false* w.r.t. I otherwise. The rule r is *satisfied* (or *true*) w.r.t. I if its head is true w.r.t. I or its body is false w.r.t. I. A *model* for $\mathcal{P}$ is an interpretation M for $\mathcal{P}$ such that every rule $r \in$ *ground*$(\mathcal{P})$ is true w.r.t. M. A model M for $\mathcal{P}$ is *minimal* if no model N for $\mathcal{P}$ exists such that N is a proper subset of M. The set of all minimal models for $\mathcal{P}$ is denoted by $\mathrm{MM}(\mathcal{P})$.

Given a ground program $\mathcal{P}$ and an interpretation I, the *reduct* of $\mathcal{P}$ w.r.t. I is the subset $\mathcal{P}^I$ of $\mathcal{P}$, which is obtained from $\mathcal{P}$ by deleting rules in which a body literal is false w.r.t. I. Note that the above definition of reduct, proposed in [17], simplifies the original definition of Gelfond-Lifschitz (GL) transform [5], but is fully equivalent to the GL transform for the definition of answer sets [17].

Let I be an interpretation for a program $\mathcal{P}$. I is an *answer set* (or stable model) for $\mathcal{P}$ if $I \in \text{MM}(\mathcal{P}^I)$ (i.e., I is a minimal model for the program $\mathcal{P}^I$) [5,18]. The set of all answer sets for $\mathcal{P}$ is denoted by $ANS(\mathcal{P})$.

2.3   Knowledge Representation and Reasoning in ASP

In the following, we briefly introduce the use of ASP as a tool for knowledge representation and reasoning, and show how the fully declarative nature of ASP allows to encode a large variety of problems by means of simple and elegant logic programs.

3-COL. As a first example, consider the well-known problem of 3-colorability, which consists of the assignment of three colors to the nodes of a graph in such a way that adjacent nodes always have different colors. This problem is known to be NP-complete.

Suppose that the nodes and the arcs are represented by a set F of facts with predicates *node* (unary) and *arc* (binary), respectively. Then, the following ASP program allows us to determine the admissible ways of coloring the given graph.

$$r_1 : \quad color(X,r) \mid color(X,y) \mid color(X,g) :- node(X).$$
$$r_2 : \quad :-arc(X,Y), color(X,C), color(Y,C).$$

Rule r_1 above states that every node of the graph must be colored as **red** or **yellow** or **green**; r_2 forbids the assignment of the same color to any couple of adjacent nodes. The minimality of answer sets guarantees that every node is assigned only one color. Thus, there is a one-to-one correspondence between the solutions of the 3-coloring problem for the instance at hand and the answer sets of $F \cup \{r_1, r_2\}$: the graph represented by F is 3-colorable if and only if $F \cup \{r_1, r_2\}$ has some answer set.

We have shown how it is possible to deal with a problem by means of an ASP program such that the instance at hand has some solution if and only if the ASP program has some answer set; in the following, we show an ASP program whose answer sets witness that a property does not hold, i.e., the property at hand holds if and only if the program has no answer sets.

RAMSEY. The Ramsey Number $R(k,m)$ is the least integer n such that, no matter how we color the arcs of the complete graph (clique) with n nodes using two colors, say red and blue, there is a red clique with k nodes (a red k-clique) or a blue clique with m nodes (a blue m-clique). Ramsey numbers exist for all pairs of positive integers k and m [19].

Similarly to what already described above, let F be the collection of facts for input predicates *node* (unary) and *edge* (binary), encoding a complete graph with n nodes; then, the following ASP program $P_{R(3,4)}$ allows to determine whether a given integer n is the Ramsey Number $R(3,4)$, knowing that no integer smaller than n is $R(3,4)$.

$$r_1: \quad blue(X,Y) \mid red(X,Y) :- edge(X,Y).$$
$$r_2: \quad :- red(X,Y), red(X,Z), red(Y,Z).$$
$$r_3: \quad :- blue(X,Y), blue(X,Z), blue(Y,Z),$$
$$blue(X,W), blue(Y,W), blue(Z,W).$$

Intuitively, the disjunctive rule r_1 guesses a color for each edge. The first constraint r_2 eliminates the colorings containing a red complete graph (i.e., a clique) on 3 nodes; the second constraint r_3 eliminates the colorings containing a blue clique on 4 nodes. The program $P_{R(3,4)} \cup F$ has an answer set if and only if there is a coloring of the edges of the complete graph on n nodes containing no red clique of size 3 and no blue clique of size 4. Thus, if there is an answer set for a particular n, then n is <u>not</u> $R(3,4)$, that is, $n < R(3,4)$. The smallest n such that no answer set is found is the Ramsey Number $R(3,4)$.

3   The EMBASP Framework

The herein adopted framework, called EMBASP, has been conceived in order to obtain an abstract core that can be adapted, in principle, to any declarative logic formalism, programming language, or target platform. The main goal of the present work is to foster the application of ASP to new scenarios, and, in particular, as already stated, in the mobile setting. Hence, for the sake of presentation, in the following we refer to a specific version of EMBASP which has been specialized to DLV and Android, and has been actually employed for the development of *GuessAndCheckers*; the full framework is subject of an ongoing work, and is already available on the official web page [16].

Among the main-stream ASP systems, DLV is the first which is undergoing an industrial exploitation by a spin-off company called DLVSystem Ltd.[2], thus fostering the interest of several industries in ASP and DLV itself [9,10,20].

Android is by far the most used mobile operating system worldwide; the official version has been originally released, and is currently developed, by Google. In the latest years, it become very popular among developers, also thanks to its open source nature, that allows any manufacturer to customize and adapt it to an incredibly wide range of devices, not just smartphones.

In the following, we introduce the architecture of EMBASP and briefly describe its implementation.

3.1   Framework Architecture

The framework consists of three different layers, as reported in Fig. 1: ASPHANDLER acts like a façade to the user of the framework; SOLVER HANDLER is meant as a middleware between the façade and the actual solver, and manages the solver invocation; ASPSOLVER actually runs the solver.

---

[2] http://www.dlvsystem.com.

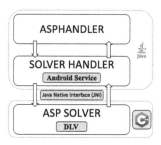

Fig. 1. The EMBASP framework architecture; overshadowed components highlight the specialization for DLV on Android.

Basically, the execution works as follows: given a logic program P, the ASPHANDLER (asynchronously) invokes the SOLVER HANDLER, providing it with P, along with proper options for the solver at hand; then, the SOLVER HANDLER starts the ASPSOLVER by invoking the native functions of the solver. Eventually, due to the asynchronous execution mode, the answer set(s) found by the ASPSOLVER are fetched by means of a callback mechanism.

3.2 A Framework Implementation for DLV on Android

Figure 2 illustrates a Java[3] implementation of the framework, tailored on DLV and Android.

ASPHANDLER LAYER. The abstract class `ASPHandler` provides all means for preparing the ASP input program(s), in several ways (simple strings, files, Java Objects) and managing the setting of all options for the solver at hand. In addition, it features proper methods to start the solver. Once that the reasoning task has produced the answer set(s), a callback function is automatically fired: it can be specified by implementing the `AnswerSetCallback` interface. The `AnswerSets` class parses the output, and collects each answer sets by means of `AnswerSet` objects. The `DLVHandler` and `DLVAnswerSets` classes are the specialization of the respective abstract classes designed for DLV. These classes provide a proper implementation tailored for the specific functionalities of the solver, i.e. the setting of its peculiar options, and the parsing of its output format.

The layer features also a `Mapper` class to convert the output from the solvers into Java Objects. Such translations are guided by Java `Annotations`[4], a form of meta-code that allows to mark classes with information that do not constitute actual Java instructions, and thus are not executed. In order to map a class to a predicate, two types of annotations can be specified: `@Predicate(string_name)`, whose target must be a class representing a predicate, and `@Term(integer_position)`, to map a term and its position among the predicate's arguments (the target must be a field of a class annotated via `@Predicate`).

---

[3] https://www.oracle.com/java.

[4] https://docs.oracle.com/javase/tutorial/java/annotations/.

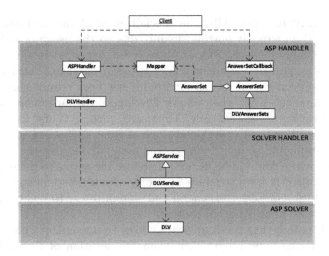

Fig. 2. Class diagram of the specialization of the EMBASP framework for DLV on Android

For instance, assume that we use EMBASP to solve the 3-coloring problem, presented in the Sect. 2.3. The following example shows how a class `Arc`, representing the concept of *arc*, can be annotated:

```
@Predicate("arc")
public class Arc {

    @Term(0)
    private String node0;

    @Term(1)
    private String node1;

    [...]
}
```

Basically, via `@Predicate("arc")` we express that the class is meant at representing atoms, containing a binary predicate *arc*, while via `@Term(N)`, with N=0,1, we express that the terms contained in these atoms are two nodes.

SOLVER HANDLER LAYER. The abstract class `ASPService` is in charge if managing invocations to the actual reasoner, and gathering the results.

The `DLVService` class extends `ASPService` and enables the invocation of DLV on Android by means of an Android's Service[5], a native component aimed at performing long-running operations in the background. Notably, it allows the applications using EMBASP to asynchronously execute other tasks while waiting for the output from the ASP solver.

---

[5] https://developer.android.com/guide/components/services.html.

ASPSOLVER LAYER. Eventually, the specialization of the ASPSOLVER layer to DLV required an explicit porting, mainly due to the usage of different programming languages. Indeed, DLV is natively implemented in C++, while on Android the standard development model is based on Java. Thus, in order to obtain a DLV version executable on Android it has been rebuilt using the NDK (Native Development Kit)[6] that allows developers to implement parts of an Android application as "native-code" languages, such as C and C++, and it is well-suited for CPU-intensive workloads such as game engines, signal processing, physics simulation, and so on. Moreover, the use of JNI grants the access to the API provided by the Android NDK, and therefore the access to the exposed DLV functionalities directly from the Java code of an Android application. Basically, these technologies represent the general and standard way to realize the porting of a C++ software in an Android context.

It is worth noting that the framework allows to dynamically change at runtime the underlying logic program(s) (from the addition of new rules, to the activation of different optimization statements, to the update of the facts base, towards brand new program(s)), and also to change solver options; this enables the change of the reasoner behaviour at runtime.

Moreover, EMBASP architecture is general enough in order to fit any kind of ASP solver by properly extending the abstract classes described above, similarly to how has been done for the DLV case.

Finally, a fruitful feature is offered by the annotation-based mapping. Indeed, it is meant to help the developer at dividing an application into two separated modules: an ASP reasoning module and a Java module. Indeed, the mapper acts like a middle-ware that enables the communication among the modules, and facilitates the developer's job by providing her with an explicit and ready-made mapping between Java objects and the logic part, without the need for any translation from string to objects and vice-versa. Such separation helps at keeping things simple when developing complex application: think about a scenario in which different figures are involved, such as Android/Java developers and KRR experts; both figures can benefit from the fact that the knowledge base and the reasoning modules can be designed and developed independently from the rest of the Java-based application.

4 A Full Native ASP-Based Android App: *GuessAndCheckers*

GuessAndCheckers is a native mobile application which has been designed and implemented by means of the specialization of EMBASP for DLV on Android discussed above. It works as a helper for users that play "live" games of the Italian checkers (i.e., by means of physical board and pieces); it has been initially developed for educational purposes. The app, that runs on Android, can help a player at any time; by means of the device camera a picture of the physical

---

[6] https://developer.android.com/tools/sdk/ndk.

board is taken: the information about the current status of the game is properly inferred from the picture by means of a vision module which relies on OpenCV[7], an open source computer vision and machine learning software.

A proper ASP representation of the state of the board is created, and eventually an ASP-based reasoning module suggests the move. Although a thorough discussion about the declarative approach to knowledge representation and reasoning supported by ASP is out of the scope of this work, it is worth noting that, as introduced in Sect. 2.3, it is significantly different from a "classic" algorithmic approach. Basically, besides solid theoretical bases that lead to declarative specifications which are already executable, there is no need for algorithm design or coding, and easy means for explicitly encoding the knowledge of an expert of the domain. For space reasons, we refer the reader to the material available online [16], including full encodings; in the following, we will illustrate how we mixed them in order to obtain the artificial prompter. The goal was to avoid ad-hoc behaviours for each situation, and develop a strategy which turns to be "good" in general. The strategy has been tuned via careful studies and a significant amount of experimental activities. It consists of several rules that can be defined as well-known by a human expert of the domain; they have been evaluated and then filtered out from a large set of hints, thus constituting a real knowledge base of good practices for real checkers players. The fully-declarative approach of ASP made easy to define and implement such knowledge base; furthermore, it was possible to develop and improve several different strategies, also experimenting with many combinations thereof.

The reasoning module consists of a Manager, developed in Java, that does not have decision-making powers; instead, it is in charge of setting up different ASP logic programs (i.e. builds and selects the ASP program to be executed at a given time) and makes use of EMBASP to communicate with DLV. At first, an ASP program (the *Capturing Program*) is invoked to check if captures are possible; if a possible capturing move is found, the *Capturing Program* provides the cells of the "jump(s)", the kind of captured pieces and the order of the capture; another logic program is then invoked to select the best move (note that each piece can capture many opponent pieces in one step). A graph-based representation of the board is used, along with a set of rules of the Italian checkers: *(a)* capture the maximum number of pieces (observing the rule "If a player is faced with the prospect of choosing which captures to make, the first and foremost rule to obey is to capture the greatest number of pieces"); *(b)* capture with a king (observing the rule "If a player may capture an equal number of pieces with either a man or king, he must do so with the king"); *(c)* if more than one capture are still available, choose the one having the lowest quantity of men pieces ("If a player may capture an equal number of pieces with a king, in which one or more options contain a number of kings, he must capture the greatest number of kings possible") *(d)* finally, if the previous constraints have not succeeded in filtering out only one capturing, select those captures where a king occurs first ("If a

---

[7] http://opencv.org.

player may capture an equal number of pieces (each series containing a king) with a king, he must capture wherever the king occurs first").

On the other hand, if no captures are possible, a different logic program (the *All Moves Program*) is invoked to find all the legal moves for a player (each Answer Set, here, represents a legal move). When more than one move can be performed (i.e., there are no mandatory moves), another logic program (*Best Game Configuration Program*) is invoked. Unlike the other programs, that only implement mandatory actions, the *Best Game Configuration Program* implements the actual strategy of our checkers prompter. The logic rules perform a reasoning task over the checkerboard configurations, and the best one is chosen by means of *weak constraints*, an ASP extension for expressing preferences [21].

The strategy of the *Best Game Configuration Program* is mainly based on the well-known strategic rule: "pieces must move in a compact formation, strongly connected to its own king-row with well-guarded higher pawns". More in detail, our reasoner guarantees that the user can move its pawns without overbalancing its formation. Pawns attack moderately, trying to remain as compact as possible, preserving at the same time a good defence of its own king row: technically, a *fully guarded back rank* (roughly speaking, prevent the opponent from getting kings unless she sacrifices some pieces).

The availability of DLV brought to Android via EMBASP allowed us to take great advantage from the declarative KRR capabilities of ASP; indeed, as already mentioned, we experimented with different logic programs, in order to find a version which plays "smartly", yet spending an acceptable (from the point of view of a typical human player) time while reasoning. We think of the design of the AI as the most interesting part of the app; we have been able to "implement" some classic strategies, a task that is typically not straightforward when dealing with the imperative programming, in a rather simple and intuitive way. Moreover, we had the chance to test the AI without the need for rebuilding the application each time we made an update, thus observing "on the fly" the impact of changes: this constitutes one of the most interesting features granted by the explicit knowledge representation. The source code of *GuessAndCheckers*, along with the APK and the benchmark results, are available online [16].

5   Related Work

The problem of embedding ASP reasoning modules into external systems and/or externally controlling an ASP system has been already investigated in the literature; to our knowledge, the more widespread solutions are the DLV Java Wrapper [22], JDLV [12], Tweety [23], and the scripting facilities featured by clingo4 [14], which allow, to different extents, the interaction and the control of ASP solvers from external applications.

In clingo4, the scripting languages *lua* and *python* enable a form of control over the computational tasks of the embedded solver clingo, with the main purpose of supporting also dynamic and incremental reasoning; on the other hand, EMBASP, similarly to the *Java Wrapper* and *JDLV*, acts like a versatile "wrapper" wherewith the developers can interact with the solver. However, differently

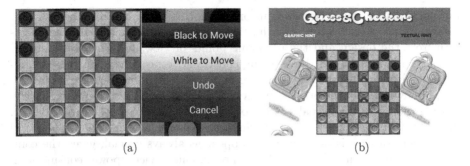

Fig. 3. Screenshots from the *GuessAndCheckers* app: recognized board (a) and suggested move for the black (b).

from the Java Wrapper, EMBASP makes use of Java Annotations, a form of meta-data that can be examined at runtime, thus allowing an easy mapping of input/output to Java Objects; and differently from JDLV, that uses JPA annotations for defining how Java classes map to relations similarly to ORM frameworks, EMBASP straightforwardly uses custom annotations, almost effortless to define, to deal with the mapping.

Moreover, our framework is not specifically bound to a single or specific solver; rather, it can be easily extended for dealing with different solvers, and to our knowledge, the specialization of EMBASP for DLV on Android is the first actual attempt reported in literature to port ASP solvers to mobile systems.

Tweety, an open source framework, is aimed at experimenting on logical aspects of artificial intelligence; it consists of a set of Java libraries that allow to make use of several knowledge representation systems supporting different logic formalisms, including ASP. Both Tweety and EMBASP provide libraries to incorporate proper calls to external declarative systems from within "traditional" applications. Tweety implementation is already very rich, covering a wide range of KR formalisms; EMBASP is mainly focused on fostering the use of ASP in a widest range of contexts, as evidenced by the offered support for the mobile setting which is currently missing in Tweety. Nevertheless, the EMBASP architecture has been designed in order to allow the creation of libraries for different programming languages, platforms and formalisms.

Several ways of taking advantage from ASP capabilities have been explored, and, interestingly, not all of them require to natively port an ASP system on the device of use. In particular, it is possible to let the reasoning tasks take place somewhere else, and use internet connections in order to communicate between the "reasoning service" and the actual application, according to a cloud computing paradigm, to some extent.

Both the approaches, native and cloud, are interesting, and each of them has pro and cons depending on the scenario it is facing. The cloud-based approach grants great computational power to low-end devices, without the need for actually porting a system to the final user's device, and completely preventing any performance issue. However, in order for this to take place, there is first the need for a proper application hosting, which requires non-negligible efforts both from the design and the economic points of view; furthermore, a steady internet connection might be a strong constraint, especially when the communication between the end user's device and the cloud infrastructure requires a large bandwidth. On the other hand, a native-based approach might involve significant efforts for the actual porting of pieces of software on the target device, which, in turn, might lead to performance or power consumption issues; and even if performance issues might not appear as always crucial, given the computational power which is available even on mobile devices, power consumption is sometimes decisive. Notably, the main idea behind this work is to embed an ASP solver directly in a mobile context, however this possibility is not hindered by the framework. In fact, due to the structure of the middleware layer (SOLVER HANDLER), it is possible to hide the details of the solver invocation, so that it can also be carried out using a cloud/server solution.

Nevertheless, in our showcase scenario, *GuessAndCheckers* shows that the development of applications that natively runs ASP-based reasoning tasks on mobile devices does not necessarily suffer from the discussed drawbacks. Indeed, DLV is invoked only on demand, when the reasoning task(s) need to be performed, and its execution requires a small amount of time; for most of the time, the user just interacts with the interface and the rest of the app: no ASP solver is running or waiting, thus preventing both performance issues and battery drain (Fig. 3).

6   Conclusions

In this preliminary work we introduced EMBASP, a general framework for embedding the reasoning capabilities of ASP into external systems; in order to assess the framework capabilities, we presented also a specialization for the mobile setting that is tailored for making use of DLV within Android apps. In addition, we presented the Android App *GuessAndCheckers*, a full-native mobile application featuring an ASP-based AI module that works as a helper for players of a "live" game of the Italian checkers (i.e., with a physical board and pieces). The framework, the app and further details are freely available online [16].

As future work, we want to test the framework over different platforms and solvers and properly evaluate performances; in addition, we are working at tuning the generalization of EMBASP in order to allow the integration of ASP in external systems for generic applications. Indeed, as ongoing work we are carrying out a redesign of the framework aimed at easily allowing for proper specializations, in principle, to any declarative logic formalism, programming language, or target platform.

References

1. Baral, C.: Knowledge Representation, Reasoning and Declarative Problem Solving. Cambridge University Press, Cambridge (2003)
2. Brewka, G., Eiter, T., Truszczynski, M.: Answer set programming at a glance. Commun. ACM **54**(12), 92–103 (2011)
3. Eiter, T., Faber, W., Leone, N., Pfeifer, G.: Declarative problem-solving using the DLV system. In: Minker, J. (ed.) Logic-Based Artificial Intelligence. The Springer International Series in Engineering and Computer Science, vol. 597, pp. 79–103. Springer, Heidelberg (2000). doi:10.1007/978-1-4615-1567-8_4
4. Eiter, T., Ianni, G., Krennwallner, T.: Answer set programming: a primer. In: Proceedings of Reasoning Web. Semantic Technologies for Information Systems, 5th International Summer School - Tutorial Lectures, Brixen-Bressanone, Italy, August-September 2009, pp. 40–110 (2009)
5. Gelfond, M., Lifschitz, V.: Classical negation in logic programs and disjunctive databases. New Gener. Comput. **9**, 365–385 (1991)
6. Marek, V.W., Truszczyński, M.: Stable models and an alternative logic programming paradigm. In: Apt, K.R., Marek, V.W., Truszczyński, M., Warren, D.S., et al. (eds.) A 25-Year Perspective. Artificial Intelligence, pp. 375–398. Springer, Heidelberg (1999)
7. Niemelä, I.: Logic programming with stable model semantics as constraint programming paradigm. Ann. Math. Artif. Intell. **25**(3–4), 241–273 (1999)
8. Calimeri, F., Gebser, M., Maratea, M., Ricca, F.: The design of the fifth answer set programming competition. CoRR abs/1405.3710 (2014)
9. Calimeri, F., Ricca, F.: On the application of the answer set programming system DLV in industry: a report from the field. Book Rev. **2013**(03) (2013)
10. Leone, N., Ricca, F.: Answer set programming: a tour from the basics to advanced development tools and industrial applications. In: Faber, W., Paschke, A. (eds.) Reasoning Web 2015. LNCS, vol. 9203, pp. 308–326. Springer, Heidelberg (2015). doi:10.1007/978-3-319-21768-0_10
11. Calimeri, F., Cozza, S., Ianni, G.: External sources of knowledge and value invention in logic programming. Ann. Math. Artif. Intell. **50**(3–4), 333–361 (2007)
12. Febbraro, O., Grasso, G., Leone, N., Ricca, F.: JASP: a framework for integrating answer set programming with java. In: Principles of Knowledge Representation and Reasoning: Proceedings of the Thirteenth International Conference, KR 2012, Rome, Italy. AAAI Press (2012)
13. Febbraro, O., Reale, K., Ricca, F.: ASPIDE: integrated development environment for answer set programming. In: Delgrande, J.P., Faber, W. (eds.) LPNMR 2011. LNCS (LNAI), vol. 6645, pp. 317–330. Springer, Heidelberg (2011). doi:10.1007/978-3-642-20895-9_37
14. Gebser, M., Kaminski, R., Kaufmann, B., Schaub, T.: Clingo = ASP + control: Preliminary report. In: Leuschel, M., Schrijvers, T. (eds.) Technical Communications of the Thirtieth International Conference on Logic Programming (ICLP 2014) (2014). arXiv:1405.3694v1. (Theory and Practice of Logic Programming, Online Supplement)
15. Leone, N., Pfeifer, G., Faber, W., Eiter, T., Gottlob, G., Perri, S., Scarcello, F.: The DLV system for knowledge representation and reasoning. ACM Trans. Comput. Logic **7**(3), 499–562 (2006)
16. Calimeri, F., Fuscà, D., Germano, S., Perri, S., Zangari, J.: EMBASP (2015)

17. Faber, W., Leone, N., Pfeifer, G.: Recursive aggregates in disjunctive logic programs: semantics and complexity. In: Alferes, J.J., Leite, J. (eds.) JELIA 2004. LNCS (LNAI), vol. 3229, pp. 200–212. Springer, Heidelberg (2004). doi:10.1007/978-3-540-30227-8_19

18. Przymusinski, T.C.: Stable semantics for disjunctive programs. New Gener. Comput. **9**, 401–424 (1991)

19. Radziszowski, S.P.: Small Ramsey numbers. Electron. J. Comb. **1** (1994). Revision 9: July 15, 2002

20. Balduccini, M., Son, T.C. (eds.): Logic Programming, Knowledge Representation, and Nonmonotonic Reasoning. LNCS (LNAI), vol. 6565. Springer, Heidelberg (2011)

21. Buccafurri, F., Leone, N., Rullo, P.: Enhancing disjunctive datalog by constraints. IEEE Trans. Knowl. Data Eng. **12**(5), 845–860 (2000)

22. Ricca, F.: The DLV java wrapper. In: de Vos, M., Provetti, A. (eds.) Proceedings ASP03 - Answer Set Programming: Advances in Theory and Implementation, Messina, Italy, pp. 305–316, September 2003. http://CEUR-WS.org/Vol-78/

23. Thimm, M.: Tweety: a comprehensive collection of java libraries for logical aspects of artificial intelligence and knowledge representation. In: Principles of Knowledge Representation and Reasoning: Proceedings of the Fourteenth International Conference, KR 2014, Vienna, Austria, 20–24 July 2014, pp. 528–537 (2014)

Relationships and Events: Towards a General Theory of Reification and Truthmaking

Nicola Guarino[1(✉)] and Giancarlo Guizzardi[1,2]

[1] ISTC-CNR Laboratory for Applied Ontology, Trento, Italy
nicola.guarino@cnr.it
[2] Federal University of Espírito Santo, Vitória, Brazil
gguizzardi@inf.ufes.br

Abstract. We propose a novel ontological analysis of relations and relationships based on a re-visitation of a classic problem in the practice of knowledge representation and conceptual modeling, namely *relationship reification*. Our idea is that a relation holds *in virtue of* a relationship's existence. Relationships are therefore *truthmakers* of relations. In this paper we present a general theory or reification and truthmaking, and discuss the interplay between events and relationships, suggesting that relationships are the *focus* of events, which emerge from the context (the *scene*) they occur in.

Keywords: Ontology · Relationships · Reification · Truthmaking · Events · Scenes

1 Introduction

In a recent paper [1], building on previous work by Guizzardi [2] on the notion of 'relator', we proposed an ontological analysis of relations and relationships based on the re-visitation of a classic problem in the practice of conceptual modeling, namely relationships reification. First, we argued that a relationship is not a tuple (i.e., an ordered set of objects), but rather an object in itself, that needs to exist in the world in order for a relation to hold: relations *hold* (that is, relational propositions are true) in virtue of the *existence* of a relationship; relationships are therefore *truthmakers*[1] of relations (more exactly, they are truthmakers of relational propositions). Then, considering the ontological nature of such truthmakers, we dismissed the idea (suggested by an early Chen's paper [3]) that they are events[2], proposing instead to consider relationships similarly to objects that can genuinely change in time. Yet, we acknowledged that reifying relationships as events may make a lot of sense in several practical cases, especially when there is no need to take change aspects into account.

---

[1] The notion of truthmaking will be further discussed and refined.
[2] For the time being, we are using here the term 'event' in its most general sense, as a synonym of what in the DOLCE ontology are called *perdurants* (note that also states and processes are considered as perdurants). In the rest of the paper, we propose a more restricted notion of event.

© Springer International Publishing AG 2016
G. Adorni et al. (Eds.): AI*IA 2016, LNAI 10037, pp. 237–249, 2016.
DOI: 10.1007/978-3-319-49130-1_18

In this paper we maintain our position that relationships are similar to objects, but we discuss the interplay between events and relationships in more detail. In short, the need to have events (in addition to relationships) in the domain of discourse is motivated by the fact that, when we describe the dynamics of a single relationship, we may want to add details concerning its spatiotemporal context, i.e., the *scene* that hosts the relationship (which may involve many *other* relationships); conversely, when we describe a complex scene, we may want to focus on a single relationship that is present in the scene. Indeed, as we shall see, we propose a view according to which events *emerge* from scenes as a result of a cognitive process that focuses on relationships: relationships are therefore the *focus* of events, which in turn can be seen as *manifestations* of relationships. So, referring to the relationship (which maintains its identity during the event) is unavoidable when we need to describe what changes in time, while referring to the event is unavoidable when we need to describe contextual aspects that go beyond the relationship itself. For instance, consider the classic example of a *works-for* relation holding between an employee and a company: we may refer to a particular employment relationship while describing how duties and claims (say, concerning the salary) vary in time, while we refer to one or more events while talking, say, of the location where the work occurs, or the weekly schedule or the activities performed in the framework of the work agreement.

The paper is structured as follows. In Sect. 2 we introduce the key notions of reification and truthmaking. First we characterize the class of relations that deserve being reified by revisiting Guizzardi's earlier distinction between *formal* and *material relations* and isolating the class of *descriptive relations,* which hold in virtue of some qualities of their relata. Then we introduce the notion of *weak truthmaker,* and, generalizing our analysis to the case of descriptive properties (monadic descriptive relations) we show how *individual qualities* such as those adopted in the DOLCE and UFO ontologies can be understood as their reifications. In Sect. 3 we discuss the crucial case of the reification of comparative relations, treated in an unsatisfactory manner in the earlier Guizzardi's work. In Sect. 4, we shift our attention to events, discussing the interplay between events and relationships, and arguing that relationships are the *focus* of events. Finally, in Sect. 5 we present our conclusions.

2   Reification and Truthmaking

Before illustrating our theory, let us briefly clarify some terminological issues concerning reification and relationships. In general, reification is the process of including a certain entity in the domain of discourse. For example, Davidson's move of putting events in the domain of discourse [6] is a reification move. Also, when we ascribe meta-properties like symmetry to a binary relation such as *married with*, we are reifying the whole relation (intended as a set of tuples). This is different from reifying a single *instance* of a relation, say the single tuple *<John, Mary>*, and is also different from reifying the result of a *nominalization* process of the relation's predicate holding for that tuple, namely the *marriage* between John and Mary. The latter, *and not the tuple*, is what we call a *relationship*.

Note that such understanding of a relationship deviates from the mainstream, since Chen defines a *relationship type* as a mathematical relation (i.e. a set of tuples), and a relationship as one of such tuples. So, under the mainstream approach relations (relationship types) and relationships are *extensional* notions. Yet, Chen admits that different relationship types (say, *friend of* and *colleague of*) may involve the same tuples, so each relationship type seems to have a unique "meaning" (its *intension*) conveyed by its *name*[3]. We claim that it is this intensional aspect of a relationship that people have in mind, when they talk of relationship reification. This is why, for the sake of clarity, we prefer to use the term 'relationship' only in its intensional meaning, adopting 'tuple' to refer to the extension. The rest of the paper is devoted to understanding when it is useful and legitimate to consider such 'intensional meaning' as an element of our domain of discourse, and, if so, what its ontological nature is.

2.1 Which Relations Deserve Reification?

In the past, a problem we encountered while developing our approach to relationships reification concerned the kinds of relation that deserve reification[4]. In the early Guizzardi's work, a crisp distinction was assumed between *formal relations*, which *"hold between two or more entities directly without any further intervening individual"*, and *material relations*, which require the existence of an intervening individual. The modeling proposal was to systematically introduce –for all material relations– a specific construct, called the *relator*, standing for such intervening individual. Note that *comparative relations* such as *taller than* were considered by Guizzardi as formal, because they hold just in virtue of the intrinsic properties of the relata.

In the philosophical literature, however, the formal/material distinction varies significantly among different authors, and overlaps with other distinctions. The definition of *formal* (vs. *material*) relations adopted by Guizzardi is indeed equivalent to one of the various definitions proposed in the literature to account for *internal* (vs. *external*) relations. We report here the one by Peter Simons, based in turn on Moore [10]: *"If it is possible that a and b both exist and it not be the case that aRb, then if aRb we say the relational predication is true externally. If it is not possible that a and b both exist and it not be the case that aRb, then where aRb we say the relational predication is true internally"* ([8], p. 203).

According to this definition, as Simons observes, comparative relational predications go across the internal/external distinction: some of them turn out to be internal (and therefore formal, in Guizzardi's terminology), but others turn out to be external, and therefore material. For instance, the mere existence of an electron e and a proton p is

---

[3] Another way to capture this meaning is to add attributes to the original tuple, which somehow express the properties the relationship has. This is the approach followed by Thalheim [7], who defines a relationship type as a sequence of entity types followed by a set of attributes.

[4] By 'a relation that deserves reification' we mean a relation that deserves reification of its relationships. Informally, we talk of reification of a relation to mean reification of its relationships.

enough to conclude that *heavier(p, e)* holds (since both of them have that particular mass *essentially*), but the mere existence of John and Mary is not enough to conclude that *taller(John, Mary)* holds, since they don't have that particular height essentially, so something else is required. Moreover, notice that, within the same relation, some predications –like *heavier(p, e)*– may be true internally, while others –like *heavier(John, Mary)*[5]– may be true externally. So, the picture is rather complicated, and this is the reason why, in our previous paper, we decided to abandon the formal/material distinction and just focus on a relevant class of relations that certainly deserve reification, those we called *extrinsic* relations, without taking a position on the reification of *intrinsic* relations, to which comparative relations belong.

In short, an intrinsic relation is a relation that can be derived from the intrinsic properties of its relata[6]. This clearly applies to all comparative relations (whether or not they depend on the mere existence of their relata), as well as to all relations that just depend on the mere existence of their relata. Extrinsic relations are just the relations that are not intrinsic: for example, you can't decide whether *married(John, Mary)* holds just on the basis of the intrinsic properties of John and Mary.

Interestingly enough, in philosophy there is another way of defining the *internal/external* distinction, owed to Armstrong [9], according to which 'internal' and 'external' become synonymous, respectively, of 'intrinsic' and 'extrinsic'[7]. So, in retrospective, we can conclude that, although Guizzardi's definition was following Simons, what he actually had in mind –especially while insisting on considering comparative relations as formal– was Armstrong's distinction.

That said, still we have to answer our original question: which relations deserve reification? Elsewhere [1, 2, 12] we have discussed extensively the practical advantages of the *relator* construct in the practice of conceptual modeling, so no doubt that most *extrinsic* relations (i.e., those Guizzardi labeled *material* relations) deserve reification. But are we sure that comparative relations don't deserve reification? For instance, one may want to keep track of the difference in height between him and his son, or may measure the extent of a temperature difference between two bodies. In general, we may be interested in talking of *qualitative relationships* among things: we can have temperature relationships, size relationships, mass relationships, so that we can say that the mass relationship between the Earth and the Moon is responsible of the way they move around.

In the light of these examples, our position is that, besides extrinsic relations, also comparative relations may deserve reification. But what do they have in common? Our

---

[5] Of course, one could consider *heavier(John@t, Mary@t)* to be a relation between the states (snapshots) of John and Mary at *t*. In this case, the relation would hold internally if at all.

[6] We shall take the notion of intrinsic property as primitive. Intuitively, an intrinsic property is a property that holds for an entity independently of the existence of any other entity.

[7] See the recent overview by MacBride [11] of the various philosophical positions on relations. According to him, there are three ways of understanding the internal/external distinction: "internal relations are determined by the mere existence of the things they relate, or internal relations are determined by the intrinsic characters of the things they relate, or internal relations supervene upon the intrinsic characters of the things they relate". The first position (adopted by Simons) is due to Moore. The second one is due to Armstrong, while the third one to Lewis.

answer is that they both belong to the broad class of what we call *descriptive relations*, which hold in virtue of *some particular aspect* (some *individual qualities*) of their relata. Under this view, both *taller than* and *being in love with* count as descriptive.

Note that *descriptive/non-descriptive* and *intrinsic/extrinsic* are orthogonal distinctions. The general picture is reported in Fig. 1, where we have shown how internal relations (in Moore's sense) go across the descriptive/non-descriptive distinction, while being included in the class of intrinsic relations. Let us briefly discuss the four quadrants shown in this figure. *Intrinsic descriptive relations* include all comparative relations holding among objects and events, plus for example all relations of mutual spatial location (at least as long we consider spatial location as an intrinsic property). *Intrinsic non-descriptive relations* include internal relations such as *existential dependence* and *resemblance,* which hold in virtue of the mere existence of their relata, as well as all comparative relations holding among tropes and qualities. Altogether, exactly because they do not depend on particular aspects of their relata, they may be called *formal* relations (abandoning therefore the notion of 'formal' as synonyms of 'internal'), in the sense that their domain and range are not limited to specific domains. The figure shows that not all formal relations are internal: for instance, *necessary part of* would be internal, while *contingent part of* would be external. *Extrinsic descriptive relations* include relations such as *works for* and *married to* that hold in virtue of some *actual* qualities of their relata, but also historical *relations* such as *author of,* which hold in virtue of some *past* quality (of the author). Finally, *extrinsic non-descriptive relations* include merely historical relations such as *born in*, that holds in virtue of an event occurred in the past, and the so-called *Cambridge relations* such as *being both observed by somebody,* which hold in virtue of something external that doesn't affect the relata.

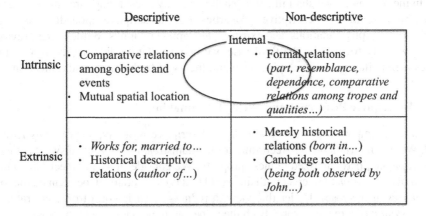

Fig. 1. Kinds of relations.

In conclusion, to decide whether a relation deserves or not reification we need to check whether it is a descriptive relation or not, i.e., whether it holds in virtue of some individual qualities of its relata. Individual qualities, originally introduced in the DOLCE ontology [13], are now a common feature (with minor differences) of other top-level ontologies such as UFO [2] and BFO [14]. In the following, we shall see how individual qualities capture the notion of truthmaking, accounting not only for the truth of a relational predication, but also for the way a relationship behaves in time. In short, we shall see how *individual qualities constitute relationships*.

2.2   Weak Truthmaking

In our earlier paper [1], we based our re-visitation of Guizzardi's original idea of *relators* on the philosophical notion of *truthmaking*. In that paper, we did not take a position concerning the specific nature of such notion, just assuming that truthmaking is a primitive, fundamental relation linking what is true to what exists. In general, a shared intuition is that a truthmaker for a property or a relation is an entity *in virtue of* which that property or relation holds. Several attempts have been made by philosophers to formally capture such intuition [15], i.e., to account for what *'in virtue of'* means. According to the mainstream doctrine, the truthmaker of a proposition is something *whose very existence* entails that the proposition is true. This means that the truthmaking relation holds *necessarily*. There is, however, a weaker notion of truthmaking, introduced by Parsons [16], according to which the truthmaker of a proposition is something that makes the proposition true not just because of its existence (i.e., because of its *essential* nature), but because *of the way it contingently is* (i.e., because of its *actual* nature). The truthmaking relation does therefore hold *contingently*. This notion of *weak truthmaking* is the one we shall adopt here, since –as we shall see– it seems to be the most apt to support our view of (descriptive) relationships as entities that can change in time, accounting not only for the *fact* that a relation holds, but also for the *way* it holds and develops in time.

In the following, we shall illustrate such a view by considering three main cases of descriptive relations: descriptive properties (i.e., descriptive monadic relations), intrinsic descriptive relations, and extrinsic descriptive relations. While in our previous paper we only focused on the third case, we believe that considering the former two cases is illuminating in developing a general theory of truthmaking and reification.

2.3   Descriptive Properties and Weak Truthmaking

Consider a simple proposition involving a descriptive property, such as *this rose is red*. What is its truthmaker? According to the mainstream theory, an answer[8] is that it is a *trope*, i.e., particularized redness property. We consider it as an object-like entity, a kind of *disposition* to interact with the world that is existentially dependent on the rose (it *inheres* in the rose). Under this view, a redness *event* is not a trope, but rather a *manifestation* of a redness trope. Both the trope and the redness event, because of their very existence, are such that the proposition is true. They are therefore both truthmakers of that proposition. Since the trope *participates* to the redness event we consider it as

---

[8] Another answer could be that the truthmaker is a *fact* of redness. In light of the discussion in [1] against facts as a viable interpretation for relationships, we do not consider this option here.

the *minimal* truthmaker, although we are aware that a precise account of the notion of minimal truthmaking is still under discussion[9].

Suppose now that the rose is red at time *t1*, and becomes brown after several days, at time *t2*. According to the mainstream theory, the truthmaker at *t2* will be a different one, namely a specific brownness. According to Parsons' theory, instead, the *weak truthmaker* at both times is the rose itself: it is the very same rose, because of the way it is at *t1* and at *t2*, that is a truthmaker of *'this rose is red'* at *t1* and a truthmaker of *'this rose is brown'* at *t2*. In other words, a weak truthmaker is something such that, because of the way it is, makes a proposition true.

We should observe, however, that the rose is not the *minimal* weak truthmaker of these propositions. There is something smaller –so to speak– than the whole rose: the rose's *color*. Indeed, it is exactly because of the rose's color that the rose is red at *t1* and brown at *t2*. As we mentioned above, this color is modeled as an *individual quality* in the DOLCE and UFO ontologies[10]. A peculiar characteristic of individual qualities is that they are *endurants*, i.e., they can qualitatively change in time (e.g., change their "value" from red to brown) while maintaining their identity.[11]

In conclusion, individual qualities are the minimal weak truthmakers of simple propositions involving a descriptive property. So to speak, they are *responsible* for the truth of such propositions, in the precise sense –as explained by Parsons– that the proposition can't become false without an intrinsic change of its weak truthmaker, i.e., since the weak truthmaker is an individual quality, without a movement in the space of possible values such quality can assume [13]. So, as we have seen, the same quality can be responsible for the truth of different propositions holding at different times.

Let us see now how the weak truthmaking mechanism described above can be used to establish an ontological foundation for *reification* choices in the practice of conceptual modeling. The practical rule we suggest is: *"Whenever a model includes a descriptive property, typically represented by an instance attribute of a class, one should reflect on the possibility of reifying it as an individual quality"*.

Suppose for example that, in an employment scenario, we have the attribute *mood* for the *Employee* class, with possible values *happy* or *sad*; reifying the *mood* quality as a separate class (whose instances inhere to the instances of *Employee*) would allow you the possibility to express, for instance: *(1) Further details on the reified entity:* "Mary has a pleasant mood"; *(2) Information on its temporal behavior:* "Mary's mood got much worse in the last days"; *(3) Information on its causal interactions with the world:* "Because of Mary's mood, she wasn't very productive at work". As we shall see, these are indeed the main reasons for the reification of descriptive relations of arbitrary arity, not just descriptive properties.

---

[9] For our purposes, we define a minimal truthmaker of a proposition as a truthmaker such that no entity *inhering* in it, *being part* of it or *participating* to it is itself a truthmaker of the same proposition.

[10] We shall not discuss the differences among these ontologies concerning the notion of quality. In particular, we shall ignore the fact that DOLCE does not consider qualities as endurants, and we shall collapse, for the sake of simplicity, UFO's distinction between *qualities* and *modes*.

[11] This notion of individual qualities as endurants is similar to Moltmann's *variable* tropes.

Note that, strictly speaking, we cannot say that in this way we are reifying the *happy* and the *sad* properties, since the same *mood* quality may "reify" both of them at different times. A strict reification would result in the explicit introduction of a specific happiness and a specific sadness, intended as *completely determined* particularized properties, which would be tropes and not qualities. In this case, however, as discussed by Moltmann [17] and in our earlier paper, we would loose the flexibility of expressing the additional information described above. So, we may consider the strategy outlined above as a *weak reification* strategy, which turns out, however, to be more effective in practice than a strict reification strategy.

3   Descriptive Relations and Weak Truthmaking

3.1   Intrinsic Descriptive Relations

Having described the truthmaking mechanism for descriptive properties, let us now generalize it to descriptive relations, considering first the intrinsic ones, and in particular *comparative relations*. Take for example *taller(John, Mary)*. In the light of the previous discussion, it is easy to see that its minimal weak truthmaker is the *mereological sum*[12] of two individual qualities, namely John's height and Mary's height. Similarly to what is noted above, should the *height relationship* between John and Mary change in a certain way, this would be also the weak truthmaker of *as-tall-as(John, Mary)* and *taller(Mary, John)*. In our view, this means that there is a single entity in our ontology, namely a *height relationship*, consisting of a *quality complex* having the two individual heights as proper parts. Such a relationship is an *endurant*, whose internal configuration may be such that one of the three possible propositions above is true at a given time.

In conclusion, we can say that comparative relationships (as mereological sums of intrinsic qualities) are the weak truthmakers of comparative relations. Reifying them has the same advantages we have seen above in terms of the possibility to add further details: for instance, we can express the actual *height-distance* between the two relata as a property of the relationship (which in principle may itself be reified, originating a height-distance individual quality inhering in the quality complex).

Note that the approach described above can be generalized to the case of relations expressing arbitrary configurations of intrinsic qualities, such as weights or colors. If we consider spatial position also an intrinsic quality (as done in DOLCE, although the choice might be debatable), then relations describing spatial configurations (patterns) may be also be reified in terms of quality complexes.

3.2   Extrinsic Descriptive Relations

Considering now extrinsic descriptive relations, their main difference from the intrinsic ones is that at least one of the qualities inhering in the two relata is a *relational* one, which is *existentially dependent* on the other one(s). Take for example *loves(John,*

---

[12] The mereological sum of x and y is an entity z such that whatever overlaps z also overlaps x or y (see, for instance, [2], Chap. 5).

Mary). Its weak truthmaker is a *quality complex* that includes John's love towards Mary (a mental disposition understood as a relational quality), and whatever other quality (relational or non-relational) it actually depends on (such as Mary's beauty) or depends on it (such as Mary's embarrassment in reaction to John's love). The internal structure of such quality complex has been discussed in detail in [1].

Note that, as we have seen, the same *relationship* (i.e., the same weak truthmaker) may make different kinds of *relational propositions* true. Indeed, at different times, we can describe it in different ways: as a mutual love, as a non-mutual-love, as a passionate love, as a mostly inexistent love, etc. All the corresponding propositions would share the same weak truthmaker, i.e., the same love relationship exhibiting qualitative changes in time.

4   Relationships and Their Context: Scenes and Events

Let us shift now our attention to *events.* In our earlier papers, as more or less customary in the philosophical literature, we used the term 'event' in its most general sense, i.e., as a synonym of *perdurant* or *occurrent* (contrasted respectively with *endurant* and *continuant*). Here, while describing the interplay between events and relationships, we shall reserve this term for a more specific use. Shortly put, *events emerge from scenes; individual qualities and relationships are the focus of events.*

4.1   Scenes

The Cambridge dictionary defines a scene as *"a part of a play or film in which the action stays in one place for a continuous period of time".* Of course the word has several more meanings, but this is the one that fits best the technical notion we would like to introduce. For us, a *scene* is whatever happens in a suitably restricted spatiotemporal region. Our intuition is that a scene is a perdurant of a particular kind, being the object of a unitary perception act. So, its main characteristic is that it is a *whole*, from a perceptual point of view. We leave it open what the specific unity conditions for this whole are. A scene may last a few milliseconds, corresponding to a "one shot" presentation, or perhaps a whole life, if we see it as a macroscopic perception act. The important facts are: *(1) A scene cannot be instantaneous: it always has a time duration bound to the intrinsic time granularity and temporal integration mechanisms of the perception system considered* (we do not perceive the single frames of a movie, nor the internal dynamics of a sound)*; (2) A scene is located in a convex region of spacetime. It occurs in a certain place, during a continuous interval of time.*

In conclusion, we see a scene as a maximal perdurant located in a convex region of spacetime: it contains all perdurants occurring there as parts. For example, consider Davidson's example [6] of a sphere rotating and heating up during a certain time interval. What the example describes is a scene, including whatever occurs in that time interval within the spatial location delimited by the sphere's surface. In the next section we shall see how two further perdurants (events, in the strong sense defended here) can be distinguished as proper parts of that scene.

4.2   Events

The etymology of the term 'event' is from Latin: *ex-venire* (to come out). If we take this etymology seriously, we have to ask where do events come from. Our answer is that they come from scenes: they emerge from scenes through a *focusing process*. We claim that all ordinary events, like those described by most natural language verbs, have a focus. This means that their participants are not involved in the event, so to speak, in an homogeneous way, but rather there are different levels of involvement, which concern also their parts and qualities. So we can distinguish some *core participants*, and others that are not involved at all in the event, except in a very indirect way. For example, consider a person writing. Her body clearly participates to this process, but some of its parts (say, the eyes) are clearly more involved than others (say, the mouth). Should the same person be writing while eating a sandwich, the mouth would be involved in the eating and not in the writing.

Consider now a scene we can perceive from a house window: a simple one like a meadow in a sunny day, or a more complex one like a busy street market in a working day. Several events may capture our attention: a butterfly passing by, a cloud showing a particular shape for a while, a person buying some food, a vendor yelling... So, we may say that many events emerge from the same scene, each one with different *focus*.

But what is this focus, exactly? One way of seeing it is as a *minimal participant* to that event. For instance, consider Titanic's collision with an iceberg, discussed by Borghini and Varzi [18]: of course there are large parts of the ship (say, the rear part) and large parts of the iceberg that are loosely involved in the event, while other parts (a suitable part of the iceberg and a suitable part of Titanic) are definitely involved. These would be the minimal participants. However, as discussed there, serious problems of vagueness and indeterminacy would emerge: how to select such minimal participants? As they put it, *"Exactly what parts of Titanic hit what parts of the iceberg?"*

In our view, a way to address this problem is to shift our attention from the participants to their *qualities*: for example, we can say that, *for sure*, the Titanic's mass and the iceberg's mass were involved in the event, while, for sure, the Titanic's color and the iceberg's color were not involved. Of course, vagueness and indeterminacy problems cannot be completely eliminated, since, for instance, determining the exact location of the hit event would still be a problem. However, for the purpose of extracting an event from a scene, we claim that pointing to some objects' qualities is enough to describe *exactly* the event we want to talk of, i.e., to let it emerge from the scene.

In other words, an event is determined by a couple $<r, f>$, where r is a spatiotemporal region, and f is the event's *focus*, consisting of a collection of individual qualities, which we shall call *focal qualities*. To see this, consider again the example of the sphere that rotates while heating up: assuming that r is the spatiotemporal region occupied by the sphere during this phenomenon, we can isolate the focus of the rotation event in a collection of individual qualities, namely the spatial locations of the sphere parts, while the focus of the heating event is clearly the sphere's temperature. In other words, if we have two different foci, we have two different events emerging from the same scene, each one with different sums of qualities as a focus.

In conclusion, what we suggest is to stop considering 'event' as synonymous of 'perdurant', but rather distinguish two broad categories within perdurants, depending on whether they have a focus or not. The former will be *events* (in this new strict sense), the latter *scenes*. The reason they are different lies in their different principle of individuation: two different scenes must have a different spatiotemporal location, while two different events may share the same spatiotemporal location. So, to quote an observation by Quine [19], if the sphere is rotating rapidly and heating slowly, we have an event that is rapid and *a different one* that is slow, while if we had a single event of course we couldn't say it is both rapid and slow.

This way of considering events allows therefore for a very fine-grained approach: an event is whatever happens to a suitably selected set of individual qualities in a particular spatiotemporal region. So, the simplest event we can describe (and imagine) would be a change (or a state) of a single individual quality, say a light's intensity changing from dark to bright, or remaining bright for a while. Indeed, in our view events are *manifestations* of individual qualities.

4.3   Reifying Events as the Context of Relationships

Let us now go back to relationships. We have advocated the view that the focus of an event is a sum of individual qualities. In the case of relational events, i.e., events involving multiple participants, this sum of individual qualities is typically constituted by relational qualities inhering in the multiple involved participants. These relational qualities form quality complexes that are reified relationships. Consider for instance, on one hand, the *marriage* between John and Mary as a relationship and, on the other hand, the event (the marriage process) that is the sum of the manifestations of the qualities (e.g., commitments and claims) constituting this relationship. We can see such event as *carved out* of a broader scene, involving John and Mary's lives, by having the marriage relationship as the focus. Analogously, Barack Obama's *presidential mandate* relationship would be the focus of Obama's term, while Paul's enrollment relationship to the University of Trento would be the focus of Paul's student life in the scope of that enrollment. All these events are carved out of complex extended scenes by being the manifestation of qualities that constitute their focal relationships.

Now, if every relational event such as the ones just described has relationships as focus, wouldn't be enough to just reify these relationships? In other words, what is the practical relevance of having also events, besides relationships, in our domain of discourse? A first reason is to make clear what the role of the relata is. Consider for instance a service offering relationship, which in a recent paper [20] we modeled as a complex sum of commitments and claims. Intuitively, a service offering has an *agent*, who is the *provider*, and a *beneficiary,* who is the *customer*. But the provider is not the agent *of the relationship*. He is the agent of an *offering event*. Indeed, roles are usually understood as *ways of participation*[13] to an event. So, being the agent of an offering event means having a commitment that is part of the focus of that event.

---

[13] We understand participation as a formal relation linking endurants to perdurants [13].

A second reason to have events in a conceptual model is the possibility to talk of the broader context of the relationship. Coming back to the example mentioned in the Introduction, consider a *works-for* relationship, modeled as a sum of duties and claims. If we want to express a constrain concerning the location where the work occurs (say, a particular office) we cannot just add an attribute to the relationship, since such location is not directly involved in the relationship, but rather it is a participant of the event focused on by the relationship. Indeed, the point is exactly that there are much more participants involved in a working event than those directly involved in the relationship: the job of the relationship is just to focus on the core participants (picking up some of their specific qualities). Thus, if we want to be able to talk of the other participants, we need both the relationship and the event.

A third reason is that we need events if we want to talk of specific temporal constraints concerning the way a relationship evolves in time. For instance, to express the constraints concerning the weekly schedule, we may need to introduce specific events corresponding to working slots as proper parts of the main event focused by the *works-for* relationship: duties and obligations usually hold for a continuous interval of time, while the working slots are not contiguous.

Finally, we may need to explicitly modeling events while dealing with extrinsic non-descriptive relations, especially merely historical relations such as *born-in*. Being non-descriptive, such relation does not need to be reified as such (we could say it is a relation without a relationship), but yet modeling the born event may have several practical advantages.

5   Final Considerations

In previous work [2, 4, 5], we have shown how reified relationships are essential for addressing many classical and recurrent modeling problems, how their explicit representation has a direct impact on the domain expert's understanding of the real-world semantics [12], and how they may help avoiding a number of occurrences of anti-patterns in the modeling of relations [21]. In our view, the work we presented here sheds new light both to the theory and the practice of reification, by clarifying which relationships deserve reification in the framework of a general ontological theory of reification and truthmaking, by clarifying the nature of descriptive relationships as quality complexes that can change in time, and by establishing a systematic, principled connection between relationships reification and events reification. The novel understanding of events we have proposed, where events emerge from scenes by means of a focusing mechanism based on relationships, further clarifies the whole picture, and gives us –we believe– the right tools to model complex scenes involving multiple emerging events.

We are aware that we still need a complete formal characterization of our theory. However, we believe the conceptual clarifications present in this paper are a first step to establish solid foundations for practical applications both in knowledge representation and conceptual modeling.

References

1. Guarino, N., Guizzardi, G.: "We need to discuss the *Relationship*": revisiting relationships as modeling constructs. In: Zdravkovic, J., Kirikova, M., Johannesson, P. (eds.) CAiSE 2015. LNCS, vol. 9097, pp. 279–294. Springer, Heidelberg (2015). doi:10.1007/978-3-319-19069-3_18

2. Guizzardi, G.: Ontological Foundations for Structural Conceptual Models (2005)

3. Chen, P.: English sentence structure and entity-relationship diagrams. Inf. Sci. **29**(2), 127–149 (1983)

4. Guizzardi, G., Herre, H., Wagner, G.: On the General Ontological Foundations of Conceptual Modeling. In: Spaccapietra, S., March, Salvatore, T., Kambayashi, Y. (eds.) ER 2002. LNCS, vol. 2503, pp. 65–78. Springer, Heidelberg (2002). doi:10.1007/3-540-45816-6_15

5. Guizzardi, G., Wagner, G., Herre, H.: On the foundations of UML as an ontology representation language. In: Motta, E., Shadbolt, N.R., Stutt, A., Gibbins, N. (eds.) EKAW 2004. LNCS (LNAI), vol. 3257, pp. 47–62. Springer, Heidelberg (2004). doi: 10.1007/978-3-540-30202-5_4

6. Davidson, D.: The individuation of events. In: Contemporary Readings in the Foundations of Metaphysics, p. 295 (1998)

7. Thalheim, B.: Entity-Relationship Modeling. Springer, Heidelberg (2000)

8. Simons, P.M.: Relations and truthmaking. Aristot. Soc. **84**(Suppl. 1), 199–213 (2010)

9. Armstrong, D.M.: Universals and scientific realism. In: A Theory of Universals, vol. 2, pp. 1–169. Cambridge University Press, Cambridge (1978)

10. Moore, G.E.: External and Internal Relations. In: Proceedings of the Aristotelian Society, pp. 1–15, December 1919

11. MacBride, F.: Relations. In: Stanford Encyclopedia of Philosophy, pp. 1–42 (2016)

12. da Silva Teixeira, M.D.G., de Almeida Falbo, R., Guizzardi, G.: Analyzing the behavior of modelers in interpreting relationships in conceptual models: an empirical study. In: OntoCOM 2014, pp. 1–12 (2014)

13. Borgo, S., Masolo, C.: Foundational choices in DOLCE. In: Staab, S. (ed.) Handbook on ontologies, pp. 361–381. Springer, Heidelberg (2009)

14. Neuhaus, F., Grenon, P., Smith, B.: A formal theory of substances, qualities and universals. In: 3rd International Conference on Formal Ontologies in Information Systems (FOIS 2004), Turin (2004)

15. MacBride, F.: Truthmakers. In: Stanford Encyclopedia of Philosophy, pp. 1–69 (2014)

16. Parsons, J.: There is no 'truthmaker' argument against nominalism. Australas. J. Philos. **77**(3), 325–334 (1999)

17. Moltmann, F.: Abstract Objects and the Semantics of Natural Language. Oxford University Press, New York (2013)

18. Borghini, A., Varzi, A.C.: Event location and vagueness. Philos. Stud. **128**(2), 313–336 (2006)

19. Quine, W.V.: Events and reification. In: LePore, E., McLaughlin, B.P. (eds.) Actions and Events: Perspectives on the Philosophy of Donald Davidson (1985)

20. Nardi, J.C., et al.: A commitment-based reference ontology for services. Inf. Syst. **54**(C), 263–288 (2015)

21. Salles, T.P., Guizzardi, G.: Ontological anti-patterns: empirically uncovered error-prone structures in ontology-driven conceptual models. Data Knowl. Eng. (DKE) **99**, 72–104 (2015)

A Self-Adaptive Context-Aware Group Recommender System

Reza Khoshkangini[1]([⊠]), Maria Silvia Pini[2], and Francesca Rossi[3]

[1] Dep. of Mathematics, University of Padova, Padova, Italy
khosh@math.unipd.it
[2] Dep. of Information Engineering, University of Padova, Padova, Italy
pini@dei.unipd.it
[3] IBM T.J. Watson Research Center, Yorktown Heights, NY, USA
frossi@math.unipd.it

Abstract. The importance role of contextual information on users' daily decisions led to develop the new generation of recommender systems called Context-Aware Recommender Systems (CARSs). Dependency of users preferences on the context of entities (e.g., restaurant, road, weather) in a dynamic domain, make the recommendation arduous to properly meet the users preferences and gain high level of users' satisfaction degree, especially in a group recommendation, in which several users need to take a joint decision. In these scenarios may also happen that some users have more weight/importance in the decision process. We propose a self-adaptive CARS (SaCARS) that provides fair services to a group of users who have different importance levels within their group Such services are recommended based on the conditional and qualitative preferences of the users that may change over time based on the different importance levels of the users in the group, on the context of the users, and the context of all the associated entities (e.g., restaurant, weather, other users) in the problem domain. In our framework we model users' preferences via conditional preference networks (CP-nets) and Time, we adapt Hyperspace Analogue to Context (HAC) model to handle the multi-dimensional context into the system, and sequential voting rule is used to aggregate users' preferences. We also evaluate the approach experimentally on a real-word scenario. Results show that it is promising.

Keywords: Context-Aware Recommender System · CP-net · User preferences

1 Introduction

The crucial impact of contextual information on users' preferences to provide services led to develop the next generation w.r.t traditional recommender systems (e.g., those based on Collaborative Filtering (CF) and Content-Based [2,3]) known as Context-Aware Recommender Systems (CARSs), which are designed

---

F. Rossi—(on leave from University of Padova).

© Springer International Publishing AG 2016
G. Adorni et al. (Eds.): AI*IA 2016, LNAI 10037, pp. 250–265, 2016.
DOI: 10.1007/978-3-319-49130-1_19

based on user, item and the context of the user as well. Context is basically defined as any information that can be used to characterize the situation of an entity [4]. Basically, CARSs have been categorized into two main methods: recommendation via context driven querying and search [5,6] and recommendation via contextual preferences elicitation and estimation. Although, such systems reduce the complexity of service selection tasks, they have difficulties to completely capture and understand the users' context due to the high-dimensionality of users' context and the lack of representation techniques in such system. In addition, dependency of users' preferences on the other entities' context bring the users change their preferences over time. Hence, there is a need to generate a self-adaptive context-aware framework that can accurately distinguish and carefully consider the changes in the context of entities to provide service(s).

To achieve the specified goals, in this paper we propose a self-adaptive context-aware recommender system (SaCARS)[1] that takes into account not only the context of user to recommend a service, but also the context of all entities e.g., user, restaurant, weather, etc. in our scenario, that are involved in the domain. In addition, it is able to provide the best restaurant, according to the group of users' conditional preferences even in the scenarios where the users have different importance in the group. Due to the complexity of the contextual information, we used Hyperspace Analogue to Context (HAC) [7] model to abstract, handle and represent the multi-dimensional context into the system. Moreover, we use CP-net formalism to model users' preferences, that often are qualitative and conditional (for example, if it is sunny, I prefer to take a restaurant that has a garden with table outside). CP-nets [8] is the most suitable way for representing qualitative and conditional preferences that have been used in automated decision making and modeling human preferences in real-world applications. We also take into account time since preferences can change over the time. Moreover, since users should take a joint decision, there is a need to aggregate their preferences. To do so, we use Sequential Weighted Majority rule, since it is a simple and powerful mechanism to deal with the issue of aggregating the users' CP-nets, when users have different power in the decision-making process.

We evaluate the proposed framework experimentally by modeling the real recommendation domain at different times using a real data set, consisting of 130 restaurants in Mexico city [9]. The experimental results show that our proposed SaCARS is able to handle the challenge of meeting users' conditional preferences in a recommendation domain, with high level satisfaction degree even with big changes in the context of the domain and different users' weights in the group.

The rest of the paper is organized as follows. We discuss related studies in Sect. 2 and report background in Sect. 3. The proposed framework and simulation are explained in Sects. 4 and 5, respectively. Finally, we conclude the paper in Sect. 6.

---

[1] In this paper we provide a revised and extended framework w.r.t. the one shown in [1]. We now assume that users can have different weights in the group and different priorities to order the features. Moreover, we evaluate the approach on real-data.

2   Related Work

Context-Aware recommender systems try to recommend a set of services to a user or a group of users with considering context. Simen et al. in [10] introduced a prototype of group CARS for concerts. They considered context such as Time and Location of the users, only to characterize users to recommend a proper concert by implementing different CF algorithms such as K-nearest neighbor, matrix factorization and Hybrid method (combination of the both methods). Palmisano et al. [11] introduced a hierarchy of contextual information with multi dimensions in their system, where each dimension could have sub-dimensions like time, location, etc. In [12] every feature is defined as a dimension (e.g., time, location) and they defined a rating function that specifies how likely user u prefers item i at time t. In this paper they considered quantitative preferences, while we deal with qualitative preferences. Baltrunas and Amatrian [13] introduced a method, in which the user's profile is divided into several sub profiles. Each sub profile represents a specific context of user and the prediction's procedure is carried out based on these sub profiles. Oku and his colleagues [14] used the modeling method to incorporate additional contextual dimensions (e.g., time, weather and companion as the user's context) in recommendation space, in which they used machine learning techniques (SVM) for preference classification to provide services. Wenyu et al. in [15] used quadratic polynomial to approximate conditional preferences into ListPMF for recommendation. All the studies mentioned above considered different context to find the best preference of the users for recommendation as in our method. However, the most difference between our proposed SaCARS w.r.t these studies lies in the ability of considering the conditional preference of the user(s) *On-the-fly*, which leads to adapt itself with current context of the domain entities to recommend the new service to users. In addition, our framework can recommend a fair service to a group of users, who may have different weights/importance in their group.

3   Background

We now provide some basic notions.

HAC in CARS. Hyperspace Analogue to Context (HAC) is a formal method to define multi-dimensional context in a Space^2. The method was inspired by Hyperspace Analogue to Language (HAL) [16] and is used (for the first time) in a context-aware system to define multi-dimensional context in a smart home [7]. In our proposed CARS, all entities in the Space $H = \langle h_1, h_2, ..., h_n \rangle$ are defined with different dimensions $h_i = \langle D_{i1}, D_{i2}, ..., D_{im} \rangle$, where D_i refers a type of context (e.g., location, time).

---

^{2} In this study, *Space* refers to a domain where all entities have dependencies. For example, in the space of selecting a restaurant, users, road, restaurants and weather have relations that can influence users' preferences.

CP-net. CP-net [8] is a graphical model to represent conditional and qualitative preference relations between variables (aka features). Assume a set of variables $V = \{X_1, ..., X_n\}$ with finite domains $D(X_1), ..., D(X_n)$. For each variable X_i, each user specifies a set of parents $P(X_i)$ that can affect her preferences over the value of X_i. So this defines a dependency graph such that every variable X_i has $P(X_i)$ as its immediate predecessors. They are sets of ceteris paribus preference statements (CP-statements). For instance, a CP-statement is "I prefer romance movie to action if it is rainy".

Sequential Majority. In order to aggregate the CP-nets of the users, we use Sequential Majority rule that in every step elicits the users' preferences over the variable X_i (it starts with independent variables, and then dependent variables). Then Majority rule is used to compute the chosen value for the variable X_i [17]. For dependent variables we select the preference ordering that is related to the assignment of the previous variables [18]. The value of a variable x_j is majority better than another x'_j ($x_j \succ_{maj} x'_j$) if and only if $|S_{x_j \succ x'_j}| > |S_{x_j \prec x'_j}| + |S_{x_j \bowtie x'_j}|$ [19], where $S_{x_j \succ x'_j}$ represents the set of users who prefer x_j to x'_j and $S_{x_j \bowtie x'_j}$ represents the set of users who are indifferent to select x_j or x'_j. The sequential procedure considered here has also been studied in terms of its resistance to a form of manipulation of the result called bribery[3] in [21–24] and it has been adapted also in scenarios where users express quantitative preferences via the soft constraint formalism [25]. Bribery issues have been studied also in this context [26,27].

4   Proposed System Framework

In this section we define our framework starting with a *case study*, which will drive us through the presentation. Bob and his wife (Alice) have decided to celebrate their anniversary. They have also decided to invite some of their common friends to the celebration in a restaurant. They have difficulties in selecting a proper restaurant, since they should consider their preferences w.r.t the context e.g., location, price, traffic, cuisine, quality, as well as their friends' preferences w.r.t the same context to select the best restaurant. Bob and Alice have some preferences in selecting food and restaurant. E.g., Alice prefers Italian restaurants to Chinese ones with high quality, while Bob prefers (unconditionally) a restaurant in a location close to where they are located. In addition, they need to take into account their friends' favorite cuisines and priorities (their friends may have certain constraints to select a restaurant or food). Furthermore, they are in a dynamic domain where any change in the context of an entity (e.g., restaurant, road or weather in the Space) may change the users' preferences. Therefore, they need to make sure that the restaurant can satisfy all the participants' requests under considering all aspect of the difficulties. It needs to be mentioned that

---

[3] The bribery problem is defined by an external agent (the briber) who wants to influence the result of the rule by convincing some users to change their preferences, in order to get a collective result which is more preferred to him; there is usually a limited budget to be spent by the briber to convince the users [20].

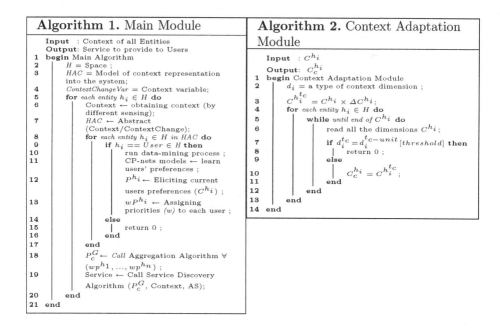

all the users (Alice and Bob's friends) are in the same recommender system such that the framework collects their data using different sensing devices (e.g., mobile devices). The system should be able to update itself to provide the right service at the right time. In other words, we need a self adaptive framework that can adapt its behavior (that is, the service it provides) based on the current context of the entities in the specific Space.

4.1 Context Integration and Abstraction to HAC

We define and extend *HAC* in our proposed framework with the following definitions.

Definition 1 (N-entity in a Space). *Space is a set of different entities* $H = \langle h_1, h_2, ..., h_n \rangle$, $h_i \in H$, *where all entities have correlations.*

For example, in a restaurant recommendation Space H; users, road, restaurant, etc. could be the entities which act as providers.

Definition 2. (N-dimensional HAC). *An n-dimensional HAC is an entity* $h_i = \langle D_{i1}, D_{i2}, ..., D_{im} \rangle$, *where each dimension* D_i *is a type of context.*

In HAC, dimensions are different attributes or context that describe an entity. These can be e.g., location or status of an entity i.e. is restaurant cheap or expensive. Thus, each entity is characterized by several different context dimensions. In this study only the most relevant context dimensions are used (with binary values for simplicity, however we generalize the binary values into non-binary values in the simulation section) in the service discovery. The most relevant context will be selected based on the users interactions' histories with the system, knowledge of domain, as well as using different machine learning techniques, e.g., Pre-filtering [12].

Definition 3 (Context Point). *Context point of an entity h_i in Space H is $C^{h_i,t_c} = \langle v_1, v_2, ..., v_n \rangle$, where, $v_i \in D_j; i = \{1, ..., n\}$ at time t. The context of an entity is defined as a point in HAC.*

For instance, in our scenario, a context point of a restaurant could be $\langle v_{location} = streetX, v_{time} = 9 : 00, v_{availability} = open \rangle$ in Space H.

Definition 4 (Context Range). *In context range that is a sub context of full H, we specify a binary value $[v_i, v_i']$ for each context dimension.*

For example, in $h_{Restaurant} = \langle d_{cuisine}, d_{queue}, d_{time}, d_{available}, d_{quality} \rangle$, the range of $d_{cuisine} = [Chinese, Italian]$, $d_{queue} = [short, long]$, $d_{time} = [9 : 00, 10 : 00]$, $d_{available} = [open, close]$ and $d_{quality} = [low, high]$. Therefore, we define $v_i \in d_i \Longleftrightarrow \forall i, d_i \in h_i$.

Definition 5 (Context Adaptation). *Context adaptation $\times \Delta C = \langle \Delta d_1, \Delta d_2, ..., \Delta d_n \rangle$, is a function/module (see Algorithm 2) to find the changes between e.g., restaurants' current context at time t_c and past context points at time t_{c-unit} (unit is defined by designer). Δd_i points the new value for a context dimension, $d_i^{t_c} = d_i \times \Delta d_i$. d_i does not change, if $\Delta d_i = \emptyset$.*

Context Adaptation module plays a crucial role in the proposed framework, since the function keeps the framework updated with the current context of all entities in the domain to provide service(s). Any change in context may have influence on the users' preferences. In this function, we use certain thresholds whose values depend on the type of the features [28].

Definition 6 (CP-net with Time and Context in HAC). *CP-net with Time and Context in HAC is a tuple P=(T,CD), where CD points the set of context dimensions (or features) $\mathbb{D} = \langle D_1^{h_1}, ..., D_n^{h_1}, D_1^{h_2}, ..., D_k^{h_2}, ..., D_1^{h_m}, ..., D_l^{h_m} \rangle$ that users prefers in Space H. T points the time that dimension $d \in D^{h_i}$ is preferred to $d' \in D^{h_i}$ in context CD.*

Each entity (h_i) may have a different set of $CDs \langle D_1, D_2, ..., D_n \rangle$, in which every CD has the binary ranges (or more) d, d'. For each context dimension D_i, there may be a set of parent feature $P(D_i)$ that can influence on preferences over the values of D_i. So this defines a dependency graph such that every node D_i has $P(D_i)$ with specific time as its immediate predecessors. Each context dimension CD is annotated with a conditional preference table (CPT) and time

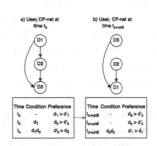

Fig. 1. Users' different CP-nets.

t, which specifies the user's preferences over a feature given the values of all parent context and time. For example, at time t_c Bob prefers unconditionally a restaurant with high quality ($d_1 \in D_1$ in Fig. 1a), while at time t_{c+unit} he may completely change his mind because of a change in context of other entities (e.g., something happened in the road and it is difficult to reach the recommended restaurant), so he may decide to completely change his preference (which may lead to have different dependencies). E.g., he may prefer a restaurant (uncon-ditionally) in a location close to where they are located ($d_3 \in D_3$ in Fig. 1b). Hence, the other conditions will appear with the new context. The refined profile of user's preferences (we consider only the most preferable one at current time) in HAC is $P^{h_{user}} = \langle d_1^{h_i,t_c}, d_2^{h_i,t_c}, d_3^{h_j,t_c} \rangle$ over $\mathbb{Z}$.

4.2   Different Importance of Users

In some cases, users may have different weights in order to select or accept a service or an item within the group. It is worth clear the users who have high weights can affect more the result, and even on the other users' preferences with less importance (but it is not guaranty). For example, in a restaurant selection guide Space, a user may have certain limitations (e.g., allergy) to select some types of restaurants (which provide specific food e.g., an Indian restaurant usually provides spicy food), thus other users may express their preferences based on the users who have high weights. Hence, the framework should consider the importance of the users in the group to provide a fair service that is acceptable for all users. In the following, we define the users' importance in *HAC* model.

Definition 7 (Users' Importance). *Every user's profile has an associated weight (w) as follows:* $P = \langle wp^{h_1}, wp^{h_2}, ..., wp^{h_n} \rangle$.

The system sends the users' profiles with users importance to aggregation module.

4.3   Preference Aggregation Module

Since CP-net is used to model the users' conditional preferences, there is a need to use a suitable mechanism to aggregate these CP-nets in a group. It is worth clear that different users have different preferences among multiple features (CP-net with different dependencies). For instance, price could be important for a user and time could be important for other users in a group. Since, in CP-net "everything else being equal", we use the concept of *independent* and *dependent* features in order to handle this challenge, such that we consider the independent features as the most important features, while the most dependent features as the less important ones. Let us consider an example (Fig. 2a), three users with four features (*D1,D2, D3* and *D4*). In User 1*D1* is the parent of *D2* and *D2* is the parent of *D3* and *D4*. So, *D1* (which is independent feature) is the most important feature for User 1, *D2* is the second most important feature, and *D3*

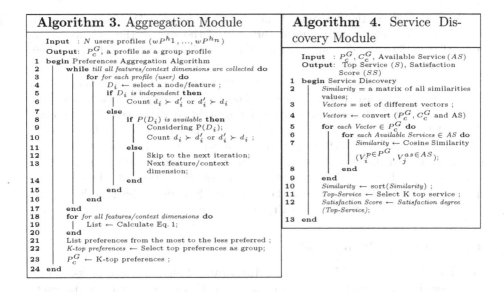

| **Algorithm 3.** Aggregation Module | **Algorithm 4.** Service Discovery Module |

Algorithm 3. Aggregation Module

```
Input   : N users profiles (wPʰ1 , ..., wPʰn )
Output: P_c^G, a profile as a group profile
1  begin Preferences Aggregation Algorithm
2  |   while till all features/context dimensions are collected do
3  |   |   for for each profile (user) do
4  |   |   |   D_i ← select a node/feature ;
5  |   |   |   if D_i is independent then
6  |   |   |   |   Count d_i ≻ d'_i or d'_i ≻ d_i
7  |   |   |   else
8  |   |   |   |   if P(D_i) is available then
9  |   |   |   |   |   Considering P(D_i);
10 |   |   |   |   |   Count d_i ≻ d'_i or d'_i ≻ d_i ;
11 |   |   |   |   else
12 |   |   |   |   |   Skip to the next iteration;
13 |   |   |   |   |   Next feature/context
                        dimension;
14 |   |   |   end
15 |   |   end
16 |   end
17 |   end
18 |   for for all features/context dimensions do
19 |   |   List ← Calculate Eq. 1;
20 |   end
21 |   List preferences from the most to the less preferred ;
22 |   K-top preferences ← Select top preferences as group;
23 |   P_c^G ← K-top preferences ;
24 end
```

Algorithm 4. Service Discovery Module

```
Input   : P_c^G, C_c^G, Available Service (AS)
Output: Top Service (S), Satisfaction
          Score (SS)
1  begin Service Discovery
2  |   Similarity = a matrix of all similarities
       values;
3  |   Vectors = set of different vectors ;
4  |   Vectors ← convert (P_c^G, C_c^G and AS)
5  |   for each Vector ∈ P_c^G do
6  |   |   for each Available Services ∈ AS do
7  |   |   |   Similarity ← Cosine Similarity
               (V_i^{p∈P^G} , V_j^{as∈AS});
8  |   |   end
9  |   end
10 |   Similarity ← sort(Similarity) ;
11 |   Top-Service ← Select K top service ;
12 |   Satisfaction Score ← Satisfaction degree
       (Top-Service);
13 end
```

and $D4$ are in the same importance. The degree-of-importance of features could be changed over time as mentioned before.

In this framework we use Sequential Weighted Majority rule to aggregate users' CP-nets. We first start with a context/feature, which is independent for each user e.g., $D1$ for User 1, $D3$ for User 2, etc. (see Fig. 2a and b). The algorithm obtains the user' information who prefers d_i over $d'_i \in D_i$ in first iteration (Fig. 2b). Thereafter, the system moves to the other features that are dependent (e.g., $D2$ for User 1, $D1$ for User 2, etc.) considering the preferences of its parents. In each iteration of collecting preference of users, the system fills missing values that are not collected from the previous iterations. This process is repeated until all the features are collected. The weight of each user is used to select the most preferable feature/context dimension as preference of the group.

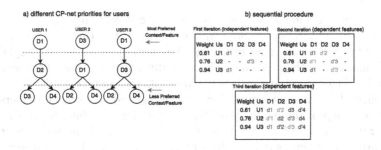

Fig. 2. (a) Different CP-net priorities for users. (b) Sequential procedure

Assume d, d' are the domains of the context dimension D_i, d is preferred to d' by preference aggregation rule if and only of the sum of the weights of the users that prefers d to d' is greater than the sum of the weights of the users that prefers d' to d (see Algorithm 3).

Since we use a concrete scenario there may be missing information within some context dimensions e.g., price for both Bob and Alice, quality for Bob, location for Alice, etc. Instead of considering indifferent the context where we do not have information, in our aggregation mechanism we use a classical Collaborative Filtering technique to find the preference of other users who have the same context dimensions on those context as unconditional preference. E.g., Alice does not have any idea about the price of the restaurant, therefore, the framework uses the preference of the other involved user who has the similar context of Alice (e.g., budget, favorite cuisine, etc.).

4.4   Recommendation Module

Service Discovery and Context Matching. This module uses k-Nearest Neighbor (kNN) algorithm [29] to find and match the best service (restaurant) with respect to the users' preferences. In this module, our algorithm does not look only for the full matches. For instance, the users' request (as a group preferences) has n dimensions, and the service can only fulfill k (*where* $k < n$) dimensions of the users' request. Thus, the system selects the best service that can satisfy more the users request. Although accuracy metrics are mostly used in RSs to show how the recommendation is suitable, there are properties of user satisfaction that are unable to obtain, such as diversity, similarity, coverage and serendipity [30]. In this study we use Cosine Similarity method [31] to find out the similarity between users' preferences and recommended services (or items), and represented as users satisfaction degree (the high similarity, the high satisfaction degree) that is explained in Sect. 5. We omit a detailed description of service discovery module, which is explained by Algorithm 4, due to space limit.

4.5   Proposed Framework's Workflow

We model a real recommendation scenario in 7 different times $(t_1, t_2, ..., t_7)$, in which at each time (t_i) one/more context of the entities (in the domain) could be changed and this may lead to change the users' preferences. We explain the whole framework in the following. The procedure of the proposed framework is explained in Algorithm 1 and it is shown in Fig. 3. The context of different entities will be elicited by different types of sensors (connected to service providers, e.g., whether service provider, traffic provider, etc.) at time t_1. Then, entities' context are abstracted to *HAC* model using the aforementioned methods (Sect. 4.2). Users' current preferences (with different weights w) are elicited (as it mentioned above, based on the users' interaction with system and the current context of the different entities in the Space), and then sent to the aggregation module as users' preferences profiles $(wP_1, wP_2, ..., wP_n)$. In this step, we aggregate the users' CP-nets sequentially and we interact with the Service Discovery module to find a proper service. Consequently, the module finds' and delivers the best possible service (e.g., restaurant in our scenario) to the group.

Meanwhile, from the time that service is delivered at t_1 to be used, to the time the service is used by users (t_2), if the relevant context of entities change at t_2, which causes changes on users' preferences (e.g., the availability of restaurant changes from open to close, or light traffic changes to heavy traffic in the road close to the recommended restaurant), *Context Change Detector* discovers that a change has happened and triggers the discovery module to find a proper service (a new restaurant) for the current time. That is because, the features of the recommended restaurant may change, which lead to decrease the satisfaction degree of users.

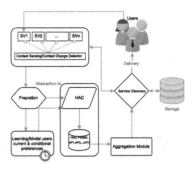

Fig. 3. Architecture of the proposed context-aware framework.

We implement the framework on a real data-set (130 restaurants in Mexico city) with contextual information, which is collected from UCI repository [9]. In this data-set, every restaurant is described and characterized with several features, from location, cuisine, price to working days and time. Hence, we use some important context of the restaurants to evaluate our framework, which are explain in the next section.

5   Simulation

We simulate the proposed self-adaptive framework using C++ and MATLAB (on a 2.7 GHz Intel Core i5 machine with 8 GB of 1867 MHz DDR3 RAM), where the proposed method is compared to a regular CARS. in two parts. The simulation's results describe the effect of the context change on users' preferences and satisfaction degree and the effect of users' weights on recommendation and their satisfaction degrees.

In this experiment we use the aforementioned data set with multiple contextual information. Then, we select (using knowledge-based domain [32]) 7 context dimensions $N^{CD} = 7$ with different domains to characterize the 130 restaurants as follows: (1) type of restaurant/cuisines, (2) serving alcohol in the restaurant, (3) smoking area, (4) accessibility, (5) price, (6) outdoor or indoor restaurant, (7) parking. Each context dimension has different domain that are explained in Table 1. We use GenCPnet [33] to model and generate $N = 7$ CP-nets as the users' conditional preferences.

As it mentioned above in this scenario, users have different weights within their group, who can influence more/less on aggregation part. We then use Sequential Weighted Majority rule to aggregate the CP-nets where we obtained a set of preferences, from the most to the least preferable. We randomly assigned

Table 1. Parameters.

Parameters	Value	Parameters	Value
Number of iteration	100	Numbers of users	7
Nof CP-nets	7	Nof available restaurant	130
Aggregation rules	Sequential Weighted Majority	Weight	0 to 1 (0 has the least and 1 has the highest importance)
Nof restaurant features	5	NoF other entities' features	2
Domain of each feature	different	Satisfaction degree	0 (minimum) 1 (maximum)
Features	Domains	Features	Domains
Type of restaurant (cuisines)	1. seafood, 2. Italian, 3. Sushi, 4. American 5. Chinese, 6. French 7. Mediterian, 8. German, 9. others	Serving alcohol	1. no, 2. wine-beer, 3. full bar
Smoking area	1. no, 2. section, 3. only at bar	Accessibility	1. completely, 2. partially, 3. no
Price	1. low, 2. medium, 3. high	Restaurant atmosphere	1. indoor, 2. outdoor
Parking	1. no, 2. yes with fee, 3. public and street	Time	$t_1, t_2, t_3, t_4, t_5, t_6, t_7$

different weights (from 0 to 1, which 0 has least and 1 has highest weight) to each user as user's importance in the group. Hence, the system selects the most preferable context dimension as the group preferences as explained in Sect. 4.4. Then, aggregation module sends a set of preferences from the most to the least group's preferences to the *Service Discovery* module, in which the most preferable preference (top one) will be considered as the group preference. In Service Discovery module, we convert all the features (context dimensions), which are used in CP-nets and restaurants into vectors (features in users' CP-nets and restaurants are the same) and we use *Cosine Similarity* to calculate the similarity between group's preferences and all the available restaurants in the Space at each time, and then we try to find the best restaurant with high similarity value. We run the framework at 7 different times from t_1 (when the service is delivered) to t_7 (when the service is used by users) to model the real scenario. Then we repeat this implementation 100 times to get the precise satisfaction value. Table 1 shows the used parameters in this simulation. In the simulation we compare regular CARS with our SaCARS in terms of users' satisfaction degrees both when the users have the same importance in the group and when they have different levels of importance.

Running Service Discovery. At t_1 the system recommends the specific restaurant, which has the nearest similarity value (highest value) with respect to the group's preferences as the best restaurant. Then we randomly change one context dimension (feature) of the recommended restaurant at t_2 e.g., the point of

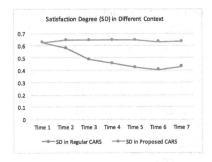
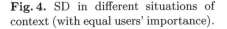

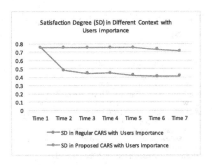

Fig. 4. SD in different situations of context (with equal users' importance).

Fig. 5. SD in different situations of context (with users who have different importance in their group).

the entity "restaurant" is changed to another point "Definition 3", thereafter, we calculate the satisfaction degree of the group. We normalized the similarity value between 0 to 1, which are the worst and the best (full match) satisfaction degree, respectively. At t_3 we change 2 context dimensions of the recommended restaurant, and then again repeating the same procedure to calculate the satisfaction degree. This procedure will continue until all the context are randomly changed. The reason that we gradually increase the number of context change (till all the context) is that to see what will happen in the worst case to evaluate if context changes influence on satisfaction degree of users. This is a regular CARS where we did not run *Context Change Detector/Context Adaptation module*. The satisfaction degree in each time is shown with blue line in Fig. 4.

Running Context Adaptation Module. At t_1 the RS provides a proper service based on the group's preferences (top one) and calculates the satisfaction degree. At t_2 we randomly change 1 context of the recommended restaurant. Thereafter, *Context Change Detector*, detects that a change has happened and reports to the discovery module to find a proper service at t_2 based on the current context of the entities. Since, the current features of the recommended restaurant might not meet the group's preferences, the satisfaction degree will be evaluated with the group's preferences. Therefore, the new service (restaurant) will replace the previous one (recommended restaurant at t_1) at t_2, if it has higher satisfaction degree upon the group's preferences. This process will continue until a threshold is reached (threshold value depends on the type of the feature [28] or the service is used by the group. The concept of different times $(t_1, ..., t_7)$ in the simulation is to model different possible situations in the specific domain, between providing a service by the framework and the time in which the service might be used by users. Moreover, we show what if/if not, the *Context Change Detector/Context Adaptation* module works during the service recommendation. We increase the number of changed context at each time and calculate the satisfaction degree, which are shown with the red line in the Fig. 4. Y-axis shows the satisfaction degree, and X-axis indicates different times (in

Table 2. Time, context changes and SD.

Time	t_1	t_2	t_3	t_4	t_5	t_6	t_7
NCC	Initial context no change	1	2	3	4	5	6
SRS	0.62	0.58	0.48	0.45	0.42	0.40	0.43
SPRS	0.62	0.64	0.64	0.64	0.64	0.63	0.63
SRSW	0.75	0.48	0.44	0.44	0.42	0.40	0.41
SPRSW	0.75	0.75	0.74	0.74	0.73	0.72	0.71

NCC: Number of context changes
SRS: Satisfaction degree (SD) in Regular RS
SPRS: Satisfaction degree in Proposed RS
SRSW:SD in Regular RS with different weights
SPRSW: SD in Proposed RS with different weights

Table 3. SD of individual user at each time.

		Satisfaction degree of each user w.r.t RR						
W	Us	RR1/t_1	RR2/t_2	RR3/t_3	RR4/t_4	RR5/t_5	RR6/t_6	RR7/t_7
0.61	U1	0.43	0.44	0.43	0.43	0.42	0.45	0.50
0.76	U2	0.49	0.51	0.501	0.50	0.49	0.54	0.62
0.94	U3	0.62	0.64	0.63	0.62	0.62	0.65	0.73
0.71	U4	0.47	0.44	0.44	0.44	0.44	0.42	0.43
0.69	U5	0.42	0.48	0.49	0.47	0.48	0.50	0.57
0.76	U6	0.43	0.45	0.44	0.44	0.43	0.45	0.50
0.51	U7	0.32	0.34	0.35	0.35	0.34	0.32	0.31
0.53	U8	0.38	0.38	0.39	0.38	0.39	0.38	0.37
0.81	U9	0.52	0.47	0.46	0.47	0.47	0.47	0.49
0.67	U10	0.44	0.45	0.44	0.45	0.44	0.46	0.50

W: Weight
RR: Recommended restaurant at each time

Figs. 4 and 5). Although, the two methods start with the same satisfaction level 0.62, at t_1, recommendation of our method has higher satisfaction degree, when there is a context change, even with high number of changes. It needs to be mentioned that at some points the change of context leads to the improvement of satisfaction degree. Having said that, on average, the more the context changes, the more satisfaction degree drops. Table 2 and Fig. 4 indicate the experiment result.

Running the proposed method with users' different importance. We have then simulated the approach assuming that the users have different levels of importance/weights in their group (see Fig. 5). We can notice a similar behavior. However, although we have a fair recommendation, the satisfaction degree of each user (individually) can decrease w.r.t the group when there are no equal weights (e.g., *U7* and *U8*, see Table 3). This is not always true, since the distribution of satisfaction degree is also related to the similarity between the features. Table 3 represents the satisfaction degree and weight of each user when we assume different weights for the users. *U3* and *U9*, who have the highest weights (0.94 and 0.81 respectively) in the group, have almost highest satisfaction degree on every recommendation at each time. As it mentioned above, the satisfaction of the users are not regularly (always) distributed according to their weights. A user with low importance could have similar satisfaction degree of a user with high importance if they have similar preferences on the highest features E.g., *U4* and *U9* have different power in the group, but they have almost same satisfaction degree.

6 Conclusion and Future Work

In this study we proposed a self-adaptive CARS that is able to deal with qualitative and conditional preferences of a group of users, that may have different importance in the decision process, also when preferences can change over time. Preferences are modeled by *CP-nets* and *Time*, and then represented via *HAC* to handle the dependencies between users' preferences and the complexity of multi-dimensional entities' context, respectively. *Sequential Weighted Majority* rule is used to aggregate such preferences. Thereafter, the recommendation module finds the best possible service that can fulfill the users' requests. Meanwhile, the system can provide new service(s) to the users, if their (one or more) most relevant context of entities changes by affecting the preferences of the group of users. The experimental evaluation showed that the changes in the context dramatically decreases the user satisfaction degree in regular CARS, while the proposed framework can keep users at a high level of satisfaction also when the users have different levels of importance. In the future work, we will investigate how extracting and learning users' conditional preferences from their interactions with the system and how combining our SaCARS with a reputation system as done in [34].

References

1. Khoshkangini, R., Pini, M.S., Rossi, F.: A design of context-aware framework for conditional preferences of group of users. In: Lee, R. (ed.) Software Engineering, Artificial Intelligence, Networking and Parallel/Distributed Computing. SCI, vol. 653, pp. 97–112. Springer, Heidelberg (2016). doi:10.1007/978-3-319-33810-1_8

2. De Gemmis, M., Iaquinta, L., Lops, P., Musto, C., Narducci, F., Semeraro, G.: Preference learning in recommender systems. Prefer. Learn. **41** (2009)

3. Lu, J., Wu, D., Mao, M., Wang, W., Zhang, G.: Recommender system application developments: a survey. Decis. Support Syst. **74**, 12–32 (2015)

4. Dey, A.K.: Understanding and using context. Pers. Ubiquit. Comput. **5**(1), 4–7 (2001)

5. Ono, C., Kurokawa, M., Motomura, Y., Asoh, H.: A context-aware movie preference model using a Bayesian network for recommendation and promotion. In: Conati, C., McCoy, K., Paliouras, G. (eds.) UM 2007. LNCS (LNAI), vol. 4511, pp. 247–257. Springer, Heidelberg (2007). doi:10.1007/978-3-540-73078-1_28

6. Setten, M., Pokraev, S., Koolwaaij, J.: Context-aware recommendations in the mobile tourist application COMPASS. In: Bra, P.M.E., Nejdl, W. (eds.) AH 2004. LNCS, vol. 3137, pp. 235–244. Springer, Heidelberg (2004). doi:10.1007/978-3-540-27780-4_27

7. Rasch, K., Li, F., Sehic, S., Ayani, R., Dustdar, S.: Context-driven personalized service discovery in pervasive environments. World Wide Web **14**, 295–319 (2011)

8. Boutilier, C., Brafman, R.I., Domshlak, C., Hoos, H.H., Poole, D.: Cp-nets: a tool for representing and reasoning with conditional ceteris paribus preference statements. J. Artif. Intell. Res. (JAIR) **21**, 135–191 (2004)

9. Lichman, M.: UCI machine learning repository (2013)

10. Smaaberg, S.F., Shabib, N., Krogstie, J.: A user-study on context-aware group recommendation for concerts. In: HT (Doctoral Consortium/Late-breaking Results/Workshops) (2014)

11. Palmisano, C., Tuzhilin, A., Gorgoglione, M.: Using context to improve predictive modeling of customers in personalization applications. IEEE Trans. Knowl. Data Eng. **20**(11), 1535–1549 (2008)

12. Adomavicius, G., Sankaranarayanan, R., Sen, S., Tuzhilin, A.: Incorporating contextual information in recommender systems using a multidimensional approach. ACM Trans. Inf. Syst. (TOIS) **23**, 103–145 (2005)

13. Baltrunas, L., Amatriain, X.: Towards time-dependant recommendation based on implicit feedback. In: Workshop on Context-Aware Recommender Systems (CARS 2009) (2009)

14. Oku, K., et al.: A recommendation system considering users past/current/future contexts. In: Proceedings of CARS (2010)
15. Liu, W., Wu, C., Feng, B., Liu, J.: Conditional preference in recommender systems. Expert Syst. Appl. **42**, 774–788 (2015)
16. Lund, K., Burgess, C.: Producing high-dimensional semantic spaces from lexical co-occurrence. Behav. Res. Meth. Instrum. Comput. **1996**, 203–208 (1996)
17. Lang, J.: Graphical representation of ordinal preferences: languages and applications. In: Croitoru, M., Ferré, S., Lukose, D. (eds.) ICCS-ConceptStruct 2010. LNCS (LNAI), vol. 6208, pp. 3–9. Springer, Heidelberg (2010). doi:10.1007/978-3-642-14197-3_3
18. Lang, J., Xia, L.: Sequential composition of voting rules in multi-issue domains. Math. Soc. Sci. **57**, 304–324 (2009)
19. Rossi, F., Venable, K.B., Walsh, T.: mCP nets: representing and reasoning with preferences of multiple agents. AAAI **4**, 729–734 (2004)
20. Faliszewski, P., Hemaspaandra, E., Hemaspaandra, L.A.: How hard is bribery in elections? JAIR **35**, 485–532 (2009)
21. Maudet, N., Pini, M.S., Venable, K.B., Rossi, F.: Influence and aggregation of preferences over combinatorial domains. In: Proceedings of AAMAS 2012, pp. 1313–1314 (2012)
22. Maran, A., Maudet, N., Pini, M.S., Rossi, F., Venable, K.B.: A framework for aggregating influenced CP-nets and its resistance to bribery. In: Proceedings of AAAI 2013 (2013)
23. Mattei, N., Pini, M.S., Venable, K.B., Rossi, F.: Bribery in voting with CP-nets. Ann. Math. Artif. Intell. **68**, 135 (2013)
24. Mattei, N., Pini, M.S., Venable, K.B., Rossi, F.: Bribery in voting over combinatorial domains is easy. In: Proceedings of AAMAS 2012, pp. 1407–1408 (2012)
25. Dalla Pozza, G., Pini, M.S., Rossi, F., Venable, K.B.: Multi-agent soft constraint aggregation via sequential voting. In: Proceedings of IJCAI, pp. 172–177 (2011)
26. Pini, M.S., Rossi, F., Venable, K.B.: Resistance to bribery when aggregating soft constraints. In: Proceedings of AAMAS 2013, pp. 1301–1302 (2013)
27. Pini, M.S., Rossi, F., Venable, K.B.: Bribery in voting with soft constraints. In: Proceedings of AAAI 2013 (2013)
28. Rasch, K., Li, F., Sehic, S., Ayani, R., Dustdar, S.: Automatic description of context-altering services through observational learning. In: Kay, J., Lukowicz, P., Tokuda, H., Olivier, P., Krüger, A. (eds.) Pervasive 2012. LNCS, vol. 7319, pp. 461–477. Springer, Heidelberg (2012). doi:10.1007/978-3-642-31205-2_28
29. Zhang, H., Berg, A.C., Maire, M., Malik, J.: SVM-KNN: discriminative nearest neighbor classification for visual category recognition. In: 2006 IEEE Computer Society Conference on Computer Vision and Pattern Recognition (CVPR 2006). vol. 2, pp. 2126–2136. IEEE (2006)
30. Ziegler, C.N., McNee, S.M., Konstan, J.A., Lausen, G.: Improving recommendation lists through topic diversification. In: Proceedings of the 14th International Conference on World Wide Web, pp. 22–32. ACM (2005)
31. Qian, G., Sural, S., Gu, Y., Pramanik, S.: Similarity between euclidean and cosine angle distance for nearest neighbor queries. In: Proceedings of the 2004 ACM Symposium on Applied Computing, pp. 1232–1237. ACM (2004)
32. Adomavicius, G., Tuzhilin, A.: Context-aware recommender systems. In: Ricci, F., Rokach, L., Shapira, B., Kantor, P.B. (eds.) Recommender Systems Handbook, pp. 217–253. Springer, New York (2011). doi:10.1007/978-0-387-85820-3_7

33. Allen, T.E., Goldsmith, J., Justice, H.E., Mattei, N., Raines, K.: Generating CP-nets uniformly at random. In: Proceedings of the 30th AAAI Conference on Artificial Intelligence (AAAI) (2016)

34. Jøsang, A., Guo, G., Pini, M.S., Santini, F., Xu, Y.: Combining recommender and reputation systems to produce better online advice. In: Torra, V., Narukawa, Y., Navarro-Arribas, G., Megías, D. (eds.) MDAI 2013. LNCS (LNAI), vol. 8234, pp. 126–138. Springer, Heidelberg (2013). doi:10.1007/978-3-642-41550-0_12

A Model+Solver Approach to Concept Learning

Francesca Alessandra Lisi[(✉)]

Dipartimento di Informatica, Università degli Studi di Bari "Aldo Moro", Bari, Italy
`francesca.lisi@uniba.it`

Abstract. Many Concept Learning problems can be seen as Constraint Satisfaction Problems (CSP). In this paper, we propose a model+solver approach to Concept Learning which combines the efficacy of Description Logics (DLs) in conceptual modeling with the efficiency of Answer Set Programming (ASP) solvers in dealing with CSPs.

Keywords: Concept learning · Declarative modeling · Description logics · Answer set programming · Constraint satisfaction

1 Introduction

Ideally, the Machine Learning (ML) task is to discover an operational description of a *target function* $f : X \rightarrow Y$ which maps elements in the *instance space* X to the values of a set Y. This function is unknown, meaning that only a set $\mathcal{D}$ (the *training data*) of points of the form $(x, f(x))$ is provided. However, it may be very difficult in general to learn such a description of f perfectly. In fact, ML algorithms are often expected to acquire only some approximation $\hat{f}$ to f by searching a very large space $\mathcal{H}$ of possible hypotheses (the *hypothesis space*) which depend on the representation chosen for f (the *language of hypotheses*). The output approximation is the one that best fits $\mathcal{D}$ according to a *scoring function* $score(f, \mathcal{D})$. It is assumed that any hypothesis $h \in \mathcal{H}$ that approximates f well w.r.t. a large set of training cases will also approximate it well for new unobserved cases. Summing up, given $\mathcal{H}$ and $\mathcal{D}$, ML algorithms are designed to find an approximation $\hat{f}$ of a target function f s.t.:

1. $\hat{f} \in \mathcal{H}$;
2. $\hat{f}(\mathcal{D}) \approx f(\mathcal{D})$; and/or
3. $\hat{f} = argmax_{f \in \mathcal{H}} score(f, \mathcal{D})$.

These notions have been mathematically formalized in computational learning theory within the Probably Approximately Correct (PAC) learning framework [1]. It has been recently stressed that the first two requirements impose constraints on the possible hypotheses, thus defining a *Constraint Satisfaction Problem* (CSP), whereas the third requirement involves the optimization step, thus turning the CSP into an *Optimization Problem* (OP) [2]. We shall refer to the ensemble of constraints and optimization criteria as the model of the learning task. Models are almost by definition declarative and it is useful to distinguish

G. Adorni et al. (Eds.): AI*IA 2016, LNAI 10037, pp. 266–279, 2016.
DOI: 10.1007/978-3-319-49130-1_20

the CSP, which is concerned with finding a solution that satisfies all the constraints in the model, from the OP, where one also must guarantee that the found solution be optimal w.r.t. the optimization function. Examples of typical CSPs in the ML context include variants of so-called *Concept Learning* which deals with inferring the general definition of a category based on members (positive examples) and nonmembers (negative examples) of this category. Here, the target is a Boolean-valued function $f : X \rightarrow \{0, 1\}$, *i.e.* a *concept*. When examples of the target concept are available, the resulting ML task is said *supervised*, otherwise it is called *unsupervised*. The positive examples are those instances with $f(x) = 1$, and negative ones are those with $f(x) = 0$. In Concept Learning, the key inferential mechanism for induction is *generalization as search* through a partially ordered space $(\mathcal{H}, \preceq)$ of hypotheses [3].

Research in ML has traditionally focussed on designing effective algorithms for solving particular tasks. However, there is an increasing interest in providing the user with a means for specifying what the ML problem in hand actually is rather than letting him struggle to outline how the solution to that problem needs to be computed (see the recent note by De Raedt [4]). This corresponds to a *model+solver* approach to ML, in which the user specifies the problem in a *declarative modeling language* and the system automatically transforms such models into a format that can be used by a *solver* to efficiently generate a solution. In this paper, we propose a *model+solver* approach to Concept Learning which combines the efficacy of *Description Logics* (DLs) [5] in conceptual modeling with the efficiency of *Answer Set Programming* (ASP) solvers (see [6] for an updated overview) in dealing with CSPs. The approach consists of a declarative modeling language based on second-order DLs under Henkin semantics, and a mechanism for transforming second-order DL formulas into a format processable by ASP solvers. This paper completes the work reported in [7]. In particular, it elaborates more on the modeling of one of the variants of the Concept Learning problem discussed in [7] (more precisely, the CSP version of the problem variant called Concept Induction), and provides a substantial contribution to the solver part which was left as future work in [7].

The paper is structured as follows. Section 2 is devoted to preliminaries on DLs and ASP. Section 3 introduces a case study from Concept Learning in DLs which is of interest to this paper. Section 4 describes our *model+solver* approach to the case being studied. Section 5 discusses related work. Section 6 summarizes the contributions of the paper and outlines directions of future work.

2   Preliminaries

2.1   Description Logics

DLs are a family of decidable First Order Logic (FOL) fragments that allow for the specification of structured knowledge in terms of classes (*concepts*), instances (*individuals*), and binary relations between instances (*roles*) [5]. Let $\mathsf{N_C}$, $\mathsf{N_R}$, and $\mathsf{N_O}$ be the alphabet of concept names, role names and individual names, respectively. Complex concepts can be defined from atomic concepts and roles

Table 1. Syntax and semantics of some typical DL constructs.

Bottom (resp. top) concept	$\bot$ (resp. $\top$)	$\emptyset$ (resp. $\Delta^{\mathcal{I}}$)
Atomic concept	A	$A^{\mathcal{I}} \subseteq \Delta^{\mathcal{I}}$
Role	R	$R^{\mathcal{I}} \subseteq \Delta^{\mathcal{I}} \times \Delta^{\mathcal{I}}$
Individual	a	$a^{\mathcal{I}} \in \Delta^{\mathcal{I}}$
Nominals	$\{a_1, \ldots, a_n\}$	$\{a_1^{\mathcal{I}}, \ldots, a_n^{\mathcal{I}}\}$
Concept negation	$\neg C$	$\Delta^{\mathcal{I}} \setminus C^{\mathcal{I}}$
Concept intersection	$C_1 \sqcap C_2$	$C_1^{\mathcal{I}} \cap C_2^{\mathcal{I}}$
Concept union	$C_1 \sqcup C_2$	$C_1^{\mathcal{I}} \cup C_2^{\mathcal{I}}$
Value restriction	$\forall R.C$	$\{x \in \Delta^{\mathcal{I}} \mid \forall y \, (x,y) \in R^{\mathcal{I}} \rightarrow y \in C^{\mathcal{I}}\}$
Existential restriction	$\exists R.C$	$\{x \in \Delta^{\mathcal{I}} \mid \exists y \, (x,y) \in R^{\mathcal{I}} \wedge y \in C^{\mathcal{I}}\}$
Self concept	$\exists R.Self$	$\{x \in \Delta^{\mathcal{I}} \mid (x,x) \in R^{\mathcal{I}}\}$
Qualified	$\leqslant n \, R.C$	$\{x \in \Delta^{\mathcal{I}} \mid \#\{y \in C^{\mathcal{I}} \mid (x,y) \in R^{\mathcal{I}}\} \leq n\}$
Number restriction	$\geqslant n \, R.C$	$\{x \in \Delta^{\mathcal{I}} \mid \#\{y \in C^{\mathcal{I}} \mid (x,y) \in R^{\mathcal{I}}\} \geq n\}$
Concept inclusion axiom	$C \sqsubseteq D$	$C^{\mathcal{I}} \subseteq D^{\mathcal{I}}$
Concept equivalence axiom	$C \equiv D$	$C^{\mathcal{I}} = D^{\mathcal{I}}$
Role inclusion axiom	$R_1 \circ \ldots \circ R_n \sqsubseteq S$	$R_1^{\mathcal{I}} \circ \ldots \circ R_n^{\mathcal{I}} \subseteq S^{\mathcal{I}}$
Concept assertion	$a : C$	$a^{\mathcal{I}} \in C^{\mathcal{I}}$
Role assertion	$\langle a, b \rangle : R$	$(a^{\mathcal{I}}, b^{\mathcal{I}}) \in R^{\mathcal{I}}$
Equality assertion	$a \doteq b$	$a^{\mathcal{I}} = b^{\mathcal{I}}$
Inequality assertion	$a \not\doteq b$	$a^{\mathcal{I}} \neq b^{\mathcal{I}}$

by means of constructors. The syntax of some typical DL constructs is reported in Table 1. A DL knowledge base (KB) $\mathcal{K} = (\mathcal{T}, \mathcal{A})$ consists of a so-called *terminological box* (TBox) $\mathcal{T}$ and a so-called *assertional box* (ABox) $\mathcal{A}$. The TBox is a finite set of *axioms* which represent either is-a relations (denoted with $\sqsubseteq$) or equivalence (denoted with $\equiv$) relations between concepts, whereas the ABox is a finite set of *assertions* (or *facts*) that represent instance-of relations between individuals (resp. couples of individuals) and concepts (resp. roles). DLs provide logical foundations to the W3C *Web Ontology Language* (OWL) [8]. Thus, when a DL-based ontology language is adopted, an ontology is nothing else than a TBox, and a populated ontology corresponds to a whole DL KB (*i.e.*, encompassing also an ABox). In particular, $\mathcal{SROIQ}$ [9] is the logical counterpart of OWL 2.[1] A distinguishing feature of $\mathcal{SROIQ}$ is that it admits inverse roles.

The semantics of DLs can be defined directly with set-theoretic formalizations as shown in Table 1 or through a mapping to FOL as shown in [10]. An *interpretation* $\mathcal{I} = (\Delta^{\mathcal{I}}, \cdot^{\mathcal{I}})$ for a DL KB $\mathcal{K}$ consists of a domain $\Delta^{\mathcal{I}}$ and a mapping function $\cdot^{\mathcal{I}}$. Under the *Unique Names Assumption* (UNA) [11], individuals are mapped to elements of $\Delta^{\mathcal{I}}$ such that $a^{\mathcal{I}} \neq b^{\mathcal{I}}$ if $a \neq b$. However UNA does not hold by default in DLs. An interpretation $\mathcal{I}$ is a *model* of $\mathcal{K}$ iff it satisfies all axioms and assertions in $\mathcal{T}$ and $\mathcal{A}$. In DLs a KB represents many different inter-

---

[1] http://www.w3.org/TR/2009/REC-owl2-overview-20091027/.

pretations, i.e. all its models. This is coherent with the *Open World Assumption* (OWA) that holds in FOL semantics. A DL KB is *satisfiable* if it has at least one model. An ABox assertion α is a *logical consequence* of a KB $\mathcal{K}$, written $\mathcal{K} \models \alpha$, if all models of $\mathcal{K}$ are also models of α.

The main reasoning task for a DL KB $\mathcal{K}$ is the *consistency check* which tries to prove the satisfiability of $\mathcal{K}$. This check is performed by applying decision procedures mostly based on tableau calculus. The *subsumption check* aims at proving whether a concept is included in another one according to the subsumption relationship. Another well known reasoning service in DLs is *instance check*, *i.e.*, the check of whether an ABox assertion is a logical consequence of a DL KB. A more sophisticated version of instance check, called *instance retrieval*, retrieves, for a DL KB $\mathcal{K}$, all (ABox) individuals that are instances of the given (possibly complex) concept expression C, *i.e.*, all those individuals a such that $\mathcal{K}$ entails that a is an instance of C. All these reasoning tasks support so-called *standard inferences* and can be reduced to the consistency check. Besides the standard ones, additional so-called *non-standard inferences* have been investigated in DL reasoning [12].

When reasoning in DLs, models can be of arbitrary cardinality. In many applications, however, the domain of interest is known to be finite. This is, *e.g.*, a natural assumption in database theory. In *finite model reasoning* [13], models have a finite yet arbitrary, unknown size. Even more interesting from the application viewpoint is the case where the domain has an a priori known cardinality, more precisely, when the domain coincides with the set of named individuals mentioned in the KB. In their proposal of *bounded model reasoning*, Gaggl *et al.* [14] refer to such models as bounded models. Also, they argue that in many applications this modification of the classical DL semantics represents a more intuitive definition of what is considered and expected as model of some KB. In fact, OWL is often "abused" by practitioners as a constraint language for an underlying fixed domain.

2.2 Answer Set Programming

Based on the stable model (answer set) semantics [15], ASP is a logic programming paradigm oriented towards difficult search problems [16]. ASP solvers (see [6] for an updated overview) are indeed powerful systems especially designed not only to find one solution to such problems but also (in some cases) to enumerate all solutions. In the following we give a brief overview of the syntax and semantics of disjunctive logic programs in ASP.

Let U be a fixed countable set of (domain) elements, also called *constants*, upon which a total order $\prec$ is defined. An *atom* α is an expression $p(t_1, \ldots, t_n)$, where p is a predicate of arity $n \geq 0$ and each t_i is either a variable or an element from U (*i.e.*, the resulting language is function-free). An atom is *ground* if it is free of variables. We denote the set of all ground atoms over U by B_U. A *(disjunctive) rule* r is of the form

$$a_1 \vee \ldots \vee a_n \leftarrow b_1, \ldots, b_k, not\ b_{k+1}, \ldots, not\ b_m$$

with $n \geq 0$, $m \geq k \geq 0$, $n + m > 0$, where $a_1, \ldots, a_n, b_1, \ldots, b_m$ are atoms, or a *count expression* of the form $\#count\{l : l_1, \ldots, l_i\} \bowtie u$, where l is an atom and l_j is a literal (*i.e.*, an atom which can be negated or not), $1 \geq j \geq i$, $\bowtie \in \{\leq, <, =, >, \geq\}$, and $u \in \mathbb{N}$. Moreover, "not" denotes *default negation*. The *head* of r is the set $head(r) = \{a_1, \ldots, a_n\}$ and the *body* of r is $body(r) = \{b_1, \ldots, b_k, notb_{k+1}, \ldots, notb_m\}$. Furthermore, we distinguish between $body^+(r) = \{b_1, \ldots, b_k\}$ and $body^-(r) = \{b_{k+1}, \ldots, b_m\}$. A rule r is *normal* if $n \leq 1$ and a *constraint* if $n = 0$. A rule r is *safe* if each variable in r occurs in $body^+(r)$. A rule r is *ground* if no variable occurs in r. A *fact* is a ground rule with $body(r) = \emptyset$ and $|head(r)| = 1$. An *(input) database* is a set of facts. A *program* is a finite set of rules. For a program Π and an input database D, we often write $\Pi(D)$ instead of $D \cup \Pi$. If each rule in a program is normal (resp. ground), we call the program normal (resp. ground).

Given a program Π, let U_Π be the set of all constants appearing in Π. $Gr(\Pi)$ is the set of rules $r\sigma$ obtained by applying, to each rule $r \in \Pi$, all possible substitutions σ from the variables in r to elements of U_Π. For count-expressions, $\{l : l_1, \ldots, l_n\}$ denotes the set of all ground instantiations of l, governed through $l_1, \ldots, l_n$. An interpretation $I \subseteq B_U$ satisfies a ground rule r iff $head(r) \cap I \neq \emptyset$ whenever $body^+(r) \subseteq I$, $body^-(r) \cap I = \emptyset$, and for each contained count-expression, $N \bowtie u$ holds, where $N = |\{l|l_1, \ldots, l_n\}|$, $u \in \mathbb{N}$ and $\bowtie \in \{\leq, <, =, >, \geq\}$. A ground program Π is satisfied by I, if I satisfies each $r \in \Pi$. A non-ground rule r (resp., a program Π) is satisfied by an interpretation I iff I satisfies all groundings of r (resp., $Gr(\Pi)$). A subset-minimal set $I \subseteq B_U$ satisfying the *Gelfond-Lifschitz reduct* $\Pi^I = \{head(r) \leftarrow body^+(r)|I \cap body^-(r) = \emptyset, r \in Gr(\Pi)\}$ is called an *answer set* of Π. We denote the set of answer sets for a program Π by $AS(\Pi)$.

3   The Case Study

Concept Learning in DLs has been paid increasing attention over the last decade. Notably, algorithms such as [17] have been proposed that follow the *generalization as search* approach by extending the methodological apparatus of ILP to DL languages. In [7], we formally defined three variants of the Concept Learning problem in the DL setting. The variants share the following two features:

1. The background knowledge is in the form of a DL KB $\mathcal{K} = (\mathcal{T}, \mathcal{A})$, and
2. The target theory is a set of DL concept definitions, *i.e.* concept equivalence axioms having an atomic concept in the left-hand side.

but differ in the requirements that an induced concept definition must fulfill in order to be considered as a correct (or valid) solution. The variant we consider in this paper is the supervised one. It is the base for the other variants being introduced in [7]. In the following, the set of all individuals occurring in $\mathcal{A}$ and the set of all individuals occurring in $\mathcal{A}$ that are instance of a given concept C w.r.t. $\mathcal{K}$ are denoted by $\mathsf{Ind}(\mathcal{A})$ and $\mathsf{Retr}_\mathcal{K}(C)$, respectively.

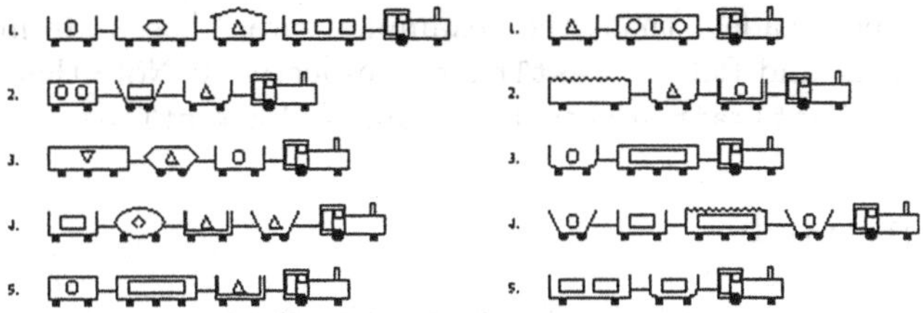

Fig. 1. Michalski's example of eastbound (left) and westbound (right) trains (illustration taken from [18]).

Definition 1 (Concept Induction - CSP version). *Let* $\mathcal{K} = (\mathcal{T}, \mathcal{A})$ *be a DL KB. Given:*

- *a (new) target concept name* C
- *a set of positive and negative examples* $\mathsf{Ind}_C^+(\mathcal{A}) \cup \mathsf{Ind}_C^-(\mathcal{A}) \subseteq \mathsf{Ind}(\mathcal{A})$ *for* C
- *a concept description language* $\mathcal{DL}_{\mathcal{H}}$

the CSP version of the Concept Induction *(CI-CSP) problem is to find a concept definition* $C \equiv D$ *with* $D \in \mathcal{DL}_{\mathcal{H}}$ *such that*

Completeness $\mathcal{K} \models (a : D) \quad \forall a \in \mathsf{Ind}_C^+(\mathcal{A})$ *and*
Consistency $\mathcal{K} \models (b : \neg D) \quad \forall b \in \mathsf{Ind}_C^-(\mathcal{A})$

Here, the sets of positive and negative examples are defined as follows

- $\mathsf{Ind}_C^+(\mathcal{A}) = \{a \in \mathsf{Ind}(\mathcal{A}) \mid (a : C) \in \mathcal{A}\} \subseteq \mathsf{Retr}_{\mathcal{K}}(C)$
- $\mathsf{Ind}_C^-(\mathcal{A}) = \{b \in \mathsf{Ind}(\mathcal{A}) \mid (b : \neg C) \in \mathcal{A}\} \subseteq \mathsf{Retr}_{\mathcal{K}}(\neg C)$

These sets can be easily computed by resorting to instance retrieval inference services usually available in DL systems.

Example 1. For illustrative purposes throughout the paper, we choose a very popular learning task in ILP proposed 20 years ago by Ryszard Michalski [18] and illustrated in Fig. 1. Here, 10 trains are described, out of which 5 are eastbound and 5 are westbound. The aim of the learning problem is to find the discriminating features between these two classes.

For the purpose of this case study, we have considered an $\mathcal{ALCO}$ ontology, *trains2*, encoding the original Trains data set[2] and distributed with the DL-Learner system.[3] The ontology encompasses 345 logical axioms, 32 classes, 5 object properties and 50 individuals. With reference to *trains2* (which therefore will play the role of $\mathcal{K}$ as in Definition 1), we might want to induce a $\mathcal{SROIQ}$ concept definition for the target concept name $C = \texttt{EastTrain}$ from the following positive and negative examples:

- $\mathsf{Ind}_{\texttt{EastTrain}}^+(\mathcal{A}) = \{\texttt{east1}, \ldots, \texttt{east5}\}$
- $\mathsf{Ind}_{\texttt{EastTrain}}^-(\mathcal{A}) = \{\texttt{west6}, \ldots, \texttt{west10}\}$

---

[2] http://archive.ics.uci.edu/ml/datasets/Trains.
[3] http://dl-learner.org/Projects/DLLearner.

We remind the reader that the examples are chosen from the sets $\mathsf{Retr}_\mathcal{K}(\texttt{EastTrain})$ and $\mathsf{Retr}_\mathcal{K}(\neg\texttt{EastTrain})$, respectively. Note that the 5 positive examples for $\texttt{EastTrain}$ are negative examples for $\texttt{WestTrain}$ and viceversa.

4   The Approach

The proposed approach consists of a declarative modeling language based on second-order DLs under Henkin semantics (see Sect. 4.1), and a mechanism for transforming second-order DL formulas into a format processable by ASP solvers (see Sect. 4.2). In particular, the transformation is a two-stage process. Second-order DL formulas are instantiated, then encoded as answer set programs.

4.1   Modeling with Second-Order DLs

As extensively discussed in [7], a major source of inspiration for the modeling part of our approach was the unified framework proposed by Colucci *et al.* [19] for non-standard DL reasoning services. In the following we summarize the results reported in [7] by revising them and adding clarifying remarks when necessary.

Let $\mathcal{DL}$ be any DL with syntax $(\mathsf{N}_\mathcal{C}, \mathsf{N}_\mathcal{R}, \mathsf{N}_\mathcal{O})$. Since we are interested in second-order formulas, we need to introduce a set $\mathsf{N}_\mathcal{X} = \{X_0, X_1, X_2, \ldots\}$ of so-called *concept variables*, *i.e.* second-order variables that can be quantified. Let then $\mathcal{DL}_\mathcal{X}$ be the second-order DL language obtained by extending $\mathcal{DL}$ with $\mathsf{N}_\mathcal{X}$.

Definition 2 (Concept term). *A concept term in $\mathcal{DL}_\mathcal{X}$ is a concept formed according to the specific syntax rules of $\mathcal{DL}$ augmented with the additional rule $C \longrightarrow X$ for $X \in \mathsf{N}_\mathcal{X}$.*

To the purpose of this work, we restrict our language to particular existentially quantified second-order formulas involving several concept subsumptions and concept assertions, but just one concept variable. Note that this fragment of second-order logic extends the one considered in [19] with concept assertions.

Definition 3 (Second-order concept expression). *Let $a_1, \ldots, a_k \in \mathcal{DL}$ be individuals, $C_1, \ldots, C_m, D_1, \ldots, D_m \in \mathcal{DL}_\mathcal{X}$ be concept terms containing just one concept variable X. A concept expression γ in $\mathcal{DL}_\mathcal{X}$ is a conjunction*

$$(C_1 \sqsubseteq D_1) \wedge \ldots \wedge (C_l \sqsubseteq D_l) \wedge (C_{l+1} \not\sqsubseteq D_{l+1}) \wedge \ldots \wedge (C_m \not\sqsubseteq D_m) \wedge$$
$$(a_1 : D_1) \wedge \ldots \wedge (a_j : D_l) \wedge (a_{j+1} : \neg D_{l+1}) \wedge \ldots \wedge (a_k : \neg D_m) \tag{1}$$

of (negated or not) concept subsumptions and concept assertions with $1 \leq l \leq m$ and $1 \leq j \leq k$.

Definition 4 (Second-order formula). *A formula ϕ in $\mathcal{DL}_\mathcal{X}$ has the form*

$$\exists X.\gamma \tag{2}$$

where γ is a concept expression of the form (1) and X is a concept variable.

We adopt the *General Semantics* (also called Henkin semantics [20]), instead of the Standard Semantics, for interpreting concept variables. A nice feature of the Henkin style is that the expressive power of the language actually remains first-order. This is due to the fact that, as opposed to the Standard Semantics where a concept variable could be interpreted as *any* subset of $\Delta^{\mathcal{I}}$, concept variables in the General Semantics can be interpreted only by *some subsets* among all the ones in the powerset of the domain $2^{\Delta^{\mathcal{I}}}$. In our case, we impose that the interpretation $X^{\mathcal{I}}$ of a concept variable $X \in \mathcal{DL}_{\mathcal{X}}$ must coincide with the interpretation $E^{\mathcal{I}}$ of some concept $E \in \mathcal{DL}$. The interpretations we refer to in the following definition are of this kind.

Definition 5 (Satisfiability of second-order concept expressions). *A concept expression γ of the form (1) is satisfiable in $\mathcal{DL}$ iff there exist a concept $E \in \mathcal{DL}$ such that, extending the semantics of $\mathcal{DL}$ for each interpretation $\mathcal{I}$, with: $(X)^{\mathcal{I}} = (E)^{\mathcal{I}}$, it holds that*

1. *for each $j = 1, \ldots, l$, and every $\mathcal{I}$, $(C_j)^{\mathcal{I}} \subseteq (D_j)^{\mathcal{I}}$ and $(a_j)^{\mathcal{I}} \in (D_j)^{\mathcal{I}}$, and*
2. *for each $j = l+1, \ldots, m$, there exists an interpretation $\mathcal{I}$ s.t. $(C_j)^{\mathcal{I}} \nsubseteq (D_j)^{\mathcal{I}}$ and $(a_j)^{\mathcal{I}} \in (\neg D_j)^{\mathcal{I}}$*

Otherwise, γ is said to be unsatisfiable in $\mathcal{DL}$.

Definition 6 (Solution for a second-order concept expression). *Let γ be a concept expression of the form (1). If γ is satisfiable in $\mathcal{DL}$, then E is a solution for γ.*

Definition 7 (Satisfiability of second-order formulas). *A formula ϕ of the form (2) is true in $\mathcal{DL}$ if there exist at least a solution for γ, otherwise it is false.*

The fragment of Second-Order DLs just introduced can be used as a declarative modeling language for Concept Learning problems in DLs. Following Definition 1, we assume that $\mathsf{Ind}_C^+(\mathcal{A}) = \{a_1, \ldots, a_m\}$ and $\mathsf{Ind}_C^-(\mathcal{A}) = \{b_1, \ldots, b_n\}$. A concept $D \in \mathcal{DL}_{\mathcal{H}}$ is a correct concept definition for the target concept name C w.r.t. $\mathsf{Ind}_C^+(\mathcal{A})$ and $\mathsf{Ind}_C^-(\mathcal{A})$ iff it is a solution for the following second-order concept expression:

$$\gamma_{\mathsf{CI\text{-}CSP}} := (a_1 : X) \wedge \ldots \wedge (a_m : X) \wedge (b_1 : \neg X) \wedge \ldots \wedge (b_n : \neg X) \qquad (3)$$

that is, iff D can be an assignment for the concept variable X. The CI-CSP problem can be modeled with the following second-order formula

$$\phi_{\mathsf{CI\text{-}CSP}} := \exists X.\gamma_{\mathsf{CI\text{-}CSP}} \qquad (4)$$

The solvability of a CI-CSP problem is therefore based on the satisfiability of the second-order formula being used for modeling the problem.

Definition 8 (Solvability of CI-CSP problems). *A CI-CSP problem $\boldsymbol{P}$ is solvable if $\phi_{CI\text{-}CSP}$ is true in $\mathcal{DL}_{\mathcal{H}}$. Otherwise, the problem is not solvable. If D is a solution for $\gamma_{CI\text{-}CSP}$, then $C \equiv D$ is a solution for $\boldsymbol{P}$.*

Example 2. According to (3), the intended CI-CSP problem of Example 1 corresponds to the following second-order concept expression $\gamma_{\text{EastTrain}}$:

$$(\text{east1}:X) \wedge \ldots \wedge (\text{east5}:X) \wedge (\text{west6}:\neg X) \wedge \ldots \wedge (\text{west10}:\neg X) \tag{5}$$

The problem is then solvable if the following second-order formula:

$$\phi_{\text{EastTrain}} := \exists X.\gamma_{\text{EastTrain}} \tag{6}$$

is true in $\mathcal{SROIQ}$, *i.e.*, if there exists a solution to $\gamma_{\text{EastTrain}}$ in $\mathcal{SROIQ}$.

4.2   Solving with ASP

In order to solve the problems modeled with the second-order concept expressions introduced in the previous section we need mechanisms for generating and evaluating candidate solutions.

How to Generate Candidate Solutions. The CI-CSP problem statement reported in Definition 1 mentions a concept description language $\mathcal{DL}_{\mathcal{H}}$ among its inputs. It is the language of hypotheses and allows for the generation of concept definitions in any $\mathcal{DL}$ according to some declarative bias. It can be considered as a generative grammar.

The concept expressions generated from $\mathcal{DL}_{\mathcal{H}}$ can be organized according to the concept subsumption relation $\sqsubseteq$. Note that $\sqsubseteq$ is a reflexive and transitive binary relation, *i.e.* a *quasi-order*. Thus, according to the *generalization as search* approach in Mitchell's vision [3], $(\mathcal{DL}_{\mathcal{H}}, \sqsubseteq)$ is a quasi-ordered set of $\mathcal{DL}$ concept definitions which defines a search space to be traversed either top-down or bottom-up by means of so-called *refinement operators*. More formally, a *downward* refinement operator is a mapping $\rho : \mathcal{DL}_{\mathcal{H}} \to 2^{\mathcal{DL}_{\mathcal{H}}}$ such that

$$\forall C \in \mathcal{DL}_{\mathcal{H}} \quad \rho(C) \subseteq \{D \in \mathcal{DL}_{\mathcal{H}} \mid D \sqsubseteq C\} \tag{7}$$

An *upward* refinement operator is dual to the downward one w.r.t. $\sqsubseteq$.

Note that there is an infinite number of generalizations and specializations in a given $(\mathcal{DL}, \sqsubseteq)$. Usually one tries to define refinement operators that can traverse efficiently the hypothesis space in pursuit of one of the correct definitions (w.r.t. the examples that have been provided). An extensive analysis of properties of refinement operators for DLs can be found in [17].

Example 3. Let us assume that the language of hypotheses allows for the generation of $\mathcal{SROIQ}$ concept expressions starting from the atomic concept and role names occurring in *trains2* (except, of course, for the target concept name). Among the concepts (instantiations of X) satisfying $\gamma_{\text{EastTrain}}$, there is

$$\geqslant 3\,\text{hasCar}.(\neg\text{JaggedCar}) \tag{8}$$

which describes the set of trains composed by at least three cars that are not jagged. It provides a correct concept definition for EastTrain w.r.t. the given examples, *i.e.*, the following concept equivalence axiom

$$\text{EastTrain} \equiv \geqslant 3 \, \text{hasCar}.(\neg \text{JaggedCar}) \tag{9}$$

is a solution for the CI-CSP problem in hand.

How to Evaluate Candidate Solutions. The choice of the solver is a critical aspect in any *model+solver* approach. In our case the use of a second-order modeling language does not necessarily imply the use of a second-order solver. Indeed, the Henkin semantics paves the way to the use of first-order solvers. Once instantiated, concept expressions of the kind (3) are just first-order DL conjunctive queries. However, the CSP nature of the problem in hand should not be neglected. This led us to assume the bounded model semantics for the instantiated concept expressions instead of the classical one.

As already mentioned in Sect. 2.1, Gaggl *et al.* [14] modify the modelhood condition by restricting the domain to a finite set of bounded size, induced by the named individuals occurring in the given $\mathcal{SROIQ}$ KB (denoted as $N_{\mathcal{O}}(\mathcal{K})$). This means that - as claimed by the proposers of the bounded model semantics - if one assumes the set of domain elements fixed and known, then there is a one-to-one correspondence between interpretations and sets of ground facts. In other words, ABoxes can be used as representations of models.

Definition 9 (Bounded model semantics [14]). *Let $\mathcal{K}$ be a $\mathcal{SROIQ}$ KB. An interpretation $\mathcal{I} = (\Delta^{\mathcal{I}}, \cdot^{\mathcal{I}})$ is said to be* individual-bounded *w.r.t. $\mathcal{K}$, if all of the following holds:*

1. *$\Delta^{\mathcal{I}} = \{a | a \in N_{\mathcal{O}}(\mathcal{K})\}$,*
2. *for each individual $a \in N_{\mathcal{O}}(\mathcal{K})\}$, $a^{\mathcal{I}} = a$.*

Accordingly, an interpretation $\mathcal{I}$ is an (individual-)bounded model *of $\mathcal{K}$, if $\mathcal{I}$ is an individual-bounded interpretation w.r.t. $\mathcal{K}$ and $\mathcal{I} \models \mathcal{K}$ holds.*

Also, $\mathcal{K}$ is called bm-satisfiable *if it has a bounded model.*

We say that $\mathcal{K}$ bm-entails *an axiom α (written $K \models_{bm} \alpha$) if every bounded model of $\mathcal{K}$ is also a model of α.*

The benefits of bounded model semantics are manifolded.

First, this non-classical semantics is computationally advantageous. Indeed, while reasoning in OWL under the classical semantics is N2ExpTime-complete [21], reasoning under the bounded model semantics is merely NP-complete [14].

Second, an arbitrary $\mathcal{SROIQ}$ KB $\mathcal{K}$ can be encoded into an answer set program $\Pi(\mathcal{K})$, such that the set of answer sets $AS(\Pi(\mathcal{K}))$ coincides with the set of bounded models of the given KB [14]. The rules for transforming $\mathcal{SROIQ}$ concept expressions into ASP are reported in Table 2. Here, O_a is a new concept name unique for the individual a. Also, $ar(r, X, Y)$ is defined as follows:

$$ar(r, X, Y) := \begin{cases} R(X, Y) & \text{if } R \text{ is an atomic role} \\ S(Y, X) & \text{if } R \text{ is an inverse role and } R = S^- \end{cases}$$

Table 2. Translation of $\mathcal{SROIQ}$ concept expressions into ASP (adapted from [14]).

C	$trans(C)$
A	$not\ \mathsf{A}(X)$
$\neg A$	$\mathsf{A}(X)$
$\{a\}$	$\{not\ \mathsf{O_a}(X)\}, \{\mathsf{O_a}(X)\}$
$\forall R.A$	$\{not\ \mathsf{A}(Y_A), ar(r, X, Y_A)\}$
$\forall R.(\neg A)$	$\{ar(r, X, Y_A), \mathsf{A}(Y_A)\}$
$\exists R.Self$	$not\ ar(r, X, X)$
$\neg \exists R.Self$	$ar(r, X, X)$
$\geqslant n\ R.A$	$\#count\{ar(r, X, Y_A) : \mathsf{A}(Y_A)\} < n$
$\geqslant n\ R.(\neg A)$	$\#count\{ar(r, X, Y_A) : not\ \mathsf{A}(Y_A)\} < n$
$\leqslant n\ R.A$	$\#count\{ar(r, X, Y_A) : \mathsf{A}(Y_A)\} > n$
$\leqslant n\ R.(\neg A)$	$\#count\{ar(r, X, Y_A) : not\ \mathsf{A}(Y_A)\} > n$

The translation into ASP requires a KB to be in the so-called *normalized form* (see [22] for details) which can be obtained however by an easy syntactic transformation. The encoding turns out to be a more effective alternative to the axiomatization, since existing OWL reasoners struggle on bounded model reasoning, due to the heavy combinatorics involved.

Last, but not least, OWL can be also used for modeling typical CSPs.

Example 4. Let $\gamma'_{\texttt{EastTrain}}$ be the first-order concept expression obtained by instantiating the unique second-order variable in $\gamma_{\texttt{EastTrain}}$ with the concept (8) generated by some refinement operator for $\mathcal{SROIQ}$. It can be encoded in ASP under bounded model semantics. The resulting ASP program can be then checked for satisfiability by any ASP solver.

5   Related Work

The *model+solver* approach, especially based on Constraint Programming, has been promoted by De Raedt *et al.* [2,23] and successfully applied to one of the most popular Data Mining (DM) tasks: Constraint-based pattern mining [24–27]. Along this line, Guns *et al.* [28] introduce MiningZinc, a general framework for constraint-based pattern mining which consists of two key components: a language component and a toolchain component. The language allows for high-level and natural modeling of DM problems, such that MiningZinc models closely resemble definitions found in the DM literature. It is inspired by the Zinc family of languages and systems and supports user-defined constraints and optimization criteria. The toolchain allows for finding solutions to the models. It ensures the solver independence of the language and supports both standard constraint solvers and specialized DM systems. Automatic model transformations enable the efficient use of different solvers and systems. The combination of both components allows one to rapidly model constraint-based DM problems and execute these with a wide variety of methods.

Bruyonooghe *et al.* [29] suggest that predicate logic can be useful as a modeling language and show how to model and solve ML and DM problems with IDP3. The core of IDP3 is a finite model generator that supports FOL enriched with types, inductive definitions, aggregates and partial functions. It offers its users a modeling language that allows them to solve a wide range of search problems. Apart from a small introductory example, applications are selected from problems that arose within ML/DM research. These research areas have recently shown a strong interest in declarative modeling and constraint solving as opposed to algorithmic approaches. The paper illustrates that the IDP3 system can be a valuable tool for researchers with such an interest.

The Meta-Interpretive Learning (MIL) framework [30] somehow follows a *model+solver* approach. It uses descriptions in the form of meta-rules with procedural constraints incorporated within a meta-interpreter. Also, it can be implemented as a simple Prolog programm or within an ASP solver. The work presented in [31] extends the theory, implementation and experimental application of MIL from grammar learning to the higher-order dyadic Datalog fragment.

6   Summary and Directions of Future Work

In this paper we have carried on the work reported in [7] by studying in more depth the case of Concept Induction, the basic case of Concept Learning in DLs which can be naturally reformulated as a CSP. In particular, we have proposed a *model+solver* approach to Concept Induction which consists of a declarative modeling language based on second-order DLs under Henkin semantics, and a mechanism for transforming second-order DL formulas into a format processable by ASP solvers. The transformation is possible under bounded model semantics, a non-standard model-theoretic semantics for DLs which has been recently proposed in order to correctly address CSPs in OWL. Overall, as suggested by Definition 8, this paper moves a step towards a new form of learnability of concepts based on the existence of solutions for the second-order concept expressions used for declaratively modeling the corresponding concept learning problems. Finally, and from a broader perspective, our proposal contributes to the current shift in AI from programming to solving as recently argued by Geffner [32].

In the future, we plan to implement and test the approach by relying on available tools. In particular, we expect full support from Wolpertinger[4] which is intended to implement the encoding proposed by Gaggl *et al.* [14] but currently can not deal properly with some OWL 2 constructs. Besides evaluation, we intend also to investigate how to express optimality criteria such as the information gain function within the second-order concept expressions.

---

[4] https://github.com/wolpertinger-reasoner/.

Acknowledgements. We would like to thank the proposers of the bounded model semantics for the fruitful discussions about their work during a visit to Dresden and for the kind remote assistance in using `Wolpertinger`.

References

1. Valiant, L.: A theory of the learnable. Commun. ACM **27**(11), 1134–1142 (1984)
2. De Raedt, L., Guns, T., Nijssen, S.: Constraint programming for data mining and machine learning. In: Fox, M., Poole, D. (eds.) Proceedings of the Twenty-Fourth AAAI Conference on Artificial Intelligence, AAAI 2010, 11–15 July 2010. AAAI Press, Atlanta (2010)
3. Mitchell, T.M.: Generalization as search. Artif. Intell. **18**, 203–226 (1982)
4. De Raedt, L.: Languages for learning and mining. In: Proceedings of the Twenty-Ninth AAAI Conference on Artificial Intelligence, 25–30 January 2015, Austin, Texas, USA, pp. 4107–4111 (2015)
5. Baader, F., Calvanese, D., McGuinness, D., Nardi, D., Patel-Schneider, P. (eds.): The Description Logic Handbook: Theory, Implementation and Applications, 2nd edn. Cambridge University Press, New York (2007)
6. Calimeri, F., Gebser, M., Maratea, M., Ricca, F.: Design and results of the fifth answer set programming competition. Artif. Intell. **231**, 151–181 (2016)
7. Lisi, F.A.: A declarative modeling language for concept learning in description logics. In: Riguzzi, F., Železný, F. (eds.) ILP 2012. LNCS (LNAI), vol. 7842, pp. 151–165. Springer, Heidelberg (2013). doi:10.1007/978-3-642-38812-5_11
8. Horrocks, I., Patel-Schneider, P.F., Harmelen, F.: From SHIQ and RDF to OWL: the making of a web ontology language. J. Web Semant. **1**(1), 7–26 (2003)
9. Horrocks, I., Kutz, O., Sattler, U.: The even more irresistible SROIQ. In: Doherty, P., Mylopoulos, J., Welty, C.A. (eds.) Proceedings, Tenth International Conference on Principles of Knowledge Representation and Reasoning, Lake District of the United Kingdom, 2–5 June 2006, pp. 57–67. AAAI Press (2006)
10. Borgida, A.: On the relative expressiveness of description logics and predicate logics. Artif. Intell. **82**(1–2), 353–367 (1996)
11. Reiter, R.: Equality and domain closure in first order databases. J. ACM **27**, 235–249 (1980)
12. Küsters, R. (ed.): Non-Standard Inferences in Description Logics. LNCS (LNAI), vol. 2100. Springer, Heidelberg (2001)
13. Lutz, C., Sattler, U., Tendera, L.: The complexity of finite model reasoning in description logics. Inf. Comput. **199**(1–2), 132–171 (2005)
14. Gaggl, S.A., Rudolph, S., Schweizer, L.: Bound your models! how to make OWL an ASP modeling language. CoRR abs/1511.00924 (2015)
15. Gelfond, M., Lifschitz, V.: Classical negation in logic programs and disjunctive. New Gener. Comput. **9**(3/4), 365–386 (1991)
16. Brewka, G., Eiter, T., Truszczynski, M.: Answer set programming at a glance. Commun. ACM **54**(12), 92–103 (2011)
17. Lehmann, J., Hitzler, P.: Concept learning in description logics using refinement operators. Mach. Learn. **78**(1–2), 203–250 (2010)
18. Michalski, R.: Pattern recognition as a rule-guided inductive inference. IEEE Trans. Pattern Anal. Mach. Intell. **2**(4), 349–361 (1980)

19. Colucci, S., Di Noia, T., Di Sciascio, E., Donini, F.M., Ragone, A.: A unified framework for non-standard reasoning services in description logics. In: Coelho, H., Studer, R., Wooldridge, M. (eds.) ECAI 2010–19th European Conference on Artificial Intelligence, Lisbon, Portugal, 16–20 August 2010, Proceedings, Frontiers in Artificial Intelligence and Applications, vol. 215, pp. 479–484. IOS Press (2010)
20. Henkin, L.: Completeness in the theory of types. J. Symbolic Logic **15**(2), 81–91 (1950)
21. Kazakov, Y.: $\mathcal{RIQ}$ and $\mathcal{SROIQ}$ are harder than $\mathcal{SHOIQ}$. In: Principles of Knowledge Representation and Reasoning: Proceedings of the Eleventh International Conference, KR 2008, Sydney, Australia, September 16-19, 2008, pp. 274–284 (2008)
22. Motik, B., Shearer, R., Horrocks, I.: Hypertableau reasoning for description logics. J. Artif. Intell. Res. **36**, 165–228 (2009)
23. Raedt, L., Nijssen, S., O'Sullivan, B., Hentenryck, P.: Constraint programming meets machine learning and data mining (Dagstuhl seminar 11201). Dagstuhl Rep. **1**(5), 61–83 (2011)
24. Nijssen, S., Guns, T., De Raedt, L.: Correlated itemset mining in ROC space: a constraint programming approach. In: Elder IV, J.F., Fogelman-Soulié, F., Flach, P.A., Zaki, M.J. (eds.) Proceedings of the 15th ACM SIGKDD International Conference on Knowledge Discovery and Data Mining, Paris, France, June 28 - July 1, 2009, pp. 647–656. ACM (2009)
25. Guns, T., Nijssen, S., Raedt, L.: Evaluating pattern set mining strategies in a constraint programming framework. In: Huang, J.Z., Cao, L., Srivastava, J. (eds.) PAKDD 2011. LNCS (LNAI), vol. 6635, pp. 382–394. Springer, Heidelberg (2011). doi:10.1007/978-3-642-20847-8_32
26. Guns, T., Nijssen, S., Raedt, L.: Itemset mining: a constraint programming perspective. Artif. Intell. **175**(12–13), 1951–1983 (2011)
27. Guns, T., Nijssen, S., Raedt, L.: k-pattern set mining under constraints. IEEE Trans. Knowl. Data Eng. **25**(2), 402–418 (2013)
28. Guns, T., Dries, A., Tack, G., Nijssen, S., De Raedt, L.: MiningZinc: a modeling language for constraint-based mining. In: Rossi, F. (ed.) IJCAI 2013, Proceedings of the 23rd International Joint Conference on Artificial Intelligence, 3–9 August 2013, IJCAI/AAAI, Beijing, China (2013)
29. Bruynooghe, M., Blockeel, H., Bogaerts, B., Cat, B., Pooter, S., Jansen, J., Labarre, A., Ramon, J., Denecker, M., Verwer, S.: Predicate logic as a modeling language: modeling and solving some machine learning and data mining problems with IDP3. Theory Pract. Logic Program. **15**(6), 783–817 (2015)
30. Muggleton, S.H., Lin, D., Pahlavi, N., Tamaddoni-Nezhad, A.: Meta-interpretive learning: application to grammatical inference. Mach. Learn. **94**(1), 25–49 (2014)
31. Muggleton, S.H., Lin, D., Tamaddoni-Nezhad, A.: Meta-interpretive learning of higher-order dyadic datalog: predicate invention revisited. Mach. Learn. **100**(1), 49–73 (2015)
32. Geffner, H.: Artificial intelligence: from programs to solvers. AI Commun. **27**(1), 45–51 (2014)

Machine Learning

A Comparative Study of Inductive and Transductive Learning with Feedforward Neural Networks

Monica Bianchini[1], Anas Belahcen[2], and Franco Scarselli[1(✉)]

[1] Department of Information Engineering and Mathematics,
University of Siena, Siena, Italy
{monica,franco}@diism.unisi.it
[2] LeRMA ENSIAS, Mohammed V University, Rabat, Morocco
belahcen.anas@gmail.com

Abstract. Traditional supervised approaches realize an inductive learning process: A model is learnt from labeled examples, in order to predict the labels of unseen examples. On the other hand, transductive learning is less ambitious. It can be thought as a procedure to learn the labels on a training set, while, simultaneously, trying to guess the best labels on the test set. Intuitively, transductive learning has the advantage of being able to directly use training patterns while deciding on a test pattern. Thus, transductive learning faces a simpler problem with respect to inductive learning. In this paper, we propose a preliminary comparative study between a simple transductive model and a pure inductive model, where the learning architectures are based on feedforward neural networks. The goal is to understand how transductive learning affects the complexity (measured by the number of hidden neurons) of the exploited neural networks. Preliminary experimental results are reported on the classical two spirals problem.

Keywords: Feedforward neural networks · Inductive learning · Transductive learning

1 Introduction

In the inductive learning framework, a model is learnt from labeled examples, in order to predict the labels of unknown examples. Formally, the model $I_{\mathbf{w}}$, whose parameters $\mathbf{w}$ are set based on the training set, takes in input a pattern and returns a class or, more generally, any predicted properties of the input pattern. More precisely, the input pattern belongs to the training set $\mathcal{L}$ during learning and it belongs to test set $\mathcal{T}$ in the test phase. On the other hand, the transductive learning framework does not imply the construction of a common predictive model. Transductive learning exploits the labels on the training set, while, simultaneously, trying to predict the best labels on the test set. Formally,

© Springer International Publishing AG 2016
G. Adorni et al. (Eds.): AI*IA 2016, LNAI 10037, pp. 283–293, 2016.
DOI: 10.1007/978-3-319-49130-1_21

a transductive algorithm makes a prediction on a pattern of the test set $\mathbf{x} \in \mathcal{T}$ by a procedure T, which takes the whole training set $\mathcal{L}$ and the pattern $\mathbf{x}$ as its input.

Interestingly, transduction is considered simpler than induction, just because it can use directly the training set patterns to decide on test patterns. In fact, transduction is not forced to pass through the intermediate task of the construction of the predictive model. As firstly argued by Vapnik [1], who coined the term *transduction*, inferring a general predictive rule from a limited set of data, sampled from an unknown stochastic process, may be a too complex target, especially when we are only interested in obtaining predictions on a limited set of domain patterns. Just citing Vapnik words, "when solving a problem of interest, do not solve a more general problem as an intermediate step", which means that a learning system tuned to a specific set should outperform a general predictive one.

Moreover, transductive learning has the advance of being suitable also for semi–supervised learning. In this case, the algorithm is provided with some supervision information, but not necessarily for all the examples. Often, this information is constituted by the targets associated with a subset of the examples, or with some constraints among examples, or it is based on a partial knowledge of the data distribution. Such a transduction capability is particularly important in real–world applications, where there are intrinsic difficulties in collecting labeled data — at least enough to train a supervised model —, which is a time–consuming and expensive procedure.

Probably, the most popular transductive approaches are graph–based methods [2–4]. They are founded on some smoothness assumptions (cluster assumption or manifold assumption), which means that data points belonging to the same topological region should have the same label. By using a graph as a discrete approximation of such manifold, graph–based transductive learning methods learn a classification of the data that should produce a low classification error on the labeled subset, whereas, it should be smooth with respect to the neighboring relations codified in the graph connectivity.

Recently, transductive learning has been successfully applied to many practical scenarios, where labeled examples are scarce, while unlabeled ones are easy to be collected. A partial list of applications of semi–supervised learning includes text and web page classification [5–8], image processing and compression [9], surveillance [10], natural language processing [11,12], intrusion detection and computer security [13], graph reconstruction (for the prediction of protein interaction networks, gene regulatory networks, and metabolic networks) [14], protein classification based on phylogenetic profiles [15], drug design [16], cancer diagnostics [17], and ECG classification [18].

2   Motivations

In this paper, we compare inductive learning with a simplified version of transductive learning[1], in order to gain some insights into the claim that the former is more difficult than the latter. As mentioned above, the main advantage of transduction lies in its capability of exploiting the training patterns directly in the prediction process. Thus, the following can be considered a very simplified approach to transduction.

> Construct a procedure $T(\mathbf{x}, \mathbf{x}_p, \mathbf{y}_p)$ that takes in input a pattern $\mathbf{x}_p$, called a prototype, with its label $\mathbf{y}_p$, and an unlabeled pattern $\mathbf{x}$, and return the predicted label of the latter one. In order to construct such a procedure, during the learning phase, both $\mathbf{x}$ and $\mathbf{x}_p$ belong to the training set. Instead, during testing, $\mathbf{x}$ is a test pattern, whereas the prototype $\mathbf{x}_p$ is a training pattern.

The idea underlying this method is that T can take a decision on a pattern $\mathbf{x}$ using the information available on the prototype $\mathbf{x}_p$. By comparing the performance achieved by T with respect to the performance of a standard inductive model $I_{\mathbf{w}}$, we can obtain a simple evaluation on how much the direct use of a training pattern helps the prediction process. Provided that T and $I_{\mathbf{w}}$ are implemented with the same method, the comparison is not affected by other differences usually existing between inductive and transductive approaches. Moreover, the complexity of the two models can be easily compared. In this paper, T (actually, $T_{\mathbf{w}}$ in this case) and $I_{\mathbf{w}}$ are implemented by standard feedforward neural networks and their complexity is measured by the number of hidden neurons.

Furthermore, notice that the transduction algorithm implemented by T is not completely defined yet, since we have not explained how a pattern and its prototype are coupled, whether a single or many prototypes are used for each pattern and, in the latter case, how the outputs of T are combined. Actually, with the simplest strategy (described in the following section), the transduction algorithm consists of a K–nearest–neighbor classifier that outputs the label of the prototype closest to the input pattern. On the other hand, with more complex strategies, the transduction algorithm relies both on the classification capability of T and on the information provided by the employed prototypes. Thus, by studying different selection strategies for prototypes, we will gain insights into the role they play in the decision process.

Finally, it is worth mentioning that the proposed method can also be considered as a mixed inductive–transductive approach. Intuitively, an inductive method should extract all the useful information from the training set and store

---

[1] Actually, the main goal of transductive learning, as proposed in the present work, is to diffuse information coming from neighbor data to improve the whole classification accuracy. Technically speaking, we face a fully supervised problem, defining first the concept of data vicinity, and then training a feedforward neural network also on the base of the target information on the neighbors. Such a simplification is required in order to compare learning by induction and learning by transduction.

such an information into the model parameters; on the other hand, a transductive method does not apply any pre–processing to the training set, since the prediction is obtained by comparing the input with the other patterns in the training set, and by diffusing the information from the training patterns to the input pattern. In our approach, some of the information is stored in the parameters of the procedure $T_\mathbf{w}$, while $T_\mathbf{w}$ has also the training set pattern(s) as input. Such a characteristic of $T_\mathbf{w}$ allows us to make studies at the crossroad between inductive and transductive models.

In this paper, a preliminary comparative experimentation is carried out between pure inductive neural networks, and several versions of the simple transductive procedure explained above. The focus is mainly on the quality of the results produced by different classifiers, without considering the respective computational costs, which can constitute a matter of research for future activities. The results are encouraging and seem to confirm the greater ease of transductive learning with respect to pure inductive approaches.

The rest of the paper is organized as follows. In Sect. 3, the transductive approach is described, based on appropriately choosing the *prototype subset*, which collects the data in relation to which the concept of proximity is defined, and combining the output of the procedure $T_\mathbf{w}$. Section 4 shows comparative experimentation and discussion on the obtained results. Finally, in Sect. 5, some conclusions are drawn.

3   The Transductive Method

In this section, the proposed transductive algorithm and the related strategies for prototype selection are described in details. Let $\mathcal{L} = \{(\mathbf{x}_i, \mathbf{y}_i), i = 1, \ldots, \ell\}$ be the training set, where ℓ denotes its cardinality, $\mathbf{x}_i$ are n–dimensional real vectors, and $\mathbf{y}_i$ are real valued labels. The test set $\mathcal{T}$ is analogously defined.

The main goal of transductive learning, as proposed in the present work, is to leverage information from neighbor data to improve the whole prediction performance. Formally, the problem of transductive learning is defined as that of constructing a predictor $T(\mathbf{x}, \mathcal{S}_\mathbf{x})$, where $\mathbf{x}$ is the input and $\mathcal{S}_\mathbf{x}$ is a subset of the learning set, i.e., $\mathcal{S}_\mathbf{x} \subseteq \mathcal{L}$, containing a neigbourood of $\mathbf{x}$. In the following, $\mathcal{S}_\mathbf{x}$ will be referred as the set of the transductive *prototypes* for $\mathbf{x}$.

T is implemented by an artificial neural network — actually $T_\mathbf{w}$ since we use a parametric model —, which is trained to predict the label of a pattern $\mathbf{x}$ based on the pattern itself, on one or more of its prototypes $\mathbf{x}_s$, together with their corresponding label(s) $\mathbf{y}_s$. More precisely, the network has $2n + 1$ inputs and one output, and it will be denoted by $N_\mathbf{w}$, where $\mathbf{w}$ represents the network parameters.

In the following, three alternative strategies will be described, which differ for: the definition of the set of prototypes $\mathbf{x}_s \in \mathcal{S}_\mathbf{x}$ associated to each pattern $\mathbf{x}$; how the neural network outputs are combined by the predictor $T_\mathbf{w}$; how the neural network is trained.

Strategy 1: The closest prototype

In this strategy, the training set is randomly split into two disjoint subsets S and $\mathcal{L} \setminus S$, such that the former is a set of candidate *prototypes*, whereas the latter is exploited to learn the transductive function. Then, for each input pattern $\mathbf{x}$, the neighborhood $S_{\mathbf{x}}$ contains just the prototype that is closest to $\mathbf{x}$, i.e., $S_{\mathbf{x}} = \{x_m\}$, with $m = \arg\min_{\mathbf{x}_k \in S} \|\mathbf{x} - \mathbf{x}_k\|$, where $\|\cdot\|$ denotes the Euclidean distance.

Thus, the network $N_{\mathbf{w}}$ is trained on the dataset:

$$\mathcal{L}_1 = \{(\mathbf{x}, \mathbf{x}_s, \mathbf{y}_s) \mid \mathbf{x} \in \mathcal{L} \setminus S, \ s = \arg\min_{\mathbf{x}_k \in S} \|\mathbf{x} - \mathbf{x}_k\| \},$$

using the original target of $\mathbf{x}$ to learn the triples $(\mathbf{x}, \mathbf{x}_s, \mathbf{y}_s)$. During the test phase, $N_{\mathbf{w}}$ is directly applied on the closest prototype: the transductive procedure returns the output of the network.

Strategy 2: The set of closest prototypes (maximum output)

In this strategy, instead of using only a single prototype for each pattern, a set of prototypes is exploited: starting with an initial set S, an ad hoc set $S_{\mathbf{x}} \in S$ is assigned to each input $\mathbf{x}$. More precisely, the Euclidean distance between the input and the prototype data is evaluated and then the k nearest patterns are chosen as prototypes for such an input. Formally, if the problem deals with c classes, then $S_{\mathbf{x}}$ contains the $\lfloor \frac{k}{c} \rfloor$–closest patterns from each class. The training set for $N_{\mathbf{w}}$ is now constructed as:

$$\mathcal{L}_2 = \{(\mathbf{x}, \mathbf{x}_s, \mathbf{y}_s) \mid \mathbf{x} \in \mathcal{L}, \ \mathbf{x}_s \text{ is one of the } \lfloor \tfrac{k}{c} \rfloor - \text{closest patterns to } \mathbf{x} \text{ from any class}\}.$$

Therefore, $N_{\mathbf{w}}$ is applied on all the prototypes in $S_{\mathbf{x}}$, and the largest obtained output is considered to be the output of the whole transductive procedure and used to measure the performance of $T_{\mathbf{w}}$ during training and testing[2]. For example, in the case of two classes, where $\mathbf{y}_s \in \{-1, 1\}$

$$T_{\mathbf{w}}(\mathbf{x}, S_{\mathbf{x}}) = N_{\mathbf{w}}(\mathbf{x}, \mathbf{x}_s, \mathbf{y}_s), \text{ where } s = \arg\max_{\mathbf{y}_i} |N_{\mathbf{w}}(\mathbf{x}, \mathbf{x}_i, \mathbf{y}_i)|, \ \mathbf{x}_i \in S_{\mathbf{x}}$$

Strategy 3: The set of closest prototypes (the label of maximum output prototype)

The last strategy coincides with the previous one, except that the target of the prototype that has produced the maximum output, instead of the network output, is considered to be the output of the transductive procedure. Thus, in the case of two classes, we have

$$T_{\mathbf{w}}(\mathbf{x}, S_{\mathbf{x}}) = \arg\max_{\mathbf{y}_s} |N_{\mathbf{w}}(\mathbf{x}, \mathbf{x}_s, \mathbf{y}_s)|, \ \mathbf{x}_s \in S_{\mathbf{x}}.$$

---

[2] Notice that even if several prototypes are used for each pattern, a single network $N_{\mathbf{w}}$ is trained.

It is worth mentioning that the computational costs of all the above strategies may be larger than the cost of a standard inductive approach for several reasons: the network $N_{\mathbf{w}}$, exploited in the transductive approaches has more inputs; the prototypes closest to the input pattern have to be computed; the network $N_{\mathbf{w}}$ is applied k times in strategies 2,3. In this paper, we focus only on comparing the quality of the results produced by the above–mentioned methods with respect to the dimension of the network, measured by the number of hidden units. We do not propose, and defer to future activities, a complete analysis of the computational issues.

4   Experiments and Discussion

The experiments were carried out on the two spirals benchmark. We have chosen such a benchmark, because it is both simple and particularly suited for a preliminary experimentation, and, at the same time, it is theoretically complex in the sense that it can be approached only with neural architectures with a large number of hidden neurons[3]. For the dataset construction, three thousands patterns were randomly generated, half positives and half negatives, with a spiral length of 1000 and an average noise of 0.2. Throughout the experiments, we use balanced training, validation and test sets, which contain 70 %, %15 and 15 % of the generated patterns, respectively. The learning architecture $N_{\mathbf{w}}$ was implemented with a multilayer feedforward neural network, composed by a hidden layer of variable dimension $(1, 3, 5, 7, 10, 15, 20, 30$ units$)$ and a unique output neuron. The activation function of the hidden neurons is the hyperbolic tangent, whereas the output unit is linear[4]. The network is trained by the scaled conjugated gradient procedure for 500 epochs, where such a value has been heuristically chosen on the base of preliminary experiments.

First of all, let us consider the results obtained with the first strategy, in which only one prototype is considered for each pattern. Several experiments have been carried out varying the dimension h of the hidden layer and the cardinality of $\mathcal{S}$, the subset of the learning set from which the closest prototype is selected. The results are reported in Table 1, where the column labeled with "Induction" describes the accuracy of a neural network used as a standard inductive classifier.

The results show that the proposed transductive approach clearly outperforms the standard inductive network only for small values of h and large values of k. Actually, transduction is theoretically simpler than induction, because transduction can directly exploit the information available in prototypes. Nevertheless, the advantage of transduction is effective only if the information provided

---

[3] It can be easily shown that the required number of hiddens increases with the length of the spirals and with the noise in the generation of the patterns.

[4] Linear–output classifiers are experimentally proved to work well in many practical problems, especially for high dimensional input spaces, reaching accuracy levels comparable to non–linear classifiers while taking less time to be trained and used [19]. Moreover, they are not affected by the saturation problems, which can arise in sigmoid neurons.

Table 1. Accuracies achieved with strategy 1 (the closest prototype) using a different number h of neurons in the hidden layer and different dimensions k for the set $\mathcal{S}$, from which the prototypes are extracted.

Number of neurons h	Induction	Transduction with $k = |\mathcal{S}|$					
		k=2	k=4	k=10	k=20	k=50	k=100
1	58.9	59.8	54.8	56.3	63.5	80.4	83.9
3	62.9	64.4	69.7	66.6	66.4	86.2	89.4
5	64.9	70.0	72.6	72.6	79.4	87.8	90.6
7	68.7	73.6	79.7	80.0	84.6	93.7	92.4
10	75.1	78.9	86.6	90.6	89.0	90.5	93.6
15	87.1	82.2	91.8	93.8	91.1	95.7	97.9
20	97.6	91.1	98.9	91.6	96.6	96.8	94.9
30	100	95.3	98.9	97.3	96.2	96.4	96.8

by the prototypes is useful for the classification of the input pattern. Intuitively, prototypes should be close to the input pattern, otherwise they do not help. This is particularly true in the two spirals problem, where the separation surface between positive and negative patterns is tangibly easy only in a neighborhood of a prototype. Thus, transduction performs worse when $|\mathcal{S}|$ is small, because, in this case, there are few prototypes in the input domain, mostly far from the input patterns.

Moreover, the network $N_{\mathbf{w}}$, used in the transduction algorithm, has a larger input dimension with respect to the one used for induction: the additional input (the prototype and its label) may even confuse the learning process. Plausibly, using strategy 1, the prototype selection is not sufficiently smart to be helpful for all the patterns. Since, in the two spirals problem, very close patterns may have a different classification, even a very large number of prototypes is not always useful. Such a consideration may explain why the transductive algorithm never reaches a 100 % precision, not even with $|\mathcal{S}| = 100$, while the inductive network is capable to reach such a limit at $h = 30$.

The results obtained using strategy 2 (a set of closest prototypes, maximum output) are summarized in Table 2. In this case, the whole set of prototypes $\mathcal{S}$ contains 75 % of the training patterns, which means that 1575 patterns (out of the 2100 composing the training set) form the initial prototype subset. For each pattern $\mathbf{x}$ to be classified, the Euclidean distance is calculated with all transductive patterns, choosing the 2, 4, 10 nearest ones (half positives and half negatives) and constructing ad hoc the set $\mathcal{S}_{\mathbf{x}}$ of the prototypes to be used for $\mathbf{x}$.

The results in Table 2 show that using a "smarter" strategy in the selection of the prototypes, transduction can achieve a 100 % accuracy. Actually, in strategy 2, the prototype set is tailored on the particular input pattern. Moreover, from all the outputs produced by $N_{\mathbf{w}}$, only the largest one is considered, so that the prototype that is finally used has undergone an accurate selection.

Table 2. Accuracies achieved with strategy 2 (a set of closest prototypes, maximum output) using a different number h of neurons in the hidden layer and different dimensions k for the set $\mathcal{S}_\mathbf{x}$ of prototypes associated to each pattern.

Number of neurons h	Induction	Transduction with $k = |\mathcal{S}_x|$		
		k=2	k=4	k=10
1	58.9	83.3	79.6	70.7
3	62.9	95.1	96.0	91.6
5	64.9	96.4	95.8	92.7
7	68.7	100	96.0	99.8
10	75.1	100	100	100
15	87.1	100	100	100
20	97.6	100	100	100
30	100	100	100	100

Therefore, the obtained results seem to support the claim on the greater simplicity of transductive with respect to inductive learning, or, more precisely, the idea that using the training set prototypes can simplify the decision process on the test set patterns. However, notice that, from a practical point of view, Table 2 does not clarify whether and when transduction is better than induction. The presented induction and transduction algorithms are not computationally comparable. For example, induction requires that $N_\mathbf{w}$ is used k times, one for each prototype. On the other hand, induction requires larger networks. A more fair comparison can be carried out by comparing the transduction results with k prototypes and h hidden neurons, with the inductive classifier with kh neurons. However, such a comparison does not provide a clear winner.

Table 3. Accuracies achieved with strategy 3 (a set of closest prototypes, the label of maximum output prototype) using a different numbers h of neurons in the hidden layer and different dimensions k for the set $\mathcal{S}_\mathbf{x}$ of prototypes associated to each pattern.

Number of neurons	Induction	Transduction, $k = |\mathcal{S}_x|$		
		k=2	k=4	k=10
1	58.9	52.2	51.1	51.1
3	62.9	98.2	97.6	94.9
5	64.9	96.9	92.4	100
7	68.7	98.4	100	100
10	75.1	100	100	100
15	87.1	100	100	100
20	97.6	100	100	100
30	100	100	100	100

Table 4. Accuracies achieved with strategy 3 (a set of closest prototypes, the label of maximum output prototype), when S_x is constructed from a set S containing only 5 % of the training set.

Number of neurons	Induction	Transduction, $k = |S_x|$		
		k=2	k=4	k=10
1	58.9	51.1	57.6	58.2
3	62.9	80.4	59.8	64.9
5	64.9	85.3	75.1	61.8
7	68.7	88.0	79.3	68.9
10	75.1	96.7	91.3	70.4
15	87.1	96.9	88.7	74.0
20	97.6	97.3	94.9	75.3
30	100	97.6	94.7	89.1

Finally, Table 3 shows the result obtained with the strategy 3 (a set of closest prototypes, the label of maximum output prototype).

In strategies 2 and 3, the classifier N_w is trained in the same way; moreover, both the strategies select the same prototype to produce the output. The difference between them stands in the way the output is produced: strategy 2 uses directly the output of the classifier, whereas strategy 3 exploits the target of the prototype. Thus, in theory, strategy 2 should be better than strategy 3, since the classifier is just trained to produce the correct result.

In fact, notice that N_w can recognize two cases: when the pattern $\mathbf{x}$ and the prototype $\mathbf{x}_s$ belong to the same class, and when they do not belong to the same class. An accurate recognition of both the cases is useful for strategy 2, whereas strategy 3 can take advantage only from the former case. Thus, comparing the strategies is useful to understand which of the two cases are really recognized by the classifier. Actually, the results in Table 3 are not largely different from those in Table 2, which seems to suggest that, at least in this experiment, the prototypes that belong to the same class of the input pattern play a more important role than those that belong to different classes. Conclusively, it is worth noting that, for both strategies, a 100 % accuracy can be reached using a significantly smaller number of hidden neurons, which stands for simpler architectures usable to face problems in the transductive learning framework.

Finally, Tables 4 and 5 display the results obtained with strategy 3, when the dimension of the set S is 5 % and 10 % of the training set, respectively.

The set S is used to construct $S_{\mathbf{x}}$, which contains the prototypes in S close to $\mathbf{x}$. Even if the construction of $S_{\mathbf{x}}$ is computationally expensive for a large S, obtained results show that the dimension of the initial set of prototypes seriously affects the performance of the transductive algorithm and a larger S, as expected, improves the achieved accuracy, probably due to a better selection of the prototypes.

Table 5. Accuracies achieved with strategy 3 (a set of closest prototypes, the label of maximum output prototype), when $\mathcal{S}_{\mathbf{x}}$ is constructed from a set $\mathcal{S}$ containing only 10 % of the training set.

Number of neurons	Induction	Transduction, $k = |\mathcal{S}_x|$		
		k=2	k=4	k=10
1	58.9	51.1	42.2	57.6
3	62.9	72.2	80.0	71.8
5	64.9	96.7	98.0	80.9
7	68.7	97.3	98.2	83.3
10	75.1	99.3	98.4	96.9
15	87.1	98.2	99.8	92.7
20	97.6	99.6	100	95.3
30	100	99.6	100	97.1

5 Conclusions

In this paper, we presented a preliminary experimental comparison between pure inductive learning and a simplified version of transductive learning based on neural network models. The experimentation has been carried out on a simple benchmark, namely the two spirals dataset. Meeting obvious expectations, the results have shown that transductive learning allows to solve the problem using a smaller number of neurons. However, the way in which the prototypes are chosen plays an important role and largely affects the performance. A simple random selection of the prototypes does not help or even may decrease the classification accuracy. On the other hand, the performance is improved if the selection of the prototypes is based on the correlation or the distance of the prototypes with respect to the input patterns. In this latter case, even the computational cost is close to that of the inductive approach.

Future matters of research will cover a wider experimentation, including a study of the computational costs of the algorithms. A theoretical analysis can be also carried out using a recently proposed measure of complexity for classification problems which allows to relate a problem with the resources (number of neurons and layers) employed by neural networks that can solve it [20, 21].

References

1. Vapnik, V.: The Nature Of Statistical Learning Theory. Springer Science & Business Media, New York (2013)
2. Belkin, M., Niyogi, P., Sindhwani, V.: Manifold regularization: a geometric framework for learning from labeled and unlabeled examples. J. Mach. Learn. Res. **7**, 2399–2434 (2006)
3. Zhou, D., Bousquet, O., Lal, T.N., Weston, J., Schölkopf, B.: Learning with local and global consistency. Adv. Neural Inf. Process. Syst. **16**, 321–328 (2004)

4. Zhu, X., Ghahramani, Z., Lafferty, J., et al.: Semi-supervised learning using Gaussian fields and harmonic functions. ICML **3**, 912–919 (2003)
5. Blum, A., Mitchell, T.: Combining labeled and unlabeled data with co-training. In: Proceedings of the 11th Annual Conference on Computational Learning Theory, pp. 92–100. ACM (1998)
6. El-Yaniv, R., Pechyony, D., Vapnik, V.: Large margin vs. large volume in transductive learning. Mach. Learn. **72**, 173–188 (2008)
7. Ifrim, G., Weikum, G.: Transductive learning for text classification using explicit knowledge models. In: Fürnkranz, J., Scheffer, T., Spiliopoulou, M. (eds.) PKDD 2006. LNCS (LNAI), vol. 4213, pp. 223–234. Springer, Heidelberg (2006). doi:10. 1007/11871637_24
8. Nigam, K., McCallum, A.K., Thrun, S., Mitchell, T.: Text classification from labeled and unlabeled documents using EM. Mach. Learn. **39**, 103–134 (2000)
9. Blake, A., Rother, C., Brown, M., Perez, P., Torr, P.: Interactive image segmentation using an adaptive GMMRF model. In: Pajdla, T., Matas, J. (eds.) ECCV 2004. LNCS, vol. 3021, pp. 428–441. Springer, Heidelberg (2004). doi:10.1007/ 978-3-540-24670-1_33
10. Balcan, M., Blum, A., Choi, P., Lafferty, J., Pantano, B., Rwebangira, M., Zhu, X.: Person identification in webcam images: an application of semi-supervised learning. In: Proceedings of the 22nd International Conference on Machine Learning (ICML05), Workshop on Learning with Partially Classified Training Data, pp. 1–9 (2005)
11. Duh, K., Kirchhoff, K.: Lexicon acquisition for dialectal Arabic using transductive learning. In: Proceedings of the 2006 Conference on Empirical Methods in Natural Language Processing, pp. 399–407. Association for Computational Linguistics (2006)
12. Ueffing, N., Haffari, G., Sarkar, A., et al.: Transductive learning for statistical machine translation. In: Annual Meeting-Association for Computational Linguistics, vol. 45, p. 25 (2007)
13. Lane, T.: A decision-theoritic, semi-supervised model for intrusion detection. In: Malo, M.A. (ed.) Machine Learning and Data Mining for Computer Security, pp. 157–177. Springer, Heidelberg (2006)
14. Vert, J.P., Yamanishi, Y.: Supervised graph inference. In: Advances in Neural Information Processing Systems, pp. 1433–1440 (2004)
15. Craig, R.A., Liao, L.: Transductive learning with EM algorithm to classify proteins based on phylogenetic profiles. Int. J. Data Min. Bioinf. **1**, 337–351 (2007)
16. Weston, J., Pérez-Cruz, F., Bousquet, O., Chapelle, O., Elisseeff, A., Schölkopf, B.: Feature selection and transduction for prediction of molecular bioactivity for drug design. Bioinformatics **19**, 764–771 (2003)
17. Bair, E., Tibshirani, R.: Semi-supervised methods to predict patient survival from gene expression data. PLoS Biol. **2**, e108 (2004)
18. Hughes, N.P., Roberts, S.J., Tarassenko, L.: Semi-supervised learning of probabilistic models for ECG segmentation. In: 26th Annual International Conference of the IEEE Engineering in Medicine and Biology Society, IEMBS 2004, vol. 1, pp. 434–437. IEEE (2004)
19. Yuan, G.X., Ho, C.H., Lin, C.J.: Recent advances of large-scale linear classification. Proc. IEEE **100**(9), 2584–2603 (2012)
20. Bianchini, M., Scarselli, F.: On the complexity of neural network classifiers: a comparison between shallow and deep architectures. IEEE Trans. Neural Netw. Learn. Syst. **25**, 1553–1565 (2014)
21. Kurková, V., Sanguineti, M.: Model complexities of shallow networks representing highly varying functions. Neurocomputing **171**, 598–604 (2016)

Structural Knowledge Extraction
from Mobility Data

Pietro Cottone, Salvatore Gaglio, Giuseppe Lo Re, Marco Ortolani$^{(\boxtimes)}$,
and Gabriele Pergola

DICGIM – Università degli Studi di Palermo, Palermo, Italy
{Pietro.Cottone,Salvatore.Gaglio,giuseppe.lore,
Marco.Ortolani,Gabriele.Pergola}@unipa.it

Abstract. Knowledge extraction has traditionally represented one of the most interesting challenges in AI; in recent years, however, the availability of large collections of data has increased the awareness that "measuring" does not seamlessly translate into "understanding", and that more data does not entail more knowledge. We propose here a formulation of knowledge extraction in terms of Grammatical Inference (GI), an inductive process able to select the best grammar consistent with the samples. The aim is to let models *emerge* from data themselves, while inference is turned into a search problem in the space of consistent grammars, induced by samples, given proper generalization operators. We will finally present an application to the extraction of structural models representing user mobility behaviors, based on public datasets.

Keywords: Structural knowledge · Grammatical inference · Mobility data

1 Introduction

Massive collections of data regarding the most disparate aspects of users' lives are produced nowadays at unprecedented rates and are readily available for machine processing. For instance, automated systems for the acquisition and processing of users' movements in everyday life have attracted growing attention, also thanks to the wide diffusion of cheap and commonly available devices (e.g. smartphones or GPS loggers) that can reliably provide the location of their owners [1]. This has an impact on the traditional approach to knowledge extraction, in that the main concern now is not just the demand for accurate predictive models, but rather the provision of reliable insights to experts. The main issue regards the choice of the most appropriate tools and features to extract information from high-dimensional, incomplete and noisy datasets. Researchers have become increasingly more aware that "measuring" does not seamlessly translate into

This work was partially supported by the Italian Ministry of Education, University and Research on the "StartUp" call funding the "BIGGER DATA" project, ref. PAC02L1_0086 – CUP: B78F13000700008.

© Springer International Publishing AG 2016
G. Adorni et al. (Eds.): AI*IA 2016, LNAI 10037, pp. 294–307, 2016.
DOI: 10.1007/978-3-319-49130-1_22

"understanding", and their primary goal is to make sense of data by letting models *emerge* from the collected samples, rather than deducing them from pre-set assumptions.

In this context, most traditional approaches to data mining are not viable to handle the complexity of data especially because they fail to provide useful insight into the real nature of the samples [2]. It has thus been claimed [3] that the availability of *qualitative information* might ease the problem: at the cost of decreasing accuracy, the user can obtain a better understanding of the data, being free to focus on the overall organization at a larger scale; once a first insight is obtained, the process can be repeated at a smaller scale, considering only a subset of the original dataset, or a projection with lower dimensionality.

Our system paves the way for employing symbolic methods through an effective strategy to develop mobility models; it explicitly uses symbolic encoding for user movements, subsequently processing them in order to extract the most relevant paths. In particular, we make use of *Grammatical Inference* (GI) [4], an inductive process able to select the best grammar (according to a metric) that is consistent with the samples. Instead of being represented in a vectorial space, we thus regard our input as strings generated by an unknown formal grammar; our claim is that GI can be successfully applied in order to get relevant insights about the hidden structure embedded in large collections of data [5], enabling the user to pose new kinds of questions, taking advantage of the generative models obtained by the inductive process. Thanks to their recursive nature, grammars are also able to make recurrent relations among data explicit at different granularities [6]. We customize *Blue**, a well-known algorithm for GI which is guided by a heuristic that makes it suitable to cope with noise in data, and to take into account the relative importance of paths. We then specifically address the issue of comparing different mobility models to estimate similarity among users, and to highlight emergent behaviours. To this end, we express the similarity measure between two languages, through their descriptions; we adapt a state-of-the-art metric [7] to solve this problem in the scenario of mobility models.

The remainder of the paper is organized as follows. Section 2 briefly discusses the relevant scientific context for the use of structural approaches in knowledge extraction, while Sect. 3 presents our approach based on GI. Section 4 will present the experimental assessment of the proposed approach and, finally, we will discuss our conclusions in Sect. 5.

2   Scientific Background

Learning from experience is a key point in the design of intelligent agents. Over the years, this issue has been addressed in different ways, depending on the available devices, algorithms, and data, beginning with expert systems, probabilistic graphical models, and other statistical approaches. It soon became apparent, however, that one of the most relevant challenges was the selection of features from unlabeled data, so a lot of effort has been devoted during the last decade to the creation of systems able to perform this task automatically. Notable examples of this class of methods fall under the name of *Deep Learning*, and it has been shown that their finest performance is comparable to the best hand-engineered systems [8]. A strong theoretical limitation, however, is represented by the well-known *No Free Lunch* theorem; one of its formulations states that "any two optimization algorithms are equivalent when their performance is averaged across all possible problems" [9]; in other words, there is no possible general criterion for choosing the optimal parameters of a method when absolutely no prior knowledge about the problem is available, except raw data [10]. If models are to be regarded as "black-boxes", there is no reasonably efficient choice among several of them, when all choices fit the data comparably well.

The most recent technological advances have once more complicated the nature of the problem; it is now possible to perform measurements regarding the most disparate aspects of users' lives at previously unconceivable rates; moreover such data are highly heterogeneous, so the obtained datasets are typically high-dimensional and possibly incomplete. One of the most common examples is the massive volume of data with diverse features collected in smart environments, where pervasive networks of sensing devices are deployed, in order to support users in controlling the monitored environments [11]. The peculiar challenges related to the analysis of this kind of data has given rise to a specific branch of AI named *Ambient Intelligence* (AmI), specifically aiming at exploiting the information about the environment state in order to personalize it, adapting the environment to users' preferences [12–14]. Very high dimensionality is hardly manageable by a human mind so, lacking support from the machine, designers are effectively prevented from grasping the most important features to consider [2]. As already mentioned, the availability of *qualitative* information might help to improve insight on data.

In this paper, we claim that qualitative information can provide very useful and compact guidelines to designers, in the preliminary set-up of systems for automatic data analysis. Also, recent findings [15] show that neural processes activated by human comprehension hint toward a grammar-based inner construction of knowledge representation; hence, modeling data in the form of grammars might help users to figure out the main structure behind relevant information. Grammar representations have been devised for *syntactic pattern recognition* [16]; in this work, we use some ideas pertaining to this research area, adapting and updating them with recent advances in data analysis.

3   Model Extraction as Grammatical Inference

We propose a system for modelling user mobility habits from frequent paths expressed as sequences of locations; our goal is to exploit the available data in order to infer a model that closely represents users' behavior. Data collected in a real-life scenario is often in a numerical form, often embedded in a geometric space, whose dimensions are the features selected by the designer. As our approach aims to formulate model extraction as a GI problem, data is required in symbolic form, so a preprocessing step is needed; as will be described in the following section, we rely on a specific encoding, namely geohash, to turn locations into symbols. Once this step is accomplished, we can express user paths as strings of a formal language. In other words, we assume that a (yet unknown) language describing our data exists; admittedly, this language may be extremely complex and, data may be corrupted by noise, so that reconstructing the original language from raw data is likely to be very challenging. However, relying on formal languages to represent, organize and process knowledge is advantageous as they naturally provide a description of the relations between elements, which may be regarded as their *hidden structure*. A formal language is a set of sentences, each finite in length and made up of a finite set of symbols [17].

Generally speaking, two different descriptions can be associated to a language, namely a *generative* description, and a *recognition-based* one. In this paper, we focus on regular languages, so the corresponding representations are *regular grammars* and *Deterministic Finite Automata* (DFAs), respectively. Inferring a language through a grammar is by all means a learning process which may be characterized by its capability of *generalizing*. Identifying a language is the main concern of *Grammatical Inference* (GI) [4], which may be defined as the process of searching for a hidden grammar by exploiting the scarce available information, often consisting in just a set of strings; as such, GI belongs to the broader framework of *Algorithmic Learning Theory* (ALT), whose central concept is that of a *language learnability model*. In this context, learnability is expressed by the paradigm of *identification in the limit* formulated by Gold [18]: the learning algorithm should identify the correct hypothesis on every possible data sequence consistent with the problem space. This idea is a non-probabilistic equivalent of statistical consistency, where the learner can fail on data sequences whose probability measure is 0; in this case, a learner (an algorithm) will identify a language in the limit if, after a number of presented strings, *its hypothesis no longer changes*. The language learnability paradigm has some theoretical limitations; in particular, while it may be proven that the class of primitive recursive languages (which includes regular languages) can be identified in the limit by a complete presentation [18], finding the minimum automaton consistent with a set of samples is an NP-hard problem; therefore, some heuristic is needed to carry out this search in an efficient way.

We will characterize the search space for our problem through its basic elements, namely: the *initial node* (an "acceptable" DFA), the *successor function* (a set of successors of an automaton generated by pairwise state merging), and the *target* (the minimum automaton that is consistent with the samples I). This search space may thus be described as a Boolean lattice [19], whose initial node is a tree automaton accepting only the positive examples I_+; this is the so-called *Prefix Tree Acceptor* (PTA) which, by construction, is the most specific DFA for the positive examples. Starting from there, we want to explore the space moving toward the minimum consistent automaton, with the negative examples as our bounds. The complexity of the search can be eased by exploiting some general-to-specific ordering of the nodes; intuitively, in grammatical induction, this ordering is based on constraints characterizing the hypotheses, with fewer

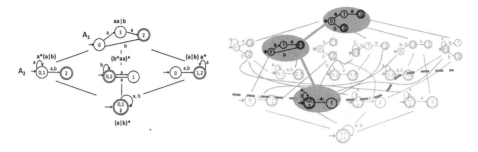

Fig. 1. An example of a merging operation (left), with a sketch of a search in the induced Boolean lattice (right).

constraints entailing more general hypotheses, and vice versa. The set of successors of a node in the lattice (an automaton) is generated by *pairwise merging* operations: two states of the original automaton are chosen for merging, resulting into a new automaton with one less state with respect to the original, as shown in the leftmost side of Fig. 1 which depicts an excerpt of a lattice. Pairwise merging may be formally defined as a partition of the set of states of the original automaton, and preserves the property of *language inclusion*, as shown in [19], which means that the application of the merging operator either (a) causes the number of states to decrease, but the recognized language is preserved, or (b) it also implies a change in the language recognized by the resulting automaton, but such language is more general, and properly includes the original one. The Boolean lattice $Lat(PTA(I_+))$ is thus completely defined by its initial node, i.e. $PTA(I_+)$, and the nodes obtained by repeatedly applying merging operations included in the partition set of $PTA(I_+)$; the deepest node in the lattice is the *Universal Automaton* (UA), that accepts all the strings defined over the alphabet. The inference of regular languages, provided a presentation from an informant, can be turned into the search for an automaton $A' \in Lat(PTA(I_+))$, given the additional hypothesis of structural completeness of I_+[1]. It may be proven that if I_+ is a structurally complete sample with respect to the minimal automaton A accepting a regular language L, then A belongs to $Lat(PTA(I_+))$, so inference in this case can be turned into the search for an automaton in the space defined by that Boolean lattice [19].

The definition of minimal DFA consistent with the sample set I can also be visualized through the so-called *Border Set* in the lattice, which establishes the limit of generalization in the search process under the control of negative samples I_-, as graphically shown by the dotted line in the rightmost side of Fig. 1.

---

[1] A I_+ sample set is said to be structurally complete with respect to an automaton A, if every transition of A is used by at least a string in I_+, and every final state in A corresponds to at least one string in I_+.

3.1   Guiding Search by Statistical Information

A well-known iterative algorithm to perform the search within the lattice of automata is *Evidence-Driven State Merging* (EDSM) [20]; it makes use of a heuristic that computes a score for all possible merges by counting the number of strings that would end in the same state; then it rejects merges making the automaton inadmissible (i.e. an element of I_- would be accepted), and selects the pair of states with the highest score. An interesting feature of the algorithm is that it implements a generalization process by producing automata that accept also strings that were not present in I_+ and in I_-; these "new" strings are able to predict unseen examples. One main issue, however, consists in the imbalance between positive and negative samples; typically, the cardinality of I_- is greater than that of I_+, thus it is important to identify and use only the most significant elements of I_-. To this end, the *frequencies* of input strings should be considered, but regular languages are not well-suited for this. In our proposal, we accounted for frequencies by employing a variation of classical EDSM, known as *Blue** [21], that incorporates a mechanism to consider only *statistically relevant* examples, without changing the nature of resulting models.

Blue* is based on a clever strategy to deal with a high amount of data and whose key insight is a statistical distinction between relevant and irrelevant information, which is treated as noise. Among the different types of noise that can be observed (e.g. noisy labels, incomplete data, fuzzy data, etc.), the case of mislabeled data is addressed; in other words, the algorithm assumes that some positive examples might be mistakenly regarded as negative, and vice versa. Its authors initially proposed a modification to the merge operator so as to make it statistically robust to misclassified examples, and later added what they called "statistical promotion". Their approach is based on the comparison of misclassified sample proportions, and aims at verifying that such proportions do not increase significantly after a state merging; the resulting reduction in the size of the produced DFA is accepted when the error does not exceed some chosen threshold. Let *population* p denote the unknown complete set of strings identifying the target language, and *population sampling*, or *sampling*, $\hat{p}$ the set of positive and negative examples $I_+ \cup I_-$. Since the error variations might depend on the particular sampling of target language, a simple comparison of misclassified proportion is not sufficient and a statistical method is necessary to deal with error variability. The proposed rule considers a merge as *statistically acceptable* "if and only if the proportion of misclassified examples in a DFA after a merging is not significantly higher than the proportion computed before the merging" [21]. Many tests are available in statistical inference theory to assess two proportions; among them, *hypothesis testing* [22] was adopted for *Blue**.

3.2   A Metric Accounting for Similarity in Structure and Language

Rather than merely comparing the structures of the obtained models, the case study in the next section will show that we are interested in capturing the language of the mobility paths, or, in other words, in computing the similarity between habits of users. We thus chose to refer to the similarity measure described in [7], whose authors propose two approaches. The first one aims at comparing two automata by looking at the languages described by the respective DFAs, whereas the second one provides an alternative insight by analyzing the structural dissimilarity. Those strategies have a complementary nature: while the latter aims at understanding of how states and transitions impact the final   . behavior, the former is language-based and well fits a deep comprehension of automata as language recognizers, so it appears more suitable for our purposes.

In order to compute the metric, one of the two automata is designated as "target", and the other as "subject"; the aim is to assess how similar the former is to the latter. The main underlying idea is the identification of a significant set of strings to be used as probes, in order to assess how both automata behave when processing them; the similarity score will depend on how many strings are identically classified by both the target and the subject. The choice of the set of probes, however, is not trivial. First of all, randomly selecting them (e.g. by random walks across the transition structure) is not a viable solution as it would result into intolerable bias towards some portion of the language. Secondarily, merely measuring the proportion of sequences in the sample that are classified equivalently by both machines may not convey any significant insight as regards the relative generalization capability of the two machines. Care must be taken during the generation of the set of probes to (1) encode every reachable state, (2) trigger every transition and (3) preserve the correct arrangement of states. To this end, the W-method [23] has been proposed as a way to compute similarity, coupled with the use of techniques of information retrieval. The author of [23] defines the *cover set* as the set of string guaranteeing that every state is reached at least one time, whereas concatenation of strings belonging to what is called the *characterization set* ensures that every possible sequences of symbols start from each state, and furthermore that every unique state from the reference automaton is explored in the subject machine. With these definitions, the *probe set* is constructed as cross product of both sets.

Merely computing a ratio of strings treated identically by both automata would likely result into an unreliable similarity score, which might be biased due to a significant asymmetry between the amount of accepted and rejected examples from the probe set (a common situation in application scenarios for the W-method and similar algorithms). To address this issue, the generated sequences are fed to the DFAs, and the outcome of their classifications are categorized in a confusion matrix; then, the similarity measure is computed by means of an F-measure [24], defined as $F = 2 * Precision * Recall/(Precision + Recall)$, which corresponds to computing the harmonic mean between the two classic measures of statistical relevance, namely Precision and Recall. Similarity computed in this way naturally emphasizes the importance of capturing the language of the reference machine, rather than ensuring accuracy with respect to language complements. Finally, in order to guarantee symmetry of the similarity measure, we compute it as the average of the scores obtained by switching the roles of target and subject automaton.

4 Inference of User Mobility Patterns

Figure 2 depicts a high-level representation of our approach. The initial step of the process requires translating paths into a symbolic representation; as mentioned in the previous section, we selected an encoding system for geographical coordinates known as *geohash* [25]. Geohash assigns a string to each *(latitude, longitude)* pair, and is based on a hierarchical spatial data structure that recursively subdivides the whole globe into "buckets" according to a grid. The space is partitioned according to a 4×8 grid; each cell is identified by an alphanumerical character and can be recursively divided into 32 smaller cells, and so on, thus providing a hierarchical structure. This process can be iterated until the desired spatial accuracy is obtained: the longer the geohash string, the smaller the area. Any path of a user within a cell may thus be represented by a string, whose symbols represent the 32 sub-cells. Geohash encoding allows for a *multi-scale* mobility model; thanks to its recursive representation, a mobility model made of a pool of 32 regular languages for each cell can be inferred. These languages may be collectively thought of as *language of paths* of the user. Moreover, the recursive partitioning of space eases the computational cost related to the model extraction: rather than inferring only one language encoding user behaviour at all possible grains, a language for each level of precision can be obtained.

Given such symbolic representation, the *Blue** algorithm can be used to infer the hierarchy of DFAs representing the user mobility model. The transition table of every DFA learned for a specific geohash cell represents user movements inside that cell; furthermore, user behavior in any sub-cell is also described by the DFA for the corresponding symbol. Those DFAs are inferred by adopting as positive samples I_+ the paths performed by the analyzed user, while as negative ones I_- the paths of all the other users within the same areas; in this fashion, a mobility model is based on the distinctive user habits emerging with respect to the other users' behaviors. This representation allows a *hierarchical* navigation among the pool of automata; increasing the spatial scale corresponds to looking into a geohash sub-cell, or in other words to "explode" a transition symbol into the correspondent language, i.e. DFA, in order to obtain a more complex

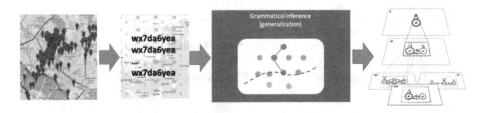

Fig. 2. From mobility data to automata: an overview of the proposed approach.

and detailed automaton. This is equivalent to concatenating a new symbol to the geohash string, and inspecting the movements of a new level of detail. The hierarchy of obtained DFAs is the result of our system, as shown in Fig. 2.

To assess the effectiveness of our approach, we used "GI-learning"[2], our software library for grammatical inference [26], to examine location data from two publicly available datasets, both collecting information about users mostly moving in metropolitan areas, measured by sensors embedded in carried-on devices, such as mobile phones; the sequences of geospatial coordinates identify the paths followed by each user. The first dataset collects fine-grained mobility data from the mobile phones of students at *Yonsei* University in Seoul, Korea [27]; the dataset contains location information (latitude and longitude) observed over a period of two months with 2–5 min rate, and user-defined types of places (workplace, cafeteria, etc.). Overall, 10 grad students with rather irregular patterns were monitored, resulting in traces about 1,848 places and 9,418 stays. For a more representative case study, we also considered the *Geolife* dataset [28], which is a collection of time-stamped tuples of the form *(latitude, longitude*, representing the spatial behavior of 182 users monitored for 5 years, collected by Microsoft Research Asia. Most trajectories took place in China, near Beijing, but routes crossing USA and Europe are also present. More than $17,000$ trajectories are contained into the dataset, for a total of approximately $50,000$ h of tracked movements. GPS loggers and smartphones acted as acquisition devices, providing a high sampling rate ($1-5$ s in time, and $5-10$ m in space) for more than 90% of the data.

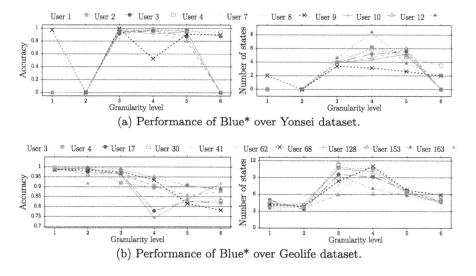

(a) Performance of Blue* over Yonsei dataset.

(b) Performance of Blue* over Geolife dataset.

Fig. 3. Accuracy and no. of states for DFAs inferred on the considered datasets.

---

[2] Software available at: https://github.com/piecot/GI-learning.

Accuracy of the Structural Models. We started by computing the accuracy and complexity of the obtained models, measured as the number of states of the resulting DFAs. Mobility models were tested at increasing scales, corresponding to increasing lengths of the geohash strings. In particular, prefixes from length 1 up to 6 have been used for the tests, corresponding to a maximum resolution for the geographical areas of about $153\,m^2$ (for geohash strings with length 7). Tests were conducted via *k-fold cross validation*, in order to prevent bias during the training process. The results of our tests will be shown for representative users within each dataset; in particular, for *Geolife*, we selected the users with the highest amount of paths. *Blue** confirms our intuition about its ability to address the unbalance between positive and negative examples, and to account for the frequencies of relevant trajectories, while preserving the determinism of the provided model. Figure 3a reports the results obtained for *Blue** over *Yonsei*, as the mean of accuracy obtained on each cell at a given prefix length. Outcomes show good performances for granularities 3 and above; the first two levels of granularities are not representative in this case as users' movements happened to be all confined in one geohash cell, so their mobility model at this granularity is trivial, consisting only in one string. Only users 2 and 12 show a slightly different behavior, moving infrequently across a wider area, as proved by the respective DFAs (see the number of states at prefix 1).

Figure 3b reports the results obtained for *Blue** over *Geolife*, clearly showing that models are very effective with coarse granularities, corresponding to the intuition that high regularities in users' movements are bound to appear if observed at wider scales. Accuracy decreases for granularities 4 and 5 (areas of about 5 km), due to higher variability in user habits as more points of interest can be identified at this scale. Notably, plotting the resulting DFA states does

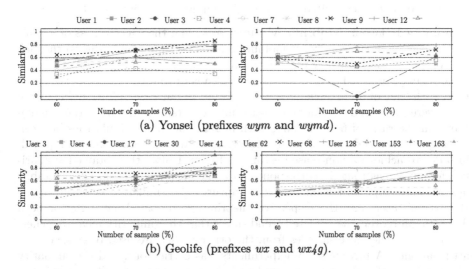

(a) Yonsei (prefixes *wym* and *wymd*).

(b) Geolife (prefixes *wx* and *wx4g*).

Fig. 4. Similarity of each user wrt his artificial copies.

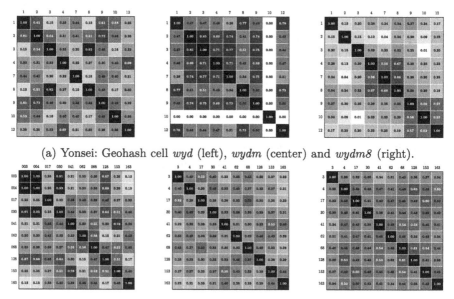

(a) Yonsei: Geohash cell *wyd* (left), *wydm* (center) and *wydm8* (right).

(b) Geolife: Geohash cell *wx* (left), *wx4* (center) and *wx4g* (right).

Fig. 5. Similarity matrices of users at three different scales.

not precisely follow the trend for the accuracy, and shows an increase for prefixes of length 3 and 4, and a decrease for lengths 5 and 6. This is likely due to the more complex models needed to express behaviors at intermediate scales, while at extreme ones (i.e. very large or small regions), it is understandably easier to predict the next place visited by a user since much fewer possible choices exist.

Similarity between Mobility Models. Assessing the effectiveness of a similarity measure in the context of mobility data is not immediate, due to the lack of a proper ground truth. We chose to overcome this difficulty by following the same strategy adopted in [29], whose basic idea is to compare the model of a real user, with that of an artificially generated one. In our case, additional artificial users were generated by randomly selecting subsets of paths from the original users (in the ratio of 60 %, to 80 % with increase of 10 %). Similarities between an original user and each of its artificially generated offspring were computed, in order to achieve a non-biased estimate. Figure 4 plots the obtained similarities for two areas roughly corresponding to the ones analyzed in [29]. Notably, our measure appears reasonable as it captures the similarity between any of the original users and the respective offspring, if compared with [29] for both datasets. Moreover, the depicted trend shows that the similarity measure generally increases the more samples from the original user are used to generate the artificial ones. A further set of experiments was carried out, and the similarity for every pair of users was computed and compared to the results presented in [29]. A partial report of the results for the second test is shown in Fig. 5, where

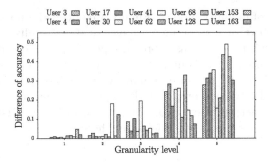

Fig. 6. Assessing accuracy in a *cold start* scenario.

the similarity for 9 users of *Yonsei* (10 users of *Geolife*) is depicted, at three different scales, with geohash length encoding of 3, 4 and 5 (2, 3, and 4 for *Geolife*), which are the prefixes representing the most visited cells. The results obtained are in accordance to the ones presented in [29] for users considered in both works, thus confirming that the reliability of the proposed similarity is comparable to the state-of-the-art metrics in literature.

A "cold start" Scenario. Finally, in order to provide an example of a practical use case for our system, we tested the proposed similarity measure in the scenario of a *recommender system*, when a previously unseen user needs to be coherently modeled, and obviously we do not possess sufficient information about them. This is the issue commonly known as *cold start*, where the only choice is to use models of others users to improve suggestions, by selecting the most similar ones according to an approximate, and necessarily imprecise model for the newcomer. To address a cold-start scenario in our context, we selected one of the users as the "newcomer" and considered only 1/4 of all available data to generate an approximate model for his mobility. We then proceed to compute the similarity between such approximate model and all the other users, in order to identify the most similar one. We can thus improve the newcomer's model by adding the (arguably coherent) mobility data from the already available user. In Fig. 6 we compare the accuracy of the approximate model with just 1/4 of the data with that of the improved model, both tested against the 3/4 data that had previously been set aside for the newcomer. As can be seen by the plot in Fig. 6, a significant improvement is achieved, which implicitly confirms the validity of our proposed similarity measure.

5   Conclusion

In this paper we proposed a novel approach to extract and compare mobility models relying on theory and methods borrowed from Grammatical Inference. Our aim was to enable multi-scale analysis so as to build models suitable to identify the most relevant relations at different spatial granularities. Moreover, we adapted a state-of-the-art metric for regular languages to the specific scenario of the comparison of mobility models. Finally, we outlined a possible use of such metric in the context of a recommender system. Our results showed that the proposed approach is promising as it provides performances comparable to other state-of-the-art approach, without the additional requirement of embedding a-priori knowledge into the model. On-going work includes the improvement of the proposed similarity measure, and the generalization of the framework to the issue of automatic extraction of *Points of Interest* (POIs), emerging by comparing several user mobility models over the same area.

References

1. De Paola, A., Gaglio, S., Lo Re, G., Ortolani, M.: An ambient intelligence architecture for extracting knowledge from distributed sensors. In: Proceedings of the 2nd International Conference on Interaction Sciences: Information Technology, Culture and Human, Seoul, Korea, pp. 104–109. ACM (2009)
2. Rajaraman, A., Ullman, J.D.: Mining of Massive Datasets. Cambridge University Press, New York (2011)
3. Carlsson, G.: Topology and data. Bull. Am. Math. Soc. **46**(2), 255–308 (2009)
4. Higuera, C.: Grammatical Inference: Learning Automata and Grammars. Cambridge University Press, New York (2010)
5. Cottone, P., Gaglio, S., Lo Re, G., Ortolani, M.: Gaining insight by structural knowledge extraction. In: Proceedings of the Twenty-Second European Conference on Artificial Intelligence, August 2016
6. Cottone, P., Ortolani, M., Pergola, G.: Detecting similarities in mobility patterns, in STAIRS 2016 - Proceedings of the 8th European Starting AI Researcher Symposium, The Hague, Holland, 26 August–2 September 2016
7. Walkinshaw, N., Bogdanov, K.: Automated comparison of state-based software models in terms of their language and structure. ACM Trans. Softw. Eng. Methodol. **22**(2), 1–37 (2013)
8. Bengio, Y.: Learning deep architectures for AI. Found. Trends Mach. Learn. **2**(1), 1–127 (2009)
9. Wolpert, D.H., Macready, W.G.: Coevolutionary free lunches. IEEE Trans. Evol. Comput. **9**(6), 721–735 (2005)
10. Whitley, D., Watson, J.P.: Search Methodologies: Introductory tutorials in optimization and decision support techniques, pp. 317–339. Springer, New York (2005)
11. Lo Re, G., Peri, D., Vassallo, S.D.: Urban air quality monitoring using vehicular sensor networks. In: Gaglio, S., Lo Re, G. (eds.) Advances onto the Internet of Things: How Ontologies Make the Internet of Things Meaningful, pp. 311–323. Springer, Cham (2014)
12. De Paola, A., La Cascia, M., Lo Re, G., Morana, M., Ortolani, M.: User detection through multi-sensor fusion in an AmI scenario. In: Proceedings of the 15th International Conference on Information Fusion, pp. 2502–2509 (2012)
13. Lo Re, G., Morana, M., Ortolani, M.: Improving user experience via motion sensors in an ambient intelligence scenario. In: Pervasive and Embedded Computing and Communication Systems (PECCS), pp. 29–34 (2013)
14. Gaglio, S., Lo Re, G., Morana, M.: Human activity recognition process using 3-d posture data. IEEE Trans. Hum. Mach. Syst. **45**(5), 586–597 (2015)

15. Ding, N., Melloni, L., Zhang, H., Tian, X., Poeppel, D.: Cortical tracking of hierarchical linguistic structures in connected speech. Nat. Neurosci. **19**(1), 158–164 (2016)
16. Fu, K.S.: Syntactic Methods in Pattern Recognition. Mathematics in Science and Engineering, vol. 112. Academic, New York (1974)
17. Chomsky, N.: Syntactic Structures. Walter de Gruyter, Berlin (2002)
18. Gold, E.M.: Language identification in the limit. Inf. Control **10**(5), 447–474 (1967)
19. Dupont, P., Miclet, L., Vidal, E.: What is the search space of the regular inference? In: Carrasco, R.C., Oncina, J. (eds.) ICGI 1994. LNCS, vol. 862, pp. 25–37. Springer, Heidelberg (1994). doi:10.1007/3-540-58473-0_134
20. Lang, K.J., Pearlmutter, B.A., Price, R.A.: Results of the Abbadingo one DFA learning competition and a new evidence-driven state merging algorithm. In: Honavar, V., Slutzki, G. (eds.) ICGI 1998. LNCS, vol. 1433, pp. 1–12. Springer, Heidelberg (1998). doi:10.1007/BFb0054059
21. Sebban, M., Janodet, J.-C., Tantini, F.: Blue: a blue-fringe procedure for learning dfa with noisy data
22. Black, K.: Business Statistics: For Contemporary Decision Making. Wiley, Hoboken (2011)
23. Chow, T.S.: Testing software design modeled by finite-state machines. IEEE Trans. Softw. Eng. **4**(3), 178–187 (1978)
24. Sokolova, M., Lapalme, G.: A systematic analysis of performance measures for classification tasks. Inf. Process. Manage. **45**(4), 427–437 (2009)
25. Balkić, Z., Šoštarić, D., Horvat, G.: GeoHash and UUID identifier for multi-agent systems. In: Jezic, G., Kusek, M., Nguyen, N.-T., Howlett, R.J., Jain, L.C. (eds.) KES-AMSTA 2012. LNCS (LNAI), vol. 7327, pp. 290–298. Springer, Heidelberg (2012). doi:10.1007/978-3-642-30947-2_33
26. Cottone, P., Ortolani, M., Pergola, G.: GI-learning: an optimized framework for grammatical inference. In: Proceedings of the 17th International Conference on Computer Systems and Technologies (Compsystech 2016). ACM, Palermo (2016)
27. Chon, J., Cha, H.: Lifemap: a smartphone-based context provider for location-based services. IEEE Perv. Comput. **2**, 58–67 (2011)
28. Zheng, Y., Liu, L., Wang, L., Xie, X.: Learning transportation mode from raw Gps data for geographic applications on the web. In: Proceedings of the 17th International Conference on World Wide Web, pp. 247–256. ACM (2008)
29. Chen, X., Pang, J., Xue, R.: Constructing and comparing user mobility profiles. ACM Trans. Web **8**(4), 21:1–21:25 (2014)

Predicting Process Behavior in WoMan

Stefano Ferilli[1,2]([✉]), Floriana Esposito[1,2], Domenico Redavid[3],
and Sergio Angelastro[1]

[1] Dipartimento di Informatica, Università di Bari, Bari, Italy
{stefano.ferilli,floriana.esposito,sergio.angelastro}@uniba.it
[2] Centro Interdipartimentale per la Logica e sue Applicazioni,
Università di Bari, Bari, Italy
[3] Artificial Brain S.r.l., Bari, Italy
redavid@abrain.it

Abstract. In addition to the classical exploitation as a means for checking process enactment conformance, process models may be precious for
making various kinds of predictions about the process enactment itself
(e.g., which activities will be carried out next, or which of a set of candidate processes is actually being executed). These predictions may be
much more important, but much more hard to be obtained as well, in less
common applications of process mining, such as those related to Ambient
Intelligence. Also, the prediction performance may provide indirect indi-
cations on the correctness and reliability of a process model. This paper
proposes a way to make these kinds of predictions using the WoMan
framework for workflow management, that has proved to be able to handle complex processes. Experimental results on different domains suggest
that the prediction ability of WoMan is noteworthy and may be useful
to support the users in carrying out their processes.

Keywords: Process mining · Activity prediction · Process prediction

1 Introduction

A classical application domain of process management is the industrial one,
where the activities of a production process must be monitored and checked for
compliance with a desired behavior. If a formal model of the desired behavior
is available, the supervision task may be automated, provided that the events
related to the process enactment can be detected and delivered to the automatic
system. Given an intermediate status of a process execution, knowing how the
execution will proceed might allow the (human or automatic) supervisor to take
suitable actions that facilitate the next activities. However, it may be expected
that in an industrial environment the rules that determine how the process must
be carried out are quite strict. So, the emphasis is more on conformance checking, while prediction of process evolution is more trivial. This situation changes
significantly if we move toward other, less traditional application domains for
process management, that have been introduced more recently. For instance,
considering the daily routines of people, at home or at work, as a process, much

© Springer International Publishing AG 2016
G. Adorni et al. (Eds.): AI*IA 2016, LNAI 10037, pp. 308–320, 2016.
DOI: 10.1007/978-3-319-49130-1_23

more variability is involved, and obtaining reliable predictions becomes both more complicated and more useful.

Another shortcoming in this landscape is that, due to the complexity of some domains, manually building the process models that are to be used for supervision, compliance checking, and prediction is very complex, costly, and error-prone. The solution of automatically learning these models from examples of actual execution, which is the task of Process Mining [1,2], opens the question of how to validate the learned models to ensure that they are correct and effective. Indeed, models are learned automatically exactly because the correct desired model is not available, and thus such a validation can be carried out only empirically by applying the learned model to future process enactments. In addition to check *a posteriori* the correctness of new process enactments based on the learned model, an investigation of how good it is in providing hints about what is going on in the process execution, and what will happen next, is another, very relevant and interesting, way to assess the quality of the model.

This paper proposes two approaches for process-related predictions. The former tackles the more classical setting in which the process (and the corresponding model) that is being enacted is known, and one wishes to predict which activities will be carried out next at any moment during the process execution. The latter is more original. It assumes that one is supervising the enactment of an unknown process, and that a set of candidate processes (and corresponding models) is available, among which one aims at predicting which one is being actually enacted. The proposed solutions work in the WoMan framework for process management [3], that introduced significant advances to the state-of-the-art. Specifically, it adopts a representation formalism that, while being more powerful than Petri/Workflow Nets, makes the above predictions more complicated. This paper investigates the reasons for such an increased complexity, and leverages their peculiarities for obtaining useful hints for making the predictions.

It is organized as follows. The next section recalls some basics on process management and the WoMan framework. Then, the proposed approaches to making predictions are presented in Sect. 4 and evaluated in Sect. 5. Finally, in the last section, we draw some conclusions and outline future work issues.

2   Process Management Basics and the WoMan Framework

A *process* consists of actions performed by agents (humans or artifacts) [4,5]. A *workflow* is a formal specification of how these actions can be composed to result in valid processes. Allowed compositional schemes include sequential, parallel, conditional, or iterative execution [6]. A process execution can be described in terms of *events*, i.e. identifiable, instantaneous actions (including decisions upon the next activity to be performed). A *case* is a particular execution of actions compliant to a given workflow. Case *traces* consist of lists of events associated to *steps* (time points) [7]. A *task* is a generic piece of work, defined to be executed for many cases of the same type. An *activity* is the actual execution of a task by a *resource* (an agent that can carry it out). Relevant events are the start and end of process executions, or of activities [5].

The WoMan framework [3,8] lies at the intersection between *Declarative Process Mining* [9] and Inductive Logic Programming (ILP) [10]. It introduced some important novelties in the process mining and management landscape. A fundamental one is the pervasive use of First-Order Logic as a representation formalism, that provides a great expressiveness potential and allows one to describe contextual information using relationships. Experiments proved that it is able to handle efficiently and effectively very complex processes, thanks to its powerful representation formalism and process handling operators. In the following, we briefly and intuitively recall its fundamental notions.

WoMan's learning module, **WIND** (Workflow INDucer), allows one to learn or refine a process model according to a case, after the case events are completely acquired. The refinement may affect the structure and/or the probabilities. Differently from all previous approaches in the literature, it is *fully incremental*: not only can it refine an existing model according to new cases whenever they become available, it can even start learning from an empty model and a single case, while others need a (large) number of cases to draw significant statistics before learning starts. This is a significant advance with respect to the state-of-the-art, because continuous adaptation of the learned model to the actual practice can be carried out efficiently, effectively and transparently to the users [8].

According to [4,11], WoMan takes as input trace elements consisting of 7-tuples $\langle T, E, W, P, A, O, R \rangle$, where T is the event timestamp, E is the type of the event (one of 'begin_process', 'end_process', 'begin_activity', 'end_activity'), W is the name of the reference workflow, P is the case identifier, A is the name of the activity, O is the progressive number of occurrence of that activity in that case, and R (optional) specifies the resource that is carrying out the activity.

A model describes the structure of a workflow using the following elements:

tasks: the kinds of activities that are allowed in the process;
transitions: the allowed connections between activities (also specifying the involved resources).

The core of the model, carrying the information about the flow of activities during process execution, is the set of transitions. A transition $t : I \Rightarrow O$, where I and O are multisets of tasks, is enabled if all input tasks in I are active; it occurs when, after stopping (in any order) the concurrent execution of all tasks in I, the concurrent execution of all output tasks in O is started (again, in any order). For analogy with the notions of 'token' and 'marking' in Petri Nets, during a process enactment we call a *token* an activity that has terminated and can be used to fire a transition, and a *marking* the set of current tokens. Both tasks and transitions are associated to the multiset C of training cases in which they occurred (a multiset because a task or transition may occur several times in the same case, if loops or duplicate tasks are present in the model). It can be exploited both during the conformance check of new cases (to ensure that the flow of activities in the new case was encountered in at least one training case) and for computing statistics on the use of tasks and transitions. In particular, it allows us to compute the probability of occurrence of a task/transition in a model learned from n training cases as the relative frequency $|C|/n$.

As shown in [3, 8], this representation formalism is more powerful than Petri or Workflow Nets [1], that are the current standard in Process Mining. It can smoothly express complex task combinations and models involving invisible or duplicate tasks, which are problematic for those formalisms. Indeed, different transitions can combine a given task in different ways with other tasks, or ignore a task when it is not mandatory for a specific passage. Other approaches, by imposing a single component for each task, route on this component all different paths passing from that task, introducing combinations that were never seen in the examples.

The increased power of WoMan's representation formalism for workflow models raises some issues that must be tackled. In Petri Nets, since a single graph is used to represent the allowed flow(s) of activities in a process, at any moment in time during a process enactment, the supervisor knows which tokens are available in which places, and thus he may know exactly which tasks are enabled. So, the prediction of the next activities that may be carried out is quite obvious, and checking the compliance of a new activity with the model means just checking that the associated task is in the set of enabled ones. Conversely, in WoMan the activity flow model is split into several 'transitions', and different transitions may share some input and output activities, which allows them to be composed in different ways with each other. As a consequence, many transitions may be eligible for application at any moment, and when a new activity takes place there may be some ambiguity about which one is actually being fired. Such an ambiguity can be resolved only at a later time. Let us see this through an example. Suppose that our model includes, among others, the following transitions:

$$t_1 : \{x\} \Rightarrow \{a, b\} \quad ; \quad t_2 : \{x, y\} \Rightarrow \{a\} \quad ; \quad t_3 : \{w\} \Rightarrow \{d, a\}$$

and that the current marking (i.e., the set of the latest activities that were terminated in the current process enactment, without starting any activity after their termination) is $\{x, y, z\}$. Now, suppose that activity a is started. It might indicate that either transition t_1 or transition t_2 have been fired. Also, if an activity d is currently being executed due to transition t_3, the current execution of a might correspond to the other output activity of transition t_3, which we are still waiting to happen to complete that transition. We call each of these alternatives a *status*. This ambiguity about different statuses that are compliant with a model at a given time of process enactment must be properly handled when supervising the process enactment, as we will show later.

3   Workflow Supervision

The supervision module allows one to check whether new cases are compliant with a given model. **WEST** (Workflow Enactment Supervisor and Trainer) takes the case events as long as they are available, and returns information about their compliance with the currently available model for the process they refer to. The output for each event can be 'ok', 'error' (e.g., when closing activities that had never begun, or terminating the process while activities are still running), or a set of warnings denoting different kinds of deviations from the model (e.g., unexpected task or transition, preconditions not fulfilled, unexpected resource running a given activity, etc.).

Now, as we have pointed out in the previous section, given an intermediate status of the process enactment and a new activity that is started, there may be different combinations of transitions that are compliant with the new activity, and one may not know which is the correct one until a later time. Thus, all the corresponding alternate evolutions of the status must be carried on by the system. Considering again the previous example, each of the three proposed options would change in a different way the status of the process, as follows:

t_1: firing this transition would consume x, yielding the new marking $\{y, z\}$ and causing the system to wait for a later activation of b;

t_2: firing this transition would consume x and y, yielding the new marking $\{z\}$ and causing the completion of transition t_2;

t_3: firing this transition would not consume any element in the marking, but would cause the completion of transition t_3.

When the next event is considered, each of these evolutions is a possible status of the process enactment. On one hand, it poses again the same ambiguity issues; on the other hand, it may point out that some current alternate statuses were wrong. So, as long as the process enactment proceeds, the set of alternate statuses that are compliant with the activities carried out so far can be both expanded with new branches, and pruned of all alternatives that become incompatible with the activities carried out so far.

Note also that each alternative may be compliant with a different set of training cases, and may rise different warnings (in the previous example, one option might be fully compliant with the model, another might rise a warning for task preconditions not fulfilled, and the other might rise a warning for an unexpected agent running that activity). WEST takes note of the warnings for each alternative and carries them on, because they might reflect secondary deviations from the model that one is willing to accept. Wrong alternatives will be removed when they will be found out to be inconsistent with later events in the process enactment. So, the question arises about how to represent each alternative status. As suggested by the previous example, we may see each status as a 5-tuple

$$\langle M, R, C, T, W \rangle$$

recording the following information:

M the marking, i.e., the set terminated activities that have not been used yet to fire a transition, each associated with the agent that carried it out and to the transition in which it was involved as an output activity;

R (for 'Ready') the set of activities that are ready to start, i.e., the output activities of transitions that have been fired in the status, and that the system is waiting for in order to complete those transitions;

C the set of training cases that are compliant with that status;
T the set of (hypothesized) transitions that have been fired to reach that status;
W the set of warnings raised by the various events that led to that status.

The system also needs to remember, at any moment in time, the set *Running* of currently running activities and the list *Transitions* of transitions actually carried out so far in the case. The set of statuses is maintained by WEST, as long as the events in a case are processed, according to Algorithm 1.

Algorithm 1. Maintenance of the structure recording valid statuses in WEST

Require: $\mathcal{M}$: process model
Require: *Statuses* : set of currently compliant statuses compatible with the case
Require: *Running* : set of currently running activities
Require: *Transitions*: list of transitions actually carried out so far
Require: $\langle T, E, W, P, A, O, R \rangle$: log entry
  if E = begin_activity **then**
    $Running \leftarrow Running \cup \{A\}$
    for all $S = \langle M, R, C, T, P \rangle \in Statuses$ **do**
      $Statuses \leftarrow Statuses \setminus \{S\}$
      if $A \in R$ **then**
        $Statuses \leftarrow Statuses \cup \{\langle M, R \setminus \{A\}, C, T, P \rangle\}$
      end if
      for all $p : I \Rightarrow O \in \mathcal{M}$ **do**
        if $I \subseteq Marking \wedge A \in O$ **then**
          $C' \leftarrow C \cap C_p$ /* C_p training cases involving p */
          $P' \leftarrow P \cup P_{A,p,S}$ /* $P_{A,p,S}$ warnings raised by running A in p given S */
          $Statuses \leftarrow Statuses \cup \{\langle M \setminus I, R \cup (O \setminus \{A\}), C', T\&\langle t \rangle, P' \rangle\}$
        end if
      end for
    end for
  end if
  if E = end_activity **then**
    if $A \notin Running$ **then**
      Error
    else
      $Running \leftarrow Running \setminus \{A\}$
      for all $S = \langle M, R, C, T, P \rangle \in Statuses$ **do**
        $S \leftarrow \langle M \cup \{A\}, R, C, T, P \rangle$
      end for
      if a transition t has been fully carried out **then**
        $Transitions \leftarrow Transitions \& \langle t \rangle$
        for all $S = \langle M, R, C, T, P \rangle \in Statuses$ **do**
          if $T \neq Transitions$ **then**
            $Statuses \leftarrow Statuses \setminus \{S\}$
          end if
        end for
      end if
    end if
  end if

---

4   Prediction Strategy

While in supervision mode, the prediction modules allow one to foresee which activities the user is likely to perform next, or to understand which process is being carried out among a given set of candidates. We recall that, due to the discussed set of alternate statuses that are compliant with the activities carried out at any moment, differently from Petri Nets it is not obvious to determine which are the next activities that will be carried out. Indeed, any status might be associated to different expected evolutions. The good news is that, having several alternate statuses, we can compute statistics on the expected activities in the different statuses, and use these statistics to determine a ranking of those that most likely will be carried out next.

Specifically, **SNAP** (Suggester of Next Action in Process) hypothesizes which are the possible/appropriate next activities that can be expected given the current intermediate status of a process execution, ranked by confidence. Confidence here is not to be interpreted in the mathematical sense. It is determined based on a heuristic combination of several parameters associated with the possible alternate process statuses that are compliant with the current partial process execution. Specifically, the activities that can be carried out next in a given status are those included in the *Ready* component of that status, or those belonging to the output set of transitions that are enabled by the *Marking* component of that status. The status parameters used for the predictions are the following:

1. frequency of activities across the various statuses (activities that appear in more statuses are more likely to be carried out next);
2. number of cases with which each status is compliant (activities expected in the statuses supported by more training cases are more likely to be carried out next);
3. number of warnings raised by the status (activities expected in statuses that raised less warnings are more likely to be carried out next);
4. confidence of the tasks and transitions as computed by the multiset of cases supporting them in the model (activities supported by more confidence, or belonging to transitions that are associated to more confidence, are more likely to be carried out next).

Finally, given a case of an unknown workflow, **WoGue** (Workflow Guesser) returns a ranking (by confidence) of a set of candidate process models. Again, this prediction is based on the possible alternate statuses identified by WEST when applying the events of the current process enactment to the candidate models. In this case, for each model, the candidate models are ranked by decreasing performance of their 'best' status, i.e. the status reporting best performance in (one or a combination of) the above parameters.

5   Evaluation

In the following, we evaluate the performance of the proposed prediction approaches on several datasets, concerning different kinds of processes. Two are related to Ambient Intelligence, and specifically to the Ambient Assisted Living domain. The other concerns chess playing. In these domains, the prediction of the next activity that will be carried out is more complex than in industrial processes, because there is much more variability and subjectivity in the users' behavior, and there is no 'correct' underlying model, just some kind of 'typicality' can be expected. The chess domain also allows us to evaluate the performance of the process prediction approach.

5.1 Datasets

Our experiments were run on the following datasets:

Aruba was taken from the CASAS benchmark repository[1]. This dataset concerns people's daily routines, and includes continuous recordings of home activities of an elderly person, visited from time to time by her children, in a time span of 220 days. We considered each day as a case of the process representing the daily routine of the elderly person. Transitions, here, correspond to terminating some activities and starting new activities. The resources (i.e., which person performs which activity) are unknown.

GPItaly was built by extracting the data from one of the Italian use cases of the GiraffPlus project[2] [12]. It concerns, again, an elderly person in her home, but focusing on the movements of the home inhabitant(s) in the various rooms of the home. The dataset considered 253 days, each of which was, again, a case of the process representing the typical movements of people in the home. Tasks, here, correspond to rooms, while transitions correspond to leaving a room and entering another. The resource (i.e., the person that moves between the rooms) is always the same.

Chess consists of 400 reports of actual top-level matches played by Anatolij Karpov and Garry Kasparov (200 matches each) downloaded from the Italian Chess Federation website[3]. In this case, playing a chess match corresponds to enacting a process, where a task is the occupation of a specific square of the chessboard by a specific kind of piece (e.g., "black rook in a8" denotes the task of occupying the top-leftmost square with a black rook), and the corresponding activities are characterized by the time at which a piece starts occupying a square and the time at which the piece leaves that square. Matches are initialized by starting the activities corresponding to the initial positions of all pieces on the chessboard. Transitions correspond to moves: indeed, each move of a player terminates some activities (since it moves pieces away from the squares they currently occupy) and starts new activities (that is, the occupation by pieces of their destination squares)[4]. Given this perspective, the involved resources are the two players: 'white'

[1] http://ailab.wsu.edu/casas/datasets.html.
[2] http://www.giraffplus.eu.
[3] http://scacchi.qnet.it.
[4] Usually, each transition terminates one activity and starts another one. Special cases, such as captures and castling, may involve more pieces.

Table 1. Dataset statistics

	Cases	Events		Activities		Tasks		Transitions	
		Overall	Avg	Overall	Avg	Overall	Avg	Overall	Avg
Aruba	220	13788	62.67	6674	30.34	10	0.05	92	0.42
GPItaly	253	185844	369.47	92669	366.28	8	0.03	79	0.31
White	158	36768	232.71	18226	115.35	681	4.31	4083	25.84
Black	87	21142	243.01	10484	120.51	663	7.62	3006	34.55
Draw	155	32422	209.17	16056	103.59	658	4.25	3434	22.15

and 'black'. In this dataset we distinguished three processes, corresponding to the possible match outcomes: *white* wins, *black* wins, or *draw*.

Table 1 reports some statistics on the experimental datasets: number of cases, events, activities, tasks and transitions. The average number of events, activities, tasks, and transitions per case is also reported. It is apparent that each dataset is characterized by a peculiar mix of features. There are more cases for the Ambient Intelligence datasets than for the chess ones. However, the chess datasets involve many more different tasks and transitions, many of which are rare or even unique. In facts, the number of tasks and transitions is much less than the number of cases in the Ambient Intelligence datasets, while the opposite holds for the chess datasets. As regards the number of events and activities, GPItaly is the most complex one, followed by the chess datasets, while Aruba is the less complex one. The datasets are different also from a qualitative viewpoint. In Aruba, the cases feature many short loops (involving 1 or 2 activities), optional and duplicated activities. Concurrency is usually limited to at most two activities. In GPItaly, again, many short loops, optional and duplicated activities are present, but no concurrency. Finally, the chess datasets are characterized by very high parallelism: each game starts with 32 concurrent activities (a number which is beyond the reach of many current process mining systems [3]). This number progressively decreases as long as the game proceeds and pieces are taken, but remains still high (about 10 concurrent activities) even at the end of the game. Short and nested loops, optional and duplicated tasks are present as well[5].

5.2   Activity Prediction

The experimental procedure for the activity prediction task was as follows. First, each dataset was translated from its original representation to the input format of WoMan. Then, each dataset was split into training and test sets using a *k*-fold

---

[5] A loop consists of a piece, after a number of moves, going back to a square that it had occupied in the past. In short loops this happens within a few moves. Optional activities are involved in that a player might make the same move with or without taking another piece. Duplicate tasks are needed when a piece occupies the same square at different stages of the match, and thus in different contexts.

cross-validation procedure (see column *Folds* in Table 2 for the values of k: for GPItaly, 3 folds used due to the very large number of cases and activities in this dataset; for the others, 5 folds were used to provide the system with more learning information; both splits were used on Aruba to allow a comparison). Then, WIND (the learning functionality of WoMan) was used to learn models from the training sets. Finally, each model was used as a reference to call WEST and SNAP on each event in the test sets: the former checked compliance of the new event and suitably updated the set of statuses associated to the current case, while the latter used the resulting set of statuses to make a prediction about the next activity that is expected in that case. In these experiments, the predictions were based exclusively on the number of cases supporting each status.

Table 2. Prediction statistics

	Folds	Activity					Process					
		Pred	Recall	Rank	(Tasks)	Quality	Pos	(%)	C	A	U	W
Aruba	3	0.85	0.97	0.92	6.06	0.78	—	—	—	—	—	—
Aruba	5	0.87	0.97	0.91	5.76	0.78	—	—	—	—	—	—
GPItaly	3	0.99	0.97	0.96	8.02	0.92	—	—	—	—	—	—
black	5	0.42	0.98	1.0	11.8	0.42	2.06	(0.47)	0.20	0.00	0.15	0.66
white	5	0.55	0.97	1.0	11.27	0.54	1.60	(0.70)	0.44	0.00	0.15	0.40
draw	5	0.64	0.98	1.0	10.95	0.62	1.78	(0.61)	0.29	0.01	0.18	0.52
chess	5	0.54	0.98	1.0	11.34	0.53	1.81	(0.59)	0.31	0.00	0.16	0.53

Table 2 (section 'Activity') reports performance averages concerning the different processes (row 'chess' refers to the average of the chess sub-datasets). Column *Pred* reports in how many cases SNAP returned a prediction. Indeed, when tasks or transitions not present in the model are executed in the current enactment, WoMan assumes a new kind of process is enacted, and avoids making predictions. Among the cases in which WoMan makes a prediction, column *Recall* reports in how many of those predictions the correct activity (i.e., the activity that is actually carried out next) is present. Finally, column *Rank* reports how close it is to the first element of the ranking (1.0 meaning it is the first in the ranking, possibly with other activities, and 0.0 meaning it is the last in the ranking), and *Tasks* is the average length of the ranking. The *Quality* index is a mix of these values, obtained as

$$Quality = Pred \cdot Recall \cdot Rank \in [0, 1]$$

It is useful to have an immediate indication of the overall performance in activity prediction. When it is 0, it means that predictions are completely unreliable; when it is 1, it means that WoMan always makes a prediction, and that such a prediction is correct (i.e., the correct activity is at the top of the ranking).

First, note that, when WoMan makes a prediction, it is extremely reliable. The correct activity that will be performed next is almost always present in the ranking (97–98 % of the times), and always in the top section (first 10 % items) of the ranking. For the chess processes it is always at the very top. This shows that WoMan is effective under very different conditions as regards the complexity of the models to be handled. In the Ambient Intelligence domain, this means that it may be worth spending some effort to prepare the environment in order to facilitate that activity, or to provide the user with suitable support for that activity. In the chess domain, this provides a first tool to make the machine able to play autonomously. The number of predictions is proportional to the number of tasks and transitions in the model. This was expected, because, the more variability in behaviors, the more likely it is that the test sets contain behaviors that were not present in the training sets. WoMan is almost always able to make a prediction in the Ambient Intelligence domain, which is extremely important in order to provide continuous support to the users. The percentage of predictions drops significantly in the chess domain, where however it still covers more than half of the match. Interestingly, albeit the evaluation metrics are different and not directly comparable, the *Quality* is at the same level or above the state-of-the-art performance obtained using Deep Learning [13] and Neural Networks [14]. The nice thing is that WoMan reaches this percentage by being able to distinguish cases in which it can make an extremely reliable prediction from cases in which it prefers not to make a prediction at all. The worst performance is on 'black', possibly because this dataset includes less training cases.

5.3   Process Prediction

The process prediction task was evaluated on the chess dataset, because it provided three different kinds of processes based on the same domain. So, we tried to predict the match outcome (white wins, black wins, or draw) as long as match events were provided to the system. The experimental procedure was similar to the procedure used for the activity prediction task. We used the same folds as for the other task, and the same models learned from the training sets. Then, we merged the test sets and proceeded as follows. On each event in the test set, WoGue was called using as candidate models *white*, *black*, and *draw*. In turn, it called WEST on each of these models to check compliance of that event and suitably update the corresponding sets of statuses associated to the current case. Finally, it ranked the models by increasing number of warnings in their 'best' status, where the 'best' status was the status that raised less warnings. Note that, in this case, WoMan always returns a prediction.

Table 2 (section 'Process') summarizes the performance on the process prediction task. Column *Pos* reports the average position of the correct prediction in the ranking (normalized in parentheses to [0, 1], where 1 represents the top of the ranking, and 0 its bottom). The last columns report, on average, for what percentage of the case duration the prediction was correct (*C*: the correct process was alone at the top of the ranking), approximately correct (*A*: the correct process shared the top of the ranking with other, but not all, processes), undefined (*U*: all processes were ranked equal), or wrong (*W*: the correct process was not at the top of the ranking). All kinds of chess processes show the same behavior. The overall percentage of correct predictions (C) is above 30 %, which is a good result, with A being almost null. Studying how the density of the different predictions is distributed along the match, we discovered that, on average, the typical sequence of outcomes is: U-A-C-W. This could be expected: total indecision happens at the beginning of the match (when all possibilities are still open), while wrong predictions are made towards the end (where it is likely that each single match has a unique final, never seen previously). In the middle of the game, where the information compiled in the model is still representative, correct or approximately correct predictions are more dense, which is encouraging. This also explains the high percentage of wrong predictions (53 %): since every match becomes somehow unique starting from $1/2$–$3/4$ of the game, the learned model is unable to provide useful predictions in this phase (and indeed, predicting the outcome of a chess match is a really hard task also for humans). So, the percentage of correct predictions should be evaluated only on the middle phase of the match, where it is much higher than what is reported in Table 2.

6   Conclusions and Future Work

While traditionally exploited for checking process enactment conformance, process models may be used to predict which activities will be carried out next, or which of a set of candidate processes is actually being executed. The prediction performance may also provide clues about the correctness and reliability of the model. In some applications, such as Ambient Intelligence ones, there is more flexibility of behaviors than on industrial applications, which makes these predictions both more relevant and harder to be obtained. This paper proposes a way to make these kinds of predictions using the WoMan framework for workflow management. Experimental results on different tasks and domains suggest that the prediction ability of WoMan is noteworthy.

Given the positive results, we plan to carry out further work on this topic. First of all, we plan to check the prediction performance on other domains. Also, we will investigate how to use and mix different parameters to improve the prediction accuracy. Finally, we would like to embed the prediction module in other applications, in order to guide their behavior.

Acknowledgments. Thanks to Amedeo Cesta and Gabriella Cortellessa for providing the GPItaly dataset, and to Riccardo De Benedictis for translating it into WoMan format. This work was partially funded by the Italian PON 2007-2013 project PON02_00563_3489339 'Puglia@Service'.

References

1. Weijters, A., van der Aalst, W.: Rediscovering workflow models from event-based data. In: Hoste, V., Pauw, G.D. (eds.) Proceedings of the 11th Dutch-Belgian Conference of Machine Learning (Benelearn 2001), pp. 93–100 (2001)

2. Aalst, W., et al.: Process mining manifesto. In: Daniel, F., Barkaoui, K., Dustdar, S. (eds.) BPM 2011. LNBIP, vol. 99, pp. 169–194. Springer, Heidelberg (2012). doi:10.1007/978-3-642-28108-2_19

3. Ferilli, S., Esposito, F.: A logic framework for incremental learning of process models. Fund. Inform. **128**, 413–443 (2013)

4. Agrawal, R., Gunopulos, D., Leymann, F.: Mining process models from workflow logs. In: Schek, H.-J., Alonso, G., Saltor, F., Ramos, I. (eds.) EDBT 1998. LNCS, vol. 1377, pp. 467–483. Springer, Heidelberg (1998)

5. Cook, J., Wolf, A.: Discovering models of software processes from event-based data. Technical report CU-CS-819-96, Department of Computer Science, University of Colorado (1996)

6. van der Aalst, W.: The application of petri nets to workflow management. J. Circ. Syst. Comput. **8**, 21–66 (1998)

7. van der Aalst, W., Weijters, T., Maruster, L.: Workflow mining: Discovering process models from event logs. IEEE Trans. Knowl. Data Eng. **16**, 1128–1142 (2004)

8. Ferilli, S.: Woman: Logic-based workflow learning and management. IEEE Trans. Syst. Man Cybern. Syst. **44**, 744–756 (2014)

9. Pesic, M., van der Aalst, W.M.P.: A Declarative Approach for Flexible Business Processes Management. In: Eder, J., Dustdar, S. (eds.) BPM 2006. LNCS, vol. 4103, pp. 169–180. Springer, Heidelberg (2006)

10. Muggleton, S.: Inductive logic programming. New Gener. Comput. **8**, 295–318 (1991)

11. Herbst, J., Karagiannis, D.: An inductive approach to the acquisition and adaptation of workflow models. In: Proceedings of the IJCAI 1999 Workshop on Intelligent Workflow and Process Management: The New Frontier for AI in Business, pp. 52–57 (1999)

12. Coradeschi, S., Cesta, A., Cortellessa, G., Coraci, L., Gonzalez, J., Karlsson, L., Furfari, F., Loutfi, A., Orlandini, A., Palumbo, F., Pecora, F., von Rump, S., Štimec, Ullberg, J., tslund, B.: Giraffplus: Combining social interaction and long term monitoring for promoting independent living. In: Proceedings of the 6th International Conference on Human System Interaction (HSI), pp. 578–585. IEEE (2013)

13. Lai, M.: Giraffe: using deep reinforcement learning to play chess. CoRR abs/1509.01549 (2015)

14. Oshri, B., Khandwala, N.: Predicting moves in chess using convolutional neural networks. In: Stanford University Course Project Reports - CS231n: Convolutional Neural Networks for Visual Recognition (Winter 2016)

On-line Learning on Temporal Manifolds

Marco Maggini and Alessandro Rossi[✉]

Department of Information Engineering and Mathematical Sciences,
University of Siena, Siena, Italy
{marco.maggini,rossi111}@unisi.it

Abstract. We formulate an online learning algorithm that exploits the temporal smoothness of data evolving on trajectories in a temporal manifold. The learning agent builds an undirected graph whose nodes store the information provided by the data during the input evolution. The agent's behavior is based on a dynamical system that is derived from the temporal coherence assumption for the prediction function. Moreover, the graph connections are developed in order to implement a regularization process in both the spatial and temporal dimensions. The algorithm is evaluated on a benchmark based on a temporal sequence obtained from the MNIST dataset by generating a video from the original images. The proposed approach is compared with standard methods when the number of supervisions decreases.

Keywords: Graph regularization · Temporal manifold · Dissipative system

1 Introduction

In this work we propose a learning paradigm that is rooted in the classic Regularization Theory based on differential operators first proposed in [1]. A fundamental assumption to achieve good generalization properties is that the input is distributed on a smooth manifold, connecting the problem to approximation and regularization theory in smooth hyper-surface reconstruction. The idea of this paper is to transfer the smoothness hypothesis in the dimension of time, assuming that the input data are presented in an high-dimensional feature space without a well-defined spatial description, but are distributed following a trajectory on a temporal manifold. This should allow the learning process to aggregate possibly different descriptions and grab invariances which are representable only with very complex structures. This is for example the case of standard Computer Vision problems such as object recognition or classification. The driving principles of this theory have been formulated in [2], into a framework blending Variational Calculus, Classical Mechanics and Dissipative Hamiltonian Systems [3]. They formulate the regularization process on the parameters of a general learning function, modeled as a set of mechanical particles. We maintain the assumption that the input trajectory lies on a temporal manifold that models the evolution of the function thus reducing the algorithm complexity. In fact,

© Springer International Publishing AG 2016
G. Adorni et al. (Eds.): AI*IA 2016, LNAI 10037, pp. 321–333, 2016.
DOI: 10.1007/978-3-319-49130-1_24

the integration of the differential equations is moved from space to time, without assuming the output function to have a specific structure. The obtained solution describes a system that is continuous in time, but that can be easily discretized to fit the data sampling. The input-output mapping is modeled by a graph whose nodes are visited in time; these nodes store the information extracted from data during the learning process. The access policy to the graph nodes allows an easy on–line update during the input processing and a fast extraction of additional knowledge from the input evolution and provided supervisions. We implemented spatial and temporal constraints on the function mapping, but other generic constraints can be embedded into the learning process, as for example logical and statistical constraints [4]. We evaluate the proposed learning scheme on a benchmark created from the MNIST dataset comparing the classification performance with other approaches as the number of supervisions decreases.

2   Regularization Theory on Temporal Manifolds

A solid formulation of the concept of regularization in Machine Learning is provided in [1] by creating a link with standard Approximation Theory. In this framework the problem of learning an input–output mapping from a set of data is cast as the approximation of a multidimensional function. A typical issue in these kind of problems is the fact that there are regions of the space in which data are not available or are affected by noise. The assumption of *smoothness* in the input space is applied to deal with these problems. This consists in requiring that small changes in the input should produce small changes in the output. This formulation can be naturally interpreted as a variational problem. Given a set of labeled data $\mathcal{L} = \{(x_i, y_i) \in \mathbb{R}^d \times \mathbb{R}, \ i = 1, \ldots, N\}$, the task is to learn the function $f : \mathbb{R}^d \to \mathbb{R}$ minimizing a cost functional of the form:

$$\Phi(f) = \lambda \|Pf\|^2 + \sum_{i=1}^{N} (f(x_i) - y_i)^2 \tag{1}$$

where the term $\|Pf\|^2$ implements the smoothness assumption. P is a generic differential operator and $\|\cdot\|$ could be for example the L^2 norm. The coefficient λ is a regularization parameter balancing the quality of interpolation on the given data with respect to smoothness. The solution to this variational problem can be found by satisfying the associated distributional Euler-Lagrange equation:

$$\hat{P}P f(x) = \frac{1}{\lambda} \sum_{i=1}^{N} (y_i - f(x))\delta(x - x_i) \tag{2}$$

where $\hat{P}$ is the formal adjoint operator of P. The solution lies in the N dimensional subspace generated by the combination of the set of basis functions obtained by centering the Green's Function $G(x; y)$ of the operator $\hat{P}P$ in the input points x_i, such that $\hat{P}P G(x; x_i) = \delta(x - x_i)$. The presence of the Dirac's

Delta and its translation property allows us to express f as the superposition of the corresponding responses centered around the supervised points:

$$f(x) = \frac{1}{\lambda} \sum_{i=1}^{N} (y_i - f(x_i)) G(x; x_i) \tag{3}$$

according to the classic representer theorem of kernel machines (see [5,6]). The final solution is then obtained by solving a linear system. Notice that in the expansion (3) a term lying in the null space of the operator $\hat{P}P$ has been discarded. The shape of this term depends on the chosen stabilizer P and on the boundary conditions, but it does not affect the global solution.

2.1   Regularization in the Temporal Domain

In this work we extend this approach to learning when dealing with temporal data characterized by a strong local correlation in time. For instance, in a video, we have portions of data with a spatial description that do not change too much among consecutive frames. This motivates the assumption that the output sequence is characterized by a smooth variation in the temporal domain. The learning process can be carried out in the time domain, possibly in a on–line never-ending learning scenario. Even though this hypothesis is consistent, it is convenient to consider problems in which the information eventually becomes redundant after a certain period T. This allows us to restrict the analysis to the equivalent case of periodic data, even if we will see that this restriction is not necessary in practice.

The input is assumed to be evolving in a feature space $\mathbb{R}^d$ as function of time $x : [0, T) \rightarrow \mathbb{R}^d$, with $0 < T < +\infty$. External teaching signals are provided by supervised pairs $(x(t_i), y_i)$ coming at t_i, where $0 < t_i < T$. The task is to learn a function $f : [0, T) \times \mathbb{R}^d \rightarrow \mathbb{R}$, whose output represents some kind of prediction on data. The assumption that the input has a strong coherence at local level can be mathematically expressed by the condition $x \in C^{k+1}([0, T), \mathbb{R}^d)$, so that we can reformulate the functional in (1) as:

$$\Phi(f) = \lambda \int_0^T (Pf)^2 \psi(t) \sqrt{1 + |x'(t)|^2} dt + \sum_{i=1}^{N} \psi(t_i)|f(t_i) - y_i|^2 \tag{4}$$

where we assume P as a general linear differential operator $P\left(\frac{d}{dt}\right) = \sum_{j=0}^{k} \alpha_j \frac{d^j}{dt^j}$ with adjoint $\hat{P}\left(\frac{d}{dt}\right) = \sum_{j=0}^{k} (-1)^k \alpha_j \frac{d^j}{dt^j}$. The introduction of the function $\psi(t)$ is related to the studies on dissipative Hamiltonian systems [3]. It represents a positive monotonically increasing function, that introduces the idea of energy dissipation. The interpretation of this concept in Machine Learning is exhaustively formulated in [2]. The whole learning process has been transformed into a dynamical system and the introduction of a dissipation term has a crucial impact on its global stability. In other words, $\psi(t)$ can be interpreted as the implementation in the functional Φ of the idea that we want to assign a growing importance to the satisfaction of the constraints as time evolves (i.e. errors in the agent's response are progressively less tolerated as its life proceeds). The Euler-Lagrange equations, formulated in a distributional sense, require a stationary point of Φ to satisfy the necessary condition:

$$\lambda \hat{P}(\bar{\psi}\sqrt{1+|x'(t)|^2}\,Pf) + \sum_{i=1}^{N} \psi(t_i)\,[f(t_i) - y_i]\,\delta(t - t_i) = 0. \tag{5}$$

The resolution of this formula is strictly dependent on the form of $x'(t)$. However, to deal with a more manageable version of (5), a substantial simplification can be achieved by choosing $\psi(t) = e^{\theta t}(\sqrt{1+|x'(t)|^2})^{-1}$. The function $\varphi(t) = e^{\theta t}$ depicts the concept of dissipation, measured by the parameter $\theta > 0$. The term $b(t) = \sqrt{1+|x'(t)|^2}$ derives from the definition of the problem in the temporal domain and represents the magnitude of the distance in local variations of the input function $x(t)$. We can rewrite (5) more concisely as:

$$\hat{P}(\varphi\,Pf) + \frac{1}{\lambda}\sum_{i=1}^{N} \psi(t_i)\,[f(t_i) - y_i]\,\delta(t - t_i) = 0 \tag{6}$$

in a form more similar to the one in (2). The condition (6) yields a non-homogeneous linear differential equation of order $\ell = 2k$, being k the highest order of the derivatives in the operator P. Also in this case, the solution is composed by the sum of two terms. The first one corresponds to the solution of the associated homogeneous equation and lies in the null space of the differential operator Q such that $Qf = \hat{P}(\varphi\,Pf)$. This solution can be expressed by a summation $\sum_{q=1}^{\ell} c_q e^{r_q t}$, where the r_q's are the roots of the characteristic polynomial of the associated homogeneous equation. The coefficients c_q depend on the boundary conditions assigned to the solutions. Obviously, to have a formulation coherent with the problem of learning the function in an on-line setting, we assume that the conditions of f at the end of the period T are not known. Hence, we have to impose the initial conditions for the correspondent Cauchy's problem, that, in general, affect the global trend of the solution. However, under the hypothesis of asymptotic stability, this part of the solution can be neglected as in the classic case. This occurs when the Routh-Hurwitz conditions are satisfied, i.e. the roots of the characteristic polynomial have negative real parts. The solution of (6) is then asymptotically given by:

$$f(t) \simeq \frac{\nu}{\lambda \alpha_k^2 \varphi(t)} \sum_{i=1}^{N} \psi(t_i)\,[f(t_i) - y_i]\,G(t; t_i). \tag{7}$$

Hence we can arbitrarily assume null Cauchy's Conditions to obtain an explicit solution of (6). Again, the function G represents the Green's Function of Q. Interestingly, G can be viewed as a *temporal kernel*, since it closely resembles classic kernels in the feature space (see [5,6]). Since we ended up in a dynamical system, the function G will be properly referred hereinafter as the *Impulse Response*, as in classical Control Theory. The shape of G (depending on $\theta, \alpha_0, \ldots, \alpha_k$) regulates the response of the system describing f to each impulse at t_i generated by the incoming supervision. The presence in (7) of the coefficients $\nu = (-1)^{k+1}$ is introduced by the sign flip in the odd order coefficients of the adjoint operator $\hat{P}$. This sign flip leads to divergent systems in the case of an odd order operator P. Given the resemblance of (7) with the classic *Gradient Descent* methods, this observation suggests an implementation with a differential operator of even order. In [7] the parallel between (7) and the Gradient Descent method in an on–line back-propagation setting is introduced using a similar framework. In fact, it is easy to see that the terms $[f(t_i) - y_i]$ represent the derivatives of the quadratic loss function appearing in the second term of the cost functional $\Phi(f)$, and the other elements in (7) compose a sort of learning rate when $\nu = -1$.

2.2   On-line Discrete Solution

The solution of equation (7) clearly depends on the Green function $G(\cdot)$, that can be written as $G(t) = \sum_{q=1}^{\ell} g_q e^{r_q t}$. Again $r_q, q = 1, \ldots, \ell$ are the roots of the characteristic polynomial associated to the differential equation (6), whereas the coefficients g_q are set to yield a unitary Dirac delta function when computing $QG(t)$. The design of the learning process in time reflects the concept of temporal causality. This means that each supervision $(x(t_i), y_i)$ does not affect the solution for $t < t_i, i \in \mathbb{N}$, that requires to exploit the *causal* impulse response with $G(t) = 0$ when $t \le 0$. The shape of $G(t)$ in the positive semi-plane depends on the parameters $\{\theta, \alpha_0, \ldots, \alpha_k\}$. Indeed, these parameters determine the coefficients of the associated characteristic polynomial and then its roots r_q, affecting the Impulse Response as in classic dynamical systems theory. When the Routh-Hurwitz conditions are satisfied, the system is stable and $G(t)$ exponentially decays to 0 after the response. The smoothness of $G(t)$ in $t = 0$ depends on ℓ, since all its derivatives up to order $\ell - 2$ are continuous and null in $t = 0$. Instead, the magnitude of the peak of $G(t)$ is related to the parameter θ. We will refer to the time needed to reach this peak as the *delay* of the response (the time necessary to let the effect of each impulse to be significant). The decay rate can be associated to the *memory* of the model, since the contribution of each example $(x(t_i), y_i)$ in (7) decreases as $G(t - t_i) \to 0$. The decay rate depends directly on the roots closest to the origin. A root in the origin generates a constant term in $G(t)$, thus yielding "infinite memory", that is in fact incompatible with the concept of dissipation. Hence, the model parameters can be set so as to define the desired behavior of the learning process. For practical reasons (numerical errors), we can skip the computation of the roots of the characteristic polynomial of equations (6), as it is more convenient to directly assign them, provided that they are coherent with a configuration of the differential operator parameters[1]. Furthermore, the direct implementation of the solution using the superposition of the impulse responses as in (7) is unfeasible in practice. An increasing amount of memory

---

[1] For instance, by observing the second coefficient of the characteristic polynomial, we have $\theta = -(r_1 + r_2)$ when $k = 1$ and $2\theta = -(r_1 + r_2 + r_3 + r_4)$ when $k = 2$.

would be necessary to store the terms $\psi(t_i)\,[f(t_i) - y_i]$ at each time instant t_i. They represent the gradients of the supervision cost function calculated at each supervised instant t_i. Since, the value of $f(t)$ depends on $\psi(t_i)\,[f(t_i) - y_i]\,G(t - t_i)$, $\forall t_i \leq t$, we would need to save all the gradients of the supervision calculated in the past and for each one update the correspondent value of $G(t - t_i)$. An approximate solution can be based on a buffer of these values, discarding each term once we decide to consider $G(t - t_i) \approx 0$.

However, it is possible to obtain the exact solution without storing the gradients and computing explicitly all the values $G(t - t_i)$. We can derive a discretized version of the solution that allows the computation of $f(t)$ at a certain time sampling-step τ. The only hypothesis needed is that supervisions are provided at time steps t_i multiple of the given τ, i.e. $t_i = \tau K, K \in \mathbb{N}$. Let us define $\mathbf{f}(t) = \left[f(t), \ldots, D^{\ell-1}f(t)\right]'$. The linear differential equation in (6) of order ℓ can be rewritten as

$$\dot{\mathbf{f}}(t) = A\mathbf{f}(t) + B\,U(t) \tag{8}$$

where A is a $\ell \times \ell$ companion matrix of the associated system of first order equations, whereas the non-homogeneous part is expressed by $B = \left[0, \ldots, \frac{\nu}{\lambda \alpha_k^2}\right]'$ and the summation of external input $U(t) = (1/\varphi(t)) \sum_{i=1}^{N} \bar{\psi}(t_i)\,[f(t_i) - y_i]\,\delta(t - t_i)$. By using the correspondent Lagrange formula, the solution is given by:

$$\mathbf{f}(t) = e^{At}\,\mathbf{f}(0) + \int_{t_0}^{t} e^{A(t-s)}\,B\,U(s)\,ds\,. \tag{9}$$

Again, the solution at time t still depends on all the supervisions that are provided at the instants $t_i \leq t$. If we select an appropriate sampling step τ, we can compute the discretized solution for $\mathbf{f}$ by using a recurrent implementation, by considering an equally spaced discretization of time $t_K = \tau K$ and $t_0 = 0$. Given the initial Cauchy conditions $\mathbf{f}[0] = \mathbf{f}(0)$, if we indicate as $\mathbf{f}[K]$ the solutions calculated at the step K, it is easy to see that we have:

$$\mathbf{f}[K+1] = e^{A\tau K}\mathbf{f}[K] + \int_{\tau K}^{\tau(K+1)} e^{A[\tau(K+1)-s]} \cdot B\,U(s)ds. \tag{10}$$

Because of the form of $U(t)$, the integral term in (10) can be rewritten as a sum of integrals, each one containing the term correspondent to one supervision instant. We can arbitrarily assume that each supervision comes in the middle of two adjacent updating steps, so that $\forall i\,\exists K : t_i = t_{\tilde{K}} = \left[K + \frac{1}{2}\right]\tau$. With this hypothesis, at the step $K+1$, because of the translation property of δ, each term of the sum of integrals is 0 apart from (potentially) the one corresponding to the index i such that $t_i = t_{\tilde{K}}$, that becomes $(1/b(t_i))\,[f(t_i) - y_i]$. Then, we can generally write the discretized updating formula for the solution as:

$$\mathbf{f}[K+1] = e^{A\tau}\mathbf{f}[K] + e^{A\tau/2} \cdot B\Delta_K \tag{11}$$

where we pose $\Delta_K = 1/b(t_i))\,[f(t_i) - y_i]$ if $\exists i : t_i = t_{\tilde{K}}$ and 0 otherwise. By using this update rule, we do not need to store any gradient term and compute

$G(t - t_i)$, since the impulses are implicitly computed by the evolution of the system. The matrices $e^{A\tau}$ and $e^{A\frac{\tau}{2}}$ can be precomputed, given the model parameters $\{\theta, \alpha_0, \ldots, \alpha_k\}$ or the roots of the characteristic polynomial.

The expression of f obtained in (11) clearly shows the dependences describing its evolution. We assumed a general form of f, neglecting the structure of the function that processes the input, bringing to a collapse of the dimensionality that simplifies the learning process. At each discrete instant of time, the prediction of f depends on the incoming input $x(t)$ only via the term $b(t)$ that modulates the contribution taking into account the variations in the input space. The input-output relation is derived by the provided targets y_i. When the system crosses unsupervised regions, the input-output coherence is kept by the imposed regularization, following the assumption that the input is distributed on a trajectory describing a smooth temporal manifold. The term $b(t)$ helps to keep a connection between the two sources of information. Indeed, when a supervision is given, the impulses received by the system are inversely proportional to the input derivative, and, hence, to the distance between the two points. That is, even in presence of a supervision, the system avoids abrupt changes when the input crosses high variability regions. This constraint should reduce problems due to noise in the input data. With this formulation, f seems to be able to simply track the provided target, performing good predictions in constant regions and possibly inaccurate predictions in the transition phase among regions with different targets. To circumvent this limitation, in the remainder we show how to exploit these regularization abilities by blending different techniques in a more sophisticated learning structure.

3   Temporal Regularization in a Learning Structure

The proposed approach is based on a learning structure that can be developed in an on–line setting and, possibly, suited for an infinite learning process. Such a structure must be capable of storing information on the regions explored in the input trajectory and the corresponding predictions, so as to improve its behavior when the input evolves on the temporal manifold. The learning agent exploits an ϵ-*Radial Network*, storing the information into the elements of a graph $\mathcal{G} = (V_G, E_G)$, where V_G and E_G represent the set of nodes and the set of edges of $\mathcal{G}$, respectively. This structure is aimed at modeling the concept of *neighborhood* of samples, by generating links among the examples which are similar in terms of spatial-features and temporal distribution. We still assume that the input is defined as a time function in a feature space, but, referring to practical cases, we consider that the input is available as a discrete stream of samples (frames for video, sampling for audio tracks and so on). This scenario fits the proposed formulation when both f and $\mathcal{G}$ are updated at a discrete time sampling step τ. When an input point $x_K = x(t_K) = x(\tau K)$ becomes available, the ϵ-*Radial Network* updating algorithm decides to add a new node into V_G or to associate the input to an existing one. Then, the function f is updated by taking in account the spatial and temporal information for the given node, together with eventual external supervisions. The computed value of the function is saved into the node and updated in future visits. When a supervision is provided, its value is stored into the correspondent node and its value is updated only if another supervision is provided for the same node (averaging the values). In general at t_K the node $v_K \in V_G$ to be assigned to the input x_K is determined in a way similar to the one proposed in [8] as:

$$
v_K = \begin{cases} v_{K-1}, & \text{if } \|x_K - x[v_{K-1}]\| \leq \epsilon_{K-1}, \ \epsilon_{K-1} \in \mathbb{R} \\ v^* = \underset{v_j \in R_G}{\arg\min} \ \|x_K - x[v_j]\|, & \text{if } \exists j : \ \|x_K - x[v_j]\| \leq \epsilon_j, \ \epsilon_j \in \mathbb{R} \quad (12) \\ v_{new} \ s.t. \ x[v_{new}] \leftarrow x_K \ \text{otherwise}, \end{cases}
$$

where $x[v], v \in V_G$ refers to the input feature vector stored in the node v. In this work, the norm used to compute the distance is the Euclidean norm in the input feature space $\mathbb{R}^d$. The parameter ϵ represents the radius of the spheres of the graph. The index assigned to ϵ means that, in general, we could assume a different radius depending on the density of the region around each node. It is worth noticing that the update rule implements a sort of on–line Nearest Neighbor to speed up the computation. Indeed, at each step we do not search V_G for the closest node, but we first try to assign the new input to previous one, provided that they are close enough. We indicate the value of $f(t)$ saved in the node v_K as f_K. The prediction function $f(t)$ evolves following the dynamical system defined in the previous Section. It is calculated on–line as the exact discretized solution of the Cauchy's problem associated with the solution of the Euler–Lagrange equations, obtained from a functional similar to the one in (4). However, to include the spatial and temporal information available in $\mathcal{G}$, we consider a modified cost functional:

$$
\Phi_S(f) = \Phi(f) + (1 - \eta)\, S(f) + \eta\, T(f) \tag{13}
$$

again integrated in $[0, T]$, where $\eta \in [0, 1]$ is used as λ to tune the global contribution of each term. The terms S and T include respectively spatial and temporal constraints encoded by the edges in $\mathcal{G}$. The first one can be expressed as:

$$
S(f) = \sum_{K=1}^{K_T - 1} \frac{\psi(t_K)}{\rho} \sum_{s=1}^{\rho} w_{Kn_s} |f(t_{\tilde{K}-1}) - f_{n_s}|^2 \tag{14}
$$

where $\mathcal{N} = \{v_{n_1}, ..., v_{n_\rho}\}$, $\rho \in \mathbb{N}_{|V_G|}$, represents the set of nodes included in the neighborhood, sorted in descending order according to the correspondent weights w_{Kn_s}, and $K_T = T/\tau$ (arbitrarily assuming that T is a multiple of τ). The w's represent the elements of the *spatial adjacency matrix* $W_G \in \mathbb{R}^{|V_G| \times |V_G|}$, that contains the weighted distances between each pair of nodes in V_G. In the remainder, we will consider the Gaussian distance calculated between the nodes v_i and v_j as:

$$
w_{ij} = e^{-\frac{\|x[v_i] - x[v_j]\|^2}{2\sigma^2}}
$$

where σ is an hyper-parameter that is computed as the mean of the Euclidean distances among the nodes in V_G. $\mathcal{S}(f)$ is related to the classic spatial regularization on graphs [8]. It gives a contribution to the prediction in a node as an average of the predictions on nodes in a neighborhood, weighted by their relative distances. The sum does not include $K = 0$ and $K = K_T$, because the value of $f(0)$ is given, whereas the instant $T = \tau K_T$ at the end of the interval is theoretically outside the optimization horizon. In our implementation the matrix W_G is computed on–the–fly, updating only the weights related to the last added node. The constraint of temporal coherence is introduced by adding the term:

$$\mathcal{T}(f) = \sum_{K=1}^{K_T-1} a_K \, \psi(t_K) \sum_{j=1}^{|V_G|} a_{Kj} |f(t_{\tilde{K}-1}) - f_j|^2. \tag{15}$$

The coefficient a_{Kj} is an element of the *temporal adjacency matrix* $A_G \in \mathbb{R}^{|V_G| \times |V_G|}$, whose entries represent the temporal correlations between nodes. The elements are intialized to zero and, then, the element a_{Kj} is incremented by one every time the input moves from node v_j to node v_K in two adjacent time steps. Each row is then normalized by the value $a_K = \max_j(a_{Kj})/(max_{ij}(a_{ij}) \cdot \sum_{j=1}^{|R_G|} a_{Kj})$. The sum term averages the contributions of the elements in the row K. The element $max_{ij}(a_{ij})$ weights the global temporal correlation over the nodes connected to node v_K $(\max_j(a_{Kj}))$, so as to assign weight 1 to the most significant row. This normalization is also useful to balance the wrong correlation that is created in the transition zones of the data sequence. For instance, when dealing with a classification task, if the input function $x(t)$ crosses a class boundary, a contribution is added into A_G carrying a wrong coherence information in practice. However, we can assume that the temporal coherence makes this kind of events quite rare. Because of their rarity, the proposed normalization should reduce the effect of the contributions in the class transitions. Both the spatial and temporal contributions at time step $K + 1$ are supposed to be placed at $t_{\tilde{K}}$ to add these terms to the eventual supervision under the effect of the Dirac's Delta in the updating formula. The idea is that the supervision instants and the updating of $\mathcal{G}$ happen with the same time-sampling step τ, but they are shifted by $\tau/2$ from the computation of the value f to be stored into $\mathcal{G}$, so as to exploit the translation property of δ and obtain a rule similar to (11). Again the function $\psi(t)$ represents the dissipation effect. We search the solution $f(t)$ among the stationary points of (13) satisfying the following distributional Euler–Lagrange equation:

$$\lambda \, \hat{P}(\varphi \, Pf) + \sum_{i=1}^{N} \psi(t_i) \, [f(t_i) - y_i] \, \delta(t - t_i)$$

$$+(1 - \eta) \sum_{K=1}^{K_T-1} \frac{\psi(t_{\tilde{K}})}{\rho} \sum_{s=1}^{\rho} w_{Kn_s}[f(t_{\tilde{K}}) - f_{n_s}]\delta(t - t_{\tilde{K}}) \tag{16}$$

$$+\eta \sum_{K=1}^{K_T-1} a_K \, \psi(t_{\tilde{K}}) \sum_{j=1}^{|V_G|} a_{Kj}[f(t_{\tilde{K}}) - f_j]\delta(t - t_{\tilde{K}}) = 0.$$

As described the previous Section for (11), the discretized solution of the previous Euler-Lagrange equation can be computed on–line with the update formula:

$$\mathbf{f}[K+1] = e^{\mathbf{A}\tau}\mathbf{f}[K] + e^{\mathbf{A}[\tau/2]} \cdot \mathbf{B}\frac{1}{b(t_{\tilde{K}})}\mathbf{E}[K+1] \tag{17}$$

where

$$\mathbf{E}[K+1] = \Delta_K + \frac{(1-\eta)}{\rho}\sum_{s=1}^{\rho} w_{Kn_s}[f(t_{\tilde{K}-1}) - f_{n_s}] + \eta\, a_K \sum_{j=1}^{|V_G|} a_{Kj}[f(t_{\tilde{K}-1}) - f_j].$$

4   Experiments

The proposed algorithm was evaluated on a video created from the MNIST dataset[2]. We randomly selected 5240 points (equally distributed per class) from the original training data (60000 examples of 28×28 images of handwritten digits). We pre-processed the data by the `mnistHelper`[3] Matlab functions and generated a sequence by applying 60 consecutive small transformations on each sample. In this way we obtained a video of 32400 frames in which each sample appears for 2 s (at 30 frames per second). The sequence of transformations is created by randomly selecting among: rotation (angle in $[-20°, 20°]$ with a step of $4°$), translation (maximum 3 pixels up, down, left or right), scaling (only few pixels bigger or smaller), blurring (Gaussian filter with parameter in $\{0.25, 0.5, 0.75, 1\}$). Since the images are 28×28 pixels and we feed the classifier with the 1-D vector of grey levels (normalized in $[0, 1]$), these small transformations are sufficient to yield a quite different descriptor, as shown in the few consecutive images in Fig. 1. The changes are not so visually evident, but they are relevant at the feature level, as shown by the reported Euclidean distances.

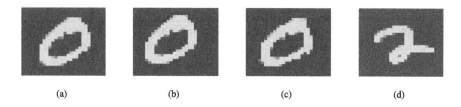

| (a) | (b) | (c) | (d) |

Fig. 1. Examples of consecutive training images generated by random transformations: (a) original image of the digit 0, (b) a left horizontal translation of 1 pixel ($\|x_b - x_a\| = 10.93$), (c) clockwise rotation of $4°$ ($\|x_c - x_b\| = 8.51$), (d) frame of the next class ($\|x_d - x_c\| = 11.33$).

We trained the agent with this sequence in different settings, by varying the number of supervisions. In order to perform the comparisons, the final accuracy is evaluated on the MNIST test set by a simple Nearest Neighbor search

---

[2] http://yann.lecun.com/exdb/mnist/.
[3] http://ufldl.stanford.edu/wiki/index.php/Using_the_MNIST_Dataset.

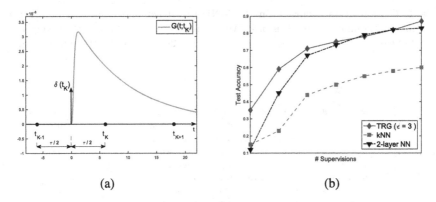

Fig. 2. (a) Impulse Response and sample spacing. (b) Test accuracy (only the setting $\epsilon = 3$ is shown).

on the nodes of the trained graph, using the stored value of f as output. We report the comparison with the plain Nearest Neighbor and a 2-layer Artificial Neural Network[4]. We set $k = 10$ (which we found to achieve the best test performance in $\{1, 5, 10, 25, 50, 100, 1000\}$) in the first one and 300 hidden units in the second one (as reported in [9] for the same architecture). For the proposed algorithm we set $\theta = 10$ (in practice the absolute value of θ is related to the sampling step τ since the dissipation depends on their product) and a fourth order system with roots $r_1 = -0.1$, $r_2 = -6$, $r_3 = -6.5$, $r_4 = -7.4$, which give the impulsive response of Fig. 2(a). The idea is that an impulse should propagate the information over few incoming samples before vanishing. The radius of the ϵ-Net spheres is tested in $\{10^{-4}, 1, 3, 5, 8\}$. Smaller values of ϵ generate graphs with more nodes, improving the prediction performance but increasing the computational cost. A good tradeoff was found at $\epsilon = 3$, that generates a graph with about 8000 nodes ($\approx 25\%$ of the available samples). The remaining parameters are chosen in a validation phase on the first 10 % of the sequence. A small regularization parameter λ amplifies the response (improving the performance) but leads to divergence (because of an accumulation of the delay of the impulsive response). A good balance is found at $\lambda = 0.01$. We selected $\sigma = 3$ from $\{0.5, 1, 2, 3\}$, $\tau = 12$ from $\{1, 2, 4, 6, 12, 18\}$, and $\rho = 5$ from $\{1, 5, 10, 25, 50\}$ so as to achieve the best accuracy on the validation part. The value of η was tried in $\{0, 0.5\}$ but there were not significant differences and we report the case $\eta = 0.5$. The results are reported in Table 1 and Fig. 2(b) with respect to the number of provided supervisions. We report also the running time for each experiment (for a correct reading notice that the proposed algorithm exploits also the unsupervised samples). Each test is carried out 3 times with supervisions on different samples (randomly chosen) and the accuracy is averaged[5].

---

[4] We used the available Matlab functions (`fitcknn` and `patternnet`).
[5] MATLAB scripts and a part of the generated video from the MNIST dataset are available at https://github.com/alered87/Graph-Regularization.

Table 1. Classification accuracy on the MNIST test set and training running time decreasing the number of supervisions. We compare our algorithm (TRG), varying ϵ, with a k-Nearest Neighbors (kNN) classifier and a 2-layer Artificial Neural Network (2-NN).

	Supervisions								Nodes generated
	0.1	1	5	10	25	50	100	%	
	30	320	1620	3240	8100	16200	32400	#	
TRG ($\epsilon = 10^{-4}$)	0.30	0.58	0.72	0.75	0.78	0.82	0.87	Accuracy	32400
	5638.3	5860.8	5821.7	5590.6	5799.5	5395.4	5326.6	Time (sec.)	
TRG ($\epsilon = 1$)	0.25	0.59	0.71	0.75	0.79	0.81	0.87		16500
	1018.2	1337.5	1041.0	998.9	1423.6	1463.3	573.3		
TRG ($\epsilon = 3$)	**0.35**	**0.59**	**0.71**	**0.75**	**0.78**	**0.82**	**0.87**		8000
	360.95	306.91	329.42	412.67	431.24	414.35	320.05		
TRG ($\epsilon = 5$)	0.34	0.55	0.67	0.73	0.77	0.79	0.85		3500
	258.05	255.24	254.93	226.35	254.59	258.48	228.81		
TRG ($\epsilon = 8$)	0.27	0.50	0.61	0.66	0.67	0.69	0.70		300
	13.3	11.31	11.68	10.87	10.59	10.77	10.95		
kNN ($k = 10$)	0.15	0.23	0.44	0.50	0.55	0.58	0.60		–
	0.44	1.26	4.81	13.64	34.70	62.43	98.11		
2-NN (HU= 300)	0.12	0.45	0.67	0.73	0.79	0.82	0.83		–
	2.14	4.05	17.09	73.98	204.87	373.59	640.01		

5   Conclusions

We presented a learning algorithm that exploits the temporal coherence of data, based on a regularization approach in both spatial and temporal dimensions. The proposed algorithm is designed for data distributed in time, that are characterized by smooth temporal variations. The results obtained on an artificially generated problem based on the MNIST dataset encourage the development of the proposed approach to deal with applications on video streams. The flexibility of the presented agent should allow a straightforward application to more complex tasks, exploiting also other kinds of information available as constraints on data.

Acknowledgements. We wish to thank Fondazione Bruno Kessler, Trento - Italy, (http://www.fbk.eu) research center that partially funded this research. We also thank Marco Gori, Duccio Papini and Paolo Nistri for helpful discussions.

References

1. Poggio, T., Girosi, F.: A theory of networks for approximation and learning. Technical report, MIT (1989)
2. Betti, A., Gori, M.: The principle of least cognitive action. Theor. Comput. Sci. **633**, 83–99 (2016). Biologically Inspired Processes in Neural Computation
3. Herrera, L., Núñez, L., Patiño, A., Rago, H.: A variational principle and the classical and quantum mechanics of the damped harmonic oscillator. Am. J. Phys. **54**, 273–277 (1986)

4. Gnecco, G., Gori, M., Melacci, S., Sanguineti, M.: Foundations of support constraint machines. Neural Comput. **27**, 388–480 (2015)
5. Schoelkopf, B., Smola, A.: From regularization operators to support vector kernels. In: Kaufmann, M. (ed.) Advances in Neural Information Processing Systems. MIT Press, Cambridge (1998)
6. Gnecco, G., Gori, M., Sanguineti, M.: Learning with boundary conditions. Neural Comput. **25**, 1029–1106 (2013)
7. Frandina, S., Gori, M., Lippi, M., Maggini, M., Melacci, S.: Variational foundations of online backpropagation. In: Mladenov, V., Koprinkova-Hristova, P., Palm, G., Villa, A.E.P., Appollini, B., Kasabov, N. (eds.) ICANN 2013. LNCS, vol. 8131, pp. 82–89. Springer, Heidelberg (2013). doi:10.1007/978-3-642-40728-4_11
8. Frandina, S., Lippi, M., Maggini, M., Melacci, S.: On–line laplacian one–class support vector machines. In: Mladenov, V., Koprinkova-Hristova, P., Palm, G., Villa, A.E.P., Appollini, B., Kasabov, N. (eds.) ICANN 2013. LNCS, vol. 8131, pp. 186–193. Springer, Heidelberg (2013). doi:10.1007/978-3-642-40728-4_24
9. Lecun, Y., Bottou, L., Bengio, Y., Haffner, P.: Gradient-based learning applied to document recognition. Proc. IEEE **86**, 2278–2324 (1998)

Learning and Reasoning with Logic Tensor Networks

Luciano Serafini[1(✉)] and Artur S. d'Avila Garcez[2]

[1] Fondazione Bruno Kessler, Trento, Italy
serafini@fbk.eu
[2] City University London, London, UK
a.garcez@city.ac.uk

Abstract. The paper introduces real logic: a framework that seamlessly integrates logical deductive reasoning with efficient, data-driven relational learning. Real logic is based on full first order language. Terms are interpreted in n-dimensional feature vectors, while predicates are interpreted in fuzzy sets. In real logic it is possible to formally define the following two tasks: (i) learning from data in presence of logical constraints, and (ii) reasoning on formulas exploiting concrete data. We implement real logic in an deep learning architecture, called logic tensor networks, based on Google's TENSORFLOW™ primitives. The paper concludes with experiments on a simple but representative example of knowledge completion.

Keywords: Knowledge representation · Relational learning · Tensor networks · Neural-symbolic computation · Data-driven knowledge completion

1 Introduction

The availability of large repositories of resources that combines multiple modalities, (image, text, audio and sensor data, social networks), has fostered various research and commercial opportunities, underpinned by machine learning methods and techniques [1–4]. In particular, recent work in machine learning has sought to combine logical services, such as knowledge completion, approximate inference, and goal-directed reasoning with data-driven statistical and neural network-based approaches [5]. We argue that there are great possibilities for improving the current state of the art in machine learning and artificial intelligence (AI) through the principled combination of knowledge representation, reasoning and learning. Guha's recent position paper [6] is a case in point, as

---

The first author acknowledges the Mobility Program of FBK, for supporting a long term visit at City University London. He also acknowledges NVIDIA Corporation for supporting this research with the donation of a GPU. We also thank Prof. Marco Gori and his group at the University of Siena for the generous and inspiring discussions on the topic of integrating logical reasoning and statistical machine learning.

© Springer International Publishing AG 2016
G. Adorni et al. (Eds.): AI*IA 2016, LNAI 10037, pp. 334–348, 2016.
DOI: 10.1007/978-3-319-49130-1_25

it advocates a new model theory for real-valued numbers. In this paper, we take inspiration from such recent work in AI, but also less recent work in the area of neural-symbolic integration [7–9] and in semantic attachment and symbol grounding [10] to achieve a vector-based representation which can be shown adequate for integrating machine learning and reasoning in a principled way.

This paper proposes a framework called *logic tensor networks (LTN)* which integrates learning based on tensor networks [5] with reasoning using first-order many-valued logic [11], all implemented in TENSORFLOW^{TM} [12]. This enables, for the first time, a range of knowledge-based tasks using rich (full first-order logic (FOL)) knowledge representation to be combined with data-driven, efficient machine learning based on the manipulation of real-valued vectors. Given data available in the form of real-valued vectors, logical soft and hard constraints and relations which apply to certain subsets of the vectors can be compactly specified in first-order logic. Reasoning about such constraints can help improve learning, and learning from new data can revise such constraints thus modifying reasoning. An adequate vector-based representation of the logic, first proposed in this paper, enables the above integration of learning and inference, as detailed in what follows.

We are interested in providing a computationally adequate approach to implementing learning and reasoning in an integrated way within an idealized agent. This agent has to manage knowledge about an unbounded, possibly infinite, set of objects $O = \{o_1, o_2, \dots\}$. Some of the objects are associated with a set of quantitative attributes, represented by an n-tuple of real values $\mathcal{G}(o_i) \in \mathbb{R}^n$, which we call *grounding*. For example, the constant "john" denoting a person may have a grounding into an n-tuple of real numbers describing the "measures" of John, e.g., his height, weight, and number of friends in some social network. ....

Object tuples can participate in a set of relations $\mathcal{R} = \{R_1, \dots, R_k\}$, with $R_i \subseteq O^{\alpha(R_i)}$, where $\alpha(R_i)$ denotes the arity of relation R_i. We presuppose the existence of a latent (unknown) relation between the above numerical properties, i.e. groundings, and partial relational structure $\mathcal{R}$ on O. Starting from this partial knowledge, an agent is required to: (i) infer new knowledge about the relational structure $\mathcal{R}$ on the objects of O; (ii) predict the numerical properties or the class of the objects in O.

Classes and relations are not usually independent. For example, it may be the case that if an object x is of class C, written as $C(x)$, and it is related to another object y through relation $R(x, y)$ then this other object y should be in the same class $C(y)$. In logic: $\forall x \exists y ((C(x) \land R(x, y)) \rightarrow C(y))$. Whether or not $C(y)$ holds will depends on the application: through reasoning, one may derive $C(y)$ where otherwise there might not have been evidence of $C(y)$ from training examples only; through learning, one may need to revise such a conclusion once training counterexamples become available. The vectorial representation proposed in this paper permits both reasoning and learning as exemplified above and detailed in the next section.

The above forms of reasoning and learning are integrated in a unifying framework, implemented within tensor networks, and exemplified in relational domains combining data and relational knowledge about the objects. It is expected that, through an adequate integration of numerical properties and relational knowledge, differently from the immediate related literature [13–15], the framework introduced in this paper will be capable of combining in an effective way full first-order logical inference on open domains with efficient relational multi-class learning using tensor networks.

The main contribution of this paper is two-fold. It introduces a novel framework for the integration of learning and reasoning which can take advantage of the representational power of full (multi-valued) first-order logic, and it instantiates the framework using tensor networks into an efficient implementation which shows that the proposed vector-based representation of the logic offers an adequate mapping between symbols and their real-world manifestations, which is appropriate for both rich inference and learning from examples.

2   Real Logic

We start from a first order language $\mathcal{L}$, whose signature contains a set $\mathcal{C}$ of constant symbols, a set $\mathcal{F}$ of functional symbols, and a set $\mathcal{P}$ of predicate symbols. The sentences of $\mathcal{L}$ are used to express relational knowledge, e.g. the atomic formula $R(o_1, o_2)$ states that objects o_1 and o_2 are related to each other through the binary relation R; $\forall xy.(R(x, y) \rightarrow R(y, x))$ states that R is a reflexive relation, where x and y are variables; $\exists y.R(o_1, y)$ states that there is an (unknown) object which is related to object o_1 through R. We assume that all sentences of $\mathcal{L}$ are in prenex conjunctive, skolemised normal form [16], e.g. a sentence $\forall x(A(x) \rightarrow \exists y R(x, y))$ is transformed into an equivalent clause $\neg A(x) \lor R(x, f(x))$, where f is a new (Skolem) function symbol.

As for the semantics of $\mathcal{L}$, we deviate from the standard abstract semantics of FOL, and we propose a *concrete* semantics with sentences interpreted as tuples of real numbers. To emphasise the fact that $\mathcal{L}$ is interpreted in a "real" world we use the term *(semantic) grounding*, denoted by $\mathcal{G}$, instead of the more standard *interpretation*[1].

- $\mathcal{G}$ associates an n-tuple of real numbers $\mathcal{G}(t)$ to any closed term t of $\mathcal{L}$; intuitively $\mathcal{G}(t)$ is the set of numeric features of the object denoted by t.
- $\mathcal{G}$ associates a real number in the interval $[0, 1]$ to each formula/clause ϕ of $\mathcal{L}$. Intuitively, $\mathcal{G}(\phi)$ represents one's confidence in the truth of ϕ; the higher the value, the higher the confidence.

A grounding is specified only for the elements of the signature of $\mathcal{L}$. The grounding of terms and clauses is defined inductively, as follows.

---

[1] In logic, the term "grounding" indicates the operation of replacing the variables of a term/formula with constants, or terms that do not contains other variables. To avoid confusion, we use the synonym "instantiation" for this sense.

Definition 1. *Let $n > 0$. A grounding $\mathcal{G}$ for a first order language $\mathcal{L}$ is a function from the signature of $\mathcal{L}$ to the real numbers that satisfies the following conditions:*

1. *$\mathcal{G}(c) \in \mathbb{R}^n$ for every constant symbol $c \in \mathcal{C}$;*
2. *$\mathcal{G}(f) \in \mathbb{R}^{n \cdot m} \longrightarrow \mathbb{R}^n$ for every $f \in \mathcal{F}$ and m is the arity of f;*
3. *$\mathcal{G}(P) \in \mathbb{R}^{n \cdot m} \longrightarrow [0,1]$ for every $P \in \mathcal{P}$ and m is the arity of P.*

A grounding $\mathcal{G}$ is inductively extended to all the closed terms and clauses, as follows:

$$\mathcal{G}(f(t_1, \ldots, t_m)) = \mathcal{G}(f)(\mathcal{G}(t_1), \ldots, \mathcal{G}(t_m))$$
$$\mathcal{G}(P(t_1, \ldots, t_m)) = \mathcal{G}(P)(\mathcal{G}(t_1), \ldots, \mathcal{G}(t_m))$$
$$\mathcal{G}(\neg P(t_1, \ldots, t_m)) = 1 - \mathcal{G}(P(t_1, \ldots, t_m))$$
$$\mathcal{G}(\phi_1 \vee \cdots \vee \phi_k) = \mu(\mathcal{G}(\phi_1), \ldots, \mathcal{G}(\phi_k))$$

where μ is an s-norm operator (the co-norm operator associated some t-norm operator). Examples of t-norms are Lukasiewicz, product, and Gödel. Lukasiewicz s-norm is defined as $\mu_{Luk}(x,y) = \min(x+y,1)$; Product s-norm is defined as $\mu_{Pr}(x,y) = x + y - x \cdot y$; Gödel s-norm is defined as $\mu_{max}(x,y) = \max(x,y)$. See [11] for details on fuzzy logics.

Notice that the previous semantics is defined only for clauses that do not contain variables. As happens in FOL the occurrences of variables in clauses are interpreted as universally quantified. For instance the clause $\neg A(x) \vee B(x)$ is interpreted as $\forall x(\neg A(x) \vee B(x))$. While the semantic of connectives is based on fuzzy logic operators, in giving the semantics of quantifiers, we deviate from fuzzy logic, introducing a different semantic that we call aggregation semantics. Intuitively the degree of truth of $\forall x P(x)$ is obtained by applying some aggregation function (e.g., mean) to the degree of truth of $P(t_1), \ldots, P(t_n)$, where $t_1, \ldots, t_n$ are considered to be a rappresentative sample of the domain of the variable x. The semantics of universally quantified formulas are defined w.r.t. an *aggregation operator* and a set of possible interpretations of the variables. More formually:

Definition 2. *Let $\phi(\mathbf{x})$ be a clause of $\mathcal{L}$ that contains a k-tuple $\mathbf{x} = \langle x_1, \ldots, x_k \rangle$ of k distinct variables. Let T be the set of all k-tuples $\mathbf{t} = \langle t_1, \ldots, t_k \rangle$ of closed terms of $\mathcal{L}$. $\mathcal{G}(\phi(\mathbf{x})) = \alpha_{\mathbf{t} \in T} \mathcal{G}(\phi(\mathbf{t}))$, where α is an aggregation operator from $[0,1]^{|T|} \to [0,1]$.*

Examples of aggregation operator that can be adopted in real logic are the minimum, the arithmetic mean, the geometric and the harmonic mean. At this stage we stay general and we simply suppose that α is a generic function from $[0,1]^{|T|}$ to $[0,1]$,

Example 1. Let $O = \{o_1, o_2, o_3, \ldots\}$ be a set of documents on a finite vocabulary $W = \{w_1, \ldots, w_n\}$ of n words. Let $\mathcal{L}$ be the language that contains a constant (id) for every document, the binary function symbol $concat(x,y)$ denoting the document resulting from the concatenation of documents x with y and the predicate $Contains(x,y)$ that means that content of y is included in the document x. Let $\mathcal{L}$ contain also the binary predicate Sim which is supposed to be *true* if document x is deemed to be similar to document y. An example of grounding is the one that associates with each document its bag-of-words vector [17]. As a consequence, a natural grounding of the *concat* function would be the sum of the vectors, and of the Sim predicate, the cosine similarity between the vectors. More formally:

- $\mathcal{G}(o_i) = \langle n_{w_1}^{o_i}, \ldots, n_{w_n}^{o_i} \rangle$, where n_w^d is the number of occurrences of word w in document d;
- if $\mathbf{v}, \mathbf{u} \in \mathbb{R}^n$, $\mathcal{G}(concat)(\mathbf{u}, \mathbf{v}) = \mathbf{u} + \mathbf{v}$;
- if $\mathbf{v}, \mathbf{u} \in \mathbb{R}^n$, $\mathcal{G}(Sim)(\mathbf{u}, \mathbf{v}) = \frac{\mathbf{u} \cdot \mathbf{v}}{||\mathbf{u}|| ||\mathbf{v}||}$.
- if $\mathbf{v}, \mathbf{u} \in \mathbb{R}^n$, $\mathcal{G}(Contains)(\mathbf{u}, \mathbf{v})$ is a value in $[0,1]$ that provides a confidence measure that the content of a document associated with the bag of words $\mathbf{u}$ is included in the document associated with the bag of words $\mathbf{v}$.

Notice that, while in the first three cases the $\mathcal{G}$ is explicitly defined (and can be calculated algorithmically), the grounding of the predicate $Contains$ need to be "learned" from a set of examples.

Let us now see how to compute the grounding of terms and formulas. Let o_1, o_2, o_3 be the following three documents:

$$o_1 = \text{``John studies logic and plays football''}$$
$$o_2 = \text{``Mary plays football and logic games''}$$
$$o_3 = \text{``John and Mary play football and study logic together''}$$

We have that $\mathbf{W} = \{$ *John, Mary, and, football, game, logic, play, study, together* $\}$. The following are examples of the grounding of terms, atomic forulas and clauses.

$$\mathcal{G}(o_1) = \langle 1, 0, 1, 1, 0, 1, 1, 1, 0 \rangle$$
$$\mathcal{G}(o_2) = \langle 0, 1, 1, 1, 1, 1, 1, 0, 0 \rangle$$
$$\mathcal{G}(o_3) = \langle 1, 1, 2, 1, 0, 1, 1, 1, 1 \rangle$$
$$\mathcal{G}(concat(o_1, o_2)) = \mathcal{G}(o_1) + \mathcal{G}(o_2) = \langle 1, 1, 2, 2, 1, 2, 2, 1, 0 \rangle$$
$$\mathcal{G}(Sim(concat(o_1, o_2)), o_3) = \frac{\mathcal{G}(concat(o_1, o_2)) \cdot \mathcal{G}(o_3)}{||\mathcal{G}(concat(o_1, o_2))|| \cdot ||\mathcal{G}(o_3)||} \approx \frac{13}{14.83} \approx 0.88$$
$$\mathcal{G}(Sim(o_1, o_3) \vee Sim(o_2, o_3)) = \mu_{max}(\mathcal{G}(Sim(o_1, o_3)), \mathcal{G}(Sim(o_2, o_3))) \approx 0.86$$

$\mathcal{G}(Contains)$ cannot be defined directly. It should be learned from a set of examples, as the following:

- $\mathcal{G}(Contains(o_1, o_2))$ should be equal or close to 0
- $\mathcal{G}(Contains(o_1, o_3))$ should be equal or close to 1
- $\mathcal{G}(Contains(concat(o_i, o_j), o_i))$ for any o_i, o_j should be equal or close to 1
- as well as $\mathcal{G}(Contains(o_i, o_i))$.

The above desiderata (aka constraints) can be expressed by the following first order formulas $\neg Contains(o_1, o_2)$, $Contains(o_1, o_3)$, $\forall xy\, Contains$ $(concat(x, y), x)$, and $\forall x\, Contains(x, x)$.

3   Learning as Approximate Satisfiability

We start by defining grounded theory and their satisfiability.

Definition 3 (Satisfiability of clauses). *Let ϕ be a clause in $\mathcal{L}$, $\mathcal{G}$ a grounding, and $v \leq w \in [0, 1]$. We say that $\mathcal{G}$ satisfies ϕ in the confidence interval $[v, w]$, written $\mathcal{G} \models_v^w \phi$, if $\mathcal{G}(\phi) \in [v, w]$.*

A partial grounding, denoted by $\hat{\mathcal{G}}$, is a grounding that is defined on a subset of the signature of $\mathcal{L}$. A grounding $\mathcal{G}$ is said to be an extension of a partial grounding $\hat{\mathcal{G}}$ if $\mathcal{G}$ coincides with $\hat{\mathcal{G}}$ on the symbols where $\hat{\mathcal{G}}$ is defined.

Definition 4 (Grounded Theory). *A grounded theory (GT) is a pair $\langle \mathcal{K}, \hat{\mathcal{G}} \rangle$ where $\mathcal{K}$ is a set of pairs $\langle [v, w], \phi(\mathbf{x}) \rangle$, where $\phi(\mathbf{x})$ is a clause of $\mathcal{L}$ containing the set $\mathbf{x}$ of free variables, and $[v, w] \subseteq [0, 1]$, and $\hat{\mathcal{G}}$ is a partial grounding.*

Definition 5 (Satisfiability of a Grounded Theory). *A GT $\langle \mathcal{K}, \hat{\mathcal{G}} \rangle$ is satisfiable if there exists a grounding $\mathcal{G}$, which extends $\hat{\mathcal{G}}$ such that for all $\langle [v, w], \phi(\mathbf{x}) \rangle \in \mathcal{K}$, $\mathcal{G} \models_v^w \phi(\mathbf{x})$.*

From the previous definition it follows that checking if a GT $\langle \mathcal{K}, \hat{\mathcal{G}} \rangle$ is satisfiable amounts to searching for an extension of the partial grounding $\hat{\mathcal{G}}$ in the space of *all possible groundings*, such that the aggregation according to α of *all* the instantiations of the clauses in $\mathcal{K}$ are satisfied w.r.t. the specified interval. Clearly a direct approach that follows the definition is unfeasible from a practical point of view. We therefore look for a form of partial satisfiability, which is computationally sustainable. There are three dimensions along which we can approximate, namely: (i) the form of $\mathcal{G}$, (ii) the set of instantiations of the free variables $\mathbf{x}$ for each clause that need to be considered to compute the aggregation function associated to the semantics of the universal quantifier and (iii) the closeness of the truth value of the clause to the indicated interval. Let us look at them separately.

Remark 1 (Limiting to regular groundings). To check satisfiability we search in all the possible functions on real numbers. However, we have said at the beginning of the paper that a grounding captures some latent correlation between the quantitative attribute of an object and its relational properties. For instance, the fact that a document is classified in the field of Artificial Intelligence depends on its bag-of-words grounding. If the language $\mathcal{L}$ contains the unary predicate $AI(x)$, standing for x is a paper about Artificial Intelligence, then the grounding of AI (which is a function from bag-of-words to $[0, 1]$) should present some regularities such that it prefers bag-of-words vectors with high values associated to terms semantically closed to "artificial intelligence". Furthermore, if two bag-of-words are rather similar (e.g., according to the cosine similarity) their classification should agree (i.e., either they are both AI or both not). This imposes some regularity constraints on the grounding associated to the predicate AI, i.e., on $\mathcal{G}(AI)$. More in general, we are interested in groundings that present some regularities, and we search them within a specific class of functions (e.g., based on neural networks, linear/polinomial functions, gaussian mistures, etc.) In this paper we will concentrate on functions based on tensor networks, but the approach could be applied to any family of functions which can be characterised by a set θ of parameters. (For instance if we choose second order polynomial functions the parameters in θ are the coefficients of the polynom).

Remark 2 (Limiting the number of clauses instantiations). In general the set of closed terms of $\mathcal{L}$ might be infinite. It's enough to have one constant c and a functional symbol f to have an infinite set of closed terms $\{c, f(c), f(f(c)), \dots\}$. If we have one single open clause $\langle [v, w], \phi(x) \rangle$ in $\mathcal{K}$, then for every grounding $\mathcal{G}$ which is a completion of $\hat{\mathcal{G}}$, we have to perform an infinite set of tests to check if $\mathcal{G}(\phi(c)) \in [v, w]$, $\mathcal{G}(\phi(f(c))) \in [v, w]$, $\mathcal{G}(\phi(f(f(c)))) \in [v, w], \dots$ This is clearly impossible, and therefore we should implement some form of sampling. The obvious way to sample, is to consider the instantiations of ϕ to terms up to a certain complexity.

Remark 3 (Degree of satisfiability). Following some intuition described in [18], we see the satisfiability problem as an optimisation problem that try to maximize the level of satisfiability of a grounded theory, by minimising some loss function. So instead of looking for a grounding that satisfies all the formulas, we search for a grounding that satisfies *as much a as possible.*

For any $\langle [v, w], \phi(\mathbf{x}) \rangle \in \mathcal{K}$ we want to find a grounding $\mathcal{G}$ that minimizes the *satisfiability error* with respect to an approximation of the domain of $\mathbf{x}$. An error occurs when a grounding $\mathcal{G}$ assigns a value $\mathcal{G}(\phi)$ to a clause ϕ which is outside the interval $[v, w]$ prescribed by $\mathcal{K}$. The measure of this error can be defined as the minimal distance between the points in the interval $[v, w]$ and $\mathcal{G}(\phi)$. For a closed clause ϕ we define

$$\text{Loss}(\mathcal{G}, \langle [v, w], \phi \rangle) = \min_{v \leq x \leq w} |x - \mathcal{G}(\phi)| \tag{1}$$

Notice that if $\mathcal{G}(\phi) \in [v, w]$, $\text{Loss}(\mathcal{G}, \phi) = 0$. For open clauses we have also to identify a subset of terms in which the variables need to be interpreted. We therefore define

$$\text{Loss}(\mathcal{G}, \langle [v, w], \phi \rangle, T_{\phi(\mathbf{x})}) = \min_{v \leq x \leq w} \left| x - \alpha_{\mathbf{t} \in T_{\phi(\mathbf{x})}} \mathcal{G}(\phi(\mathbf{t})) \right| \tag{2}$$

where, $T_{\phi(\mathbf{x})}$ is a finite set of k-tuples of grounded terms of $\mathcal{L}$, with $|\mathbf{x}| = k$, and α is the aggregation function used for interpreting the universally quantified variables. The above gives rise to the following definition of approximate satisfiability w.r.t. a family $\mathbb{G}$ of grounding functions on the language $\mathcal{L}$.

Definition 6 (Best satisfiability problem). *Let* $\langle \mathcal{K}, \hat{\mathcal{G}} \rangle$ *be a grounded theory. Let* $\mathbb{G}$ *be a family of grounding functions. For every* $\langle [v, w], \phi(\mathbf{x}) \rangle \in \mathcal{K}$, *let* $T_{\phi(\mathbf{x})}$ *be a subset of all the k-tuples of ground terms of* $\mathcal{L}$. *We define the* best satisfiability problem *as the problem of finding an extensions* $\mathcal{G}^*$ *of* $\hat{\mathcal{G}}$ *in* $\mathbb{G}$ *that minimizes the satisfiability error on the clauses of* $\mathcal{K}$ *interpreted w.r.t the relative domain*

$$\mathcal{G}^* = \operatorname*{argmin}_{\hat{\mathcal{G}} \subseteq \mathcal{G} \in \mathbb{G}} \sum_{\langle [v,w], \phi(\mathbf{x}) \rangle \in \mathcal{K}} Loss(\mathcal{G}, \langle [v, w], \phi(\mathbf{x}) \rangle, T_{\phi(\mathbf{x})})$$

4 Implementing Real Logic in Tensor Networks

Specific instances of Real Logic can be obtained by selecting the space $\mathbb{G}$ of groundings and the specific s-norm for the interpretation of disjunction and the α aggregation function for the interpretation of universally quantified variables. In this section, we describe a realization of real logic where $\mathbb{G}$ is the space of real tensor transformations of order k (where k is a parameter). In this space, function symbols are interpreted as linear transformations. More precisely, if f is a function symbol of arity m and $\mathbf{v}_1, \ldots, \mathbf{v}_m \in \mathbb{R}^n$ are real vectors corresponding to the grounding of m terms then $\mathcal{G}(f)(\mathbf{v}_1, \ldots, \mathbf{v}_m)$ is of the form:

$$\mathcal{G}(f)(\mathbf{v}_1, \ldots, \mathbf{v}_m) = M_f \mathbf{v}^T + N_f$$

for some $n \times mn$ matrix M_f and n-vector N_f, where $\mathbf{v} = \langle \mathbf{v}_1, \ldots, \mathbf{v}_n \rangle$.

The grounding of an m-ary predicate P, $\mathcal{G}(P)$, is defined as a generalization of the neural tensor network [5] (which has been shown effective at knowledge compilation in the presence of simple logical constraints), as a function from $\mathbb{R}^{mn}$ to $[0, 1]$, as follows:

$$\mathcal{G}(P) = \sigma \left(u_P^T \tanh \left(\mathbf{v}^T W_P^{[1:k]} \mathbf{v} + V_P \mathbf{v} + B_P \right) \right) \tag{3}$$

where $W_P^{[1:k]}$ is a 3-D tensor in $\mathbb{R}^{mn \times mn \times k}$, V_P is a matrix in $\mathbb{R}^{k \times mn}$, and B_P is a vector in $\mathbb{R}^k$, and σ is the sigmoid function. With this encoding, the grounding (i.e. truth-value) of a clause can be determined by a neural network which first computes the grounding of the literals contained in the clause, and then combines them using the specific s-norm. An example of tensor network for $\neg P(x, y) \rightarrow A(y)$ is shown in Fig. 1.

This architecture has been used in combination with vector embedding, i.e., a way to associate a vectorial semantics to concepts and relations, which is, in our terms, a specific example of grounding function. In the above architecture the node shown in gray contains the parameters that need to be learned in order to minimize the loss function, or equivalently, to maximize the satisfaction degree of a grounded theory. In the above tensor network formulation, W_*, V_*, B_* and u_* with $* \in \{P, A\}$ are parameters to be learned by minimizing the loss function or, equivalently, to maximize the satisfiability of the clause $P(x, y) \rightarrow A(y)$. We

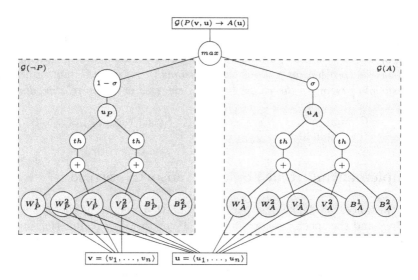

Fig. 1. Tensor net for $P(x,y) \rightarrow A(y)$, with $\mathcal{G}(x) = \mathbf{v}$ and $\mathcal{G}(y) = \mathbf{u}$ and $k = 2$.

have developed a python library that supports the definition, and the parameter learning of the structure described above. Such a library is based on Google's TensorFlow[TM]2.

5   An Example of Knowledge Completion

To test our idea, in this section we use the well-known *friends and smokers*[3] example [19] to illustrate the task of knowledge completion in LTN. There are 14 people divided into two groups $\{a, b, \ldots, h\}$ and $\{i, j, \ldots, n\}$. Within each group of people we have complete knowledge of their smoking habits. In the first group, we have complete knowledge of who has and does not have cancer. In the second group, this is not known for any of the persons. Knowledge about the friendship relation is complete within each group only if symmetry of friendship is assumed. Otherwise, it is incomplete in that it may be known that, e.g., a is a friend of b, but not known whether b is a friend of a. Finally, there is also general knowledge about smoking, friendship and cancer, namely, that smoking causes cancer, friendship is normally a symmetric and anti-reflexive relation, everyone has a friend, and that smoking propagates (either actively or passively) among friends. All this knowledge can be represented by the knowledge-bases shown in Fig. 2.

Notice that, if we adopt classical FOL semantics the knowledge base $\mathcal{K}^{SFC} = \mathcal{K}^{SFC}_{x,y} \cup \mathcal{K}^{SF}_{a,\ldots,h} \cup \mathcal{K}^{SFC}_{i,\ldots,n}$ is inconsistent. Indeed the axiom $\forall x(S(x) \rightarrow C(x))$

---

[2] https://www.tensorflow.org/.

[3] Normally, a probabilistic approach is taken to solve this problem, and one that requires instantiating all clauses to remove variables, essentially turning the problem into a propositional one; `ltn` takes a different approach.

$$\mathcal{K}_{a...h}^{SFC}$$

$$
\begin{array}{c}
S(a),\ S(e),\ S(f),\ S(g), \\
\neg S(b),\ \neg S(c),\ \neg S(d),\ \neg S(g),\ \neg S(h), \\
F(a,b),\ F(a,e),\ F(a,f),\ F(a,g),\ F(b,c), \\
F(c,d),\ F(e,f),\ F(g,h), \\
\neg F(a,c),\ \neg F(a,d),\ \neg F(a,h),\ \neg F(b,d),\ \neg F(b,e), \\
\neg F(b,f),\ \neg F(b,g),\ \neg F(b,h),\ \neg F(c,e),\ \neg F(c,f), \\
\neg F(c,g),\ \neg F(c,h),\ \neg F(d,e),\ \neg F(d,f),\ \neg F(d,g), \\
\neg F(d,h),\ \neg F(e,g),\ \neg F(e,h),\ \neg F(f,g),\ \neg F(f,h), \\
C(a),\ C(e), \\
\neg C(b),\ \neg C(c),\ \neg C(d),\ \neg C(f),\ \neg C(g),\ \neg C(h)
\end{array}
$$

$$\mathcal{K}_{i...n}^{SF}$$

$$
\begin{array}{c}
S(i),\ S(n), \\
\neg S(j),\ \neg S(k), \\
\neg S(l),\ \neg S(m), \\
F(i,j),\ F(i,m), \\
F(k,l),\ F(m,n), \\
\neg F(i,k),\ \neg F(i,l), \\
\neg F(i,n),\ \neg F(j,k), \\
\neg F(j,l),\ \neg F(j,m), \\
\neg F(j,n),\ \neg F(l,n), \\
\neg F(k,m),\ \neg F(l,m)
\end{array}
$$

$$\mathcal{K}_{x,y}^{SFC}$$

$$
\begin{array}{c}
\forall x \neg F(x,x), \quad \forall xy(S(x) \wedge F(x,y) \rightarrow S(y)), \\
\forall xy(F(x,y) \rightarrow F(y,x)), \quad \forall x(S(x) \rightarrow C(x)) \\
\forall x \exists y F(x,y),
\end{array}
$$

Fig. 2. Knowledge-bases for the friends-and-smokers example.

contained in $\mathcal{K}_{x,y}^{SFC}$ contraddicts the facts $S(f)$ and $\neg C(f)$ contained in $\mathcal{K}_{a,...,h}^{SFC}$. If instead we admit the possibility for facts in $\mathcal{K}^{SFC}$ to be true "to a certain extent", i.e., they are associated to an (unknown) degree of truth smaller than 1, then $\mathcal{K}^{SFC}$ is consistent and admits a grounding. Our main tasks are:

(i) find the maximum degree of truth of the facts contained so that $\mathcal{K}^{SFC}$ is consistent;
(ii) find a truth-value for all the ground atomic facts not explicitly mentioned in $\mathcal{K}^{SFC}$;
(iii) find the grounding of each constant symbol $a, ..., n^4$.

To answer (i)-(iii), we use LTN to find a grounding that best approximates the complete KB. To show the role of background knolwedge in the learning-inference process, we run two experiments. In the first ($exp1$), we seek to complete KB consisting of only factual knowledge: $\mathcal{K}_{exp1}^{SFC} = \mathcal{K}_{a...h}^{SFC} \cup \mathcal{K}_{i...n}^{SF}$. In the second ($exp2$), we include background knowledge, that is: $\mathcal{K}_{exp2}^{SFC} = \mathcal{K}_{exp1} \cup \mathcal{K}_{x,y}^{SFC}$.

We configure the network as follows: each constant (person, in this case) has 30 real features. The number of layers k in the tensor network equal to 10, and the regularization parameter[5] $\lambda = 1^{-10}$. We choose the Lukasiewicz t-norm[6], and use the harmonic mean as aggregation operator. An estimation of the optimal grounding is obtained by 5,000 runs of the RMSProp optimisation algorithm [20] available in TENSORFLOW[TM].

The results of the two experiments are reported in Table 1. For readability, we use boldface for truth-values greater than 0.5. The truth-values of the facts

---

[4] Notice how no grounding is provided about the signature of the knowledge-base.
[5] A smoth factor $\lambda \|\mathbf{\Omega}\|_2^2$ is added to the loss to limit the size of parameters.
[6] $\mu(a,b) = \min(1, a + b)$.

listed in a knowledge-base are highlighted with the same background color of the knowledge-base in Fig. 2. The values with white background are the result of the knowledge completion produced by the LTN learning-inference procedure. To evaluate the quality of the results, one has to check whether (i) the truth-values of the facts listed in a KB are indeed close to 1.0, and (ii) the truth-values associated with knowledge completion correspond to expectation. An initial analysis shows that the LTN associated with $\mathcal{K}_{exp1}$ produces the same facts as $\mathcal{K}_{exp1}$ itself. In other words, the LTN fits the data. However, the LTN also learns to infer additional positive and negative facts about the predicates F (friends) and C (cancer) not derivable from $\mathcal{K}_{exp1}$ by pure logical reasoning; for example: $F(c, b)$, $F(g, b)$ and $\neg F(b, a)$. These facts are derived by exploiting similarities between the groundings of the constants generated by the LTN. For instance, $\mathcal{G}(c)$ and $\mathcal{G}(g)$ happen to present a high cosine similarity measure. As a result, facts about the friendship relations of c affect the friendship relations of g and vice-versa, for instance $F(c, b)$ and $F(g, b)$. The level of satisfiability associated with $\mathcal{K}_{exp1} \approx 1$, which indicates that $\mathcal{K}_{exp1}$ is classically satisfiable.

The results of the second experiment show that more facts can be learned with the inclusion of background knowledge. For example, the LTN now predicts that

Table 1. Learning and reasoning in $\mathcal{K}_{exp1}$ and $\mathcal{K}_{exp2}$

			F							
	S	C	a	b	c	d	e	f	g	h
a	**1.00**	**1.00**	0.00	**1.00**	0.00	0.00	**1.00**	**1.00**	**1.00**	0.00
b	0.00	0.00	0.00	0.00	**1.00**	0.00	0.00	0.00	0.00	0.00
c	0.00	0.00	0.00	**0.82**	0.00	**1.00**	0.00	0.00	0.00	0.00
d	0.00	0.00	0.00	0.00	0.06	0.00	0.00	0.00	0.00	0.00
e	**1.00**	**1.00**	0.00	0.33	0.21	0.00	0.00	**1.00**	0.00	0.00
f	**1.00**	0.00	0.00	0.00	0.05	0.00	0.00	0.00	0.00	0.00
g	**1.00**	0.00	0.03	**1.00**	**1.00**	**1.00**	0.11	**1.00**	0.00	**1.00**
h	0.00	0.00	0.00	0.23	0.01	0.14	0.00	0.02	0.00	0.00

			F					
	S	C	i	j	k	l	m	n
i	**1.00**	0.00	0.00	**1.00**	0.00	0.00	**1.00**	0.00
j	0.00	0.00	0.00	0.00	0.00	0.00	0.00	0.00
k	0.00	0.00	0.10	**1.00**	0.00	**1.00**	0.00	0.00
l	0.00	0.00	0.00	0.02	0.00	0.00	0.00	0.00
m	0.00	0.03	**1.00**	**1.00**	0.12	**1.00**	0.00	**1.00**
n	**1.00**	0.01	0.00	0.98	0.00	0.01	0.02	0.00

Learning and reasoning on $\mathcal{K}_{exp1} = \mathcal{K}^{SFC}_{a\ldots h} \cup \mathcal{K}^{SF}_{i\ldots n}$

			F							
	S	C	a	b	c	d	e	f	g	h
a	**0.84**	**0.87**	0.02	**0.95**	0.01	0.03	**0.93**	**0.97**	**0.98**	0.01
b	0.13	0.16	0.45	0.01	**0.97**	0.04	0.02	0.03	0.06	0.03
c	0.13	0.15	0.02	**0.94**	0.11	**0.99**	0.03	0.16	0.15	0.15
d	0.14	0.15	0.01	0.06	**0.88**	0.08	0.01	0.03	0.07	0.02
e	**0.84**	**0.85**	0.32	0.06	0.05	0.03	0.04	**0.97**	0.07	0.06
f	**0.81**	0.19	0.34	0.11	0.08	0.04	0.42	0.08	0.06	0.05
g	**0.82**	0.19	**0.81**	0.26	0.19	0.30	0.06	0.28	0.00	**0.94**
h	0.14	0.17	0.05	0.25	0.26	0.16	0.20	0.14	**0.72**	0.01

			F					
	S	C	i	j	k	l	m	n
i	**0.83**	**0.86**	0.02	**0.91**	0.01	0.03	**0.97**	0.01
j	0.19	0.22	**0.73**	0.03	0.00	0.04	0.02	0.05
k	0.14	0.34	0.17	0.07	0.04	**0.97**	0.04	0.02
l	0.16	0.19	0.11	0.12	0.15	0.06	0.05	0.03
m	0.14	0.17	**0.96**	0.07	0.02	0.11	0.00	**0.92**
n	**0.84**	**0.86**	0.13	0.28	0.01	0.24	**0.69**	0.02

	$a, \ldots, h, i, \ldots, n$	
$\forall x \neg F(x, x)$	**0.98**	
$\forall xy(F(x, y) \rightarrow F(y, x))$	**0.90**	**0.90**
$\forall x(S(x) \rightarrow C(x))$	**0.77**	
$\forall x(S(x) \wedge F(x, y) \rightarrow S(y))$	**0.96**	**0.92**
$\forall x \exists y(F(x, y))$	**1.0**	

Learning and reasoning on $\mathcal{K}_{exp2} = \mathcal{K}^{SFC}_{a\ldots h} \cup \mathcal{K}^{SF}_{i\ldots n} \cup \mathcal{K}^{SFC}$

$C(i)$ and $C(n)$ are true. Similarly, from the symmetry of the friendship relation, the LTN concludes that m is a friend of i, as expected. In fact, all the axioms in the generic background knowledge $\mathcal{K}^{SFC}$ are satisfied with a degree of satisfiability higher than 90 %, apart from the *smoking causes cancer* axiom - which is responsible for the classical inconsistency since in the data f and g smoke and do not have cancer -, which has a degree of satisfiability of 77 %.

6   Related Work

In his recent note, Guha [6], advocates the need for a new model theory for distributed representations (such as those based on embeddings). The note sketches a proposal, where terms and (binary) predicates are all interpreted as points/vectors in an n-dimensional real space. The computation of the truth-value of the atomic formulae $P(t_1, \ldots, t_n)$ is obtained by comparing the projections of the vector associated to each t_i with that associated to P_i. Real logic shares with [6] the idea that terms must be interpreted in a geometric space. It has, however, a different (and more general) interpretation of functions and predicate symbols. Real Logic is more general because the semantics proposed in [6] can be implemented within an ltn with a single layer ($k = 1$), since the operation of projection and comparison necessary to compute the truth-value of $P(t_1, \ldots, t_m)$ can be encoded within an $nm \times nm$ matrix W with the constraint that $\langle \mathcal{G}(t_1), \ldots, \mathcal{G}(t_n) \rangle^T W \langle \mathcal{G}(t_1), \ldots, \mathcal{G}(t_n) \rangle \leq \delta$, which can be encoded easily in ltn.

Real Logic is orthogonal to the approach taken by (Hybrid) Markov Logic Networks (MLNs) and its variations [19,21,22]. In MLNs, the level of truth of a formula is determined by the number of models that satisfy the formula: the more models, the higher the degree of truth. Hybrid MLNs introduce a dependency on the real features associated to constants, which is given, and not learned. In Real Logic, instead, the level of truth of a complex formula is determined by (fuzzy) logical reasoning, and the relations between the features of different objects is learned through error minimization. Another difference is that MLNs work under the *closed world assumption*, while Real Logic is open domain. Much work has been done also on neuro-fuzzy approaches [23]. These are essentially propositional while real logic is first-order.

Bayesian logic (BLOG) [24] is open domain, and in this respect is similar to real logic and LTNs. But, instead of taking an explicit probabilistic approach, LTNs draw from the efficient approach used by tensor networks for knowledge graphs, as already discussed. LTNs can have a probabilistic interpretation but this is not a requirement. Other statistical AI and probabilistic approaches such as lifted inference fall in this same category, including probabilistic variations of inductive logic programming (ILP) [25], which are normally restricted to Horn clauses. Metainterpretive ILP [26], together with BLOG, seems closer to LTNs in what concerns the knowledge representation language, but without exploring tensor networks and its benefits in terms of computational efficiency.

Similarly to [9], LTN is a framework for learning in presence of logical constraints. The two approaches share the idea that logical constraints and training examples should be considered uniformly as supervisions of a learning algorithm. The learning process is an optimization that minimize the distance between the supervisions and the predictions. LTN introduces two novelties: First, in LTN existential quantifiers are not grounded into a finite disjunction, but it is skolemized. This means that we do not assume closed world assumption, and existential quantified formula can be satisfied by "new" individuals. Second, LTN allows to generate and predict data. For instance if a grounded theory contains the formula $\forall x \exists y R(x, y)$, LTN generates a real function (corresponding to the grounding of the skolem function intoduced in the formula) that, for every feature vector $\mathbf{v}$ returns the feature vector $f(\mathbf{v})$, which can be intuitively interpreted as the set of features of the *typical* R-related object of an object with features equal to $\mathbf{v}$.

Finally, related work in the domain of neural-symbolic computing and neural network fibring [8] has sought to combine neural networks with ILP to gain efficiency [27] and other forms of knowledge representation, such as propositional modal logic and logic programming. The above are more tightly-coupled approaches. In contrast, LTNs use a richer FOL language, exploit the benefits of knowledge compilation and tensor networks within a more loosely- coupled approach, and might even offer an adequate representation of equality in logic. Experimental evaluations and comparison with other neural-symbolic approaches are desirable though, including the latest developments in the field, a good snapshot of which can be found in [15].

7   Conclusion and Future Work

We have proposed *Real Logic*: a uniform framework for learning and reasoning. Approximate satisfiability is defined as a learning task with both knowledge and data being mapped onto real-valued vectors. With an inference-as-learning approach, relational knowledge constraints and state-of-the-art data-driven approaches can be integrated. We showed how Real Logic can be implemented in deep tensor networks, which we call Logic Tensor Networks (LTNs), and applied efficiently to knowledge completion and data prediction tasks. As future work, we will make the implementation of LTN available in TENSORFLOW^{TM} and apply it to large-scale experiments and relational learning benchmarks for comparison with statistical relational learning, neural-symbolic computing, and (probabilistic) inductive logic programming approaches.

References

1. Bengio, Y.: Learning deep architectures for AI. Found. Trends Mach. Learn. **2**, 1–127 (2009)
2. Silver, D., Huang, A., Maddison, C.J., Guez, A., Sifre, L., van den Driessche, G., Schrittwieser, J., Antonoglou, I., Panneershelvam, V., Lanctot, M., Dieleman, S., Grewe, D., Nham, J., Kalchbrenner, N., Sutskever, I., Lillicrap, T., Leach, M., Kavukcuoglu, K., Graepel, T., Hassabis, D.: Mastering the game of go with deep neural networks and tree search. Nature **529**, 484–503 (2016)

3. Kephart, J.O., Chess, D.M.: The vision of autonomic computing. Computer **36**, 41–50 (2003)
4. Kiela, D., Bottou, L.: Learning image embeddings using convolutional neural networks for improved multi-modal semantics. In: Proceedings of EMNLP 2014 (2014)
5. Socher, R., Chen, D., Manning, C.D., Ng, A.: Reasoning with neural tensor networks for knowledge base completion. In: Advances in Neural Information Processing Systems, pp. 926–934 (2013)
6. Guha, R.: Towards a model theory for distributed representations. In: 2015 AAAI Spring Symposium Series (2015)
7. Bottou, L.: From machine learning to machine reasoning. Technical report, arXiv.1102.1808 (2011)
8. d'Avila Garcez, A.S., Lamb, L.C., Gabbay, D.M.: Neural-Symbolic Cognitive Reasoning. Cognitive Technologies. Springer, Heidelberg (2009)
9. Diligenti, M., Gori, M., Maggini, M., Rigutini, L.: Bridging logic and kernel machines. Mach. Learn. **86**, 57–88 (2012)
10. Barrett, L., Feldman, J., MacDermed, L.: A (somewhat) new solution to the variable binding problem. Neural Comput. **20**, 2361–2378 (2008)
11. Bergmann, M.: An Introduction to Many-Valued and Fuzzy Logic: Semantics, Algebras, and Derivation Systems. Cambridge University Press, New York (2008)
12. TensorFlow: Large-scale machine learning on heterogeneous systems (2016). tensorflow.org
13. d'Avila Garcez, A.S., Gori, M., Hitzler, P., Lamb, L.C.: Neural-symbolic learning and reasoning (dagstuhl seminar 14381). Dagstuhl Rep. **4**, 50–84 (2014)
14. McCallum, A., Gabrilovich, E., Guha, R., Murphy, K. (eds.): Knowledge representation and reasoning: integrating symbolic and neural approaches. In: AAAI Spring Symposium, Stanford University, CA, USA (2015)
15. Besold, T.R., d'Avila Garcez, A., Marcus, G.F., Miikulainen, R. (eds.): Cognitive computation: integrating neural and symbolic approaches. In: Workshop at NIPS 2015, Montreal, Canada, CEUR-WS 1583, April 2016
16. Huth, M., Ryan, M.: Logic in Computer Science: Modelling and Reasoning About Systems. Cambridge University Press, New York (2004)
17. Blei, D.M., Ng, A.Y., Jordan, M.I.: Latent Dirichlet allocation. J. Mach. Learn. Res. **3**, 993–1022 (2003)
18. Brys, T., Drugan, M.M., Bosman, P.A., De Cock, M., Nowé, A.: Solving satisfiability in fuzzy logics by mixing CMA-ES. In: Proceedings of the 15th Annual Conference on Genetic and Evolutionary Computation, GECCO 2013, pp. 1125–1132. ACM, New York (2013)
19. Richardson, M., Domingos, P.: Markov logic networks. Mach. Learn. **62**, 107–136 (2006)
20. Tieleman, T., Hinton, G.: Lecture 6.5 - RMSProp, COURSERA: Neural networks for machine learning. Technical report (2012)
21. Wang, J., Domingos, P.M.: Hybrid markov logic networks. In: AAAI, pp. 1106–1111 (2008)
22. Nath, A., Domingos, P.M.: Learning relational sum-product networks. In: Proceedings of the Twenty-Ninth AAAI Conference on Artificial Intelligence, January 25–30, 2015, Austin, Texas, USA, pp. 2878–2886 (2015)
23. Kosko, B.: Neural Networks and Fuzzy Systems: A Dynamical Systems Approach to Machine Intelligence. Prentice-Hall Inc., Upper Saddle River (1992)
24. Milch, B., Marthi, B., Russell, S.J., Sontag, D., Ong, D.L., Kolobov, A.: BLOG: probabilistic models with unknown objects. In: IJCAI 2005, pp. 1352–1359 (2005)

25. Raedt, L.D., Kersting, K., Natarajan, S., Poole, D.: Statistical Relational Artificial Intelligence: Logic, Probability, and Computation. Synthesis Lectures on Artificial Intelligence and Machine Learning. Morgan & Claypool, San Rafael (2016)
26. Muggleton, S.H., Lin, D., Tamaddoni-Nezhad, A.: Meta-interpretive learning of higher-order dyadic datalog: predicate invention revisited. Mach. Learn. **100**, 49–73 (2015)
27. França, M.V.M., Zaverucha, G., d'Avila Garcez, A.S.: Fast relational learning using bottom clause propositionalization with artificial neural networks. Mach. Learn. **94**, 81–104 (2014)

Semantic Web and Description Logics

Probabilistic Logical Inference on the Web

Marco Alberti[1], Giuseppe Cota[2], Fabrizio Riguzzi[1], and Riccardo Zese[2(✉)]

[1] Dipartimento di Matematica e Informatica, University of Ferrara,
Via Saragat 1, 44122 Ferrara, Italy
{marco.alberti,fabrizio.riguzzi}@unife.it
[2] Dipartimento di Ingegneria, University of Ferrara,
Via Saragat 1, 44122 Ferrara, Italy
{giuseppe.cota,riccardo.zese}@unife.it

Abstract. `cplint` on SWISH is a web application for probabilistic logic programming. It allows users to perform inference and learning using just a web browser, with the computation performed on the server. In this paper we report on recent advances in the system, namely the inclusion of algorithms for computing conditional probabilities with exact, rejection sampling and Metropolis-Hasting methods. Moreover, the system now allows hybrid programs, i.e., programs where some of the random variables are continuous. To perform inference on such programs likelihood weighting is used that makes it possible to also have evidence on continuous variables. `cplint` on SWISH offers also the possibility of sampling arguments of goals, a kind of inference rarely considered but useful especially when the arguments are continuous variables. Finally, `cplint` on SWISH offers the possibility of graphing the results, for example by drawing the distribution of the sampled continuous arguments of goals.

Keywords: Probabilistic logic programming · Probabilistic logical inference · Hybrid program

1 Introduction

Probabilistic Programming (PP) [1] has recently emerged as a useful tool for building complex probabilistic models and for performing inference and learning on them. Probabilistic Logic Programming (PLP) [2] is PP based on Logic Programming that allows to model domains characterized by complex and uncertain relationships among domain entities.

Many systems have been proposed for reasoning with PLP. Even if they are freely available for download, using them usually requires a complex installation process and a steep learning curve. In order to mitigate these problems, we developed `cplint` on SWISH [3], a web application for reasoning on PLP with just a web browser: the algorithms run on a server and the users can post queries and see the results in their browser. The application is available at http://cplint.lamping.unife.it and Fig. 1 shows its interface.

`cplint` on SWISH uses the reasoning algorithms included in the `cplint` suite, including exact and approximate inference and parameter and structure learning.

© Springer International Publishing AG 2016
G. Adorni et al. (Eds.): AI*IA 2016, LNAI 10037, pp. 351–363, 2016.
DOI: 10.1007/978-3-319-49130-1_26

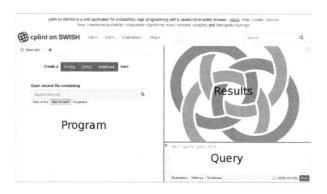

Fig. 1. Interface of `cplint` on SWISH.

In this paper we report on new advances implemented in the system. In particular, we included algorithms for computing conditional probabilities with exact, rejection sampling and Metropolis-Hasting methods. The system now also allows hybrid programs, where some of the random variables are continuous. Likelihood weighting is exploited in order to perform inference on hybrid programs and to collect evidences on continuous variables. When using such variables, the availability of techniques for sampling arguments of goals is extremely useful though it is infrequently considered. `cplint` on SWISH offers these features together with the possibility of graphing the results, for example by drawing the distribution of the sampled continuous arguments of goals. To the best of our knowledge, `cplint` on SWISH is the only online system able to cope with these problems.

A similar system is ProbLog2 [4], which has also an online version[1]. The main difference between `cplint` on SWISH and ProbLog2 is that the former currently also offers structure learning, approximate conditional inference through sampling and handling of continuous variables. Moreover, `cplint` on SWISH is based on SWISH[2], a web framework for Logic Programming using features and packages of SWI-Prolog and its Pengines library and uses a Prolog-only software stack, whereas ProbLog2 relies on several different technologies, including JavaScript, Python 3 and the DSHARP compiler. In particular, it writes intermediate files to disk in order to call external programs such as DSHARP, while we work in main memory only.

All the examples in the paper named as `<name>.pl` can be accessed online at `http://cplint.lamping.unife.it/example/inference/<name>.pl`

2   Syntax and Semantics

The distribution semantics [5] is one of the most used approaches for representing probabilistic information in Logic Programming and it is at the basis of many

---

[1] https://dtai.cs.kuleuven.be/problog/.
[2] http://swish.swi-prolog.org.

languages, such as Independent Choice Logic, PRISM, Logic Programs with Annotated Disjunctions (LPADs) and ProbLog.

We consider first the discrete version of probabilistic logic programming languages. In this version, each atom is a Boolean random variable that can assume values true or false. The facts and rules of the program specify the dependences among the truth value of atoms and the main inference task is to compute the probability that a ground query is true, often conditioned on the truth of another ground goal, the evidence. All the languages following the distribution semantics allow the specification of alternatives either for facts and/or for clauses. We present here the syntax of LPADs because it is the most general [6].

An LPAD is a finite set of annotated disjunctive clauses of the form h_{i1} : $\Pi_{i1}; \ldots; h_{in_i}$: Π_{in_i} :- $b_{i1}, \ldots, b_{im_i}$. where $b_{i1}, \ldots, b_{im_i}$ are literals, $h_{i1}, \ldots h_{in_i}$ are atoms and $\Pi_{i1}, \ldots, \Pi_{in_i}$ are real numbers in the interval $[0, 1]$. This clause can be interpreted as "if $b_{i1}, \ldots, b_{im_i}$ is true, then h_{i1} is true with probability Π_{i1} or ... or h_{in_i} is true with probability Π_{in_i}."

Given an LPAD P, the grounding $ground(P)$ is obtained by replacing variables with terms from the Herbrand universe in all possible ways. If P does not contain function symbols and P is finite, $ground(P)$ is finite as well.

$ground(P)$ is still an LPAD from which, by selecting a head atom for each ground clause, we can obtain a normal logic program, called "world", to which we can assign a probability by multiplying the probabilities of all the head atoms chosen. In this way we get a probability distribution over worlds from which we can define a probability distribution over the truth values of a ground atom: the probability of an atom q being true is the sum of the probabilities of the worlds where q is true, that can be checked because the worlds are normal programs that we assume have a two-valued well-founded model.

This semantics can be given also a sampling interpretation: the probability of a query q is the fraction of worlds, sampled from the distribution over worlds, where q is true. To sample from the distribution over worlds, you simply randomly select a head atom for each clause according to the probabilistic annotations. Note that you don't even need to sample a complete world: if the samples you have taken ensure the truth value of q is determined, you don't need to sample more clauses.

To compute the conditional probability $P(q|e)$ of a query q given evidence e, you can use the definition of conditional probability, $P(q|e) = P(q, e)/P(e)$, and compute first the probability of q, e (the sum of probabilities of worlds where both q and e are true) and the probability of e and then divide the two.

If the program P contains function symbols, a more complex definition of the semantics is necessary, because $ground(P)$ is infinite, a world would be obtained by making an infinite number of choices and so its probability, the product of infinite numbers all smaller than one, would be 0. In this case you have to work with sets of worlds and use Kolmogorov's definition of probability space [7].

Up to now we have considered only discrete random variables and discrete probability distributions. How can we consider continuous random variables and probability density functions, for example real variables following a Gaussian distribution? `cplint` allows the specification of density functions over arguments of atoms in the head of rules. For example, in

`g(X,Y): gaussian(Y,0,1):- object(X).`

X takes terms while Y takes real numbers as values. The clause states that, for each X such that `object(X)` is true, the values of Y such that `g(X,Y)` is true follow a Gaussian distribution with mean 0 and variance 1. You can think of an atom such as `g(a,Y)` as an encoding of a continuous random variable associated to term `g(a)`. A semantics to such programs was given independently in [8] and [9]. In [10] the semantics of these programs, called Hybrid Probabilistic Logic Programs (HPLP), is defined by means of a stochastic generalization STp of the Tp operator that applies to continuous variables the sampling interpretation of the distribution semantics: STp is applied to interpretations that contain ground atoms (as in standard logic programming) and terms of the form $t = v$ where t is a term indicating a continuous random variable and v is a real number. If the body of a clause is true in an interpretation I, $STp(I)$ will contain a sample from the head.

In [9] a probability space for N continuous random variables is defined by considering the Borel σ-algebra over $\mathbb{R}^N$ and a Lebesgue measure on this set as the probability measure. The probability space is lifted to cover the entire program using the least model semantics of constraint logic programs.

If an atom encodes a continuous random variable (such as `g(X,Y)` above), asking the probability that a ground instantiation, such as `g(a,0.3)`, is true is not meaningful, as the probability that a continuous random variables takes a specific value is always 0. In this case you are more interested in computing the distribution of Y of a goal `g(a,Y)`, possibly after having observed some evidence. If the evidence is on an atom defining another continuous random variable, the definition of conditional probability cannot be applied, as the probability of the evidence would be 0 and so the fraction would be undefined. This problem is resolved in [10] by providing a definition using limits.

3   Inference

Computing all the worlds is impractical because their number is exponential in the number of ground probabilistic clauses. Alternative approaches have been considered that can be grouped in exact and approximate ones.

For exact inference from discrete program without function symbols a successful approach finds explanations for the query q [11], where an explanation is a set of clause choices that are sufficient for entailing the query. Once all explanations for the query are found, they are encoded as a Boolean formula in DNF (with a propositional variable per choice and a conjunction per explanation) and the problem is reduced to that of computing the probability that a propositional formula is true. This problem is difficult (#P complexity) but converting the DNF into a language from which the computation of the probability is polynomial (knowledge compilation [12]) yields algorithm able to handle problems of significant size [11,13].

For approximate inference one of the most used approach consists in Monte Carlo sampling, following the sampling interpretation of the semantics given above. Monte Carlo backward reasoning has been implemented in [14,15] and found to give good performance in terms of quality of the solutions and of running time. Monte Carlo sampling is attractive for the simplicity of its implementation and because you can improve the estimate as more time is available. Moreover, Monte Carlo can be used also for programs with function symbols, in which goals may have infinite explanations and exact inference may loop. In sampling, infinite explanations have probability 0, so the computation of each sample eventually terminates.

Monte Carlo inference provides also smart algorithms for computing conditional probabilities: rejection sampling or Metropolis-Hastings Markov Chain Monte Carlo (MCMC). In rejection sampling [16], you first query the evidence and, if the query is successful, query the goal in the same sample, otherwise the sample is discarded. In Metropolis-Hastings MCMC [17], a Markov chain is built by taking an initial sample and by generating successor samples.

The initial sample is built by randomly sampling choices so that the evidence is true. A successor sample is obtained by deleting a fixed number of sampled probabilistic choices. Then the evidence is queried by taking a sample starting with the undeleted choices. If the query succeeds, the goal is queried by taking a sample. The sample is accepted with a probability of $\min\{1, \frac{N_0}{N_1}\}$ where N_0 is the number of choices sampled in the previous sample and N_1 is the number of choices sampled in the current sample. Then the number of successes of the query is increased by 1 if the query succeeded in the last accepted sample. The final probability is given by the number of successes over the number of samples.

When you have evidence on ground atoms that have continuous values as arguments, you can still use Monte Carlo sampling. You cannot use rejection sampling or Metropolis-Hastings, as the probability of the evidence is 0, but you can use likelihood weighting [10] to obtain samples of continuous arguments of a goal. For each sample to be taken, likelihood weighting samples the query and then assigns a weight to the sample on the basis of evidence. The weight is computed by deriving the evidence backward in the same sample of the query starting with a weight of one: each time a choice should be taken or a continuous variable sampled, if the choice/variable has already been taken, the current weight is multiplied by probability of the choice/by the density value of the continuous value.

If likelihood weighting is used to find the posterior density of a continuous random variable, you obtain a set of samples for the variables with each sample associated with a weight that can be interpreted as a relative frequency. The set of samples without the weight, instead, can be interpreted as the prior density of the variable. These two set of samples can be used to plot the density before and after observing the evidence.

You can sample arguments of queries also for discrete goals: in this case you get a discrete distribution over the values of one or more arguments of a goal. If the query predicate is determinate in each world, i.e., given values for input arguments there is a single value for output arguments that make the query true, for each sample you get a single value. Moreover, if clauses sharing an atom in the head are mutually exclusive, i.e., in each world the body of at most one clause is true, then the query defines a probability distribution over output arguments. In this way we can simulate those languages such as PRISM and Stochastic Logic Programs that define probability distributions over arguments rather than probability distributions over truth values of ground atoms.

4   Inference with cplint

cplint on SWISH uses two modules for performing inference, pita for exact inference by knowledge compilation and mcintyre for approximate inference by sampling. In this section we discuss the algorithms and predicates provided by these two modules.

The unconditional probability of an atom can be asked using pita with the predicate

```
prob(:Query:atom,-Probability:float).
```

The conditional probability of an atom query given another atom evidence can be asked with the predicate

```
prob(:Query:atom,:Evidence:atom,-Probability:float).
```

With mcintyre, you can estimate the probability of a goal by taking a given number of sample using the predicate

```
mc_sample(:Query:atom,+Samples:int,-Probability:float).
```

You can ask conditional queries with rejection sampling or with Metropolis-Hastings MCMC too.

```
mc_rejection_sample(:Query:atom,:Evidence:atom,+Samples:int,
    -Successes:int,-Failures:int,-Probability:float).
```

In Metropolis-Hastings MCMC, mcintyre follows the algorithm proposed in [17] (the non adaptive version). The initial sample is built with a backtracking meta-interpreter that starts with the goal and randomizes the order in which clauses are selected during the search so that the initial sample is unbiased. Then the goal is queried using regular mcintyre.

A successor sample is obtained by deleting a number of sampled probabilistic choices given by parameter Lag. Then the evidence is queried using regular mcintyre starting with the undeleted choices. If the query succeeds, the goal is queried using regular mcintyre. The sample is accepted with the probability indicated in Sect. 3. In [17] the lag is always 1 but the proof in [17] that the above acceptance probability yields a valid Metropolis-Hastings algorithm holds also when forgetting more than one sampled choice, so the lag is user defined in cplint.

You can take a given number of sample with Metropolis-Hastings MCMC using

```
mc_mh_sample(:Query:atom,:Evidence:atom,+Samples:int,+Lag:int,
    -Successes:int,-Failures:int,-Probability:float).
```

Moreover, you can sample arguments of queries with rejection sampling and Metropolis-Hastings MCMC using

```
mc_rejection_sample_arg(:Query:atom,:Evidence:atom,+Samples:int,
    ?Arg:var,-Values:list).
mc_mh_sample_arg(:Query:atom,:Evidence:atom,+Samples:int,
    +Lag:int,?Arg:var,-Values:list).
```

Finally, you can compute expectations with

```
mc_expectation(:Query:atom,+N:int,?Arg:var,-Exp:float).
```

that computes the expected value of Arg in Query by sampling. It takes N samples of Query and sums up the value of Arg for each sample. The overall sum is divided by N to give Exp.

To compute conditional expectations, use

```
mc_mh_expectation(:Query:atom,:Evidence:atom,+N:int,
    +Lag:int,?Arg:var,-Exp:float).
```

For visualizing the results of sampling arguments you can use

```
mc_sample_arg_bar(:Query:atom,+Samples:int,?Arg:var,-Chart:dict).
mc_rejection_sample_arg_bar(:Query:atom,:Evidence:atom,
    +Samples:int,?Arg:var,-Chart:dict).
mc_mh_sample_arg_bar(:Query:atom,:Evidence:atom,+Samples:int,
    +Lag:int,?Arg:var,-Chart:dict).
```

that return in Chart a bar chart with a bar for each possible sampled value whose size is the number of samples returning that value.

When you have continuous random variables, you may be interested in sampling arguments of goals representing continuous random variables. In this way you can build a probability density of the sampled argument. When you do not have evidence or you have evidence on atoms not depending on continuous random variables, you can use the above predicates for sampling arguments.

When you have evidence on ground atoms that have continuous values as arguments, you need to use likelihood weighting [10] to obtain samples of continuous arguments of a goal.

For each sample to be taken, likelihood weighting uses a meta-interpreter to find a sample where the goal is true, randomizing the choice of clauses when more than one resolves with the goal in order to obtain an unbiased sample. This meta-interpreter is similar to the one used to generate the first sample in Metropolis-Hastings.

Then a different meta-interpreter is used to evaluate the weight of the sample. This meta-interpreter starts with the evidence as the query and a weight of 1. Each time the meta-interpreter encounters a probabilistic choice over a continuous variable, it first checks whether a value has already been sampled. If so, it computes the probability density of the sampled value and multiplies the weight by it. If the value has not been sampled, it takes a sample and records it, leaving the weight unchanged. In this way, each sample in the result has a weight that is 1 for the prior distribution and that may be different from the posterior distribution, reflecting the influence of evidence.

The predicate

```
mc_lw_sample_arg(:Query:atom,:Evidence:atom,+N:int,?Arg:var,
   -ValList).
```

returns in ValList a list of couples V-W where V is a value of Arg for which Query succeeds and W is the weight computed by likelihood weighting according to Evidence (a conjunction of atoms is allowed here).

You can use the samples to draw the probability density function of the argument. The predicate

```
histogram(+List:list,+NBins:int,-Chart:dict).
```

draws a histogram of the samples in List dividing the domain in NBins bins. List must be a list of couples of the form [V]-W where V is a sampled value and W is its weight. This is the format of the list of samples returned by argument sampling predicates except mc_lw_sample_arg/5 that returns a list of couples V-W. In this case you can use

```
densities(+PriorList:list,+PostList:list,+NBins:int,-Chart:dict).
```

that draws a line chart of the density of two sets of samples, usually prior and post observations. The samples from the prior are in PriorList as couples [V]-W, while the samples from the posterior are in PostList as couples V-W where V is a value and W its weight. The lines are drawn dividing the domain in NBins bins.

5 Examples

5.1 Generative Model

Program arithm.pl encodes a model for generating random functions:

```
eval(X,Y) :- random_fn(X,0,F), Y is F.
op(+):0.5;op(-):0.5.
random_fn(X,L,F) :- comb(L), random_fn(X,l(L),F1),
   random_fn(X,r(L),F2), op(Op), F=..[Op,F1,F2].
random_fn(X,L,F) :- \+ comb(L), base_random_fn(X,L,F).
comb(_):0.3.
base_random_fn(X,L,X) :- identity(L).
base_random_fn(_,L,C) :- \+ identity(L), random_const(L,C).
identity(_):0.5.
random_const(_,C):discrete(C,[0:0.1,1:0.1,2:0.1,3:0.1,4:0.1,
   5:0.1,6:0.1,7:0.1,8:0.1,9:0.1]).
```

A random function is either an operator ('+' or '−') applied to two random functions or a base random function. A base random function is either an identity or a constant drawn uniformly from the integers $0, \ldots, 9$.

You may be interested in the distribution of the output values of the random function with input 2 given that the function outputs 3 for input 1. You can get this distribution with

```
?- mc_mh_sample_arg_bar(eval(2,Y),eval(1,3),1000,1,Y,V).
```

that samples 1000 values for Y in `eval(2,Y)` and returns them in V. A bar graph of the frequencies of the sampled value is shown in Fig. 2a. Since each world of the program is determinate, in each world there is a single value of Y in `eval(2,Y)` and the list of sampled values contain a single element.

5.2   Gaussian Mixture Model

Example gaussian_mixture.pl defines a mixture of two Gaussians:

```
heads:0.6;tails:0.4.
g(X): gaussian(X,0, 1).
h(X): gaussian(X,5, 2).
mix(X) :- heads, g(X).
mix(X) :- tails, h(X).
```

The argument X of `mix(X)` follows a model mixing two Gaussian, one with mean 0 and variance 1 with probability 0.6 and one with mean 5 and variance 2 with probability 0.4. The query

```
?- mc_sample_arg(mix(X),10000,X,L0), histogram(L0,40,Chart).
```

draws the density of the random variable X of `mix(X)`, shown in Fig. 2b.

5.3   Bayesian Estimation

Let us consider a problem proposed on the Anglican [18] web site[3]. We are trying to estimate the true value of a Gaussian distributed random variable, given some observed data. The variance is known (its value is 2) and we suppose that the mean has itself a Gaussian distribution with mean 1 and variance 5. We take different measurement (e.g. at different times), indexed with an integer.

This problem can be modeled with (gauss_mean_est.pl)

```
value(I,X) :- mean(M),value(I,M,X).
mean(M): gaussian(M,1.0, 5.0).
value(_,M,X): gaussian(X,M, 2.0).
```

Given that we observe 9 and 8 at indexes 1 and 2, how does the distribution of the random variable (value at index 0) changes with respect to the case of no

---

[3] http://www.robots.ox.ac.uk/~fwood/anglican/examples/viewer/?
worksheet=gaussian-posteriors.

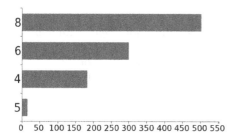

(a) Distribution of sampled values the arithm.pl example.

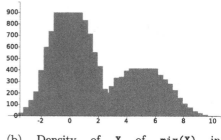

(b) Density of X of mix(X) in gaussian_mixture.pl.

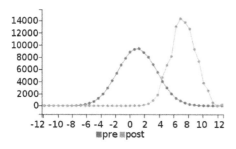

(c) Prior and posterior densities in gauss_mean_est.pl.

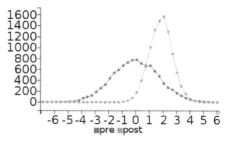

(d) Prior and posterior densities in kalman.pl.

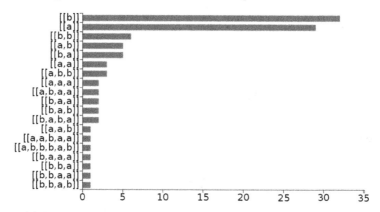

(e) Samples of sentences of the language defined in slp_pcfg.pl.

Fig. 2. Graphs of examples

observations? This example shows that the parameters of the distribution atoms can be taken from the probabilistic atom. The query

```
?- mc_sample_arg(value(0,Y),100000,Y,L0),
   mc_lw_sample_arg(value(0,X),(value(1,9),value(2,8)),1000,X,L),
   densities(L0,L,40,Chart).
```

takes 100000 samples of argument X of value(0,X) before and after observing value(1,9),value(2,8) and draws the prior and posterior densities of the samples using a line chart. Figure 2c shows the resulting graph where the posterior is clearly peaked at around 9.

5.4   Kalman Filter

Example kalman_filter.pl (adapted from [17])

```
kf(N,O, T) :- init(S), kf_part(0, N, S,0,T).
kf_part(I, N, S,[V|RO], T) :- I < N, NextI is I+1, trans(S,I,NextS),
  emit(NextS,I,V), kf_part(NextI, N, NextS,RO, T).
kf_part(N, N, S, [],S).
trans(S,I,NextS) :- {NextS =:= E + S}, trans_err(I,E).
emit(NextS,I,V) :- {V =:= NextS+X}, obs_err(I,X).
init(S):gaussian(S,0,1).
trans_err(_,E):gaussian(E,0,2).
obs_err(_,E):gaussian(E,0,1).
```

encodes a Kalman filter, i.e., a Hidden Markov model with a real value as state and a real value as output. The next state is given by the current state plus Gaussian noise (with mean 0 and variance 2 in this example) and the output is given by the current state plus Gaussian noise (with mean 0 and variance 1 in this example). A Kalman filter can be considered as modeling a random walk of a single continuous state variable with noisy observations.

Continuous random variables are involved in arithmetic expressions (in the predicates trans/3 and emit/3). It is often convenient, as in this case, to use CLP(R) constraints because in this way the same clauses can be used both to sample and to evaluate the weight the sample on the basis of evidence, otherwise different clauses have to be written.

Given that at time 0 the value 2.5 was observed, what is the distribution of the state at time 1 (filtering problem)? Likelihood weighting is used to condition the distribution on evidence on a continuous random variable (evidence with probability 0). CLP(R) constraints allow both sampling and weighting samples with the same program: when sampling, the constraint {V=:=NextS+X} is used to compute V from X and NextS. When weighting, V is known and the constraint is used to compute X from V and NextS, which is then given the density value at X as weight. The above query can be expressed with

```
?- mc_sample_arg(kf(1,_O1,Y),10000,Y,L0),
   mc_lw_sample_arg(kf(1,_O2,T),kf(1,[2.5],_T),10000,T,L),
   densities(L0,L,40,Chart).
```

that returns the graph of Fig. 2d, from which it is evident that the posterior distribution is peaked around 2.5.

5.5   Stochastic Logic Programs

Stochastic logic programs (SLPs) [19] are a probabilistic formalism where each clause is annotated with a probability. The probabilities of all clauses with the same head predicate sum to one and define a mutually exclusive choice on how to continue a proof. Furthermore, repeated choices are independent, i.e., no stochastic memorization is done. SLPs are used most commonly for defining a distribution over the values of arguments of a query. SLPs are a direct generalization of probabilistic context-free grammars and are particularly suitable for representing them. For example, the grammar

```
0.2:S->aS    0.2:S->bS    0.3:S->a    0.3:S->b
```

can be represented with the SLP

```
0.2::s([a|R]):- s(R).   0.2::s([b|R]):- s(R).
0.3::s([a]).            0.3::s([b]).
```

This SLP can be encoded in `cplint` as (slp_pcfg.pl):

```
s_as(N):0.2;s_bs(N):0.2;s_a(N):0.3;s_b(N):0.3.
s([a|R],N0):- s_as(N0), N1 is N0+1, s(R,N1).
s([b|R],N0):- s_bs(N0), N1 is N0+1, s(R,N1).
s([a],N0):- s_a(N0).    s([b],N0):- s_b(N0).
s(L):-s(L,0).
```

where we have added an argument to s/1 for passing a counter to ensure that different calls to s/2 are associated to independent random variables.

By using the argument sampling features of `cplint` we can simulate the behavior of SLPs. For example the query

```
?- mc_sample_arg_bar(s(S),100,S,L).
```

samples 100 sentences from the language and draws the bar chart of Fig. 2e.

6   Conclusions

PLP has now become flexible enough to encode and solve problems usually tackled only with other Probabilistic Programming paradigms, such as functional or imperative PP. `cplint` on SWISH allows the user to exploit these new features of PLP without the need of a complex installation process. In this way we hope to reach out to a wider audience and increase the user base of PLP. In the future we plan to explore more in detail the connection with probabilistic programming using functional/imperative languages and exploit the techniques developed in that field.

A complete online tutorial [20] is available at http://ds.ing.unife.it/~gcota/plptutorial/.

Acknowledgement. This work was supported by the "GNCS-INdAM".

References

1. Pfeffer, A.: Practical Probabilistic Programming. Manning Publications, Cherry Hill (2016)
2. De Raedt, L., Kimmig, A.: Probabilistic (logic) programming concepts. Mach. Learn. **100**(1), 5–47 (2015)
3. Riguzzi, F., Bellodi, E., Lamma, E., Zese, R., Cota, G.: Probabilistic logic programming on the web. Softw. Pract. Exper. **46**, 1381–1396 (2015)
4. Fierens, D., den Broeck, G.V., Renkens, J., Shterionov, D.S., Gutmann, B., Thon, I., Janssens, G., De Raedt, L.: Inference and learning in probabilistic logic programs using weighted Boolean formulas. Theoret. Pract. Log. Prog. **15**(3), 358–401 (2015)
5. Sato, T.: A statistical learning method for logic programs with distribution semantics. In: 12th International Conference on Logic Programming, Tokyo Japan, pp. 715–729. MIT Press, Cambridge (1995)
6. Vennekens, J., Verbaeten, S., Bruynooghe, M.: Logic programs with annotated disjunctions. In: Demoen, B., Lifschitz, V. (eds.) ICLP 2004. LNCS, vol. 3132, pp. 431–445. Springer, Heidelberg (2004). doi:10.1007/978-3-540-27775-0_30
7. Riguzzi, F.: The distribution semantics for normal programs with function symbols. Int. J. Approximate Reasoning **77**, 1–19 (2016)
8. Gutmann, B., Thon, I., Kimmig, A., Bruynooghe, M., Raedt, L.D.: The magic of logical inference in probabilistic programming. Theoret. Pract. Log. Prog. **11**(4–5), 663–680 (2011)
9. Islam, M.A., Ramakrishnan, C., Ramakrishnan, I.: Inference in probabilistic logic programs with continuous random variables. Theoret. Pract. Log. Prog. **12**, 505–523 (2012)
10. Nitti, D., De Laet, T., De Raedt, L.: Probabilistic logic programming for hybrid relational domains. Mach. Learn. **103**(3), 407–449 (2016)
11. De Raedt, L., Kimmig, A., Toivonen, H.: ProbLog: A probabilistic Prolog and its application in link discovery. In: 20th International Joint Conference on Artificial Intelligence, (IJCAI 2005), Hyderabad, India, vol. 7, pp. 2462–2467. AAAI Press, Palo Alto, California USA (2007)
12. Darwiche, A., Marquis, P.: A knowledge compilation map. J. Artif. Intell. Res. **17**, 229–264 (2002)
13. Riguzzi, F., Swift, T.: The PITA system: tabling and answer subsumption for reasoning under uncertainty. Theoret. Pract. Log. Prog. **11**(4–5), 433–449 (2011)
14. Kimmig, A., Demoen, B., De Raedt, L., Costa, V.S., Rocha, R.: On the implementation of the probabilistic logic programming language ProbLog. Theoret. Pract. Log. Prog. **11**(2–3), 235–262 (2011)
15. Riguzzi, F.: MCINTYRE: a Monte Carlo system for probabilistic logic programming. Fundam. Inform. **124**(4), 521–541 (2013)
16. Von Neumann, J.: Various techniques used in connection with random digits. Nat. Bureau Stand. Appl. Math. Ser. **12**, 36–38 (1951)
17. Nampally, A., Ramakrishnan, C.: Adaptive MCMC-based inference in probabilistic logic programs. arXiv:1403.6036 (2014)
18. Wood, F., van de Meent, J.W., Mansinghka, V.: A new approach to probabilistic programming inference. In: Proceedings of the 17th International Conference on Artificial Intelligence and Statistics, pp. 1024–1032 (2014)
19. Muggleton, S.: Learning stochastic logic programs. Electron. Trans. Artif. Intell. **4**(B), 141–153 (2000)
20. Riguzzi, F., Cota, G.: Probabilistic logic programming tutorial. Assoc. Logic Program. Newsl. **29**(1), 1 (2016)

Probabilistic Hybrid Knowledge Bases Under the Distribution Semantics

Marco Alberti[1]([✉]), Evelina Lamma[2], Fabrizio Riguzzi[1], and Riccardo Zese[2]

[1] Dipartimento di Matematica e Informatica, University of Ferrara,
Via Saragat 1, 44122 Ferrara, Italy
{marco.alberti,fabrizio.riguzzi}@unife.it
[2] Dipartimento di Ingegneria, University of Ferrara,
Via Saragat 1, 44122 Ferrara, Italy
{evelina.lamma,riccardo.zese}@unife.it

Abstract. Since Logic Programming (LP) and Description Logics (DLs) are based on different assumptions (the closed and the open world assumption, respectively), combining them provides higher expressiveness in applications that require both assumptions.

Several proposals have been made to combine LP and DLs. An especially successful line of research is the one based on Lifschitz's logic of Minimal Knowledge with Negation as Failure (MKNF). Motik and Rosati introduced Hybrid knowledge bases (KBs), composed of LP rules and DL axioms, gave them an MKNF semantics and studied their complexity. Knorr et al. proposed a well-founded semantics for Hybrid KBs where the LP clause heads are non-disjunctive, which keeps querying polynomial (provided the underlying DL is polynomial) even when the LP portion is non-stratified.

In this paper, we propose Probabilistic Hybrid Knowledge Bases (PHKBs), where the atom in the head of LP clauses and each DL axiom is annotated with a probability value. PHKBs are given a distribution semantics by defining a probability distribution over deterministic Hybrid KBs. The probability of a query being true is the sum of the probabilities of the deterministic KBs that entail the query. Both epistemic and statistical probability can be addressed, thanks to the integration of probabilistic LP and DLs.

Keywords: Hybrid knowledge bases · MKNF · Distribution semantics

1 Introduction

Complex domains are often modeled using Logic Programming (LP) or Description Logics (DLs). Both LP and DLs are based on first order logic so they share many similarities. The main and remarkable difference between them is the domain closure assumption: LP is based on the closed-world assumption (CWA) while DLs use the open-world assumption (OWA). Several authors proposed combinations of LP and DLs. Motik and Rosati [1] define Hybrid Knowledge Bases, composed of a logic program and a DL KB, with semantics based

G. Adorni et al. (Eds.): AI*IA 2016, LNAI 10037, pp. 364–376, 2016.
DOI: 10.1007/978-3-319-49130-1_27

on the logic of Minimal Knowledge with Negation as Failure (MKNF) [2]; as shown by the authors, their proposal exhibits desirable properties (faithfulness, i.e. preservation of the semantics of both formalisms when the other is absent; tightness, i.e. no layering of LP and DL; flexibility, the possibility to view each predicate under both open and closed world assumption; decidability), that each of the other existing approaches to LP and DL integration lacks at least partly.

HKBs can manage domains where different information requires different closure assumptions, such as in legal reasoning; for instance, in [3] it is shown that modeling a real world penal code requires both assumptions.

Many domains, especially those that model the real world, are often characterized by uncertain information. In LP a large number of works have appeared for allowing probabilistic reasoning, leading to the dawn of the Probabilistic Logic Programming (PLP) field. One of the most widespread approaches is the distribution semantics [4]. According to this semantics, a program defines a probability distribution over normal Logic Programs called worlds from which the probability of a query is obtained by marginalization. The distribution semantics underlies many languages such as Logic Programs with Annotated Disjunctions (LPADs), CP-logic and ProbLog. All these languages have the same expressive power as a program in one language can be translated into each of others [5].

Similarly, DLs need as well to manage uncertainty to correctly model real world domains. Some proposals for combining DLs with probability theory exploit graphical models: [6,7] exploit Bayesian networks while [8] combine DLs with Markov networks. Differently, other approaches such as [9–12] exploit Nilsson's probabilistic logic [13] to reason with intervals of probability values.

In [14] we applied the distribution semantics to DLs defining DISPONTE (for "DIstribution Semantics for Probabilistic ONTologiEs"). DISPONTE allows to associate probability values to axioms of a KB. The probability of queries is computed as for PLP languages.

In this paper we propose an approach for defining Probabilistic Hybrid KBs (PHKBs) under the distribution semantics. We combine LPADs with DLs under DISPONTE semantics, both following the distribution semantics. In a PHKB, a query is always either entailed or not entailed in the MKNF sense, so its probability can be computed as for LPADs and DISPONTE.

Halpern [15] distinguishes statistical statements from statements about degrees of belief and presents two examples: "the probability that a randomly chosen bird flies is 0.9" and "the probability that Tweety (a particular bird) flies is 0.9". The first statement captures statistical information about the world while the second captures a degree of belief. The first type of statement is called "Type 1" while the latter "Type 2". The first statement can be read as: given a randomly chosen x in the domain, if x is a bird, the probability that x flies is 0.9, or the conditional probability that x flies given that it is a bird is 0.9. DISPONTE allows to define only "Type 2" statements since the probability associated with an axiom represents the degree of belief in that axiom as a whole. Note that "Type 1" differs from statistical information on the domain such as partial concept overlapping of the form "90 % of birds fly". In fact, this second statement means that for every bird we know with certainty whether it flies or not but, of all birds, only 90 % fly. However, if each individual bird has probability 0.9 of flying, the expected number of birds that fly is 90 % of all birds, so we can model partial overlapping with "Type 2" statements, i.e., with probabilistic statements about individuals. The integration of LP and DLs in PHKBs allows to express a form of statistical probabilistic knowledge that is not permitted by DISPONTE alone: in particular, with the LP part we can express "Type 1" statements.

To understand PHKBs, one needs to first acquire background information about what they combine together. Thus in Sect. 2 we provide a description of these background notions and set the notation for the current work. In Sect. 3 we introduce our probabilistic extension to hybrid MKNF knowledge bases and we define their semantics. Section 4 discusses related work while Sect. 5 concludes the paper and presents remarks on future work.

2   Background and Notation

This section is devoted to introducing the background notions required to understand PHKBs. Hybrid KBs, presented in Sect. 2.3, combine Description Logics (DLs) and Logic Programming (LP) following the Minimal Knowledge with Negation as Failure (MKNF) semantics. Therefore, we start with these blocks to achieve the goal of introducing HKBs. Then, we will discuss probabilistic extensions to LP and DLs and in particular about Logic Programs with Annotated Disjunctions (Sect. 2.4) and DISPONTE (Sect. 2.5), both following the distribution semantics [4].

2.1   Description Logics

DLs are fragments of First Order Logic (FOL) languages used for modeling ontologies [16]. These knowledge representation formalisms differ on which information they permit to define and are usually designed to assure computational properties such as decidability and/or low complexity.

Usually, DLs' syntax is based on concepts, corresponding to sets of individuals, and roles, sets of pairs of individuals of the domain. In order to illustrate DLs, we now describe $\mathcal{SHOIQ}$ [17] as a prototype of expressive DLs.

Let consider a set of *atomic concepts* $\mathbf{C}$, a set of *atomic roles* $\mathbf{R}$ and a set of individuals $\mathbf{I}$. *Concepts* are defined by induction as follows. Each $C \in \mathbf{C}$, $\bot$ and $\top$ are concepts. If C, C_1 and C_2 are concepts and $R \in \mathbf{R}$, then $(C_1 \sqcap C_2)$, $(C_1 \sqcup C_2)$ and $\neg C$ are concepts, as well as $\exists R.C$ and $\forall R.C$. Considering again C, C_1 and C_2, if $S \in \mathbf{R} \cup \mathbf{R}^-$, then $\geq nS.C$ and $\leq nS.C$ for an integer $n \geq 0$ are also concepts. Finally, if $a \in \mathbf{I}$, then $\{a\}$ is a concept called *nominal*.

Roles are either atomic roles $R \in \mathbf{R}$ or their inverse R^- where $R \in \mathbf{R}$. The set of all inverses of roles in $\mathbf{R}$ is denoted by $\mathbf{R}^-$.

A *TBox* $\mathcal{T}$ is a finite set of *concept inclusion axioms* $C \sqsubseteq D$, where C and D are concepts. We use $C \equiv D$ to abbreviate the conjunction of $C \sqsubseteq D$ and $D \sqsubseteq C$. An *RBox* $\mathcal{R}$ consists of a finite set of *transitivity axioms* $Trans(R)$,

Table 1. Translation of $\mathcal{SHOIQ}$ axioms into predicate logic.

Axiom	Translation
$C \sqsubseteq D$	$\forall x.\pi_x(C) \rightarrow \pi_x(D)$
$R \sqsubseteq S$	$\forall x,y.R(x,y) \rightarrow S(x,y)$
$Trans(R)$	$\forall x,y,z.R(x,y) \wedge R(y,z) \rightarrow R(x,z)$
$a : C$	$\pi_a(C)$
$(a,b) : R$	$R(a,b)$
$a = b$	$a = b$
$a \neq b$	$a \neq b$

where $R \in \mathbf{R}$, and *role inclusion axioms* $R \sqsubseteq S$, where $R, S \in \mathbf{R} \cup \mathbf{R}^-$. An *ABox* $\mathcal{A}$ is a finite set of *concept membership axioms* $a : C$, *role membership axioms* $(a,b) : R$, *equality axioms* $a = b$ and *inequality axioms* $a \neq b$, where $C \in \mathbf{C}$, $R \in \mathbf{R}$ and $a, b \in \mathbf{I}$.

A $\mathcal{SHOIQ}$ KB $\mathcal{K} = (\mathcal{T}, \mathcal{R}, \mathcal{A})$ consists of a TBox $\mathcal{T}$, an RBox $\mathcal{R}$ and an ABox $\mathcal{A}$. It is usually assigned a semantics in terms of interpretations $\mathcal{I} = (\Delta^{\mathcal{I}}, \cdot^{\mathcal{I}})$, where $\Delta^{\mathcal{I}}$ is a non-empty *domain* and $\cdot^{\mathcal{I}}$ is the *interpretation function*. This function assigns an element in $\Delta^{\mathcal{I}}$ to each $a \in \mathbf{I}$, a subset of $\Delta^{\mathcal{I}}$ to each $C \in \mathbf{C}$ and a subset of $\Delta^{\mathcal{I}} \times \Delta^{\mathcal{I}}$ to each $R \in \mathbf{R}$.

The *satisfaction* of an axiom E in an interpretation $\mathcal{I} = (\Delta^{\mathcal{I}}, \cdot^{\mathcal{I}})$, denoted by $\mathcal{I} \models E$, is defined as follows: (1) $\mathcal{I} \models C \sqsubseteq D$ iff $C^{\mathcal{I}} \subseteq D^{\mathcal{I}}$, (2) $\mathcal{I} \models a : C$ iff $a^{\mathcal{I}} \in C^{\mathcal{I}}$, (3) $\mathcal{I} \models (a,b) : R$ iff $(a^{\mathcal{I}}, b^{\mathcal{I}}) \in R^{\mathcal{I}}$, (4) $\mathcal{I} \models a = b$ iff $a^{\mathcal{I}} = b^{\mathcal{I}}$, (5) $\mathcal{I} \models a \neq b$ iff $a^{\mathcal{I}} \neq b^{\mathcal{I}}$, (6) $\mathcal{I} \models Trans(R)$ iff $R^{\mathcal{I}}$ is transitive, i.e., $\forall X, Y, Z R(X,Y) \wedge R(Y,Z) \rightarrow R(X,Z)$, (7) $\mathcal{I} \models R \sqsubseteq S$ iff $R^{\mathcal{I}} \subseteq S^{\mathcal{I}}$. $\mathcal{I}$ *satisfies* a set of axioms $\mathcal{E}$, denoted by $\mathcal{I} \models \mathcal{E}$, iff $\mathcal{I} \models E$ for all $E \in \mathcal{E}$. An interpretation $\mathcal{I}$ *satisfies* a knowledge base $\mathcal{K} = (\mathcal{T}, \mathcal{R}, \mathcal{A})$, denoted $\mathcal{I} \models \mathcal{K}$, iff $\mathcal{I}$ satisfies $\mathcal{T}$, $\mathcal{R}$ and $\mathcal{A}$. In this case we say that $\mathcal{I}$ is a *model* of $\mathcal{K}$.

$\mathcal{SHOIQ}$ is decidable iff there are no number restrictions on roles which are transitive or have transitive subroles.

DLs can be directly translated into FOL by using function π_x that maps concept expressions to logical formulas. Table 1 shows the translation of each axiom of $\mathcal{SHOIQ}$ KBs:

Example 1. In a social network scenario, the following axioms

$$\begin{aligned}
\exists rejectedBy.\top &\sqsubseteq spammer \\
\geq_3 reported.\top &\sqsubseteq hasReports \\
\exists reported.trustedUser &\sqsubseteq hasReports \\
hasReports &\sqsubseteq spammer
\end{aligned}$$

model the fact that a user is considered a spammer if she or he has had at least a friend request rejected, or if she or he has reports, which means she or he has been reported at least three times, or at least once by a trusted user.

2.2 MKNF

The logic of Minimal Knowledge with Negation as Failure (MKNF) was introduced in [2]. We briefly recall its syntax and semantics, following [1]. The syntax of MKNF is the syntax of first order logic augmented with modal operators **K** and **not**. In the following, Δ is the Herbrand universe of the signature at hand.

An MKNF *structure* is a triple (I, M, N) where I as a first-order interpretation over Δ and M and N are sets of first order interpretations over Δ. Satisfaction of a closed formula by an MKNF structure is defined as follows (where p is an atom and ψ is a formula):

$$
\begin{aligned}
(I, M, N) &\models p & &\text{iff } p \in I \\
(I, M, N) &\models \neg\psi & &\text{iff } (I, M, N) \not\models \psi \\
(I, M, N) &\models \psi_1 \wedge \psi_2 & &\text{iff } (I, M, N) \models \psi_1 \text{and } (I, M, N) \models \psi_2 \\
(I, M, N) &\models \exists x : \psi & &\text{iff } (I, M, N) \models \psi[\alpha/x] \text{for some } \alpha \in \Delta \\
(I, M, N) &\models \mathbf{K}\,\psi & &\text{iff } (J, M, N) \models \psi \text{ for all } J \in M \\
(I, M, N) &\models \mathbf{not}\,\psi & &\text{iff } (J, M, N) \not\models \psi \text{ for some } J \in N
\end{aligned}
$$

An MKNF *interpretation* is a set M of interpretations over Δ. An interpretation M is an MKNF *model* of a closed formula ψ iff

– $(I, M, M) \models \psi$ for all $I \in M$
– for all $M' \supsetneq M$, for some $I' \in M' (I', M', M) \not\models \psi$

A formula ψ *entails* a formula ϕ, written $\psi \models_{\mathrm{MKNF}} \phi$, iff for all MKNF models M of ψ and for all $I \in M$ $(I, M, M) \models \phi$.

2.3 Hybrid Knowledge Bases

Let $\mathcal{DL}$ be a description logic, i.e., a fragment of first order logic such that

– a transformation π (such as the one in Table 1) exists from each knowledge base $\mathcal{O}$ of $\mathcal{DL}$ to a formula of function-free first order logic with equality;
– it supports *ABoxes* (assertions of the form $C(a_1)$ and of the form $R(a_1, a_2)$, where C is an unary predicate -a class-, P is a binary predicate -a role- and the a_i's are $\mathcal{DL}$ constants);
– satisfiability checking and instance checking are decidable.

A Hybrid Knowledge Base (HKB, [1]) is a pair $\mathcal{K} = \langle \mathcal{O}, \mathcal{P} \rangle$ where $\mathcal{O}$ is a $\mathcal{DL}$ knowledge base and $\mathcal{P}$ is a set of LP rules of the form $h \leftarrow a_1, \ldots, a_n, {\sim}b_1, \ldots, {\sim} b_m$, where a_i and b_i are atoms. Note that [1] allow disjunctions in rule heads, but we do not introduce them because they are not required for our definition of *PHKBs* (see Sect. 3).

[1] define HKB's semantics by transforming it into an MKNF formula. More precisely, the transformation π defined for $\mathcal{DL}$ is extended as follows to support LP rules:

– if C is a rule of the form $h \leftarrow a_1, \ldots, a_n, {\sim}b_1, \ldots, {\sim}b_m$ and $\mathbf{X}$ is the vector of all variables in C, $\pi(C) = \forall\mathbf{X}(\mathbf{K}\,a_1 \wedge \ldots \wedge \mathbf{K}\,a_n \wedge \mathbf{not}\,b_1 \wedge \ldots \wedge \ldots \mathbf{not}\,b_m \rightarrow \mathbf{K}\,h)$

- $\pi(\mathcal{P}) = \bigwedge_{C \in \mathcal{P}} \pi(C)$
- $\pi(\langle \mathcal{O}, \mathcal{P} \rangle) = \mathbf{K}\,\pi\mathcal{O} \wedge \pi(\mathcal{P})$

In Sect. 3, we employ grounding in order to define the semantics of Probabilistic Hybrid Knowledge Bases (PHKBs); for this purpose, it is important for a HKB to have the same MKNF models as its grounding. As shown in [1], a sufficient condition is *DL-safety*. A rule in a HKB $\mathcal{K} = \langle \mathcal{O}, \mathcal{P} \rangle$ is DL-safe if all variables in it occur in a positive atom in its body, whose predicate does not appear in $\mathcal{O}$. A HKB is DL-safe if all the rules in $\mathcal{P}$ are DL-safe. In [1], the authors also argue that non DL-safe knowledge bases can be made DL-safe by a syntactic transformation that does not affect their semantics, so in practice it can be assumed that all knowledge bases are DL-safe.

Example 2. Consider the HKB $\mathcal{K} = \langle \mathcal{O}, \mathcal{P} \rangle$, where $\mathcal{O}$ is the set of axioms defined in Example 1, except for the last one, and $\mathcal{P} =$

$$spammer(X) \quad \leftarrow hasReports(X), {\sim}trustedUser(X).$$
$$rejectedBy(X,Y) \leftarrow invited(X,Y), {\sim}accepted(Y,X).$$

The LP rules define the role *rejectedBy* which occurs in the DL axioms in terms of missing acceptance of a friend request, employing the closed world assumption, and specify that a user with reports is a spammer, but only if she or he is not a trusted user, again using default negation.

The corresponding MKNF formula is

$$
\begin{aligned}
\pi(\mathcal{K}) = \ &\forall X (\mathbf{K}\,hasReports(X) \wedge \mathbf{not}\;trustedUser(X) \rightarrow \mathbf{K}\,spammer(X)) \\
\wedge \ &\forall X \forall Y (\mathbf{K}\,invited(X,Y) \wedge \mathbf{not}\;invited(Y,X) \rightarrow \mathbf{K}\,rejectedBy(X,Y)) \\
\wedge \ &\mathbf{K}\,(\forall X (\exists Y\,rejectedBy(X,Y) \rightarrow spammer(X)) \\
\wedge \ &\forall X (\exists^{\geq 3} Y\,reported(X,Y) \rightarrow hasReports(X)) \\
\wedge \ &\forall X (\exists Y (reported(X,Y) \wedge trustedUser(Y)) \rightarrow hasReports(X)))
\end{aligned}
$$

2.4   Probabilistic Logic Programs

We consider Logic Programs with Annotated Disjunctions (LPADs) and we do not allow for function symbols; for the treatment of function symbols, see [18].

LPADs [19] consist of a finite set of annotated disjunctive clauses C_i of the form $h_{i1} : \Pi_{i1}; \ldots; h_{in_i} : \Pi_{in_i} \leftarrow b_{i1}, \ldots, b_{im_i}$. Here, $b_{i1}, \ldots, b_{im_i}$ are logical literals which form the *body* of C_i, denoted by $body(C_i)$, while $h_{i1}, \ldots h_{in_i}$ are logical atoms and $\{\Pi_{i1}, \ldots, \Pi_{in_i}\}$ are real numbers in the interval $[0,1]$ such that $\sum_{k=1}^{n_i} \Pi_{ik} \leq 1$. Note that if $n_i = 1$ and $\Pi_{i1} = 1$ the clause corresponds to a non-disjunctive clause. Otherwise, if $\sum_{k=1}^{n_i} \Pi_{ik} < 1$, the head of the annotated disjunctive clause implicitly contains an extra atom *null* that does not appear in the body of any clause and whose annotation is $1 - \sum_{k=1}^{n_i} \Pi_{ik}$. The grounding of an LPAD $\mathcal{P}$ is denoted by $ground(\mathcal{P})$.

$ground(P)$ is still an LPAD from which we can obtain a normal logic program by selecting a head atom for each ground clause. In this way we obtain a so-called "world" to which we can assign a probability by multiplying the probabilities of all the head atoms chosen. In this way we get a probability distribution over worlds. We consider only *sound* LPADs, where each possible world w has a total well-founded model, so for a query Q (a ground clause) either $w \models Q$ (Q is true in the well-founded model of w) or $w \not\models Q$, i.e. the well-founded model is two-valued. The probability of a query Q given a world w can be thus defined as $P(Q|w) = 1$ if $w \models Q$ and 0 otherwise. The probability of Q is then:

$$P(Q) = \sum_{w \in W_\mathcal{P}} P(Q, w) = \sum_{w \in W_\mathcal{P}} P(Q|w)P(w) = \sum_{w \in W_\mathcal{P}:w \models Q} P(w) \qquad (1)$$

Example 3. In the same setting of Example 2, the program

$$spammer(john) : 0.3 \leftarrow hasReports(john), \sim trustedUser(john).$$
$$spammer(john) : 0.4 \leftarrow rejectedBy(john, jack).$$
$$hasReports(john). \qquad rejectedBy(john, jack).$$

is ground and has two probabilistic clauses, so there are four worlds. The query $spammer(john)$ is true in three of them, i.e., those containing at least one probabilistic clause, and false in the world that does not contain any probabilistic clause. The probability of Q is $0.3 \times 0.4 + 0.3 \times 0.6 + 0.7 \times 0.4 = 0.58$.

2.5   Probabilistic Description Logics

DISPONTE [14] applies the distribution semantics to probabilistic ontologies [4]. In DISPONTE a *probabilistic knowledge base* $\mathbb{O}$ is a set of certain and probabilistic axioms. *Certain axioms* are regular DL axioms. *Probabilistic axioms* take the form $p :: E$, where p is a real number in $[0, 1]$ and E is a DL axiom. Probability p can be interpreted as an *epistemic probability*, i.e., as the degree of our belief in axiom E. For example, a probabilistic concept membership axiom $p :: a : c$ means that we have degree of belief p in $c(a)$. The statement that Tweety flies with probability 0.9 can be expressed as $0.9 :: tweety : flies$.

The idea of DISPONTE is to associate independent Boolean random variables with the probabilistic axioms. By assigning values to every random variable we obtain a *world*, i.e. the set of probabilistic axioms whose random variable takes on value 1 together with the set of certain axioms. DISPONTE defines a probability distribution over worlds as in PLP.

We can now assign probabilities to queries. Given a world w, the probability of a query Q is defined as $P(Q|w) = 1$ if $w \models Q$ and 0 otherwise. The probability of a query can be defined by marginalizing the joint probability of the query and the worlds, as for PLP.

Example 4. Consider the following KBs, a probabilistic version of the one in Example 1:

$$0.3 :: \exists rejectedBy.\top \sqsubseteq spammer$$
$$\geq_3 reported.\top \sqsubseteq hasReports$$
$$\exists reported.trustedUser \sqsubseteq hasReports$$
$$0.5 :: hasReports \sqsubseteq spammer$$

Given that $john : \exists rejectedBy.\top$, the query $john : spammer$ has probability 0.3. With the additional assumption $john : \exists reported.trustedUser$, the probability of $john : spammer$ is given by the sum of the probabilities of the three worlds where either of the probabilistic axioms occurs, that is $0.3 \times 0.5 + 0.3 \times 0.5 + 0.7 \times 0.5 = 0.65$.

3   Probabilistic Hybrid Knowledge Bases

In this section we formally define Probabilistic Hybrid Knowledge Bases (PHKBs), which combine a probabilistic DL knowledge base with a proabbilistic logic program, and their semantics.

Definition 1. *A PHKB is a pair $\mathbb{K} = \langle \mathbb{O}, \mathbb{P} \rangle$ where $\mathbb{O}$ is a DISPONTE knowledge base and $\mathbb{P}$ is an LPAD.*

The semantics of a PHKB, as usually in the distribution semantics approach, is given by considering possible worlds.

Definition 2. *A possible world of a PHKB $\mathbb{K} = \langle \mathbb{O}, \mathbb{P} \rangle$ is a non probabilistic HKB $w = \langle \mathcal{O}, \mathcal{P} \rangle$ where $\mathcal{O}$ is a possible world of $\mathbb{O}$ and $\mathcal{P}$ is a possible world of $\mathbb{P}$, where $\mathbb{P}$'s grounding is performed over $\mathbb{O}$'s individuals together with $\mathbb{P}$'s Herbrand universe.*

The probability distribution over worlds in the PHKB is induced by the probability distributions of worlds in its DL and LP components.

Definition 3. *If $P(\mathcal{P})$ is the probability of $\mathcal{P}$ and $P(\mathcal{O})$ is the probability of $\mathcal{O}$, the probability of $w = \langle \mathcal{O}, \mathcal{P} \rangle$ is $P(w) = P(\mathcal{P})P(\mathcal{O})$.*

It is easy to see that this is a probability distribution over the worlds of $\mathbb{K}$.

We can assign probabilities to queries as for LPAD and DISPONTE, by defining a joint probability distribution over worlds and queries, where a query's conditional probability given a world is 1 if the world entails the query in the MKNF sense (see Sect. 2.2), and 0 otherwise. The probability of a query is again defined by marginalizing the joint probability of the query and the worlds.

Definition 4. *Given a world w, the probability of a query Q is defined as $P(Q|w) = 1$ if $w \models_{MKNF} \mathbf{K} Q$ and 0 otherwise.*
   The probability of the query is its marginal probability:

$$P(Q) = \sum_w P(w) * P(Q|w) \tag{2}$$

Note that Eq. (2) is the sum of the probabilities of the worlds that entail, in the MKNF sense, the query.

A nice result of PHKBs is that they allow coping with both types of probabilities defined by Halpern in [15]. In fact, "Type 1" probabilistic statements about individuals can be expressed using the LP part of PHKB because each LP clause stands for the set of its ground instantiations and there is a different random variable for each instantiation.

Example 5. The knowledge base $\mathbb{K} = \langle \mathbb{O}, \mathbb{P} \rangle$ with

$$\mathbb{P} = soldier(X) : 0.8 \leftarrow person(X), guard(X). \tag{3}$$
$$person(pete).$$
$$person(al).$$
$$person(john).$$
$$\mathbb{O} = \forall commands.soldier \sqsubseteq commander$$
$$pete : guard$$
$$al : guard$$
$$(john, pete) : commands$$
$$(john, al) : commands$$
$$john : \forall commands.guard$$

expresses that if X is a person and a guard, then X is a soldier with probability 80 %. Moreover, we know that those who command only soldiers are commanders, that *pete* and *al* are guards and that *john* commands only guards. Note that this KB is DL-safe because the predicate *person*/1 does not appear in the DL portion. What is the probability of *commander*(*john*)?

There are four pairs of possible worlds (in each pair the worlds are identical, except for the presence of the clause $soldier(john) \leftarrow person(john), guard(john)$), which share $\mathcal{O}$ and differ for the LP clauses:

1. both instantiations of (3): $soldier(pete) \leftarrow person(pete), guard(pete)$ and $soldier(al) \leftarrow person(al), guard(al)$ (probability $0.8 \times 0.8 = 0.64$);
2. $soldier(pete) \leftarrow person(pete), guard(pete)$ (probability $0.8 \times 0.2 = 0.16$);
3. $soldier(al) \leftarrow person(al), guard(al)$ (probability $0.2 \times 0.8 = 0.16$);
4. no clause(probability $0.2 \times 0.2 = 0.04$).

All MKNF models of world 1 entail $soldier(pete)$ and $soldier(al)$, so $john$: $\forall commands.soldier$, and therefore $commander(john)$; this means that world 1 entails $commander(john)$. $soldier(al)$ is not entailed by world 2, so neither is $commander(john)$. Likewise, worlds 3 and 4 do not entail $commander(john)$. The probability of $commander(john)$ is thus the probability of world 1, i.e., 0.64.

This example shows that PHKB allows "Type 1" statements: the only LP rule (3) models the fact that an individual guard has 80 % of being a soldier. In this way PHKB highly extends the expressive power of DISPONTE.

Please note however that rule (3) is not equivalent to saying that 80 % of guards are soldiers. In this case in fact the query would be false with probability 1, as there exist guards that are not soldiers so *john* does not command only soldiers. However, the expected number of soldiers, given that there are two guards, is $2 \times 0.64 + 1 \times 0.16 + 1 \times 0.16 + 0 \times 0.04 = 1.6$, which is 80 % of 2.

Example 6. Consider the knowledge base $\mathbb{K} = \langle \mathbb{O}, \mathbb{P} \rangle$, a probabilistic version of the knowledge base in Example 2, with

$$\mathbb{P} = spammer(X) : 0.3 \leftarrow hasReports(X), \sim trustedUser(X). \tag{4}$$
$$rejectedBy(X, Y) \leftarrow invited(X, Y), \sim accepted(Y, X).$$
$$\mathbb{O} = 0.4 :: \exists rejectedBy.\top \sqsubseteq spammer \tag{5}$$
$$\geq_3 reported.\top \sqsubseteq hasReports$$
$$\exists reported.trustedUser \sqsubseteq hasReports$$

Here, the probability of a randomly chosen user reported for spamming activity that is not trusted is considered a spammer is 0.3, so the expected number of spammers is 30 % that of untrusted users reported for spamming.

With the assertion $\langle john, mary \rangle : invited$, the KB has eight worlds and the query $Q = john : spammer$ is true in four of them, those containing axiom (5). The probability of the query is $0.7 \times 0.7 \times 0.4 + 0.3 \times 0.7 \times 0.4 + 0.7 \times 0.3 \times 0.4 + 0.3 \times 0.3 \times 0.4 = 0.4$.

Adding the assertions $john :\geq_3 reported.\top$, Q is true in six worlds: those that contain axiom (5) and (4) with X instantiated to $john$. The probability of Q is thus $0.3 \times 0.7 \times 0.6 + 0.7 \times 0.7 \times 0.4 + 0.3 \times 0.3 \times 0.6 + 0.3 \times 0.7 \times 0.4 + 0.7 \times 0.3 \times 0.4 + 0.3 \times 0.3 \times 0.4 = 0.58$.

Finally, if $john : trustedUser$ holds, the rule (4) can no longer be used to entail Q, so the probability is 0.4.

Example 7. We now consider the example in Sect. 4 of [20] and make one of the DL axioms probabilistic thus obtaining $\mathbb{K} = \langle \mathbb{O}, \mathbb{P} \rangle$ with:

$$\mathbb{P} = notMarried(X) \leftarrow person(X), \sim married(X).$$
$$discount(X) \leftarrow spouse(X, Y), person(X), person(Y).$$
$$\mathbb{O} = notMarried \equiv \neg married$$
$$0.4 :: notMarried \sqsubseteq highRisk \tag{6}$$
$$\exists spouse.\top \sqsubseteq married$$
$$john : person$$

In [20], $highRisk(john)$ is entailed by the deterministic knowledge base. In the probabilistic version, there are two possible worlds: one where axiom (6) occurs, whose probability is 0.4, and one where it does not. The query $highRisk(john)$ is true only in the world where axiom (6) occurs, so its probability is 0.4.

4   Related Work

Despite the large number of proposals regarding the combination of probability and LP or DLs, the field of hybrid KBs is still in fast development. Moreover, to the best of our knowledge the definition of probabilistic hybrid KBs represents a completely new field.

FOProbLog [21] is an extension of ProbLog where a program contains a set of *probabilistic facts*, i.e. facts annotated with probabilities, and a set of general clauses which can have positive and negative probabilistic facts in their body. Each fact is assumed to be probabilistically independent. FOProbLog follows the distribution semantics and exploits Binary Decision Diagrams to compute the probability of queries. Differently from our approach, it follows only the open world assumption. Moreover, it permits to associate probability values only to facts. In our case a probability value can be associated also with implications in the LP part and with TBox axioms in the DL part of the hybrid KB. FOProbLog is a reasoner for FOL that is not tailored to DLs, so the algorithm could be suboptimal for them.

In [22] the authors extend the definition of ontologies to allow the management of information under the open world assumption. Axioms not included in the KB are called *open* and are associated with a probability interval between 0 and a fixed threshold. In this way a query can return a probability interval in the case open information is used for answering a query. Moreover, a background knowledge can be specified to restrict the worlds defined by the open KB. Similar to [21], only assertional probabilistic data is allowed.

A combination between DLs and logic programs was presented in [12] in order to integrate ontologies and rules. They use a tightly coupled approach to (probabilistic) disjunctive description logic programs. They define a description logic program as a pair (L, P), where L is a DL KB and P is a disjunctive logic program which contains rules on concepts and roles of L. P may contain probabilistic alternatives in the style of ICL [23]. The integration follows Nilsson's probabilistic logic [13] approach.

Nilsson's logic allows weaker conclusions than the distribution semantics: consider a probabilistic DISPONTE ontology composed of the axioms $0.4 :: a : C$ and $0.5 :: b : C$ and a probabilistic Nilsson KB composed of $C(a) \geq 0.4$ and $C(b) \geq 0.5$. The distribution semantics permits the derivation of $P(a : C \vee b : C) = 0.4 \times (1 - 0.5) + (1 - 0.4) \times 0.5 + 0.4 \times 0.5 = 0.7$. Differently, Nilsson's logic returns the lowest p such that Pr satisfies all the $F \geq p$ in the KB. Since in Nilsson's logic Pr satisfies $F \geq p$ iff $Pr(F) \geq p$, in this example the lowest p such that $Pr(C(a) \vee C(b)) \geq p$ holds is 0.5. This is due to the fact that in the distribution semantics the probabilistic axioms are considered as independent, which allows to make stronger conclusions, without limiting the expressive power as any probabilistic dependence can be modeled.

5   Conclusions and Future Work

In this paper we introduced Probabilistic Hybrid Knowledge Bases, an extension of Hybrid MKNF Knowledge Bases to support probabilistic reasoning, and gave them a distribution semantics.

The next step is to provide a reasoner for PHKBs. The SLG($\mathcal{O}$) procedure [24] for hybrid knowledge bases under the well founded semantics integrates a reasoner for the DL at hand with the SLG procedure in the form of an oracle: the DL reasoner returns the LP atoms that need to be true for the query to succeed. We are following a similar approach for PHKBs, integrating the TRILL probabilistic DL reasoner [25,26] with the PITA algorithm [27] for PLP reasoning. We also plan to develop a web application for using the system, similarly to what we have done for TRILL[1] [28] and PITA[2] [29].

Acknowledgement. This work was supported by the "GNCS-INdAM".

References

1. Motik, B., Rosati, R.: Reconciling description logics and rules. J. ACM **57**(5), 30:1–30:62 (2010)
2. Lifschitz, V.: Nonmonotonic databases and epistemic queries. In: Proceedings of the 12th International Joint Conference on Artificial Intelligence, IJCAI 1991, vol. 1, pp. 381–386. Morgan Kaufmann Publishers Inc., San Francisco (1991)
3. Alberti, M., Gomes, A.S., Gonçalves, R., Leite, J., Slota, M.: Normative systems represented as hybrid knowledge bases. In: Leite, J., Torroni, P., Ågotnes, T., Boella, G., Torre, L. (eds.) CLIMA 2011. LNCS (LNAI), vol. 6814, pp. 330–346. Springer, Heidelberg (2011). doi:10.1007/978-3-642-22359-4_23
4. Sato, T.: A statistical learning method for logic programs with distribution semantics. In: Sterling, L. (ed.) 12th International Conference on Logic Programming, Tokyo, Japan, pp. 715–729. MIT Press, Cambridge (1995)
5. Vennekens, J., Verbaeten, S.: Logic programs with annotated disjunctions. Technical report CW386, KU Leuven (2003)
6. d'Amato, C., Fanizzi, N., Lukasiewicz, T.: Tractable reasoning with Bayesian description logics. In: Greco, S., Lukasiewicz, T. (eds.) SUM 2008. LNCS (LNAI), vol. 5291, pp. 146–159. Springer, Heidelberg (2008). doi:10.1007/978-3-540-87993-0_13
7. Ceylan, İ.İ, Peñaloza, R.: Bayesian description logics. In: Bienvenu, M., Ortiz, M., Rosati, R., Simkus, M. (eds.) Informal Proceedings of the 27th International Workshop on Description Logics. CEUR Workshop Proceedings, 17–20 July 2014, Vienna, Austria, vol. 1193, pp. 447–458 (2014). CEUR-WS.org
8. Gottlob, G., Lukasiewicz, T., Simari, G.I.: Conjunctive query answering in probabilistic datalog+/− ontologies. In: Rudolph, S., Gutierrez, C. (eds.) RR 2011. LNCS, vol. 6902, pp. 77–92. Springer, Heidelberg (2011). doi:10.1007/978-3-642-23580-1_7
9. Giugno, R., Lukasiewicz, T.: P-$\mathcal{SHOQ}$(D): a probabilistic extension of $\mathcal{SHOQ}$(D) for probabilistic ontologies in the semantic web. In: Flesca, S., Greco, S., Ianni, G., Leone, N. (eds.) JELIA 2002. LNCS (LNAI), vol. 2424, pp. 86–97. Springer, Heidelberg (2002). doi:10.1007/3-540-45757-7_8
10. Lukasiewicz, T.: Probabilistic default reasoning with conditional constraints. Ann. Math. Artif. Intell. **34**(1–3), 35–88 (2002)
11. Lukasiewicz, T.: Expressive probabilistic description logics. Artif. Intell. **172**(6–7), 852–883 (2008)
12. Calì, A., Lukasiewicz, T., Predoiu, L., Stuckenschmidt, H.: Tightly coupled probabilistic description logic programs for the semantic web. In: Spaccapietra, S. (ed.) Journal on Data Semantics XII. LNCS, vol. 5480, pp. 95–130. Springer, Heidelberg (2009). doi:10.1007/978-3-642-00685-2_4

---

[1] http://trill.lamping.unife.it.

[2] http://cplint.lamping.unife.it.

13. Nilsson, N.J.: Probabilistic logic. Artif. Intell. **28**(1), 71–87 (1986)
14. Bellodi, E., Lamma, E., Riguzzi, F., Albani, S.: A distribution semantics for probabilistic ontologies. In: 7th International Workshop on Uncertainty Reasoning for the Semantic Web. CEUR Workshop Proceedings, vol. 778, pp. 75–86. Sun SITE Central Europe, Aachen, Germany (2011)
15. Halpern, J.Y.: An analysis of first-order logics of probability. Artif. Intell. **46**(3), 311–350 (1990)
16. Baader, F., Horrocks, I., Sattler, U.: Description logics. In: Handbook of Knowledge Representation, Chap. 3, pp. 135–179. Elsevier, Amsterdam (2008)
17. Horrocks, I., Sattler, U.: A tableau decision procedure for SHOIQ. J. Autom. Reasoning **39**(3), 249–276 (2007)
18. Riguzzi, F.: The distribution semantics for normal programs with function symbols. Int. J. Approximate Reasoning **77**, 1–19 (2016)
19. Vennekens, J., Verbaeten, S., Bruynooghe, M.: Logic programs with annotated disjunctions. In: Demoen, B., Lifschitz, V. (eds.) ICLP 2004. LNCS, vol. 3132, pp. 431–445. Springer, Heidelberg (2004). doi:10.1007/978-3-540-27775-0_30
20. Motik, B., Rosati, R.: A faithful integration of description logics with logic programming. In: Proceedings of the 20th International Joint Conference on Artifical Intelligence, IJCAI 2007, pp. 477–482. Morgan Kaufmann Publishers Inc., San Francisco (2007)
21. Bruynooghe, M., Mantadelis, T., Kimmig, A., Gutmann, B., Vennekens, J., Janssens, G., De Raedt, L.: ProBlog technology for inference in a probabilistic first order logic. In: 19th European Conference on Artificial Intelligence, ECAI 2010. Frontiers in Artificial Intelligence and Applications, 16–20 August 2010, Lisbon, Portugal, vol. 215, pp. 719–724. IOS Press (2010)
22. Ceylan, İ.İ., Darwiche, A., den Broeck, G.V.: Open-world probabilistic databases. In: Baral, C., Delgrande, J.P., Wolter, F., (eds.) Principles of Knowledge Representation and Reasoning: Proceedings of the Fifteenth International Conference, KR 2016, 25–29 April 2016, Cape Town, South Africa, pp. 339–348. AAAI Press (2016)
23. Poole, D.: The independent choice logic for modelling multiple agents under uncertainty. Artif. Intell. **94**, 7–56 (1997)
24. Alferes, J.J., Knorr, M., Swift, T.: Query-driven procedures for hybrid MKNF knowledge bases. ACM Trans. Comput. Logic **14**(2), 16:1–16:43 (2013)
25. Riguzzi, F., Bellodi, E., Lamma, E., Zese, R., Cota, G.: Learning probabilistic description logics. In: Bobillo, F., et al. (eds.) URSW 2011-2013. LNCS, vol. 8816, pp. 63–78. Springer International Publishing, Switzerland (2014). doi:10. 1007/978-3-319-13413-0_4
26. Zese, R., Bellodi, E., Riguzzi, F., Lamma, E.: Tableau reasoners for probabilistic ontologies exploiting logic programming techniques. In: Bellodi, E., Bonfietti, A. (eds.) Doctoral Consortium (DC) Co-located with the 14th Conference of the Italian Association for Artificial Intelligence (AI*IA 2015). CEUR Workshop Proceedings, vol. 1485, pp. 1–6. Sun SITE Central Europe, Aachen, Germany (2015)
27. Riguzzi, F., Swift, T.: The PITA system: tabling and answer subsumption for reasoning under uncertainty. Theoret. Pract. Log. Prog. **11**(4–5), 433–449 (2011)
28. Bellodi, E., Lamma, E., Riguzzi, F., Zese, R., Cota, G.: A web system for reasoning with probabilistic OWL. Software: Practice and Experience (2016, to appear)
29. Riguzzi, F., Bellodi, E., Lamma, E., Zese, R., Cota, G.: Probabilistic logic programming on the web. Softw. Pract. Exper. **46**, 1381–1396 (2015)

Context-Awareness for Multi-sensor Data Fusion in Smart Environments

Alessandra De Paola[(✉)], Pierluca Ferraro, Salvatore Gaglio,
and Giuseppe Lo Re

DICGIM Department, University of Palermo, Palermo, Italy
{alessandra.depaola,pierluca.ferraro,
salvatore.gaglio,giuseppe.lore}@unipa.it

Abstract. Multi-sensor data fusion is extensively used to merge data collected by heterogeneous sensors deployed in smart environments. However, data coming from sensors are often noisy and inaccurate, and thus probabilistic techniques, such as Dynamic Bayesian Networks, are often adopted to explicitly model the noise and uncertainty of data.

This work proposes to improve the accuracy of probabilistic inference systems by including context information, and proves the suitability of such an approach in the application scenario of user activity recognition in a smart home environment. However, the selection of the most convenient set of context information to be considered is not a trivial task. To this end, we carried out an extensive experimental evaluation which shows that choosing the right combination of context information is fundamental to maximize the inference accuracy.

Keywords: Multi-sensor data fusion · Dynamic Bayesian Networks · Context awareness

1 Motivations and Related Work

Nowadays, users expect end-applications to provide useful context-aware services, by exploiting the increasing number of sensors deployed in smart environments and smartphones [1]. To this end, pervasive computing applications need to accurately infer the current context, by efficiently processing large amounts of raw sensory data.

For this purpose, multi-sensor data fusion is extensively used to combine data collected by heterogeneous sensors [2]. However, since sensor data are often noisy and inaccurate, probabilistic techniques are widely adopted to explicitly model the noise and uncertainty of raw data, as described in [3]. In particular, Dynamic Bayesian Networks (DBNs) [4] take into consideration the past belief of the system, in addition to data coming from sensors, and allow to handle the dynamicity of the observed phenomena. Many works leverage DBNs to perform adaptive data fusion for different applications, such as fire detection [5], target tracking [6], and user presence detection [7,8]. A detailed survey on multi-sensor data fusion can be found in [9].

© Springer International Publishing AG 2016
G. Adorni et al. (Eds.): AI*IA 2016, LNAI 10037, pp. 377–391, 2016.
DOI: 10.1007/978-3-319-49130-1_28

Many systems presented in the literature exploit context information to improve the inference accuracy and reduce the uncertainty of unreliable sensor data [10]. Multi-attribute utility theory is exploited in [11] for modeling and merging context attributes, with the goal of achieving situation awareness. The authors of [12] propose a context aggregation framework that can recognize context information of various scale (i.e., personal, local, city-wide, and global) and combine it hierarchically. Moreover, various frameworks use context information to reduce unnecessary communications among wireless sensors, thus reducing their energy consumption [13,14].

We propose a context-aware multi-sensor data fusion system to infer high-level context information about the world, that includes low-level context information in order to refine the inference process. The output of the inference process can be further exploited by higher level reasoning modules to derive new knowledge, in a multi-layered architecture that aims to provide a symbolic description of the environment.

We demonstrate the effectiveness of the proposed approach in an Ambient Intelligence (AmI) [15] scenario, whose goal is to create smart environments which satisfy users' needs, by exploiting pervasive sensors and actuators that surround users, with a low level of intrusiveness [16]. To meet such requirement, many AmI designers prefer to use low-cost and low-impact devices, possibly already deployed in the environment, rather than developing ad-hoc sensors to specifically monitor the features of interest, and thus the collected data are usually only partially related to observed phenomena [8].

In the field of AmI, a key challenge is recognizing users' activities [17,18]. Various approaches have been proposed in the literature, depending on the kind of activities to classify. For example, to recognize activities of daily living (e.g., sleeping, working, eating), wireless sensors are often unobtrusively deployed in smart environments, so as not to bother users [3]. Conversely, inertial sensors, such as those commonly found in smartphones, are better suited to recognize activities that involve physical movements, e.g., sitting down, walking, and running [19,20].

We focus on user activity recognition in a smart home environment, and exploit context information at different levels. The inference of context information, as a high-level description of the users' activity, is the main goal of the system. Moreover, basic context attributes, such as time-related and location-related information, are used to refine the inference process. Such basic context attributes can be reliably and easily sensed, and thus do not increase the uncertainty of the system.

Unlike other works presented in the literature, we advocate that it is not always convenient to blindly include all available context information in the data fusion process. On the contrary, as we demonstrate in the experimental section, choosing the right combination of context information is fundamental to maximize the inference accuracy. To this end, we propose to exploit only context attributes which are readily available and easy to measure in a reliable way, so as not to increase the uncertainty of the system. Moreover, we prove that choosing the right combination of context information is fundamental to

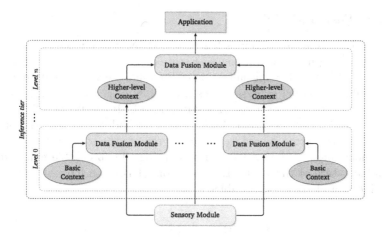

Fig. 1. Multi-layered architecture of the context-aware data fusion system.

maximize the inference accuracy, especially when only few sensors are available. In such cases, our results show that exploiting context information improves the accuracy of the system by almost 13 %.

The remainder of this paper is organized as follows. Section 2 describes the multi-layered architecture of the proposed system, focusing on the context-aware DBN that performs the inference. Section 3 discusses the context information that can be exploited to increase the accuracy of the system. Section 4 presents the experimental setting and the results of our analysis. Finally, Sect. 5 draws our conclusions and proposes directions for future work.

2   Multi-layer Architecture

This paper proposes a novel approach to multi-sensor data fusion for intelligent systems based on the use of pervasive sensors. One of the main features of the system is its capability of dealing with inaccurate and noisy data coming from sensory devices. In particular, the use of probabilistic techniques allows our system to merge information coming from multiple sensors by explicitly modeling the noise and uncertainty of data [9].

Figure 1 shows the multi-layered architecture of the system. At the lowest tier, the *Sensory* module perceives the world through the pervasive sensory infrastructure. The inference tier is composed of multiple levels: at each level, one or more *Data Fusion* modules exploit context attributes coming from lower levels to perform probabilistic inference on the pre-processed sensory data, fusing them to infer new context information which provides a higher level description of the environment. The process of knowledge abstraction continues until the context information requested by the top-level application is inferred.

In this work, we will focus on a single *Data Fusion* module, and on the impact that context information has on its inference accuracy. A more accurate description of the *Data Fusion* module is presented in the following section.

2.1   Data Fusion Module

The proposed data fusion system is based on a DBN, which models the observed phenomena taking into account the past state of the world besides current sensory readings. DBNs are a specialization of Bayesian Networks that guarantee a great flexibility in model expressiveness [21]. They pose no restrictions on conditional probability distributions, differently from Kalman filters [22], and allow for more general topologies than Hidden Markov Models [23]. A DBN is partitioned in temporal slices, where each slice represents the state of the world in a given moment, besides the evidences representing the observable manifestation of the hidden state of the world. Each slice of a DBN can have any number of state variables and evidence variables.

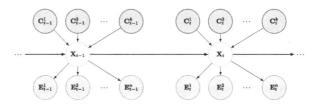

Fig. 2. Structure of the Dynamic Bayesian Network (DBN) used for the inference.

Figure 2 shows the structure of the DBN we designed. Our goal is to infer the state of the world, in the form of a given feature of interest, on the basis of a set of sensory readings, represented by the evidence nodes $E_t = (E_t^1, \ldots, E_t^n)$ at any time slice t. Differently from prior work, we also exploit a set of context information, represented by the evidence nodes $C_t = (C_t^1, \ldots, C_t^k)$ in the time slice t. We will analyze in detail the choice of which context information to use in Sect. 3.

To fully characterize the DBN, it is necessary to define the *sensor model* and the *state transition model* [4]. The probability distribution $P(E_t|X_t)$ expresses how sensory readings are affected by the state variable, and is named *sensor model*. The *state transition model*, defined as $P(X_t|X_{t-1}, C_t)$, represents the probability that the state variable takes a certain value, given its previous value and the current context information.

The *belief* of the system about a specific value of the state variable at time t is defined as:

$$Bel(x_t) = P(x_t|E_{1:t}, C_{1:t}). \tag{1}$$

By following a procedure analogous to that adopted in [24] for deriving the equation of Bayes filters, it is possible to express Eq. (1) in the following recursive formulation:

$$Bel(x_t) = \eta \cdot \prod_{e_t^i} P(e_t^i|x_t) \cdot \sum_{x_{t-1}} P(x_t|x_{t-1}, C_t) \cdot Bel(x_{t-1}), \tag{2}$$

where η is a normalizing constant. Using Eq. 2, we only need to store the last two slices of the DBN, and thus the time and space required for updating the belief do not increase over time. Calculating the belief for a single x_t has a computational complexity of $O(n+m)$, where n is the number of sensor nodes and m is the number of possible values of the state variable. The overall complexity of computing $Bel(x_t)$ for all values of X_t is therefore $O(m^2 + m \cdot n)$.

In order to populate the conditional probability tables of the DBN, several different methods can be adopted, depending on the training set. In a fully labeled dataset, we can compute sample statistics for each node. Otherwise, if the values of one or more of the variables are missing for some of the training records, we can adopt the Expectation Maximization (EM) algorithm or some form of gradient ascent [25].

3   Context-Awareness

The role of context in our system is twofold. First, our main goal is the inference of context information, intended as a high-level description of the surrounding world. In particular, as described in Sect. 1, we are interested in recognizing the activities performed by users in a smart home environment, which in turn will enable higher-level applications to provide to users the most appropriate services.

Low-level context information, such as time and location, can be exploited by our data fusion system to improve the accuracy of reasoning by refining the inference process, as demonstrated by many context-aware data fusion systems proposed by researchers over the years [11,26].

However, using too many context attributes can actually be detrimental to the inference accuracy, as will be demonstrated in Sect. 4.3, and increases the computational burden of the system, especially in the training phase. Thus, it is important to analyze the possible context information and select only the most informative attributes, which may vary depending on the application scenario.

We identify some principles that should drive the selection of context attributes. First of all, context information should be readily available in all situations, regardless of the sensors used. Therefore, we suggest to discard information provided by users manually, together with context attributes which are difficult to sense or that cannot be sensed directly and reliably, thus introducing new elements of uncertainty in the system.

The authors of [27] provide a widely accepted definition of context, which identifies the *primary* categories of contextual information, i.e., identity, activity, location, and time. Identity and activity are high level attributes, while location and time are low level attributes. Thus, according to the principles stated above, we will focus on location-related and time-related context information, analyzing the possible benefits they can provide to the system, and validating our intuitions in the experimental section.

Time-related context information is used by most context-aware systems in literature, since it is very easy to obtain (i.e., it is sufficient to check the current date and time). For activity recognition systems, in particular, time-related context information provides remarkable improvements to the accuracy [28]. First of all, intuitively, activities performed by users may vary a lot in different periods of day: for example, sleeping is the most probable activity during the night, and many users have lunch and dinner at regular time each day. Thus, exploiting this context attribute should improve the accuracy of the system, with almost no drawbacks. However, the number of periods in which a day is divided can influence the performance of the system, as we will demonstrate in Sect. 4.3. Both too coarse-grained periods (e.g., intervals of 12 h) and too fine-grained ones (e.g., intervals of 1 min) do not convey much information; hence, finding the best granularity is very important.

Similarly, activities performed by users might be influenced by the current day of the week and, to a lesser extent, by the month of the year. However, we expect the activities of users to be less correlated to these context attributes, with respect to the period of day. For example, it is possible that users will behave differently during weekends, but it is unlikely that activities will change much among the other days. We defer further considerations regarding the day of the week and month of the year to the experimental section. Other time-related context information, such as the timezone, might be interesting for different scenarios, but are irrelevant for our case study of activity recognition in a smart house.

As regards location-related context information, we focus primarily on the position of users, leaving to future work an analysis on how to exploit the position of objects to improve the awareness of the system about users' surroundings. In the case of a smart home, with no strong assumption on the kind of sensors used, we propose to exploit user location information with a room-level granularity. Regardless of the sensors used, estimating the position of users with this level of detail is required to correctly inferring their activities.

However, a system that relies primarily on location-related context information will encounter difficulties in recognizing certain activities. Intuitively, this can be explained by considering that some activities are performed in well-defined locations (e.g., sleeping in the bedroom), and therefore are well recognized using this kind on information, while other activities are more irregular (e.g., housekeeping, which may be carried out in all rooms of the smart home), and more heterogeneous context information should be exploited to recognize them with higher accuracy.

4 Experimental Analysis

In order to evaluate the possible contribute of different context information to the data fusion process, we test the performance of the proposed system while varying the type and granularity of context information.

4.1 Simulation Setting

We evaluated our system in a simulated smart home, pervaded by several sensor devices, as proposed by [29]. Sensory traces and corresponding user activities were obtained from the Aruba dataset of the CASAS Smart Home Project [3], at Washington State University. This dataset contains annotated data collected in a smart apartment with a single resident, over a period of seven months. Events are generated by 31 motion sensors, 3 door sensors, and 5 temperature sensors, deployed in 8 rooms (5 sensors per room on average).

We partitioned the sequence of sensor events into time windows of 30 s, counting how many times each sensor was activated during each slice. We noticed a low correlation between temperature readings and the activity performed by the user, and thus we decided to discard this information.

The Aruba dataset considers eleven activities of daily living (ADLs), i.e., *Bed to Toilet, Eating, Enter Home, Housekeeping, Leave Home, Meal Preparation, Relax, Resperate*[1], *Sleeping, Wash Dishes*, and *Work*. We added a new activity, named *Outside*, that takes into consideration the periods of time when the user is not at home, i.e., the intervals between *Leave Home* and *Enter Home*.

We also added another activity, named *Other*, which groups all the sensor events that do not match any of the known activities. We think it is essential to detect this activity class accurately in a real world scenario, since nearly 20 % of the sensor events in the dataset considered here belong to the *Other* class. However, considering the heterogeneity of the activities grouped by this class, it is very challenging to recognize it with good accuracy, and many approaches in the literature ignore it altogether, relying on a static list of known activities, as noted in [17].

We used the cross validation method to evaluate the system, dividing the dataset into ten parts. For each test, nine parts were used for learning the CPTs (Conditional Probability Tables) of the DBN, and the tenth was used for the test. This process was then repeated changing the test set ten times and averaging the results.

After the pre-processing phase, the dataset consisted of 633 468 sensor events. Each test of the cross validation used 570 121 sensor events as training cases, and 63 347 sensor events as test cases. All experiments have been performed on a workstation equipped with an Intel® Core™ i5-3470 CPU (4 cores, 3.20 GHz, 4 GB RAM). The training phase required 4 914 ms on average.

4.2 Performance Metrics

We adopted the average accuracy as metric to evaluate the performance of the activity recognition systems, defined as:

$$Acc = \frac{TP + TN}{TP + TN + FP + FN}, \tag{3}$$

where *TP, TN, FP*, and *FN* are, respectively, the true positives, true negatives, false positives and false negatives. However, accuracy alone is not sufficient to evaluate different approaches, since data are skewed towards the most probable activities. In fact, activities such as *Sleeping* and *Relax* account for a large number of time slices, while others like *Resperate* and *Leave Home* or *Enter Home*

---

[1] Resperate is a device used for the treatment of high blood pressure.

are much rarer and shorter. For this reason, we adopted additional metrics to provide a more detailed analysis of the performance of the systems.

To measure the uncertainty of the probabilistic reasoning performed by the systems, we used an index based on the classic definition of Shannon entropy [30]. We also calculated the average cross-entropy error function, which is defined as follows:

$$CE = -\frac{1}{N} \sum_{i=1}^{N} \sum_{j=1}^{M} y_{ij} \log p_{ij}, \tag{4}$$

where N is the number of timesteps, M is the number of activity classes, and y_{ij} and p_{ij} are, respectively, the ground truth and the predicted probability for the jth activity class at time i. The cross-entropy error is an information-theoretic measure of accuracy that incorporates the idea of probabilistic confidence, measuring the cross-entropy between the distribution of true labels and the prediction of the system. This kind of error becomes extremely large (i.e., $+\infty$ in the extreme case) if the system is over-confident about a wrong prediction, and it is thus useful to evaluate the accuracy of the belief with a fine granularity. Finally, we determined the *precision* (positive predictive value), as fidelity measure, and the *recall* (sensitivity), for measuring completeness, which are defined as follows:

$$precision = \frac{TP}{TP + FP}, \qquad recall = \frac{TP}{TP + FN}. \tag{5}$$

Precision and recall, in turn, are used to calculate the *F-score*, defined as the harmonic mean of precision and recall, as follows:

$$F\text{-}score = 2 \cdot \frac{precision \cdot recall}{precision + recall}. \tag{6}$$

4.3   Experimental Results

Time-Related Context Information. The first set of experiments we present is a detailed analysis on the importance of some time-related context attribute, i.e., *period of day*, *day of week*, and *month*.

We will begin by studying the performance of the system when changing the granularity of the *period of day* node. Figure 3a shows the accuracy, uncertainty, F-score and cross-entropy error of a system exploiting the *period of day* node, as a function of the number of periods in which a day is divided, starting from a single period (i.e., a single interval of 24 h) up to a maximum of 48 periods (i.e., 48 intervals of 30 min). We notice an increment of the accuracy and F-score when increasing the granularity up to 6 periods (i.e., intervals of 4 h). Likewise, uncertainty and cross-entropy error are very low using this granularity. However, if we divide the day in more than 6 periods, we observe a steady decrease of the F-score, as well as an increase of uncertainty, whilst accuracy and cross-entropy remain unchanged. Thus, we can conclude that increasing the time granularity is beneficial only up to a point; going further only adds to the noise, resulting in a system that performs worse with no added benefits.

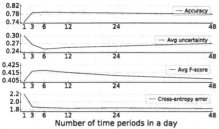

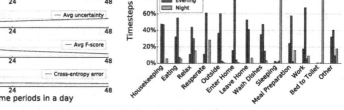

(a) System performance when varying the granularity of the *period of day* node.

(b) Activities' frequency during morning, afternoon, evening, and night.

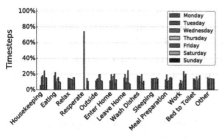

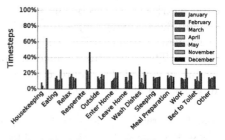

(c) Activities' frequency in different days.

(d) Activities' frequency in different months.

Fig. 3. Analysis on the importance of time-related context information.

Our experimental results show that it is possible to improve accuracy and F-score of the system even more by dividing the day manually in four periods, namely *morning* (8 AM - 12 PM), *afternoon* (12 PM - 8 PM), *evening* (8 PM - 11 PM) and *night* (11 PM - 8 AM). This way, the periods closely follow the phases of the day when the type of activities performed by typical users changes, as shown in Fig. 3b. As the figure points out, the user's behavior changes remarkably during the day. For example, the *Housekeeping* and *Wash Dishes* activities are much more probable during morning or afternoon, and it seems the user works prevalently on afternoons. As expected, activities such as *Sleeping* and *Bed to Toilet* take place mainly at night. However, some activities, such as *Other*, show less variance during the day, and are thus more difficult to identify. Results show that this granularity yields the best accuracy, F-score and uncertainty (0.793, 0.416, and 0.231, respectively), and one of the lowest cross-entropy errors, i.e., 1.858.

In order to evaluate the effect of context information concerning the day of week and the month, we analyzed the frequency of the user's activities, during the week (Fig. 3c) and among different months (Fig. 3d). It is worth noting that the user's behavior is pretty regular during the week, including weekends. The only exception appears to be the *Resperate* activity; however, this is a fairly rare activity, thus its weight when determining the accuracy of the system is limited. Even among different months, the activities are quite regular (only *Housekeeping* and *Resperate* show a remarkable variance). By analyzing such results, it is easy to predict the different impact of the *period of day* node with respect to the *day of week* and *month* ones.

In order to verify our analysis, we compared eight systems that exploit different combinations of such context information, as reported in Table 1. The difference in accuracy between the best and worst combination of context nodes is more than 10 %. We can observe that four systems out of the five with highest accuracy exploit the *period of day* node. Moreover, these systems show high F-scores and low uncertainty and cross-entropy errors. As expected, the accuracy of all systems improves significantly if the *Other* activity class is ignored, increasing by about 10 % on the average. Surprisingly, the system which includes all three context nodes performs worse than the one which excludes them. This can be explained by the interference of the *month* and *day of week* nodes. In fact, the system that exploits only these two context nodes is the worst according to all the metrics. Conversely, the system that performs better is the one which uses only the *period of day* node. Activities are too regular during the week and among months, and therefore the usefulness of the *day of week* and *month* nodes is limited. Thus, at a first glance, it appears that the *day of week* and *month* nodes are not needed to improve the performance of the data fusion system, and can, in fact, be detrimental.

Table 1. Average accuracy (Acc), uncertainty, cross-entropy error (CE), and F-score of the analyzed systems, sorted by accuracy in descending order.

Period of day	Day of week	Month	Acc	Acc w/o *Other*	Uncertainty	CE	F-score
✓	–	–	0.793	0.889	0.231	1.858	0.416
✓	✓	–	0.779	0.874	0.245	1.950	0.400
✓	–	✓	0.778	0.876	0.280	1.815	0.385
–	–	–	0.760	0.853	0.282	2.146	0.403
✓	✓	✓	0.739	0.833	0.373	1.911	0.366
–	✓	–	0.734	0.826	0.347	2.232	0.390
–	–	✓	0.714	0.800	0.429	2.222	0.363
–	✓	✓	0.690	0.772	0.562	2.347	0.349

The system which uses only the *period of day* node will be considered as baseline for comparison with other systems in next experiments. To provide a more detailed analysis of its performances, its confusion matrix, row-wise normalized, is presented in Fig. 4. Each cell C_{ij} represents the number of instances of class i predicted to be in class j by the system. Therefore, diagonal entries correspond to true positives, and non diagonal entries correspond to classification errors. To explain why some activities are more difficult to recognize than others, in the following we will analyze the location in which each activity is carried out.

Location-Related Context Information. Intuitively, we can hypothesize that some activities are performed in well-defined locations, and therefore are well recognized using only motion sensors, while other activities are more irregular. Furthermore, we supposed that some activities are performed mainly in

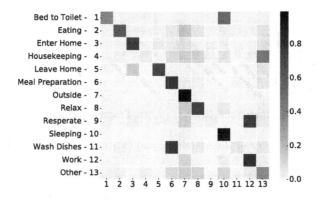

Fig. 4. Confusion matrix of the baseline data fusion system.

the same rooms (and roughly in the same time periods), such as *Wash Dishes* and *Meal Preparation*. To verify these hypotheses, we divided the smart house in rooms, and measured the variability of the association between activities and rooms, through the *diversity index*, defined as the classical Shannon entropy [30]. Figure 5a summarizes the *diversity index* of the activities, which indicates how they are carried out in different rooms. Activities performed in a well-defined location have a low diversity index, while activities carried out different rooms exhibit a high diversity index. As expected, activities which are difficult to recognize correctly, such as *Housekeeping* and *Other*, exhibit the highest diversity indices. On the other hand, activities that are easier to classify accurately, such as *Sleeping*, have low *diversity indices*. The *Wash Dishes* activity seems to contradict this statement, since it sports a low *diversity index*, but is often misclassified. However, this activity only takes place in the kitchen, since almost 80 % of sensor events associated with it comes from sensors deployed there. This accounts for its low *diversity index*, but, as Fig. 5b shows, there are much more probable activities taking place in the same room, such as *Meal Preparation* (52.9 % of sensor events) and *Other* (37.7 % of sensor events); even *Relax* is a more probable activity than *Wash Dishes*, in the kitchen. Therefore, it is understandable that a system which relies mostly on motion sensors will have a hard time identifying this kind of activity. To overcome this problem, we can exploit the information associated to the duration of each activity.

Duration of Activities. We observed that some activities exhibit a much longer average duration than others. For example, *Sleeping* has an average duration of about 4 h, while *Eating* generally takes about 10 min. Thus, it is intuitive that making use of this kind of context information should be beneficial to the system.

In order to verify the usefulness of such information, we tested a system with an additional context node exploiting the duration of activities, and compared it to our baseline system. Surprisingly, the resulting accuracy was lower than the baseline system by about 2 %, and the other metrics were unchanged or slightly

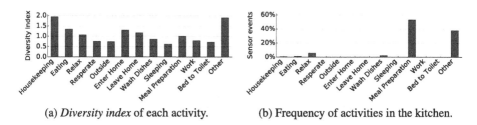

(a) *Diversity index* of each activity.   (b) Frequency of activities in the kitchen.

Fig. 5. (a) *Diversity index* of the activities, and (b) frequency of activities performed in the kitchen.

worse. A closer look at the data reveals that only a couple of activities (i.e., *Sleeping* and *Outside*) have average durations longer than one hour. Most of the other activities have durations similar to each other, generally between 10 and 30 min. It seems that this type of context information fails to help the system if we can exploit enough data coming from the sensory devices.

However, when performing data fusion, it might not always be efficient to sample all available sensors. On the contrary, it may be useful to activate only a subset of sensors, depending on the application scenario. For instance, if the sensory infrastructure is composed of devices with limited energy resources, the use of a subset of devices might increase the lifetime of the whole network.

For this reason, we repeated the comparison experiments using only a subset of sensors, discarding the rest of the data. As expected, in these conditions context information proved to be much more valuable. Using only 10 sensors (out of 34), the accuracy of the baseline system is 65.87 %, while exploiting duration information results in an accuracy of 74.25 %, with a significant improvement of 8.38 %. As it turns out, the same is true for other context information as well.

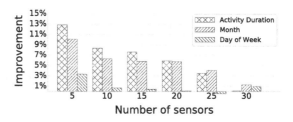

Fig. 6. Improvement of inference accuracy when exploiting context information with different number of sensors.

Figure 6 shows the improvement in accuracy of systems exploiting *activity duration*, *month* and *day of week*, with respect to the baseline system (i.e., the one exploiting only the *period of day*) as a function of the number of sensors used. It can be noted that, in the extreme case of using only 5 sensors, exploiting the *activity duration* improves the accuracy of the system by almost 13 %. Conversely, the benefits of using context information decrease when there is enough data coming from the sensory devices. The same holds true for the *month* node, whilst the improvement when using the *day of week* is negligible even with few sensors.

5   Conclusions

In this paper, we have proposed a multi-sensor data fusion system that aims to improve the accuracy of probabilistic inference by including context information in the fusion process. The key idea is that context information can be involved at different levels of the reasoning process. Basic context attributes can contribute to improve inference accuracy, as demonstrated in the experimental evaluation. At the same time, the context information inferred as result of the data fusion constitutes a high-level description of the environment, and can be exploited by other reasoning engines to better support top-level applications that provide context-aware services to users.

We have demonstrated the suitability of such approach in the application scenario of user activity recognition in a smart home environment. The experimental results have confirmed that choosing the right combination of context information is fundamental to maximize the inference accuracy, especially when only few sensors are available, and that exploiting the best context information set greatly improves the accuracy of activity recognition systems.

As future work, we are interested in studying how to further use context information to dynamically reconfigure the sensory infrastructure, by sampling a subset of sensors in order to minimize energy consumption, whilst maintaining a high degree of inference accuracy.

Moreover, in this paper, we focused on a scenario involving a single user in a smart apartment, and we will study multi-user scenarios in the future. However, recognizing activities performed by multiple users is really challenging, since users can influence each other. Several studies demonstrated that using personalized models for each user dramatically improve systems performance [31], but learning personalized models is computational expensive, and it is thus necessary to find a good trade-off between accuracy and computational efficience. Finally, we are interested in considering training and test data coming from different smart environments, so as to verify the generalization potential of the system.

References

1. Lo Re, G., Morana, M., Ortolani, M.: Improving user experience via motion sensors in an ambient intelligence scenario. In: Pervasive and Embedded Computing and Communication Systems (PECCS), 2013, pp. 29–34 (2013)
2. De Paola, A., La Cascia, M., Lo Re, G., Morana, M., Ortolani, M.: Mimicking biological mechanisms for sensory information fusion. Biol. Inspired Cogn. Archit. **3**, 27–38 (2013)
3. Cook, D.J.: Learning setting-generalized activity models for smart spaces. IEEE Intell. Syst. **2010**(99), 1 (2010)

4. Murphy, K.P.: Dynamic Bayesian networks: representation, inference and learning. PhD thesis, University of California, Berkeley (2002)
5. Cheng, N., Wu, Q.: A decision-making method for fire detection data fusion based on Bayesian approach. In: Proceedings of 4th International Conference on Digital Manufacturing and Automation (ICDMA), pp. 21–23. IEEE (2013)
6. Zhang, Y., Ji, Q.: Active and dynamic information fusion for multisensor systems with dynamic Bayesian networks. IEEE Trans. Syst. Man Cybern. Part B Cybern. **36**(2), 467–472 (2006)
7. De Paola, A., La Cascia, M., Lo Re, G., Morana, M., Ortolani, M.: User detection through multi-sensor fusion in an AmI scenario. In: Proceedings of 15th International Conference on Information Fusion (FUSION), pp. 2502–2509. IEEE (2012)
8. De Paola, A., Gaglio, S., Lo Re, G., Ortolani, M.: Multi-sensor fusion through adaptive Bayesian networks. In: Pirrone, R., Sorbello, F. (eds.) AI*IA 2011. LNCS (LNAI), vol. 6934, pp. 360–371. Springer, Heidelberg (2011). doi:10.1007/978-3-642-23954-0_33
9. Khaleghi, B., Khamis, A., Karray, F.O., Razavi, S.N.: Multisensor data fusion: a review of the state-of-the-art. Inf. Fusion **14**(1), 28–44 (2013)
10. Huebscher, M.C., McCann, J.A.: Adaptive middleware for context-aware applications in smart-homes. In: Proceedings of 2nd Workshop on Middleware for Pervasive and Ad-Hoc Computing, pp. 111–116. ACM (2004)
11. Padovitz, A., Loke, S.W., Zaslavsky, A., Burg, B., Bartolini, C.: An approach to data fusion for context awareness. In: Dey, A., Kokinov, B., Leake, D., Turner, R. (eds.) CONTEXT 2005. LNCS (LNAI), vol. 3554, pp. 353–367. Springer, Heidelberg (2005). doi:10.1007/11508373_27
12. Cho, K., Hwang, I., Kang, S., Kim, B., Lee, J., Lee, S., Park, S., Song, J., Rhee, Y.: HiCon: a hierarchical context monitoring and composition framework for next-generation context-aware services. IEEE Netw. **22**(4), 34–42 (2008)
13. Kang, S., Lee, J., Jang, H., Lee, Y., Park, S., Song, J.: A scalable and energy-efficient context monitoring framework for mobile personal sensor networks. IEEE Trans. Mob. Comput. **9**(5), 686–702 (2010)
14. Nath, S.: ACE: exploiting correlation for energy-efficient and continuous context sensing. In: Proceedings of 10th ACM International Conference on Mobile Systems, Applications, and Services (MobiSys), pp. 29–42. ACM (2012)
15. Cook, D., Augusto, J., Jakkula, V.: Ambient intelligence: technologies, applications, and opportunities. Pervasive and Mob. Comput. **5**(4), 277–298 (2009)
16. De Paola, A., Farruggia, A., Gaglio, S., Lo Re, G., Ortolani, M.: Exploiting the human factor in a WSN-based system for ambient intelligence. In: International Conference on Complex, Intelligent and Software Intensive Systems, 2009. CISIS 2009, pp. 748–753 (2009)
17. Krishnan, N.C., Cook, D.J.: Activity recognition on streaming sensor data. Pervasive Mob. Comput. **10**, 138–154 (2012)
18. Gaglio, S., Lo Re, G., Morana, M.: Human activity recognition process using 3-D posture data. IEEE Trans. Hum. Mach. Syst. **45**(5), 586–597 (2015)
19. Gao, L., Bourke, A., Nelson, J.: Evaluation of accelerometer based multi-sensor versus single-sensor activity recognition systems. Med. Eng. Phys. **36**(6), 779–785 (2014)
20. Kwapisz, J.R., Weiss, G.M., Moore, S.A.: Activity recognition using cell phone accelerometers. ACM SigKDD Explor. Newsl. **12**(2), 74–82 (2011)

21. De Paola, A., Gagliano, L.: Design of an adaptive Bayesian system for sensor data fusion. In: Gaglio, S., Lo Re, G. (eds.) Advances onto the Internet of Things. Advances in Intelligent Systems and Computing, pp. 61–76. Springer, Switzerland (2014). doi:10.1007/978-3-319-03992-3_5

22. Meinhold, R.J., Singpurwalla, N.D.: Understanding the Kalman filter. Am. Stat. **37**(2), 123–127 (1983)

23. Sanchez, D., Tentori, M., Favela, J.: Hidden Markov models for activity recognition in Ambient Intelligence environments. In: Proceedings of 8th Mexican International Conference on Current Trends in Computer Science, pp. 33–40. IEEE (2007)

24. Thrun, S., Burgard, W., Fox, D.: Probabilistic Robotics. MIT Press, Cambridge (2005)

25. Koller, D., Friedman, N.: Probabilistic Graphical Models: Principles and Techniques. MIT Press, Cambridge (2009)

26. Nimier, V.: Introducing contextual information in multisensor tracking algorithms. In: Bouchon-Meunier, B., Yager, R.R., Zadeh, L.A. (eds.) IPMU 1994. LNCS, vol. 945, pp. 595–604. Springer, Heidelberg (1995). doi:10.1007/BFb0035992

27. Dey, A.K., Abowd, G.D., Salber, D.: A conceptual framework and a toolkit for supporting the rapid prototyping of context-aware applications. Hum. Comput. Interact. **16**(2), 97–166 (2001)

28. Perera, C., Zaslavsky, A., Christen, P., Georgakopoulos, D.: Context aware computing for the Internet of Things: a survey. IEEE Commun. Surv. Tutor. **16**(1), 414–454 (2014)

29. Lalomia, A., Lo Re, G., Ortolani, M.: A hybrid framework for soft real-time WSN simulation. In: Proceedings of 13th IEEE/ACM International Symposium Distributed Simulation and Real Time Applications (DS-RT), pp. 201–207 (2009)

30. Shannon, C.E.: A mathematical theory of communication. ACM SIGMOBILE Mob. Comput. Commun. Rev. **5**(1), 3–55 (2001)

31. Weiss, G.M., Lockhart, J.W.: The impact of personalization on smartphone-based activity recognition. In: AAAI Workshop on Activity Context Representation: Techniques and Languages (2012)

Reasoning about Multiple Aspects in Rational Closure for DLs

Valentina Gliozzi$^{(\boxtimes)}$

Dipartimento di Informatica, Center for Logic, Language and Cognition,
Università di Torino, Turin, Italy
valentina.gliozzi@unito.it

Abstract. We propose a logical analysis of the concept of typicality, central in human cognition (Rosch 1978). We start from a previously proposed extension of the basic Description Logic $\mathcal{ALC}$ with a typicality operator **T** that allows to consistently represent the attribution to classes of individuals of properties with exceptions (as in the classic example (i) typical birds fly, (ii) penguins are birds but (iii) typical penguins don't fly). We then strengthen this extension in order to separately reason about the typicality with respect to different aspects (e.g., flying, having nice feather: in the previous example, penguins may not inherit the property of flying, for which they are exceptional, but can nonetheless inherit other properties, such as having nice feather).

Keywords: Description Logics · Nonmonotonic reasoning · Multipreference

1 Introduction

In [1] it is proposed a rational closure strengthening of $\mathcal{ALC}$. This strengthening allows to perform non monotonic reasoning in $\mathcal{ALC}$ in a computationally efficient way. The extension, as already the related logic $\mathcal{ALC} + \mathbf{T}_{min}$ proposed in [2] and the weaker (monotonic) logic $\mathcal{ALC} + \mathbf{T}$ presented in [3], allows to consistently represent typical properties with exceptions that could not be represented in standard $\mathcal{ALC}$.

For instance, in all the above logics one can say that:

SET 1:
Typical students don't earn money
Typical working students do earn money
Typical apprentice working students don't earn money

without having to conclude that there cannot exist working students nor apprentice working students. On the contrary, in standard $\mathcal{ALC}$ typicality cannot be represented, and these three propositions can only be expressed by the stronger ones:

© Springer International Publishing AG 2016
G. Adorni et al. (Eds.): AI*IA 2016, LNAI 10037, pp. 392–405, 2016.
DOI: 10.1007/978-3-319-49130-1_29

SET 2:

Students don't earn money (Student $\sqsubseteq \neg$ EarnMoney)

Working students do earn money (Worker $\sqcap$ Student $\sqsubseteq$ EarnMoney)

Apprentice working students don't earn money (Worker $\sqcap$ Apprentice $\sqcap$ Student $\sqsubseteq \neg$ EarnMoney)

These propositions are consistent in $\mathcal{ALC}$ only if there are no working students nor apprentice working students.

In all the extensions of $\mathcal{ALC}$ mentioned above one can represent the set of propositions in $SET1$ by means of a typicality operator $\mathbf{T}$ that, given a concept C (e.g. Student) singles out the most typical instances of C: so, for instance, $\mathbf{T}(Student)$ refers to the typical instances of the concept Student. The semantics of $\mathbf{T}$ is given by means of a preference relation $<$ that compares the typicality of two individuals: for any two x and y, $x < y$ means that x is more typical than y. Typical instances of a concept C are those minimal with respect to $<$ (formally, as we will see later, $(\mathbf{T}(C))^I = min_<(C)^I$, where $min_<(C)^I = \{x \in C^I : \not\exists y \in C^I \text{ s.t. } y < x\}$).

The operator $\mathbf{T}$ has all the properties that, in the analysis of·Kraus et al. [4] any non monotonic entailment should have. For instance, $\mathbf{T}$ satisfies the principle of cautious monotonicity, according to which if $\mathbf{T}(Student) \sqsubseteq Young$, then $\mathbf{T}(Student) = \mathbf{T}(Student \sqcap Young))$. The precise relations between the properties of $\mathbf{T}$ and preferential entailment are established in [3].

Although the extensions of $\mathcal{ALC}$ with the typicality operator $\mathbf{T}$ allow to express $SET1$ of propositions, the resulting logic is monotonic, and it does not allow to perform some wanted, non monotonic inferences. For instance, it does not allow to deal with irrelevance which is the principle that from the fact that typical students are young, one would want to derive that typical blond students also are young, since being blond is irrelevant with respect to youth. As another example, when knowing that an individual, say John, is a student, and given $SET1$ of propositions, one would want to conclude that John is a typical student and therefore does not earn money. On the other hand, when knowing that John is a working student, one would want to conclude that he is a typical working student and therefore does earn money. In other words one would want to assume that an individual is a typical instance of the most specific class it belongs to, in the absence of information to the contrary.

These stronger inferences all hold in the strengthening of $\mathcal{ALC} + \mathbf{T}$ presented in [1,2]. In particular, [1] proposes an adaptation to $\mathcal{ALC}$ of the well known mechanism of *rational closure*, first proposed by Lehmann and Magidor in [5]. From a semantic point of view, this strengthening of $\mathcal{ALC} + \mathbf{T}$ corresponds to restricting one's attention to minimal models, that minimize the height (rank) of all domain elements with respect to $<$ (i.e. that minimize the length of the $<$-chains starting from all individuals). Under the condition that the models considered are canonical, the semantic characterization corresponds to the syntactical rational closure. This semantics supports all the above wanted inferences, and the nice computational properties of rational closure guarantee that whether the above inferences are valid or not can be computed in reasonable time.

The main drawback of rational closure is that it is an **all-or-nothing** mechanism: for any subclass C' of C it holds that either the typical members of C' inherit all the properties of C or they don't inherit any property. Once the typical members of C' are recognized as exceptional with respect to C for a given aspect, they become exceptional for all aspects. Consider an enriched version of the classic birds/penguins example, expressed by propositions:

SET 3:
Typical birds have nice feather
Typical birds fly
Penguins are birds
Typical penguins do not fly

In this case, since penguins are exceptional with respect to the aspect of flying, they are *non-typical* birds, and for this reason they do not inherit any of the typical properties of birds.

On the contrary, given $SET3$ of propositions, one wants to conclude that:

– (**) **Typical penguins have nice feather**

This is to say that one wants to separately reason about the different aspects: the property of flying is not related to the property of having nice feather, hence we want to separately reason on the two aspects.

Here we propose a strengthening of the semantics used for rational closure in $\mathcal{ALC}$ [1] that only used a single preference relation $<$ by allowing, beside $<$, several preference relations that compare the typicality of individuals with respect to a given aspect. **Obtaining a strengthening of rational closure is the purpose of this work. This puts strong constraints on the resulting semantics, and defines the horizon of this work**. In this new semantics we can express the fact that, for instance, x is more typical than y with respect to the property of flying but y is more typical that x with respect to some other property, as the property of having nice feather. To this purpose we consider preference relations indexed by concepts that stand for the above mentioned aspects under which we compare individuals. So we will write that $x <_A y$ to mean that x is preferred to y for what concerns aspect A: for instance $x <_{Fly} y$ means that x is more typical than y with respect to the property of flying.

We therefore proceed as follows: we first recall the semantics of the extension of $\mathcal{ALC}$ with a typicality operator which was at the basis of the definition of rational closure and semantics in [1,6]. We then expand this semantics by introducing several preference relations, that we then minimize obtaining our new minimal models' mechanism. As we will see this new semantics leads to a strengthening of rational closure, allowing to separately reason about the inheritance of different properties.

2   The Operator T and the General Semantics

Let us briefly recall the logic $\mathcal{ALC} + \mathbf{T}_\mathsf{R}$ which is at the basis of a rational closure construction proposed in [1] for $\mathcal{ALC}$. The intuitive idea of $\mathcal{ALC} + \mathbf{T}_\mathsf{R}$ is to extend the standard $\mathcal{ALC}$ with concepts of the form $\mathbf{T}(C)$, whose intuitive meaning is that $\mathbf{T}(C)$ selects the *typical* instances of a concept C, to distinguish between the properties that hold for all instances of concept C ($C \sqsubseteq D$), and those that only hold for the typical such instances ($\mathbf{T}(C) \sqsubseteq D$). The $\mathcal{ALC} + \mathbf{T}_{\mathsf{R}}$ language is defined as follows: $C_R := A \mid \top \mid \bot \mid \neg C_R \mid C_R \sqcap C_R \mid C_R \sqcup C_R \mid \forall R.C_R \mid \exists R.C_R$, and $C_L := C_R \mid \mathbf{T}(C_R)$, where A is a concept name and R a role name. A KB is a pair (TBox, ABox). TBox contains a finite set of concept inclusions $C_L \sqsubseteq C_R$. ABox contains a finite set of assertions of the form $C_L(a)$ and $R(a, b)$, where a, b are individual constants.

The semantics of $\mathcal{ALC} + \mathbf{T}_{\mathsf{R}}$ is defined in terms of rational models: ordinary models of $\mathcal{ALC}$ are equipped with a *preference relation* $<$ on the domain, whose intuitive meaning is to compare the "typicality" of domain elements: $x < y$ means that x is more typical than y. Typical members of a concept C, instances of $\mathbf{T}(C)$, are the members x of C that are minimal with respect to $<$ (such that there is no other member of C more typical than x). In rational models $<$ is further assumed to be modular: for all $x, y, z \in \Delta$, if $x < y$ then either $x < z$ or $z < y$. These rational models characterize $\mathcal{ALC} + \mathbf{T}_{\mathsf{R}}$.

Definition 1 (Semantics of $\mathcal{ALC} + \mathbf{T}_{\mathsf{R}}$ [1]). *A model $\mathcal{M}$ of $\mathcal{ALC} + \mathbf{T}_{\mathsf{R}}$ is any structure $\langle \Delta, <, I \rangle$ where: Δ is the domain; $<$ is an irreflexive, transitive, and modular relation over Δ that satisfies the finite chain condition (there is no infinite $<$-descending chain, hence if $S \neq \emptyset$, also $min_<(S) \neq \emptyset$); I is the extension function that maps each concept name C to $C^I \subseteq \Delta$, each role name R to $R^I \subseteq \Delta^I \times \Delta^I$ and each individual constant $a \in \mathcal{O}$ to $a^I \in \Delta$. For concepts of $\mathcal{ALC}$, C^I is defined in the usual way. For the $\mathbf{T}$ operator, we have $(\mathbf{T}(C))^I = min_<(C^I)$.*

As shown in [1], the logic $\mathcal{ALC} + \mathbf{T}_{\mathsf{R}}$ enjoys the finite model property and finite $\mathcal{ALC} + \mathbf{T}_{\mathsf{R}}$ models can be equivalently defined by postulating the existence of a function $k_{\mathcal{M}} : \Delta \longmapsto \mathbb{N}$, where $k_{\mathcal{M}}$ assigns a finite rank to each world: the rank $k_{\mathcal{M}}$ of a domain element $x \in \Delta$ is the length of the longest chain $x_0 < \cdots < x$ from x to a minimal x_0 (s.t. there is no x' with $x' < x_0$). The rank $k_{\mathcal{M}}(C_R)$ of a concept C_R in $\mathcal{M}$ is $i = min\{k_{\mathcal{M}}(x) : x \in C_R^I\}$.

A model $\mathcal{M}$ satisfies a knowledge base $K = $ (TBox, ABox) if it satisfies its TBox (and for all inclusions $C \sqsubseteq D$ in TBox, it holds $C^I \subseteq D^I$), and its ABox (for all $C(a)$ in ABox, $a^I \in C^I$, and for all aRb in ABox, $(a^I, b^I) \in R^I$). A query F (either an assertion $C_L(a)$ or an inclusion relation $C_L \sqsubseteq C_R$) is logically (rationally) entailed by a knowledge base K ($K \models_{\mathcal{ALC} + \mathbf{T}_{\mathsf{R}}} F$) if F holds in all models satisfying K.

Although the typicality operator $\mathbf{T}$ itself is nonmonotonic (i.e. $\mathbf{T}(C) \sqsubseteq D$ does not imply $\mathbf{T}(C \sqcap E) \sqsubseteq D$), the logic $\mathcal{ALC} + \mathbf{T}_{\mathsf{R}}$ is monotonic: what is logically entailed by K is still entailed by any K' with $K \subseteq K'$.

In [1,6] the non monotonic mechanism of rational closure has been defined over $\mathcal{ALC} + \mathbf{T}_{\mathsf{R}}$, which extends to DLs the notion of rational closure proposed in the propositional context by Lehmann and Magidor [5]. The definition is based on the notion of exceptionality. Roughly speaking $\mathbf{T}(C) \sqsubseteq D$ holds (is included in the rational closure) of K if C (indeed, $C \sqcap D$) is less exceptional than $C \sqcap \neg D$. We briefly recall this construction and we refer to [1,6] for full details. Here we only consider rational closure of TBox, defined as follows.

Definition 2 (Exceptionality of Concepts and Inclusions). *Let T_B be a TBox and C a concept. C is said to be* exceptional *for T_B if and only if $T_B \models_{\mathcal{ALC}+\mathbf{T}_R} \mathbf{T}(\top) \sqsubseteq \neg C$. A $\mathbf{T}$-inclusion $\mathbf{T}(C) \sqsubseteq D$ is exceptional for T_B if C is exceptional for T_B. The set of $\mathbf{T}$-inclusions of T_B which are exceptional in T_B will be denoted as $\mathcal{E}(T_B)$.*

Given a DL TBox, it is possible to define a sequence of non increasing subsets of TBox ordered according to the exceptionality of the elements $E_0 \supseteq E_1, E_1 \supseteq E_2, \ldots$ by letting $E_0 = $ TBox and, for $i > 0$, $E_i = \mathcal{E}(E_{i-1}) \cup \{C \sqsubseteq D \in$ TBox s.t. $\mathbf{T}$ does not occurr in $C\}$. Observe that, being KB finite, there is an $n \geq 0$ such that, for all $m > n$, $E_m = E_n$ or $E_m = \emptyset$. A concept C has *rank i* (denoted $rank(C) = i$) for TBox, iff i is the least natural number for which C is not exceptional for E_i. If C is exceptional for all E_i then $rank(C) = \infty$ (C has no rank).

Rational closure builds on this notion of exceptionality:

Definition 3 (Rational Closure of TBox). *Let $KB = (TBox, ABox)$ be a DL knowledge base. The rational closure of TBox $\overline{TBox} = \{\mathbf{T}(C) \sqsubseteq D \mid$ either $rank(C) < rank(C \sqcap \neg D)$ or $rank(C) = \infty\} \cup \{C \sqsubseteq D \mid KB \models_{\mathcal{ALC}+\mathbf{T}_R} C \sqsubseteq D\}$, where C and D are $\mathcal{ALC}$ concepts.*

As a very interesting property, in the context of DLs, the rational closure has a very interesting complexity: deciding if an inclusion $\mathbf{T}(C) \sqsubseteq D$ belongs to the rational closure of TBox is a problem in ExpTime [1].

In [1] it is shown that the semantics corresponding to rational closure can be given in terms of minimal canonical $\mathcal{ALC} + \mathbf{T}_R$ models. With respect to standard $\mathcal{ALC} + \mathbf{T}_R$ models, in these models the rank of each domain element is as low as possible (each domain element is assumed to be as typical as possible). This is expressed by the following definition.

Definition 4 (Minimal Models of K (with Respect to $TBox$)). *Given $\mathcal{M} = \langle \Delta, <, I \rangle$ and $\mathcal{M}' = \langle \Delta', <', I' \rangle$, we say that $\mathcal{M}$ is preferred to $\mathcal{M}'$ ($\mathcal{M} < \mathcal{M}'$) if: $\Delta = \Delta'$, $C^I = C^{I'}$ for all concepts C, for all $x \in \Delta$, it holds that $k_{\mathcal{M}}(x) \leq k_{\mathcal{M}'}(x)$ whereas there exists $y \in \Delta$ such that $k_{\mathcal{M}}(y) < k_{\mathcal{M}'}(y)$.*

Given a knowledge base $K = \langle TBox, ABox \rangle$, we say that $\mathcal{M}$ is a minimal model of K (with respect to TBox) if it is a model satisfying K and there is no $\mathcal{M}'$ model satisfying K such that $\mathcal{M}' < \mathcal{M}$.

Furthermore, the models corresponding to rational closure are canonical. This property, expressed by the following definition, is needed when reasoning about the (relative) rank of the concepts: it is important to have them all represented.

Definition 5 (Canonical Model). *Given $K = (TBox, ABox)$, a model $\mathcal{M} = \langle \Delta, <, I \rangle$ satisfying K is* canonical *if for each set of concepts $\{C_1, C_2, \ldots, C_n\}$ consistent with K, there exists (at least) a domain element $x \in \Delta$ such that $x \in (C_1 \sqcap C_2 \sqcap \cdots \sqcap C_n)^I$.*

Definition 6 (Minimal Canonical Models (with Respect to TBox)).
$\mathcal{M}$ *is a canonical model of K minimal with respect to TBox if it satisfies K, it is minimal with respect to TBox (Definition 4) and it is canonical (Definition 5).*

The correspondence between minimal canonical models and rational closure is established by the following key theorem.

Theorem 1 [1]. *Let $K = (TBox, ABox)$ be a knowledge base and $C \sqsubseteq D$ a query. We have that $C \sqsubseteq D \in \overline{TBox}$ if and only if $C \sqsubseteq D$ holds in all minimal canonical models of K with respect to TBox (Definition 6).*

3   Semantics with Several Preference Relations

The main weakness of rational closure, despite its power and its nice computational properties, is that it is an all-or-nothing mechanism that does not allow to separately reason on single aspects. To overcome this difficulty, we here consider models with several preference relations, one for each aspect we want to reason about. We assume an aspect can be any concept occurring in K: we call $\mathcal{L}_{\mathcal{A}}$ the set of these aspects (observe that A may be non-atomic). For each aspect A, $<_A$ expresses the preference for aspect A : $<_{Fly}$ expresses the preference for flying, so if we know that $\mathbf{T}(Bird) \sqsubseteq Fly$, birds that do fly will be preferred to birds that do not fly, with respect to aspect fly, i.e. with respect to $<_{Fly}$. All these preferences, as well as the global preference relation $<$, satisfy the properties in Definition 7 below. We now enrich the definition of an $\mathcal{ALC} + \mathbf{T}_R$ model given above (Definition 1) by taking into account preferences with respect to all of the aspects. In the semantics we can express that for instance $x <_{A_i} y$, whereas $y <_{A_j} x$ (x is preferred to y for aspect A_i but y is preferred to x for aspect A_j).

This semantic richness allows to obtain a strengthening of rational closure in which typicality with respect to every aspect is maximized. Since we want to compare our approach to rational closure, we keep the language the same as in $\mathcal{ALC} + \mathbf{T}_R$. In particular, we only have one single typicality operator $\mathbf{T}$. However, the semantic richness could motivate the introduction of several typicality operators $\mathbf{T}_{A_1} \dots \mathbf{T}_{A_n}$ by which one might want to explicitly talk in the language about the typicality w.r.t. aspect A_1, or A_2, and so on. We leave this extension for future work.

Definition 7 (Enriched Rational Models). *Given a knowledge base K, we call an enriched rational model a structure $\mathcal{M} = \langle \Delta, <, <_{A_1}, \dots, <_{A_n}, I \rangle$, where Δ, I are defined as in Definition 1, and $<, <_{A_1}, \dots, <_{A_n}$ are preference relations over Δ, with the properties of being irreflexive, transitive, satisfying the finite chain condition, modular (for all $x, y, z \in \Delta$, if $x <_{A_i} y$ then either $x <_{A_i} z$ or $z <_{A_i} y$).*

For all $<_{A_i}$ and for $<$ it holds that $min_{<_{A_i}}(S) = \{x \in S \text{ s.t. there is no } x_1 \in S \text{ s.t. } x_1 <_{A_i} x\}$ and $min_<(S) = \{x \in S \text{ s.t. there is no } x_1 \in S \text{ s.t. } x_1 < x\}$ and $(\mathbf{T}(C))^I = min_<(C^I)$.

$<$ satisfies the further conditions that $x < y$ if:

(a) there is A_i such that $x <_{A_i} y$, and there is no A_j such that $y <_{A_j} x$ or;
(b) there is $\mathbf{T}(C_i) \sqsubseteq A_i \in K$ s.t. $y \in (C_i \sqcap \neg A_i)^I$, and for all $\mathbf{T}(C_j) \sqsubseteq A_j \in K$
    s.t. $x \in (C_j \sqcap \neg A_j)^I$, there is $\mathbf{T}(C_k) \sqsubseteq A_k \in K$ s.t. $y \in (C_k \sqcap \neg A_k)^I$ and
    $k_{\mathcal{M}}(C_j) < k_{\mathcal{M}}(C_k)$.

In this semantics the global preference relation $<$ is related to the various preference relations $<_{A_i}$ relative to single aspects A_i. Given (a) $x < y$ when x is preferred to y for a single aspect A_i, and there is no aspect A_j for which y is preferred to x. (b) captures the idea that in case two individuals are preferred with respect to different aspects, preference (for the global preference relation) is given to the individual that satisfies all typical properties of the **most specific** concept (if C_k is more specific than C_j, then $k_{\mathcal{M}}(C_j) < k_{\mathcal{M}}(C_k)$), as illustrated by Example 1 below.

We insist in highlighting that this semantics somewhat complicated is needed since we want to provide a strengthening of rational closure. For this, we have to respect the constraints imposed by rational closure. One might think in the future to study a semantics in which only (a) holds. We have not considered such a simpler semantics since it would no longer be a strengthening of the semantics corresponding to rational closure, and is therefore out of the focus of this work.

In order to be a model of K an $\mathcal{ALC}^{\mathbf{R}}\mathbf{T}_E$ model must satisfy the following constraints.

Definition 8 (Enriched Rational Models of K). *Given a knowledge base K, and an enriched rational model for K $\mathcal{M} = \langle \Delta, <, <_{A_1}, \ldots, <_{A_n}, I \rangle$, $\mathcal{M}$ is a model of K if it satisfies both its TBox and its ABox, where $\mathcal{M}$ satisfies TBox if for all inclusions $C \sqsubseteq A_i \in TBox$: if $\mathbf{T}$ does not occur in C, then $C^I \subseteq A_i{}^I$ if $\mathbf{T}$ occurs in C, and C is $\mathbf{T}(C')$, then both (i) $\min_<(C'^I) \subseteq A_i{}^I$ and (ii) $\min_{<_{A_i}}(C'^I) \subseteq A_i{}^I$. $\mathcal{M}$ satisfies ABox if (i) for all $C(a)$ in ABox, $a^I \in C^I$, (ii) for all aRb in ABox, $(a^I, b^I) \in R^I$.*

Example 1. Let $K = \{Penguin \sqsubseteq Bird, \mathbf{T}(Bird) \sqsubseteq HasNiceFeather, \mathbf{T}(Bird) \sqsubseteq Fly, \mathbf{T}(Penguin) \sqsubseteq \neg Fly\}$. $\mathcal{L}_A = \{HasNiceFeather, Fly, \neg Fly, Bird, Penguin\}$. We consider an $\mathcal{ALC}^{\mathbf{R}}\mathbf{T}_E$ model $\mathcal{M}$ of K, that we don't fully describe but which we only use to observe the behavior of two Penguins x, y with respect to the properties of (not) flying and having nice feather. In particular, let us consider the three preference relations: $<, <_{\neg Fly}, <_{HasNiceFeather}$.

Suppose $x <_{\neg Fly} y$ (because x, as all typical penguins, does not fly whereas y exceptionally does) and there is no other aspect A_i such that $y <_{A_i} x$, and in particular it does not hold that $y <_{HasNiceFeather} x$ (because for instance both have a nice feather). In this case, obviously it holds that $x < y$ (since (a) is satisfied).

Consider now a more tricky situation in which again $x <_{\neg Fly} y$ holds (because for instance x does not fly whereas y flies), (x is a typical penguin for what concerns Flying) but this time $y <_{HasNiceFeather} x$ holds (because for instance y has a nice feather, whereas x has not). So x is preferred to y for a given aspect whereas y is preferred to x for another aspect. However, x enjoys the typical properties of penguins, and violates the typical properties of birds, whereas y

enjoys the typical properties of birds and violates those of penguins. Being concept Penguin more specific than concept Bird, we prefer x to y, since we prefer the individuals that inherit the properties of the most specific concepts of which they are instances. This is exactly what we get: by (b) $x < y$ holds.

Logical entailment for $\mathcal{ALC}^\mathbf{R}\mathbf{T}_E$ is defined as usual: a query (with form $C_L(a)$ or $C_L \sqsubseteq C_R$) is logically entailed by K if it holds in all models of K, as stated by the following definition. The following theorem shows the relations between $\mathcal{ALC}^\mathbf{R}\mathbf{T}_E$ and $\mathcal{ALC} + \mathbf{T}_R$. Proofs are omitted due to space limitations.

Theorem 2. *If* $K \models_{\mathcal{ALC}+\mathbf{T}_R} F$ *then also* $K \models_{\mathcal{ALC}^\mathbf{R}\mathbf{T}_E} F$. *If* $\mathbf{T}$ *does not occur in* F *the other direction also holds: If* $K \models_{\mathcal{ALC}^\mathbf{R}\mathbf{T}_E} F$ *then also* $K \models_{\mathcal{ALC}+\mathbf{T}_R} F$.

The following example shows that $\mathcal{ALC}^\mathbf{R}\mathbf{T}_E$ alone is not strong enough, and this motivates the minimal models' mechanism that we introduce in the next section. In the example we show that $\mathcal{ALC}^\mathbf{R}\mathbf{T}_E$ alone does not allow us to perform the stronger inferences with respect to rational closure mentioned in the Introduction (and in particular, it does not allow to infer (**), that typical penguins have a nice feather).

Example 2. Consider the above Example 1. As said in the Introduction, in rational closure we are not able to reason separately about the property of flying or not flying, and the property of having or not having a nice feather. Since penguins are exceptional birds with respect to the property of flying, in rational closure which is an all-or-nothing mechanism, they do not inherit any of the properties of typical birds. In particular, they do not inherit the property of having a nice feather, even if this property and the fact of flying are independent from each other and there is no reason why being exceptional with respect to one property should block the inheritance of the other one. Does our enriched semantics enforce the separate inheritance of independent properties?

Consider a model $\mathcal{M}$ in which we have $\Delta = \{x, y, z\}$, where x is a bird (not a penguin) that flies and has a nice feather ($x \in Bird^I, x \in Fly^I, x \in HasNiceFeather^I, x \notin Penguin^I$), y is a penguin that does not fly and has a nice feather ($y \in Penguin^I, y \in Bird^I, y \notin Fly^I, y \in HasNiceFeather^I$), z is a penguin that does not fly and has no nice feather ($z \in Penguin^I, z \in Bird^I, z \notin Fly^I, z \notin HasNiceFeather^I$). Suppose it holds that $x <_{Fly} y$, $x <_{Fly} z$, $x <_{HasNiceFeather} z$, $y <_{HasNiceFeather} z$, and $x < y$, $x < z$, $y < z$. It can be verified that this is an $\mathcal{ALC}^\mathbf{R}\mathbf{T}_E$ model, satisfying $\mathbf{T}(Penguin) \sqsubseteq HasNiceFeather$ (since the only typical Penguin is y, instance of HasNiceFeather).

Unfortunately, this is not the only $\mathcal{ALC}^\mathbf{R}\mathbf{T}_E$ model of K. For instance there can be $\mathcal{M}'$ equal to $\mathcal{M}$ except from the fact that $y <_{HasNiceFeather} z$ does not hold, nor $y < z$ holds. It can be easily verified that this is also an $\mathcal{ALC}^\mathbf{R}\mathbf{T}_E$ model of K in which $\mathbf{T}(Penguin) \sqsubseteq HasNiceFeather$ does not hold (since now also z is a typical Penguin, and z is not an instance of HasNiceFeather).

This example shows that although there are $\mathcal{ALC}^\mathbf{R}\mathbf{T}_E$ models satisfying well suited inclusions, the logic is not strong enough to limit our attention to these models. We would like to constrain our logic in order to exclude models like $\mathcal{M}'$. Roughly speaking, we want to eliminate $\mathcal{M}'$ because it is not minimal: although the model as it is satisfies K, so y does not *need* to be preferred to z to satisfy K (neither with respect to $<$ nor with respect to $<_{HasNiceFeather}$), intuitively we would like to prefer y to z (with respect to the property HasNiceFeather, whence in this case with respect to the global $<$), since y does not falsify any of the inclusions with HasNiceFeather, whereas z does. This is obtained by imposing the constraint of considering only models minimal with respect to all relations $<_A$, defined as in Definition 10 below. Notice that the wanted inference does not hold in $\mathcal{ALC} + \mathbf{T_R}$ minimal canonical models corresponding to rational closure: in these models $y < z$ does never hold (the two elements have the same rank) and this semantics does not allow us to prefer y to z. By adopting the restriction to minimal canonical models, we obtain a semantics which is stronger than rational closure (and therefore enforces all conclusions enforced by rational closure) and, furthermore, separately allows to reason on different aspects.

Before we end the section, similarly to what done above, let us introduce a rank of a domain element with respect to an aspect. We will use this notion in the following section.

Definition 9. *The rank $k_{A_i\mathcal{M}}(x)$ of a domain element x with respect to $<_{A_i}$ in $\mathcal{M}$ is the length of the longest chain $x_0 <_{A_i} \cdots <_{A_i} x$ from x to a minimal x_0 (s.t. for no x' $x' <_{A_i} x_0$). To refer to the rank of an element x with respect to the preference relation $<$ we will simply write $k_{\mathcal{M}}(x)$.*

The notion just introduced will be useful in the following. Since $k_{A_i\mathcal{M}}$ and $<_{A_i}$ are clearly interdefinable (by the previous definition and by the properties of $<_{A_i}$ it easily follows that in all enriched models $\mathcal{M}$, $x <_{A_i} y$ iff $k_{A_i\mathcal{M}}(x) < k_{A_i\mathcal{M}}(y)$, and $x < y$ iff $k_{\mathcal{M}}(x) < k_{\mathcal{M}}(y)$), we will shift from one to other whenever this simplifies the exposition.

4   Nonmonotonicity and Relation with Rational Closure

We here define a minimal models mechanism starting from the enriched models of the previous section. With respect to the minimal canonical models used in [1] we define minimal models by separately minimizing all the preference relations with respect to all aspects (steps (i) and (ii) in the definition below), before minimizing $<$ (steps (iii) and (iv) in the definition below). By the constraints linking $<$ to the preference relations $<_{A_1} \cdots <_{A_n}$, this leads to preferring (with respect to the global $<$) the individuals that are minimal with respect to all $<_{A_i}$ for all aspects A_i, or to aspects of most specific categories than of more general ones. It turns out that this leads to a stronger semantics than what is obtained by directly minimizing $<$.

Definition 10 (Minimal Enriched Models). *Given two $\mathcal{ALC}^{\mathbf{R}}\mathbf{T}_E$ enriched models $\mathcal{M} = \langle \Delta, <_{A_1}, \ldots, <_{A_n}, <, I \rangle$ and $\mathcal{M}' = \langle \Delta', <'_{A_1}, \ldots, <'_{A_n}, <', I' \rangle$ we say that $\mathcal{M}'$ is preferred to $\mathcal{M}$ with respect to the single aspects (and write $\mathcal{M}' <_{Enriched_A spects} \mathcal{M}$) if $\Delta = \Delta'$, $I = I'$, and:*

(i) for all $x \in \Delta$, for all A_i: $k_{A_i{}_{\mathcal{M}'}}(x) \leq k_{A_i{}_{\mathcal{M}}}(x)$;
(ii) for some $y \in \Delta$, for some A_j, $k_{A_j{}_{\mathcal{M}'}}(y) < k_{A_j{}_{\mathcal{M}}}(y)$

We let the set $Min_{Aspects} = \{\mathcal{M} :$ there is no $\mathcal{M}'$ such that $\mathcal{M}' <_{Enriched_A spects}$ $\mathcal{M}\}$. Given $\mathcal{M}$ and $\mathcal{M}' \in Min_{Aspects}$, we say that $\mathcal{M}'$ is overall preferred to $\mathcal{M}$ (and write $\mathcal{M}' <_{Enriched} \mathcal{M}$) if $\Delta = \Delta'$, $I = I'$, and:

(iii) for all $x \in \Delta$, $k_{\mathcal{M}'}(x) \leq k_{\mathcal{M}}(x)$;
(iv) for some $y \in \Delta$, $k_{\mathcal{M}'}(y) < k_{\mathcal{M}}(y)$

We call $\mathcal{M}$ a minimal enriched model of K if it is a model of K and there is no $\mathcal{M}'$ model of K such that $\mathcal{M}' <_{Enriched} \mathcal{M}$.

K minimally entails a query F if F holds in all minimal $\mathcal{ALC}^{\mathbf{R}}\mathbf{T}_E$ models of K. We write $K \models_{\mathcal{ALC}^{\mathbf{R}}\mathbf{T}_{E min}} F$. We have developed the semantics above in order to overcome a weakness of rational closure, namely its all-or-nothing character. In order to show that the semantics hits the point, we show here that the semantics is stronger than the one corresponding to rational closure. Furthermore, Example 3 below shows that indeed we have strengthened rational closure by making it possible to separately reason on the different properties. Since the semantic characterization of rational closure is given in terms of rational *canonical* models, here we restrict our attention to enriched rational models which are *canonical*.

Definition 11 (Minimal Canonical Enriched Models of K). *An $\mathcal{ALC}^{\mathbf{R}}\mathbf{T}_E$ enriched model $\mathcal{M}$ is a minimal canonical enriched model of K if it satisfies K, it is minimal (with respect to Definition 10) and it is canonical: for all the sets of concepts $\{C_1, C_2, \ldots, C_n\}$ s.t. $K \not\models_{\mathcal{ALC}^{\mathbf{R}}\mathbf{T}_E} C_1 \sqcap C_2 \sqcap \cdots \sqcap C_n \sqsubseteq \bot$, there exists (at least) a domain element x such that $x \in (C_1 \sqcap C_2 \sqcap \cdots \sqcap C_n)^I$.*

We call $\mathcal{ALC}^{\mathbf{R}}\mathbf{T}_E + min - canonical$ the semantics obtained by restricting attention to minimal canonical enriched models. In the following we will write: $K \models_{\mathcal{ALC}^{\mathbf{R}}\mathbf{T}_E+min-canonical} C \sqsubseteq D$ to mean that $C \sqsubseteq D$ holds in all minimal canonical enriched models of K.

The following example shows that this semantics allows us to correctly deal with the wanted inferences of the Introduction, as (**). The fact that the semantics $\mathcal{ALC}^{\mathbf{R}}\mathbf{T}_E + min - canonical$ is a genuine strengthening of the semantics corresponding to rational closure is formally shown in Theorem 3 below.

Example 3. Consider any minimal canonical model $\mathcal{M}^*$ of the same K used in Example 1. It can be easily verified that in $\mathcal{M}^*$ there is a domain element y which is a penguin that does not fly and has a nice feather ($y \in Penguin^I, y \in Bird^I, y \in HasNiceFeather^I$). First, it can be verified that $y \in min_<(Penguin^I)$ (by Definition 7, and since by minimality of $<_{Fly}$ and $<_{HasNiceFeather}$, $y \in min_{<_{Fly}}(Penguin^I)$ and $y \in min_{<_{HasNiceFeather}}(Penguin^I)$). Furthermore, for any penguin z that has not a nice feather, $y < z$ (again by Definition 7, and since by minimality of $<_{Fly}$ and $<_{HasNiceFeather}$, $y <_{HasNiceFeather} z$). From this, in all minimal canonical $\mathcal{ALC}^{\mathbf{R}}\mathbf{T}_E$ models of K it holds that $\mathbf{T}(Penguin) \sqsubseteq HasNiceFeather$, i.e., $K \models_{\mathcal{ALC}^{\mathbf{R}}\mathbf{T}_E+min-canonical} \mathbf{T}(Penguin) \sqsubseteq HasNiceFeather$, which was the wanted inference (**) of the Introduction.

The following theorem is the important technical result of the paper:

Theorem 3. The minimal models semantics $\mathcal{ALC}^{\mathbf{R}}\mathbf{T}_E + min - canonical$ is stronger than the semantics for rational closure. Let $(K = TBox, ABox)$. If $C \sqsubseteq D \in \overline{TBox}$ then $K \models_{\mathcal{ALC}^{\mathbf{R}}\mathbf{T}_E+min-canonical} C \sqsubseteq D$.

Proof **(Sketch).** By contraposition suppose that $K \not\models_{\mathcal{ALC}^{\mathbf{R}}\mathbf{T}_E+min-canonical}$ $C \sqsubseteq D$. Then there is a minimal canonical enriched $\mathcal{ALC}^{\mathbf{R}}\mathbf{T}_E$ model $\mathcal{M} = \langle \Delta, <_{A_1}, \ldots, <_{A_n}, <, I \rangle$ of K and an $y \in C^I$ such that $y \notin D^I$. All consistent sets of concepts consistent with K w.r.t. $\mathcal{ALC}^{\mathbf{R}}\mathbf{T}_E$ are also consistent with K with respect to $\mathcal{ALC} + \mathbf{T}_R$, and viceversa (by Theorem 2). By definition of *canonical*, there is also a canonical $\mathcal{ALC} + \mathbf{T}_R$ model of K $\mathcal{M}_{RC} = \langle \Delta, <_R C, I \rangle$ be this model. If C does not contain the $\mathbf{T}$ operator, we are done: in $\mathcal{M}_{RC}$, as in $\mathcal{M}$, there is $y \in C^I$ such that $y \notin D^I$, hence $C \sqsubseteq D$ does not hold in $\mathcal{M}_{RC}$, and $C \sqsubseteq D \notin \overline{TBox}$. If $\mathbf{T}$ occurs in C, and $C = \mathbf{T}(C')$, we still need to show that also in $\mathcal{M}_{RC}$, as in $\mathcal{M}$, $y \in min_{<_{RC}}(C'^I)$. We prove this by showing that for all $x, y \in \Delta$ if $x <_{RC} y$ in $\mathcal{M}_{RC}$, then also $x < y$ in $\mathcal{M}$. The proof is by induction on $k_{\mathcal{M}_{RC}}(x)$.

(a) let $k_{\mathcal{M}_{RC}}(x) = 0$ and $k_{\mathcal{M}_{RC}}(y) > 0$. Since x does not violate any inclusion, also in $\mathcal{M}$ (by minimality of $\mathcal{M}$) for all preference relations $<_{A_j}$ $k_{A_{j\mathcal{M}}}(x) = 0$, and also $k_{\mathcal{M}}(x) = 0$. This cannot hold for y, for which $k_{\mathcal{M}}(y) > 0$ (otherwise $\mathcal{M}$ would violate K, against the hypothesis). Hence $x < y$ in $\mathcal{M}$.

(b) let $k_{\mathcal{M}_{RC}}(x) = i < k_{\mathcal{M}_{RC}}(y)$, i.e. $x <_{RC} y$. As $x <_{RC} y$ in $\mathcal{M}_{RC}$ and the rank of x in $\mathcal{M}_{RC}$ is i, there must be a $\mathbf{T}(B_i) \sqsubseteq A_i \in E_i - E_{i+1}$ such that $x \in (\neg B_i \sqcup A_i)^I$ whereas $y \in (B_i \sqcap \neg A_i)^I$ in $\mathcal{M}_{RC}$. Before we proceed let us notice that by definition of E_i, as well as by what stated just above on the relation between rank of a concept and $k_{\mathcal{M}_{RC}}$, $k_{\mathcal{M}_{RC}}(B_i) = k_{\mathcal{M}_{RC}}(x)$. We will use this fact below. We show that, for any inclusion $\mathbf{T}(B_l) \sqsubseteq A_l \in K$ that is violated by x, it holds that $k_{\mathcal{M}}(B_l) < k_{\mathcal{M}}(B_i)$, so that, by (b), $x < y$.

Let $\mathbf{T}(B_l) \sqsubseteq A_l \in K$ violated by x, i.e. such that $x \in (B_l \sqcap \neg A_l)^I$. Since $\mathcal{M}_{RC}$ satisfies K, there must be $x' <_{RC} x$ in $\mathcal{M}_{RC}$ with $x' \in (B_l)^I$. As $k_{\mathcal{M}_{RC}}(x') < i$, by inductive hypothesis, $x' < x$ in $\mathcal{M}$. As $x' \in B_l^I$, $k_{\mathcal{M}}(B_l) \leq k_{\mathcal{M}}(x')$. Since it can be shown that $k_{\mathcal{M}}(x') < k_{\mathcal{M}}(B_i)$, $k_{\mathcal{M}}(B_l) < k_{\mathcal{M}}(B_i)$, and by condition (b), it holds that $x < y$ in $\mathcal{M}$.

With these facts, since $y \in min_<(C'^I)$ holds in $\mathcal{M}$, also $y \in min_{<_{RC}}(C'^I)$ in $\mathcal{M}_{RC}$, hence $\mathbf{T}(C') \sqsubseteq D$ does not hold in $\mathcal{M}_{RC}$, and $C \sqsubseteq D = \mathbf{T}(C') \sqsubseteq D \notin \overline{TBox}$.

The theorem follows by contraposition.

5   Conclusions and Related Works

A lot of work has been done in order to extend the basic formalism of Description Logics (DLs) with nonmonotonic reasoning features [2,7–19]. The purpose of these extensions is to allow reasoning about *prototypical properties* of individuals or classes of individuals.

The interest of rational closure for DLs is that it provides a significant and reasonable skeptical nonmonotonic inference mechanism, while keeping the same complexity as the underlying logic. The first notion of rational closure for DLs was defined by Casini and Straccia [15]. Their rational closure construction for $\mathcal{ALC}$ directly uses entailment in $\mathcal{ALC}$ over a materialization of the KB. A variant of this notion of rational closure has been studied in [19], and a semantic characterization for it has been proposed. In [1,6] a notion of rational closure for the logic $\mathcal{ALC}$ has been proposed, building on the notion of rational closure proposed by Lehmann and Magidor [5], together with a minimal model semantics characterization.

It is well known that rational closure has some weaknesses that accompany its well-known qualities, both in the context of propositional logic and in the context of Description Logics. Among the weaknesses is the fact that one cannot separately reason property by property, so that, if a subclass of C is exceptional for a given aspect, it is exceptional "tout court" and does not inherit any of the typical properties of C. Among the strengths of rational closure there is its computational lightness, which is crucial in Description Logics. To overcome the limitations of rational closure, in [20,21] an approach is introduced based on the combination of rational closure and *Defeasible Inheritance Networks*, while in [22] a lexicographic closure is proposed, and in [23] relevant closure, a syntactic stronger version of rational closure. To address the mentioned weakness of rational closure, in this paper we have proposed a finer grained semantics of the semantics for rational closure proposed in [1], where models are equipped with several preference relations. In such a semantics it is possible to relativize the notion of typicality, whence to reason about typical properties independently from each other. We are currently working at the formulation of a syntactic characterization of the semantics which will be a strengthening of rational closure.

As the semantics we have proposed provides a strengthening of rational closure, a natural question arises whether this semantics is equivalent to the lexicographic closure proposed in [24]. In particular, lexicographic closure construction for the description logic $\mathcal{ALC}$ has been defined in [22]. Concerning our Example 3 above, our minimal model semantics gives the same results as lexicographic closure, since $\mathbf{T}(Penguin) \sqsubseteq HasNiceFeather$ can be derived from the lexicographic closure of the TBox and $\mathbf{T}(Penguin) \sqsubseteq HasNiceFeather$ holds in all the minimal canonical enriched models of TBox.

However, a general relation needs to be established.

An approach related to our approach is given in [25], where it is proposed an extension of $\mathcal{ALC} + \mathbf{T}$ with several typicality operators, each corresponding to a preference relation. This approach is related to ours although different: the language in [25] allows for several typicality operators whereas we only have a single typicality operator. The focus of [25] is indeed different from ours, as it does not deal with rational closure, whereas this is the main contribution of our paper.

Acknowledgement. This research is partially supported by INDAM-GNCS Project 2016 "Ragionamento Defeasible nelle Logiche Descrittive".

References

1. Giordano, L., Gliozzi, V., Olivetti, N., Pozzato, G.L.: Semantic characterization of rational closure: from propositional logic to description logics. Artif. Intell. **226**, 1–33 (2015)
2. Giordano, L., Gliozzi, V., Olivetti, N., Pozzato, G.L.: A nonmonotonic description logic for reasoning about typicality. Artif. Intell. **195**, 165–202 (2013)
3. Giordano, L., Gliozzi, V., Olivetti, N., Pozzato, G.L.: ALC+T: a preferential extension of description logics. Fundamenta Informaticae **96**, 1–32 (2009)
4. Kraus, S., Lehmann, D., Magidor, M.: Nonmonotonic reasoning, preferential models and cumulative logics. Artif. Intell. **44**(1–2), 167–207 (1990)
5. Lehmann, D., Magidor, M.: What does a conditional knowledge base entail? Artif. Intell. **55**(1), 1–60 (1992)
6. Giordano, L., Gliozzi, V., Olivetti, N., Pozzato, G.L.: Minimal model semantics and rational closure in description logics. In: 26th International Workshop on Description Logics (DL 2013), vol. 1014, pp. 168–180 (2013)
7. Straccia, U.: Default inheritance reasoning in hybrid KL-ONE-style logics. In: Bajcsy, R. (ed.) Proceedings of the 13th International Joint Conference on Artificial Intelligence (IJCAI 1993), Chambéry, pp. 676–681. Morgan Kaufmann, August 1993
8. Baader, F., Hollunder, B.: Priorities on defaults with prerequisites, and their application in treating specificity in terminological default logic. J. Autom. Reason. (JAR) **15**(1), 41–68 (1995)
9. Donini, F.M., Nardi, D., Rosati, R.: Description logics of minimal knowledge and negation as failure. ACM Trans. Comput. Log. (ToCL) **3**(2), 177–225 (2002)
10. Eiter, T., Lukasiewicz, T., Schindlauer, R., Tompits, H.: Combining answer set programming with description logics for the semantic web. In: Dubois, D., Welty, C., Williams, M. (eds.) Principles of Knowledge Representation and Reasoning: Proceedings of the 9th International Conference (KR 2004), Whistler, pp. 141–151. AAAI Press, June 2004
11. Giordano, L., Gliozzi, V., Olivetti, N., Pozzato, G.L.: Preferential description logics. In: Dershowitz, N., Voronkov, A. (eds.) LPAR 2007. LNCS (LNAI), vol. 4790, pp. 257–272. Springer, Heidelberg (2007). doi:10.1007/978-3-540-75560-9_20
12. Ke, P., Sattler, U.: Next steps for description logics of minimal knowledge and negation as failure. In: Baader, F., Lutz, C., Motik, B. (eds.) Proceedings of Description Logics. CEUR Workshop Proceedings, Dresden, vol. 353, May 2008. CEUR-WS.org
13. Britz, K., Heidema, J., Meyer, T.: Semantic preferential subsumption. In: Brewka, G., Lang, J. (eds.) Principles of Knowledge Representation and Reasoning: Proceedings of the 11th International Conference (KR 2008), Sidney, pp. 476–484. AAAI Press, September 2008
14. Bonatti, P.A., Lutz, C., Wolter, F.: The complexity of circumscription in DLs. J. Artif. Intell. Res. (JAIR) **35**, 717–773 (2009)
15. Casini, G., Straccia, U.: Rational closure for defeasible description logics. In: Janhunen, T., Niemelä, I. (eds.) JELIA 2010. LNCS (LNAI), vol. 6341, pp. 77–90. Springer, Heidelberg (2010). doi:10.1007/978-3-642-15675-5_9
16. Motik, B., Rosati, R.: Reconciling description logics and rules. J. ACM **57**(5), 1–62 (2010)

17. Krisnadhi, A.A., Sengupta, K., Hitzler, P.: Local closed world semantics: keep it simple, stupid! In: Proceedings of Description Logics. CEUR Workshop Proceedings, Barcelona, vol. 745, July 2011
18. Knorr, M., Hitzler, P., Maier, F.: Reconciling owl and non-monotonic rules for the semantic web. In: ECAI 2012, pp. 474–479 (2012)
19. Casini, G., Meyer, T., Varzinczak, I.J., Moodley, K.: Nonmonotonic reasoning in description logics: rational closure for the ABox. In: 26th International Workshop on Description Logics. DL 2013. CEUR Workshop Proceedings, vol. 1014, pp. 600–615 (2013). CEUR-WS.org
20. Casini, G., Straccia, U.: Defeasible inheritance-based description logics. In: Walsh, T. (ed.) Proceedings of the 22nd International Joint Conference on Artificial Intelligence (IJCAI 2011), Barcelona, pp. 813–818. Morgan Kaufmann, July 2011
21. Casini, G., Straccia, U.: Defeasible inheritance-based description logics. J. Artif. Intell. Res. (JAIR) **48**, 415–473 (2013)
22. Casini, G., Straccia, U.: Lexicographic closure for defeasible description logics. In: Proceedings of Australasian Ontology Workshop, vol. 969, pp. 28–39 (2012)
23. Casini, G., Meyer, T., Moodley, K., Nortjé, R.: Relevant closure: a new form of defeasible reasoning for description logics. In: Fermé, E., Leite, J. (eds.) JELIA 2014. LNCS (LNAI), vol. 8761, pp. 92–106. Springer, Heidelberg (2014). doi:10.1007/978-3-319-11558-0_7
24. Lehmann, D.J.: Another perspective on default reasoning. Ann. Math. Artif. Intell. **15**(1), 61–82 (1995)
25. Gil, O.F.: On the non-monotonic description logic ALC+T_{min}. In: 15th International Workshop on Non-Monotonic Reasoning (2014)

A Framework for Automatic Population of Ontology-Based Digital Libraries

Laura Pandolfo[1]([🖂]), Luca Pulina[2], and Giovanni Adorni[1]

[1] DIBRIS, Università di Genova, Via Opera Pia, 13, 16145 Genova, Italy
laura.pandolfo@edu.unige.it, adorni@unige.it
[2] Polcoming, Università di Sassari, Viale Mancini n. 5, 07100 Sassari, Italy
lpulina@uniss.it

Abstract. Maintaining updated ontology-based digital libraries faces two main issues. First, documents are often unstructured and in heterogeneous data formats, making it even more difficult to extract information and search in. Second, manual ontology population is time consuming and therefore automatic methods to support this process are needed.

In this paper, we present an ontology-based framework aiming at populating ontologies. In particular, we propose an approach for triplet extraction from heterogeneous and unstructured documents in order to automatically populate ontology-based digital libraries. Finally, we evaluate the proposed framework on a real world case study.

Keywords: Ontology population · Ontology-based digital library · Information Extraction

1 Introduction

The Web has witnessed a rapid growth in the amount of data and documents stored in a variety of digital archives and libraries. However, research and practice in digital libraries are not new concepts. As early as 1945, Vannevar Bush proposed a hypertext-like system describing it as "a device in which an individual stores all his books, records, and communications, and which is mechanized so that it may be consulted with exceeding speed and flexibility" [1]. Nowadays, a large number of digital libraries are accessible and they relate different types of documents, such as government dossiers, online news, cultural heritage data, medical and biological records.

In this field, Semantic Web technologies provide methodologies to support, e.g., the integration of multiple data sources, and the semantic annotation of digital resources. Ontologies are commonly used in several digital libraries, see, e.g., [2–4]. Maintaining updated ontology-based digital libraries faces two main issues. First, extracting data from unstructured texts is a challenging task, since processing natural language obviously requires more than a simple conversion of text in machine readable format. Second, automatic ontology population – i.e., the process of adding new instances to the ontology – is a crucial process.

© Springer International Publishing AG 2016
G. Adorni et al. (Eds.): AI*IA 2016, LNAI 10037, pp. 406–417, 2016.
DOI: 10.1007/978-3-319-49130-1_30

In fact, instantiating ontologies with entities and relations annotated in documents is a relevant step towards the provision of valuable semantic digital libraries and related services [5]. Manual ontology population requires specialized expertise, and it is also a very time-consuming and resource-intensive task. If done (semi-) automatically, this activity brings several challenges. In the last years, several approaches have been proposed in the scientific literature, most of them based on Natural Language Processing (NLP), Machine Learning (ML), and Information Extraction (IE) techniques – see, e.g., [6,7].

In this paper we present a framework for the automatic population of ontology-based applications. The main contribution of the proposed framework is represented by the triplet extractor (TE) module, devoted to detect the three components of RDF triples, namely subject, predicate, and object. We design TE by taking into account linguistic studies demonstrating that several languages, such as Italian and English, follows the Subject-Verb-Object (SVO) word order [8]. For instance, in Italian language each sentence semantically well-formed must contain at least one verb, since it regulates the syntactic relationships between the subject and the object [9]. The verb represents the dynamic element which allocates the meaning of a concept in the speaker's mind [10]. Given the considerations above, the first (and crucial) step in TE is the detection of the main verb V in the sentence. Next, subject and object are detected considering the nouns group to the left and to the right of V, respectively. In order to do that, linguistic pre-processing tasks are needed, among which sentence splitting, Part-Of-Speech (POS), and Named Entity Recognition (NER).

The approach used in the presented framework differs from [11] since we take into account unstructured and heterogeneous documents that not share common layout properties. Moreover, unlike [12,13], we start with the main verb detection and then we extract the subject and object candidates, instead to detect triples starting from the subject.

We test our framework in the context of the ontology-based digital library STOLE [14], a collection of journal articles concerning the legislative history of public administration in Italy. This case study offers several challenges concerning automatic ontology population and, in general, the whole information extraction process.

The rest of the paper is organized as follows. After some preliminaries (Sect. 2), in Sect. 3 we describe the proposed ontology-based framework for automatic ontology population. In Sect. 4 we describe the case study mentioned above, and we report the results of a preliminary experimental analysis. We conclude the paper in Sect. 5 with some final remarks and discussing future work.

2 Preliminaries

2.1 Ontologies

A *knowledge base* is a pair $\mathcal{K} = \langle \mathcal{T}, \mathcal{A} \rangle$ where $\mathcal{T}$ is the *Terminological Box* (*Tbox*) specifying known classes of data and relations among them, while $\mathcal{A}$ is the *Assertional Box* (*Abox*) specifying the extensional knowledge, i.e., factual data and their classification. The Tbox is a formal conceptualization of some domain of interest, i.e., an *ontology*, in the words of [15]. In order to describe in a formal way real domains through ontologies – and to connect them with Semantic Web based applications –, the World Wide Web Consortium's (W3C) has proposed the Web Ontology Language (OWL), whose newest version is OWL 2 — see [16].

As stated in [17], there are two alternative ways of assigning meaning to ontologies in OWL 2: using the Direct Semantics (OWL 2 DL) and the RDF-Based Semantics (OWL 2 Full). From an automated reasoning perspective, OWL 2 DL is a syntactic variant of some *Description Logics* (DLs), i.e., formal language and rules to enable well-founded inference of new facts from those explicitly stored in the *Abox* according to the structure of the *Tbox*. In several knowledge representation systems, the ontology is the central core for the representation for a specific domain. In fact, the greatly expressive power provided by OWL 2 allows to manage the structure of knowledge.

The most important building elements of OWL 2 can be grouped into three categories: *individuals, classes, properties*. Individuals, also called instances, represent actual entities from the specific domain. *Individuals* can be named with explicit name and they are usually identified using an Internationalized Resource Identifier (IRI). *Classes* represent concepts that include sets of individuals. Classes can be in relation to each other through `rdfs:subClassOf` constructor. *Properties*, also called roles, represent another basic building block of OWL. Properties are binary relations and they can be of two types, namely *object properties*, i.e., relations that connect pairs of individuals, and *data properties*, i.e., relations between instances of classes and datatypes.

2.2 Information Extraction

Information Extraction (IE) is the name given to any process that selectively structures and combines data that is found in one or more texts [18]. In the mid-to-late 1980s, the interest and rapid advances in the field of IE have been essentially influenced by the DARPA-initiated series of Message Understanding Conferences (MUCs)[1].

Unlike Information Retrieval (IR), where the aim is to match user queries with relevant information documents, the general goal of IE is to discover and extract relevant information from unstructured or semi-structured texts. Being machine-readable, the extracted information can be presented to an end-user or it can be reused by other computer systems such as search engines and database management systems to provide better services to end-users [19].

The process of extracting structured information involves identification of certain small-scale structures like noun phrases and verbs. However, some domain-specific knowledge is required in order to correctly aggregate the partially extracted information into a structured form [20]. For this purpose, the IE

---

[1] http://www-nlpir.nist.gov/related_projects/muc/.

process is based on user-defined structures called templates, that contains a set of pre-defined slots.

Recently, Ontology-based Information Extraction (OBIE) has emerged as a new sub-field of IE. The use of ontologies in these systems make domain knowledge more explicit and formal [21]. The main goal of OBIE systems is to identify, e.g., in text, concepts, properties, and relations expressed in ontologies. Computational linguistic techniques and theories are playing a strong role in this area, since the IE activity concerns often processing human language texts by means of some NLP techniques. Typically, most OBIE systems – for a detailed survey see [22] – combine different IE and NLP tasks, including the following:

- **Tokenization.** The main task of a tokenizer is to segment an input string into a sequence of token with sentential boundaries marked.
- **Sentence Splitting.** It is done by segmenting a compound sentence containing conjunctions into several simple sentences.
- **Part-Of-Speech (POS).** POS tagging is the process of going through a corpus of sentences and labeling each word in each sentence with its grammatical category, such as verbs, nouns, adjectives, adverbs, etc.
- **Named Entity Recognition (NER).** This task is high language-dependent and concerns the identification and classification of predefined types of named entities, such as organizations, persons, place names, dates, numerical and currency expressions, etc.

3   Automatic Triple Extraction and Ontology Population

In this section we present both architecture – Fig. 1 – and implementation details of the proposed framework.

Looking at Fig. 1, we can see that the framework is composed of the modules described in the following.

Sentence Detector module (SD). It is devoted to accomplish the Sentence Splitting task. It takes as input the content of a document in text format, and a sentence detector model in OpenNLP format. The output is a text file containing the computed sentences. SD has been implemented using OpenNLP 1.6.0 Sentence Detection APIs [23].

Named Entity Recognition module (NER). It aims to accomplish the Named Entity Recognition task. It receives as input the output of SD, while returns as output a text file with tagged entities. NER has been implemented on top of the OpenNLP 1.6.0 Name Finder APIs [23]. Following what suggested in the OpenNLP documentation, Named Entity Recognition Models is composed of a pool of model files, one for each entity to recognize. In our framework, entities have the same name of the ontology classes considered for the automatic ontology population. The set of classes to populate is given as input by the user.

Part-Of-Speech Tagger module (POS). Its role is to add Part-Of-Speech tags to the output of the NER module. In the current implementation we use

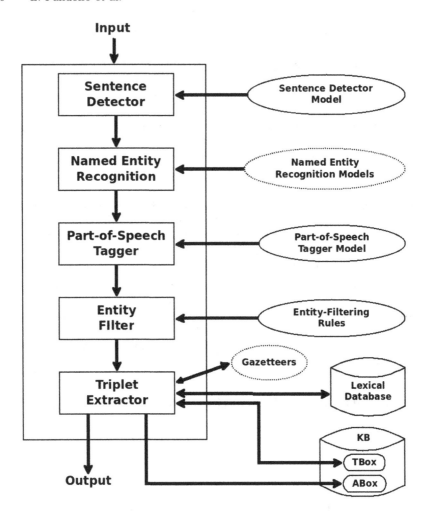

Fig. 1. Overview of the architecture of the framework.

OpenNLP 1.6.0 POS Tagger APIs [23]. As in the case of SD, it also needs a POS Tagger model in OpenNLP format.

Entity Filter module (EF). It aims to refine the results returned by NER on the basis of the output obtained by POS. As an example, e.g., if we would like to populate an ontology class named **Person**, EF will remove terms tagged by the POS such as conjunctions if they occur in a NER tagged term related to **Person**. Rules such as the one mentioned above are encoded in the module **Entity-Filtering Rules** (EFR). EFR is composed of a distinct set of rules for each entity type. Finally, EF returns a new POS and ER tagged text.

Triplet Extractor module (TE). It is the core of the framework, and it takes as input the output of EF. As the name suggests, TE is devoted to extract relevant triples from the text and add them to the KB. As mentioned in the Introduction, our approach to TE exploits the fact that the verb is the dynamic element which allocates the meaning of a concept in the speaker's mind.

Figure 2 summarizes the procedure for triple extraction in TE. Looking at the figure, we can see that TRIPLETEXTRACTOR takes the following input:

```
TRIPLETEXTRACTOR(T, TB, G)
 1  CANDIDATE ← NIL
 2  foreach t in T
 3    S ← NIL
 4    O ← NIL
 5    for i = 1 to SIZE(t)
 6      v ← t[i]
 7      if ISVERB(v) then
 8        DETECTSUBJECT(i, S)
 9        DETECTOBJECT(i, O)
10      if not EMPTY(S) and not EMPTY(O)
11        foreach s in S
12          foreach o in O
13            ADDCANDIDATE(CANDIDATE, s, v, o)
14      i ← i + 1
15  TRIPLE ← NIL
16  foreach c in CANDIDATE
17    v ← GETVERB(c)
18    op ← NIL
19    foreach g in G
20      for i = 1 to SIZE(g)
21        if v == g[i]
22          op ← NAMEOF(g)
23          break
24        GETSYNSET(SYN, g[i])
25        foreach syn in SYN
26          if (syn == v)
27            op ← NAMEOF(g)
28            break
29        i ← i + 1
30    if op ≠ NIL
31      d ← GETDOMAINNAME(TB, op)
32      r ← GETRANGENAME(TB, op)
33      s ← GETSUBJECT(c)
34      o ← GETOBJECT(c)
35      if ENTITY(s) == d and ENTITY(o) == r
36        t ← MAKETRIPLE(TB, s, op, o)
37        PUSH(TRIPLE, t)
```

Fig. 2. Pseudo-code of triplet extraction basic routine.

- An array T, in which every element is an array containing the text lines coming from the output of EF;
- TB, that denotes a data structure containing the considered $TBox$; and
- An array G, in which every element is an array containing a gazetteer of verbs (Gazetteer in Fig. 1).

The procedure in Fig. 2 first looks for a pool of candidate triples in the form SVO (Lines 2–14), and store them in an array initially empty (Line 1). Each token in t is checked by the function IsVERB (Line 7). Notice that, for compound tenses, we only consider the main verb, excluding auxiliaries. If the result of the check is true, the task of DETECTSUBJECT (Line 8) is to find NER tagged words on the left of v, and add them to the array S. DETECTSUBJECT stops when it finds a verb or reaches the begin of t.

DETECTOBJECT (Line 9) works in a similar way; it looks for NER tagged words on the right of v, and add them to the array O. DETECTOBJECT stops when it finds a verb or reaches the end of t. Then, candidate triples are stored in a dedicate array (Lines 10–13).

Next, the procedure checks if the triples contained in CANDIDATE are consistent with respect to TB (Lines 16–36). The verb v of each candidate triple c is conjugate to its infinitive form (Line 17), and it performs a syntactic check with respect to the elements g of the gazetteers G (Lines 19–21). Notice that G contains a g related to each Object Property involved in the RDF triples detection.

If the match is positive, the name of the Object Property related to g is stored in op, and the loop ends (Lines 21–23). If it is not the case, synonyms of the current element $g[i]$ of g are collected and stored in the array SYN (Line 24) – see also Lexical Database in Fig. 1. Then, v is syntactically checked against each element of SYN (Lines 25–28).

Once detected op, TB is queried in order to obtain the name of domain and range – d and r, respectively – related to op (Lines 31–32). Both d and r are compared with the entity names of subject and object of $c - s$ and o, respectively (Line 35). If the resulting triple t is consistent with respect to the ontology, it is collected in the array TRIPLE (Lines 35–37), else c is discarded. Finally, triples are added to the KB.

Concerning implementation details, we report that TE is implemented in JAVA language. The interaction with TBox is implemented on top of the OWL APIs [24] (Version 3), while as triple store we use Stardog 4 Community[2]. Triples are inserted to Stardog using SPARQL 1.1 INSERT DATA queries [25]. Lexical Database is implemented using MultiWordNet [26]; Notice that TE implements a cache mechanism in order to minimize the total amount of calls to MultiWordNet. Finally, extracted triples are also output in a Notation 3 [27] file.

---

[2] http://stardog.com.

4   Case Study: The STOLE Ontology-Based Digital Library

In this Section we test the framework described in Sect. 3 to populate the STOLE ontology-based digital library [14]. STOLE collects journal articles published in the 19th and 20th centuries concerning the history of public administration in Italy. The main goal of STOLE is to clearly model historical concepts and, at the same time, to gain insights into this specific field, e.g., supporting historians to find out some unexplored but useful aspects about a particular event or person.

The STOLE ontology [28–30] is the conceptual layer of the digital library, and its modeling language is OWL2 DL. Main classes of the STOLE ontology are briefly described in the following:

- `Article` is the class that represents the collection of journal articles.
- `Event` contains relevant events for this specific domain.
- `Institution` is the class that represents the different public institutions.
- `Journal` denotes the collection of journals.
- `LegalSystem` includes the successive systems for interpreting and enforcing the laws.
- `Person` is the class representing people cited in the articles or involved in events. This class contains one subclass, `Author`, that includes the contributors of the articles.
- `Subject` is a class representing topics tackled in journals.

Some of the most important object properties in STOLE are listed below:

- `cites` is the object property that connects individuals in `Article` to individuals in `Person`, highlighting the people that are mentioned in an article.
- `hasInstitutionEventOf` is the object property that relates individuals in `Institution` to individuals in `Event`.
- `hasLegalSystemEventOf` is a relation connecting individuals in `LegalSystem` to individuals in `Event`.
- `hasLifeEventOf` connects individuals in `Person` to individuals in `Event`.
- `mentions` highlights which historical event is mentioned in a given article. Its domain is `Article`, while its range is `Event`.
- `reportsInstitution` points out which institutions appear in an article. The domain is `Article`, while the range is `Institution`.
- `reportsLegalSystem` indicates which legal systems are reported in an article. The domain is `Article`, while the range is `LegalSystem`.

The case study here presented offers several challenging issues related to the Ontology Population task. First, most part of the historical journal articles in STOLE are low-quality scanned PDFs of photocopies; despite the progress made in Optical Character Recognition (OCR), the converted file in text format reports several typos and errors. In general, it is well-established that performance of IE process could be seriously affected by poor OCR. Second, most part of the considered journal articles are written in an old technical Italian language, and even the manual annotation by experts required a great deal of effort. The complexity of the language represents a significant barrier since there are not available models for IE and NLP tasks, neither for old Italian language nor for the investigated domain. Furthermore, the lack of specific gazetteers and lexical databases it makes this case study even trickier.

In our case study, we are mainly interested to extract from the historical journal articles triples having as object property the ones related to `Event`, namely `hasInstitutionEventOf`, `hasLegalSystemEventOf`, and `hasLifeEventOf` ("Pool A" in the following). In order to do that, with reference to Fig. 1 in Sect. 3, we set up our framework as follows:

- Models for SD and POS modules are the ones available at https://github. com/aciapetti/opennlp-italian-models.
- Concerning NER, we train[3] models for the entities related to `Event`, `Institution`, `LegalSystem`, and `Person`. Both model files and the list of documents used to train the models are available at http://visionlab.uniss.it/ 2016AIIA.tar.gz.
- Regarding EF Rules, a set of rules for each entity has been hard coded in the EF module.
- We compute a gazetteer for each "Pool A" object property. Each gazetteer contains a list of verbs (in infinitive form) suggested by the domain experts.

As by-product of the process, we can also add triples related to the object properties `cites`, `mentions`, `reportsInstitution`, and `reportsLegalSystem` ("Pool B").

We test the proposed framework on the following articles and encyclopedia entries, annotated by domain experts:

- AA.VV. "Cronaca Parlamentare. Sessione 1851". Rivista Amministrativa del Regno. Vol. 2, 1851. (Document 1, in the following)
- Carlo Boggio. "Materie generali. Progetto di legge del ministro Pinelli". Rivista Amministrativa del Regno, Vol. 3, 1853. (Document 2)
- Giacomo Curlo Spinola. "Materie generali. Necessità di migliorare la condizione degli impiegati dell'Amministrazione provinciale". Rivista Amministrativa del Regno, Vol. 7, 1856. (Document 3)
- Dizionario Zanichelli. "Cavour, Camillo Benso", http://dizionaripiuzanichelli. it. (Document 4)
- Dizionario Zanichelli. "Carlo Alberto", http://dizionaripiuzanichelli.it. (Document 5)
- Portale Storico della Camera dei Deputati. "Carlo Cadorna", http://storia. camera.it/presidenti/cadorna-carlo. (Document 6)

Table 1 shows the results of the experimentation mentioned before. Concerning "Pool A", first we notice that domain experts found a small number of triples

---

[3] We used OpenNLP default parameters, namely cutoff frequencies set to 5, number of iterations set to 100.

Table 1. Automatic triples extraction results. The table is organized as follows. The first column ("Document") reports the article name, and it is followed by two group of columns, "Pool A" and "Pool B", respectively. Each group is composed of three columns reporting values of Precision, Recall, and F-Measure (columns "P", "R", and "F", respectively.

Document	Pool A			Pool B		
	P	R	F	P	R	F
Document 1	1.00	1.00	1.00	0.85	0.55	0.74
Document 2	1.00	0.50	0.66	0.87	0.65	0.74
Document 3	0.66	1.00	0.57	0.83	0.60	0.69
Document 4	1.00	1.00	1.00	1.00	0.46	0.63
Document 5	1.00	0.40	0.57	0.74	0.46	0.57
Document 6	1.00	0.50	0.66	0.82	0.28	0.42

(equals or smaller to 5) in each document. In addition, low values of recall are mainly due to the small accuracy of NER models, especially the one related to the class Event; this is confirmed looking at the results related to "Pool B". NER models have been trained on datasets with a total amount of sentences ranging from 1000 to 2000, while OpenNLP developers suggest that training data should contain at least 15000 sentences.

5   Conclusions

In this paper we described our framework for automatic ontology population from unstructured texts in natural language. As we have seen in the previous Section, NLP tasks play a crucial role in the success of the whole process. In particular, we notice that optimized sentence splitting and POS tagging models are needed, since the only available models for Italian language report several inaccuracies and inconsistencies in this case study. For instance, in sentences containing words such as *"l'art. 3 della legge..."* the splitting model marked the end of the sentence – in a wrong way – after the full stop. This can heavily affect the performance of triple detection because the sentence could become meaningless. Even the POS tag module reports some failures, e.g., tagging nouns or verbs as conjunctions.

As part of future work, we are planning to train extended NER models, in order to better detect entities. Moreover, we intend to include in our framework the co-reference resolution analysis for resolving anaphoric references by pronouns and definite noun phrases. Finally, further experimental analysis of the case study proposed will be performed using the described framework.

Acknowledgments. The authors would like to thank Dr. Anastasia Di Nunzio for helpful discussion on linguistic typology studies.

References

1. Bush, V.: As we may think. ACM SIGPC Notes **1**(4), 36–44 (1979)
2. Zghal, H.B., Moreno, A.: A system for information retrieval in a medical digital library based on modular ontologies and query reformulation. Multimedia Tools Appl. **72**(3), 2393–2412 (2014)
3. Li, N., Zhu, L., Mitra, P., Mueller, K., Poweleit, E., Giles, C.L.: Orechem chemxseer: a semantic digital library for chemistry. In: Proceedings of the 10th Annual Joint Conference on Digital Libraries, pp. 245–254. ACM (2010)
4. Doerr, M., Gradmann, S., Hennicke, S., Isaac, A., Meghini, C., van de Sompel, H.: The Europeana data model (edm). In: World Library and Information Congress: 76th IFLA General Conference and Assembly, pp. 10–15 (2010)
5. Ruiz-Martínez, J.M., Minarro-Giménez, J.A., Castellanos-Nieves, D., García-Sánchez, F., Valencia-Garcia, R.: Ontology population: an application for the e-tourism domain. Int. J. Innovative Comput. Inf. Control (IJICIC) **7**(11), 6115–6134 (2011)
6. Bontcheva, K., Tablan, V., Maynard, D., Cunningham, H.: Evolving gate to meet new challenges in language engineering. Nat. Lang. Eng. **10**(3–4), 349–373 (2004)
7. Faria, C., Serra, I., Girardi, R.: A domain-independent process for automatic ontology population from text. Sci. Comput. Program. **95**, 26–43 (2014)
8. Antinucci, F., Cinque, G.: Sull'ordine delle parole in italiano: l'emarginazione. Studi di grammatica italiana VI, pp. 121–146 (1977)
9. Boschi, S.: La comunicazione vista dal nostro cervello. Lampi di stampa, Milan (2008)
10. Sabatini, F.: La comunicazione e gli usi della lingua. Loescher, Torino (1991)
11. Adrian, W.T., Leone, N., Manna, M.: Ontology-driven information extraction. arXiv preprint arXiv:1512.06034 (2015)
12. Benammar, R., Trémeau, A., Maret, P.: An approach for ontology population based on information extraction techniques. In: Debruyne, C., Panetto, H., Meersman, R., Dillon, T., Weichhart, G., An, Y., Agostino Ardagna, C. (eds.) On the Move to Meaningful Internet Systems: OTM 2015 Conferences. LNCS, vol. 9415, pp. 397–404. Springer, Switzerland (2015). doi:10.1007/978-3-319-26148-5_26
13. Rusu, D., Dali, L., Fortuna, B., Grobelnik, M., Mladenic, D.: Triplet extraction from sentences. In: Proceedings of the 10th International Multiconference Information Society-IS, pp. 8–12 (2007)
14. Adorni, G., Maratea, M., Pandolfo, L., Pulina, L.: An ontology-based archive for historical research. In: Proceedings of the 28th International Workshop on Description Logics. CEUR Workshop Proceedings, Athens, Greece, 7–10 June 2015, vol. 1350. CEUR-WS.org (2015)
15. Gruber, T.R.: A translation approach to portable ontology specifications. Knowl. Acquis. **5**(2), 199–220 (1993)
16. Motik, B., Patel-Schneider, P.F., Parsia, B., Bock, C., Fokoue, A., Haase, P., Hoekstra, R., Horrocks, I., Ruttenberg, A., Sattler, U., et al.: Owl 2 web ontology language: structural specification and functional-style syntax. W3C Recommendation **27**(65), 159 (2009)
17. Hitzler, P., Krötzsch, M., Parsia, B., Patel-Schneider, P.F., Rudolph, S.: Owl 2 Web Ontology Language Primer, 2nd edn. W3C Recommendation, December 2012
18. Dale, R., Moisl, H., Somers, H.: Handbook of Natural Language Processing. CRC Press, Boca Raton (2000)

19. Jiang, J.: Information extraction from text. In: Aggarwal, C.C., Zhai, C. (eds.) Mining Text Data, pp. 11–41. Springer, Heidelberg (2012). doi:10.1007/978-1-4614-3223-4_2

20. Piskorski, J., Yangarber, R.: Information extraction: past, present and future. In: Poibeau, T., Saggion, H., Piskorski, J., Yangarber, R. (eds.) Multi-source, Multilingual Information Extraction and Summarization. Theory and Applications of Natural Language Processing, pp. 23–49. Springer, Heidelberg (2013). doi:10.1007/978-3-642-28569-1_2

21. Saggion, H., Funk, A., Maynard, D., Bontcheva, K.: Ontology-Based Information Extraction for Business Intelligence. Springer, Heidelberg (2007)

22. Wimalasuriya, D.C., Dou, D.: Ontology-based information extraction: an introduction and a survey of current approaches. J. Inf. Sci. **36**, 306 (2010)

23. The opennlp project (2005). http://opennlp.apache.org. Accessed June 2016

24. Horridge, M., Bechhofer, S.: The owl API: a java api for owl ontologies. Semant. Web **2**(1), 11–21 (2011)

25. Harris, S., Seaborne, A., Prudhommeaux, E.: Sparql 1.1 Query Language, vol. 21. W3C Recommendation (2013)

26. Pianta, E., Bentivogli, L., Girardi, C.: MultiWordNet: developing an aligned multilingual database. In: Proceedings of the 1st International Conference on Global WordNet, pp. 293–302 (2002)

27. Berners-Lee, T., Connolly, D., Kagal, L., Scharf, Y., Hendler, J.: N3logic: a logical framework for the world wide web. Theor. Pract. Logic Program. **8**(03), 249–269 (2008)

28. Adorni, G., Maratea, M., Pandolfo, L., Pulina, L.: An ontology for historical research documents. In: Cate, B., Mileo, A. (eds.) RR 2015. LNCS, vol. 9209, pp. 11–18. Springer, Heidelberg (2015). doi:10.1007/978-3-319-22002-4_2

29. Adorni, G., Maratea, M., Mura, S., Pandolfo, L., Pulina, L., Soddu, F.: A domain ontology for historical research documents. In: Artificial Intelligence for Cultural Heritage, pp. 25–48 Cambridge Scholars Publishing (2016)

30. Kontchakov, R., Pandolfo, L., Pulina, L., Ryzhikov, V., Zakharyaschev, M.: Temporal and spatial OBDA with many-dimensional Halpern-Shoham logic. To appear in Proceedings of IJCAI (2016)

Reasoning About Surprising Scenarios in Description Logics of Typicality

Gian Luca Pozzato$^{(\boxtimes)}$

Dipartimento di Informatica, Università di Torino, Turin, Italy
gianluca.pozzato@unito.it

Abstract. We continue our investigation on nonmonotonic procedures for preferential Description Logics in order to reason about plausible but *surprising* scenarios. We consider an extension $\mathcal{ALC} + \mathbf{T_R^{exp}}$ of the non-monotonic logic of typicality $\mathcal{ALC} + \mathbf{T_R}$ by inclusions of the form $\mathbf{T}(C) \sqsubseteq_d D$, where d is a degree of expectedness. We consider a notion of extension of an ABox, in order to assume typicality assertions about individuals satisfying cardinality restrictions on concepts, then we define a preference relation among such extended ABoxes based on the degrees of expectedness, then we restrict entailment to those extensions that are minimal with respect to this preference relation. We propose a decision procedure for reasoning in $\mathcal{ALC} + \mathbf{T_R^{exp}}$ and we exploit it to show that entailment is in EXPTIME as for the underlying $\mathcal{ALC}$. Last, we introduce a further extension of the proposed approach in order to reason about all plausible extensions of the ABox, by restricting the attention to specific degrees of expectedness ranging from the most surprising scenarios to the most expected ones.

Keywords: Description Logics · Nonmonotonic reasoning · Typicality

1 Introduction

Nonmonotonic extensions of Description Logics (from now on, DLs for short) have been actively investigated since the early 90s [1–8] in order to tackle the problem of representing *prototypical* properties of classes and to reason about *defasible* inheritance. A simple but powerful nonmonotonic extension of DLs is proposed in [9–13]: in this approach "typical" or "normal" properties can be directly specified by means of a "typicality" operator $\mathbf{T}$ enriching the underlying DL. The semantics of the $\mathbf{T}$ operator is characterized by the core properties of nonmonotonic reasoning axiomatized by either *preferential logic* [14] or *rational logic* [15]. We focus on the Description Logic $\mathcal{ALC} + \mathbf{T_R}$ introduced in [13]. In this logic one can express defeasible inclusions such as "normally, depressed people have sleep disorders":

$$\mathbf{T}(Depressed) \sqsubseteq \exists Symptom.SleepDisorder$$

As a difference with standard DLs, one can consistently express exceptions and reason about defeasible inheritance as well. For instance, a knowledge base can

© Springer International Publishing AG 2016
G. Adorni et al. (Eds.): AI*IA 2016, LNAI 10037, pp. 418–432, 2016.
DOI: 10.1007/978-3-319-49130-1_31

consistently express that "normally, a patient affected by depression is not able to react to positive life events", whereas "mood reactivity (ability to feel better temporarily in response to positive life events) is a typical symptom of atypical depression" as follows:

$$AtypicalDepressed \sqsubseteq Depressed$$

$$\mathbf{T}(Depressed) \sqsubseteq \neg\exists Symptom.MoodReactivity$$

$$\mathbf{T}(AtypicalDepressed) \sqsubseteq \exists Symptom.MoodReactivity$$

From a semantic point of view, models of $\mathcal{ALC} + \mathbf{T_R}$ are standard models extended by a function f which selects the typical/most normal instances of any concept C, i.e. the extension of $\mathbf{T}(C)$ is defined as $(\mathbf{T}(C))^{\mathcal{I}} = f(C^{\mathcal{I}})$. The function f satisfies a set of postulates that are a restatement of Kraus, Lehmann and Magidor's axioms of rational logic $\mathbf{R}$. This allows the typicality operator to inherit well-established properties of nonmonotonic reasoning (e.g. specificity).

The logic $\mathcal{ALC} + \mathbf{T_R}$ itself is too weak in several application domains. Indeed, although the operator $\mathbf{T}$ is nonmonotonic ($\mathbf{T}(C) \sqsubseteq E$ does not imply $\mathbf{T}(C \sqcap D) \sqsubseteq E$), the logic $\mathcal{ALC} + \mathbf{T_R}$ is monotonic, in the sense that if the fact F follows from a given knowledge base KB, then F also follows from any KB' $\supseteq$ KB. As a consequence, unless a KB contains explicit assumptions about typicality of individuals, there is no way of inferring defeasible properties about them: in the above example, if KB contains the fact that Kate is a depressed woman, i.e. $Depressed(kate)$ belongs to KB, it is not possible to infer that she has sleep disorders ($\exists Symptom.SleepDisorder(kate)$). This would be possible only if the KB contained the stronger information that Kate is a *typical* depressed woman, i.e. $\mathbf{T}(Depressed)(kate)$ belongs to (or can be inferred from) KB. In order to overcome this limit and perform useful inferences, in [13] the authors have introduced a nonmonotonic extension of the logic $\mathcal{ALC} + \mathbf{T_R}$ based on a minimal model semantics, corresponding to a notion of *rational closure* as defined in [15] for propositional logic. Intuitively, the idea is to restrict our consideration to (canonical) models that maximize typical instances of a concept when consistent with the knowledge base. The resulting logic, call it $\mathcal{ALC} + \mathbf{T_R}^{RaCl}$, supports typicality assumptions, so that if one knows that Kate is depressed, one can nonmonotonically assume that she is also a *typical* depressed if this is consistent, and therefore that she has sleep disorders. From a semantic point of view, the logic $\mathcal{ALC} + \mathbf{T_R}^{RaCl}$ is based on a preference relation among $\mathcal{ALC} + \mathbf{T_R}$ models and a notion of *minimal entailment* restricted to models that are minimal with respect to such preference relation.

The logic $\mathcal{ALC} + \mathbf{T_R}^{RaCl}$ imposes to consider *all* typicality assumptions that are consistent with a given KB. This seems to be too strong in several application domains, in particular when the need arises of bounding the cardinality of the extension of a given concept, that is to say the number of domain elements being members of such a concept, as introduced in [16]. As a further example, consider the KB from the domain of sports entertainment from [17], where preliminary ideas of this approach have been outlined by considering an extension of the lightweight DL-Lite$_c$ore for surprising scenarios: $\mathbf{T}(FaceWrestler) \sqsubseteq$

$RoyalRumbleWinner$; $\mathbf{T}(Returning) \sqsubseteq RoyalRumbleWinner$; $\mathbf{T}(Predicted) \sqsubseteq$ $RoyalRumbleWinner$. The first inclusion represents that, normally, a face wrestler wins the Royal Rumble match, an annual wrestling event involving thirty athletes. The second one states that, typically, an athlete returning from an injury wins the Royal Rumble match. The third and last inclusion represents that an athlete whose victory has been predicted by wrestling web sites normally wins the Royal Rumble match. If the assertional part of the KB contains the facts: $FaceWrestler(dean)$, $Returning(seth)$, $FaceWrestler(roman)$, $Predicted(roman)$, whose meaning is that Dean is a face athlete, Seth is returning from an injury, and Roman is a face wrestler who has been predicted to win the rumble match, respectively, then in $\mathcal{ALC} + \mathbf{T}_{\mathbf{R}}^{RaCl}$ we conclude that: $\mathbf{T}(FaceWrestler)(dean)$, $\mathbf{T}(Returning)(seth)$, $\mathbf{T}(FaceWrestler)(roman)$, $\mathbf{T}(Predicted)(roman)$, and then that Dean, Seth and Roman are all winners. This happens in $\mathcal{ALC} + \mathbf{T}_{\mathbf{R}}^{RaCl}$ because it is consistent to make the three assumptions above, that hold in all minimal models, however one should be interested in three distinct scenarios that cannot be captured by $\mathcal{ALC} + \mathbf{T}_{\mathbf{R}}^{RaCl}$ as it is. One could think of extending the logic $\mathcal{ALC} + \mathbf{T}_{\mathbf{R}}^{RaCl}$ by means of *cardinality restrictions*, in the example by imposing that there is only one member of the extension of the concept $RoyalRumbleWinner$, however the resulting knowledge base would be inconsistent.

Furthermore, it is sometimes useful to restrict reasoning to *surprising* scenarios, excluding "trivial"/"obvious" ones. For instance, recently a great attention has been devoted to *serendipitous* search engines, that must be able to provide results that are "*surprising, semantically cohesive*, i.e. *relevant* to some information need of the user, or just *interesting*" [18]. In this sense, the scenario (among those satisfying cardinality restrictions) obtained by assuming the largest set of consistent typicality assumptions in $\mathcal{ALC} + \mathbf{T}_{\mathbf{R}}^{RaCl}$ corresponds to the most trivial one, whereas one could be interested in less expected ones, in which some typicality assumptions are discarded.

The solution we propose in this work, called $\mathcal{ALC} + \mathbf{T}_{\mathbf{R}}^{\mathsf{exp}}$, is based on the combination of two components. On the one hand, we allow one to express different *degrees of expectedness* of typicality inclusions, having the form $\mathbf{T}(C) \sqsubseteq_d D$ where d is a positive integer such that an inclusion with degree d is more "trivial" (or "obvious") with respect to another one with degree $d' \le d$: this allows one to describe several plausible scenarios by considering different combinations of typicality assumptions about individuals named in the ABox. Such degrees introduce a rank of expectedness among plausible scenarios, ranging from surprising to obvious ones. On the other hand, TBoxes are extended to allow restrictions about the cardinality of concepts, in order to "filter" such plausible scenarios. Finally, reasoning tasks are restricted to reasonable but "surprising enough" (or "not obvious") scenarios satisfying cardinality restrictions. We then define notions of skeptical and credulous entailment and we describe a sound and complete decision procedure that allows us to show that reasoning in $\mathcal{ALC} + \mathbf{T}_{\mathbf{R}}^{\mathsf{exp}}$ is EXPTIME-complete for both skeptical and credulous entailment.

It is worth noticing that the proposed logic $\mathcal{ALC} + \mathbf{T}_{\mathbf{R}}^{\text{exp}}$ is not intended to replace existing extensions of DLs for representing and reasoning about prototypical properties and defeasible inheritance. The idea is that, in some applications, the need of reasoning about surprising scenarios could help domain experts to achieve their goals, wherever standard reasoning is not enough to do it: as an example, in medical diagnosis, the most likely explanation for a set of symptoms is not always the solution to the problem, whereas reasoning about surprising scenarios could help the medical staff in taking alternative explanations into account. In other words, the logic $\mathcal{ALC} + \mathbf{T}_{\mathbf{R}}^{\text{exp}}$ is not intended to replace existing nonmonotonic DLs, but to tile them in order to reason about alternative, plausible scenarios when it is needed to go beyond most likely solutions.

2   Preferential Description Logics

2.1   The Monotonic Logic $\mathcal{ALC} + \mathbf{T}_{\mathbf{R}}$

The logic $\mathcal{ALC} + \mathbf{T}_{\mathbf{R}}$ is obtained by adding to standard $\mathcal{ALC}$ the typicality operator $\mathbf{T}$ [13]. The intuitive idea is that $\mathbf{T}(C)$ selects the *typical* instances of a concept C. We can therefore distinguish between the properties that hold for all instances of concept C ($C \sqsubseteq D$), and those that only hold for the normal or typical instances of C ($\mathbf{T}(C) \sqsubseteq D$).

The semantics of the $\mathbf{T}$ operator can be formulated in terms of *rational models*: a model $\mathcal{M}$ is any structure $\langle \Delta^{\mathcal{I}}, <, \cdot^{\mathcal{I}} \rangle$ where $\Delta^{\mathcal{I}}$ is the domain, $<$ is an irreflexive, transitive, well-founded and modular (for all x, y, z in $\Delta^{\mathcal{I}}$, if $x < y$ then either $x < z$ or $z < y$) relation over $\Delta^{\mathcal{I}}$. In this respect, $x < y$ means that x is "more normal" than y, and that the typical members of a concept C are the minimal elements of C with respect to this relation. An element $x \in \Delta^{\mathcal{I}}$ is a *typical instance* of some concept C if $x \in C^{\mathcal{I}}$ and there is no C-element in $\Delta^{\mathcal{I}}$ more typical than x. In detail, $\cdot^{\mathcal{I}}$ is the extension function that maps each concept C to $C^{\mathcal{I}} \subseteq \Delta^{\mathcal{I}}$, and each role R to $R^{\mathcal{I}} \subseteq \Delta^{\mathcal{I}} \times \Delta^{\mathcal{I}}$. For concepts of $\mathcal{ALC}$, $C^{\mathcal{I}}$ is defined as usual. For the $\mathbf{T}$ operator, we have $(\mathbf{T}(C))^{\mathcal{I}} = Min_<(C^{\mathcal{I}})$. A model $\mathcal{M}$ can be equivalently defined by postulating the existence of a function $k_{\mathcal{M}} : \Delta^{\mathcal{I}} \longmapsto \mathbb{N}$, where $k_{\mathcal{M}}$ assigns a finite rank to each domain element: the rank function $k_{\mathcal{M}}$ and $<$ can be defined from each other by letting $x < y$ if and only if $k_{\mathcal{M}}(x) < k_{\mathcal{M}}(y)$.

Given standard definitions of satisfiability of a KB in a model, we define a notion of entailment in $\mathcal{ALC} + \mathbf{T}_{\mathbf{R}}$. Given a query F (either an inclusion $C \sqsubseteq D$ or an assertion $C(a)$ or an assertion of the form $R(a, b)$), we say that F is entailed from a KB, written KB $\models_{\mathcal{ALC} + \mathbf{T}_{\mathbf{R}}} F$, if F holds in all $\mathcal{ALC} + \mathbf{T}_{\mathbf{R}}$ models satisfying KB.

2.2   The Nonmonotonic Logic $\mathcal{ALC} + \mathbf{T}_{\mathbf{R}}^{RaCl}$

Even if the typicality operator $\mathbf{T}$ itself is nonmonotonic (i.e. $\mathbf{T}(C) \sqsubseteq E$ does not imply $\mathbf{T}(C \sqcap D) \sqsubseteq E$), what is inferred from a KB can still be inferred from any KB' with KB $\subseteq$ KB', i.e. the logic $\mathcal{ALC} + \mathbf{T}_{\mathbf{R}}$ is monotonic. In order to perform useful nonmonotonic inferences, in [13] the authors have strengthened the above semantics by restricting entailment to a class of minimal models. Intuitively, the idea is to restrict entailment to models that *minimize the untypical instances of a concept*. The resulting logic is called $\mathcal{ALC} + \mathbf{T}_{\mathbf{R}}^{RaCl}$ and it corresponds to a notion of *rational closure* on top of $\mathcal{ALC} + \mathbf{T}_{\mathbf{R}}$. Such a notion is a natural extension of the rational closure construction provided in [15] for the propositional logic.

The nonmonotonic semantics of $\mathcal{ALC} + \mathbf{T}_{\mathbf{R}}^{RaCl}$ relies on minimal rational models that minimize the *rank of domain elements*. Informally, given two models of KB, one in which a given domain element x has rank 2 (because for instance $z < y < x$), and another in which it has rank 1 (because only $y < x$), we prefer the latter, as in this model the element x is assumed to be "more typical" than in the former. Query entailment is then restricted to minimal *canonical models*. The intuition is that a canonical model contains all the individuals that enjoy properties that are consistent with KB. A model $\mathcal{M}$ is a minimal canonical model of KB if it satisfies KB, it is minimal and it is canonical[1]. A query F is minimally entailed from a KB, written KB $\models_{\mathcal{ALC}+\mathbf{T}_{\mathbf{R}}^{RaCl}} F$, if it holds in all minimal canonical models of KB. In [13] it is shown that query entailment in $\mathcal{ALC} + \mathbf{T}_{\mathbf{R}}^{RaCl}$ is in ExpTime. Details about the construction of the rational closure and the correspondence between semantics and construction can be found in [13].

3   Between $\mathcal{ALC} + \mathbf{T}_{\mathbf{R}}$ and $\mathcal{ALC} + \mathbf{T}_{\mathbf{R}}^{RaCl}$: The Logic $\mathcal{ALC} + \mathbf{T}_{\mathbf{R}}^{exp}$

In this section we define an alternative semantics that allows us to express a degree of expectedness for the typicality inclusions and to limit the number of typicality assumptions in the ABox in order to obtain less predictable scenarios. The basic idea is similar to the one proposed in [9], where a completion of an $\mathcal{ALC}+\mathbf{T}$ ABox is proposed in order to assume that every individual constant of the ABox is a typical element of the most specific concept he belongs to, if this is consistent with the knowledge base. Here we propose a similar, algorithmic construction in order to compute only *some* assumptions of typicality of domain elements/individual constants, in order to describe an alternative, surprising but plausible scenario. Constraints about the cardinality of the extensions of concepts are also introduced in order to *filter* scenarios, allowing to define *eligible* extensions of the ABox satisfying such constraints, and entailment is restricted to *minimal* scenarios, called *perfect* extensions, with respect to an order relation among extensions: intuitively, an extension is preferred to another one if it represents a more surprising scenario. The logic $\mathcal{ALC} + \mathbf{T}_{\mathbf{R}}^{exp}$ allows one to express cardinality restrictions in the TBox. More expressive DLs allow one to specify (un)qualified number restrictions, in order to specify the number of possible elements filling a given role R. As an example, number restrictions allow one to

---

[1] In Theorem 10 in [13] the authors have shown that for any KB there exists a finite minimal canonical model of KB minimally satisfying the ABox.

express that a student attends to 3 courses. Number restrictions are therefore "localized to the fillers of one particular role" [16], for instance we can have *Student* $\sqsubseteq= 3Attends.Course$ as a restriction on the number of role fillers of the role *Attends*. However one could need to express *global* restrictions on the number of domain elements belonging to a given concept, for instance to express that in the whole domain there are exactly 3 courses. In DLs not allowing cardinality restrictions one can only express that every student must attend to three courses, but not that all must attend to the same ones. In the logic $\mathcal{ALC} + \mathbf{T}_\mathbf{R}^{exp}$, cardinality restrictions on concepts are added to the TBox: they are expressions of the form either $(\geq n\ C)$ or $(\leq n\ C)$ or $(= n\ C)$, where n is a positive integer and C is a concept. This is formally defined in the next definition, where, given a set S, $\sharp S$ is the cardinality of S.

Definition 1. *We consider an alphabet of concept names $\mathcal{C}$, of role names $\mathcal{R}$, and of individual constants $\mathcal{O}$. Given $A \in \mathcal{C}$ and $R \in \mathcal{R}$, we define:*

$$C := A \mid \top \mid \bot \mid \neg C \mid C \sqcap C \mid C \sqcup C \mid \forall R.C \mid \exists R.C$$

An $\mathcal{ALC} + \mathbf{T}_\mathbf{R}^{exp}$ knowledge base is a pair $(\mathcal{T}, \mathcal{A})$. $\mathcal{T}$ contains axioms of the form: (i) $C \sqsubseteq C$; (ii) $\mathbf{T}(C) \sqsubseteq_d C$, where $d \in \mathbb{N}^+$ is called the degree of expectedness; (iii) $(\odot n\ C)$, where $\odot \in \{=, \leq, \geq\}$ and $n \in \mathbb{N}^+$. $\mathcal{A}$ contains assertions of the form $C(a)$ and $R(a,b)$, where $a, b \in \mathcal{O}$.

Given an inclusion $\mathbf{T}(C) \sqsubseteq_d D$, the higher the degree of expectedness the more the inclusion is, in some sense, "obvious"/not surprising. Given another inclusion $\mathbf{T}(C') \sqsubseteq_{d'} D'$, with $d' < d$, we assume that this inclusion is less "obvious", more surprising with respect to the other one. As an example, let KB contain $\mathbf{T}(Student) \sqsubseteq_4 SocialNetworkUser$ and $\mathbf{T}(Student) \sqsubseteq_2 PartyParticipant$, representing that typical students make use of social networks, and that normally they go to parties; however, the second inclusion is less obvious with respect to the first one. In other words, one can think of representing the fact that both are properties of a prototypical student, however there are more exceptions of students not taking part to parties with respect to the number of exceptions of students not being part of the social media ecosphere.

It is worth noticing that using positive integers for expressing degrees of expectedness is only a way of formalizing a partial order among typicality inclusions, however all properties expressed by typicality inclusions of the form $\mathbf{T}(C) \sqsubseteq_d D$ are *typical* properties, even if n is low: the ontology engineer has still to distinguish between properties that are prototypical (even with some exceptions) and those that are not and do not deserve to be represented by a typicality inclusion. It is also worth noticing that degrees of expectedness are not intended to represent priorities among inclusions (as in circumscribed KBs), since specificity is provided for free by the preferential semantics of the logic $\mathcal{ALC} + \mathbf{T}_\mathbf{R}^{RaCl}$.

Before introducing technical details and formal definitions (Sects. 3.1 and 3.2), we provide an example in order to give an intuitive idea of what we mean for reasoning about surprising scenarios in the logic $\mathcal{ALC} + \mathbf{T}_\mathbf{R}^{exp}$.

Example 1. (Mysterious medical diagnosis). Let KB $= (\mathcal{T}, \emptyset)$ where $\mathcal{T}$ is as follows:

Depressed $\sqsubseteq$ *Condition*,

ProstateCancerPatient $\sqsubseteq$ *Condition*,

Bipolar $\sqsubseteq$ *Condition*,

AtypicalDepressed $\sqsubseteq$ *Depressed*, $\mathbf{T}$(*Depressed*) $\sqsubseteq_2$ ¬∃*Symptom.MoodReactivity*,

$\mathbf{T}$(*Bipolar*) $\sqsubseteq_5$ ∃*Symptom.MoodReactivity*,

$\mathbf{T}$(*AtypicalDepressed*) $\sqsubseteq_4$ ∃*Symptom.MoodReactivity*,

$\mathbf{T}$(*ProstateCancerPatient*) $\sqsubseteq_2$ ∃*Symptom.MoodReactivity*,

(≥ 1 *Condition*),

(≤ 2 *Condition*)

the last ones stating that we want to focus on at least one/at most two conditions determining patient's symptoms. We have that $\mathbf{T}$(*Depressed* ⊓ *AcromegalicGiant*) $\sqsubseteq$ ¬∃*Symptom.MoodReactivity* follows[2] from KB, and this is a wanted inference, since being affected by acromegaly, a rare syndrome that results when the anterior pituitary gland produces excess growth hormone, is irrelevant with respect to mood reactivity as far as we know. This is a nonmonotonic inference that does no longer follow if it is discovered that typical depressed people also affected by acromegaly are subject to mood reactivity: given $\mathcal{T}'= \mathcal{T} \cup \{\mathbf{T}$(*Depressed* ⊓ *AcromegalicGiant*) $\sqsubseteq$ ∃*Symptom.MoodReactivity*}, we have that the inclusion

$$\mathbf{T}(Depressed \sqcap AcromegalicGiant) \sqsubseteq \neg\exists Symptom.MoodReactivity$$

does not follow from the KB with $\mathcal{T}'$ in the logic $\mathcal{ALC} + \mathbf{T_R^{exp}}$. As for rational closure, the set of facts/inclusions that are entailed from a $\mathcal{ALC} + \mathbf{T_R^{exp}}$ KB is closed under the property known as *rational monotonicity*: for instance, from KB and the fact that $\mathbf{T}$(*Depressed*) $\sqsubseteq$ ¬*Elder* is not entailed from KB in $\mathcal{ALC} + \mathbf{T_R^{exp}}$, it follows that the inclusion $\mathbf{T}$(*Depressed* ⊓ *Elder*) $\sqsubseteq$ ¬∃*Symptom.MoodReactivity* is entailed in $\mathcal{ALC} + \mathbf{T_R^{exp}}$. Concerning ABox reasoning, we can think of exploiting the logic $\mathcal{ALC} + \mathbf{T_R^{exp}}$ in order to find a mysterious medical diagnosis to explain patients' symptoms and signs, a set of formulas of the form $C_r(a)$ that we call $\mathcal{P}$. For instance, let $\mathcal{P}$ describe Greg's symptom, in particular that he has mood reactivity: $\mathcal{P} = \{\exists Symptom.MoodReactivity(greg)\}$. We have that $\mathcal{P}$ is not entailed by KB, but KB $\cup \mathcal{P}$ is consistent. We are then interested in finding a diagnosis for Greg's symptoms, that is to say a set of assertions $\mathcal{D}$ such that $\mathcal{P}$ follows from KB $\cup \mathcal{D}$. The most trivial scenario suggests that Greg is affected by the bipolar disorder, i.e. $\mathcal{D} = \{Bipolar(greg)\}$, however this scenario is discarded in the logic $\mathcal{ALC} + \mathbf{T_R^{exp}}$: in this context, the condition that is taken into consideration

---

[2] As mentioned, at this point of the presentation we only want to give an intuition of the inferences characterizing the logic $\mathcal{ALC} + \mathbf{T_R^{exp}}$. Formal definitions of nonmonotonic entailment in $\mathcal{ALC} + \mathbf{T_R^{exp}}$ will be provided in Definition 10.

is prostatic cancer, i.e. $\mathcal{D} = \{ProstateCancerPatient(greg)\}$, and such a non trivial diagnosis could be confirmed by an evaluation of other typical symptoms of such a disease (e.g. nocturia).

One could object that no one would be interested in a medical diagnosis support system that discards the most likely explanation for a medical problem, however, as mentioned in the Introduction, the idea underlying the proposed approach is not to ignore the most expected explanation, rather to "go beyond" it in order to find (unexpected) alternative ones in case of a failure with the standard diagnosis. In other words: if the most likely explanation does not provide a solution, the logic $\mathcal{ALC} + \mathbf{T_R^{exp}}$ tries to provide surprising alternatives that could be taken into account for further investigations.

3.1   Extensions of ABox

Given a KB, we define the finite set $\mathbb{C}$ of concepts occurring in the scope of the typicality operator, i.e. $\mathbb{C} = \{C \mid \mathbf{T}(C) \sqsubseteq_d D \in \text{KB}\}$. These are the concepts whose atypical instances we want to minimize.

Given an individual a explicitly named in the ABox, we define the set of "plausible" typicality assumptions $\mathbf{T}(C)(a)$ that can be minimally entailed from KB *without cardinality restrictions* in the logic $\mathcal{ALC} + \mathbf{T_R^{RaCl}}$, with $C \in \mathbb{C}$. We then consider an ordered set of pairs (a, C) of all possible assumptions $\mathbf{T}(C)(a)$, for all concepts $C \in \mathbb{C}$ and all individual constants a occurring in the ABox.

Definition 2 (Assumptions in $\mathcal{ALC}+\mathbf{T_R^{exp}}$). *Given an $\mathcal{ALC}+\mathbf{T_R^{exp}}$ KB=$(\mathcal{T} \cup \mathcal{T}_{card}, \mathcal{A})$, where $\mathcal{T}_{card}$ is a set of cardinality restrictions and $\mathcal{T}$ does not contain cardinality restrictions, let $\mathcal{T}'$ be the set of inclusions of $\mathcal{T}$ without degrees of expectedness. Given a finite set of concepts $\mathbb{C}$, we define, for each individual name a occurring in $\mathcal{A}$:*

$$\mathbb{C}_a = \{C \in \mathbb{C} \mid (\mathcal{T}', \mathcal{A}) \models_{\mathcal{ALC}+\mathbf{T_R^{RaCl}}} \mathbf{T}(C)(a)\}$$

We also define $\mathbb{C}_\mathcal{A} = \{(a, C) \mid C \in \mathbb{C}_a \text{ and } a \text{ occurs in } \mathcal{A}\}$ and we impose an order on the elements of $\mathbb{C}_\mathcal{A}$: $\mathbb{C}_\mathcal{A} = [(a_1, C_1), (a_2, C_2), \ldots, (a_n, C_n)]$. Furthermore, we define the ordered multiset $d_\mathcal{A} = [d_1, d_2, \ldots, d_n]$, respecting the order imposed on $\mathbb{C}_\mathcal{A}$, where $d_i = avg(\{d \in \mathbb{N}^+ \mid \mathbf{T}(C_i) \sqsubseteq_d D \in \mathcal{T}\})$.

Intuitively, the ordered multiset $d_\mathcal{A}$ is a tuple of the form $[d_1, d_2, \ldots, d_n]$, where d_i is the degree of expectedness of the assumption $\mathbf{T}(C)(a)$, such that $(a, C) \in \mathbb{C}_\mathcal{A}$ at position i. d_i corresponds to the average[3] of all the degrees d of typicality inclusions $\mathbf{T}(C) \sqsubseteq_d D$ in the TBox.

In order to define alternative scenarios, where not all plausible assumptions are taken into account, we consider different extensions of the ABox and we

---

[3] Other aggregation functions could be used to define d_i (e.g. maximun/minimum degree). We aim at studying the impact of this choice on the reasoning machinery in future research.

introduce an order among them, allowing to range from unpredictable to trivial ones. Starting from $d_{\mathcal{A}} = [d_1, d_2, \ldots, d_n]$, the first step is to build all alternative tuples where 0 is used in place of some d_i to represent that the corresponding typicality assertion $\mathbf{T}(C)(a)$ is no longer assumed (Definition 3). Furthermore, we define the *extension* of the ABox corresponding to a string so obtained (Definition 4).

Definition 3 (Strings of plausible assumptions $\mathbb{S}$). *Given a KB=$(\mathcal{T}, \mathcal{A})$ and the set $\mathbb{C}_{\mathcal{A}}$, let $d_{\mathcal{A}} = [d_1, d_2, \ldots, d_n]$ be as in Definition 2. We define the set $\mathbb{S}$ of all the* strings of plausible assumptions *with respect to KB as*

$$\mathbb{S} = \{[s_1, s_2, \ldots, s_n] \mid \forall i = 1, 2, \ldots, n \ either \ s_i = d_i \ or \ s_i = 0\}$$

Definition 4 (Extension of the ABox). *Let KB=$(\mathcal{T}, \mathcal{A})$ and let $\mathbb{C}_{\mathcal{A}} = [(a_1, C_1), (a_2, C_2), \ldots, (a_n, C_n)]$ as in Definition 2. Given a string of plausible assumptions $[s_1, s_2, \ldots, s_n] \in \mathbb{S}$ of Definition 3, we define the extension $\widehat{\mathcal{A}}$ of $\mathcal{A}$ with respect to $\mathbb{C}_{\mathcal{A}}$ and $\mathbb{S}$*

$$\widehat{\mathcal{A}} = \{\mathbf{T}(C_i)(a_i) \mid (a_i, C_i) \in \mathbb{C}_{\mathcal{A}} \ and \ s_i \neq 0\}$$

It can be observed that, in $\mathcal{ALC} + \mathbf{T}_{\mathbf{R}}^{RaCl}$, the set of typicality assumptions that can be inferred from a KB corresponds to the extension of $\mathcal{A}$ corresponding to the string $d_{\mathcal{A}}$ (no element set to 0): all the typicality assertions of individuals occurring in the ABox, that are consistent with the KB, are assumed. On the contrary, in $\mathcal{ALC} + \mathbf{T}_{\mathbf{R}}$, no typicality assumptions can be derived from a KB, and this corresponds to extending $\mathcal{A}$ by the assertions corresponding to the string $[0, 0, \ldots, 0]$, i.e. by the empty set.

3.2 Cardinality Restrictions and Perfect Extensions

Let us now introduce models of the Description Logic $\mathcal{ALC} + \mathbf{T}_{\mathbf{R}}^{exp}$ taking cardinality restrictions into account, as well as the notion of *eligible* extension of the ABox as a set of typicality assumptions satisfying cardinality restrictions.

Definition 5. *Given a model $\mathcal{M} = \langle \Delta^{\mathcal{I}}, <, .^{\mathcal{I}} \rangle$, it satisfies: (TBox) (i) an inclusion $C \sqsubseteq D$ if $C^{\mathcal{I}} \subseteq D^{\mathcal{I}}$; (ii) a typicality inclusion $\mathbf{T}(C) \sqsubseteq_d D$ if $Min_<(C^{\mathcal{I}}) \subseteq D^{\mathcal{I}}$; (iii) a cardinality restriction of the form $(\odot \ n \ C)$ if $\sharp C^{\mathcal{I}} \odot n$, where $\odot \in \{\leq, \geq, =\}$ and $n \in \mathbb{N}^+$. (ABox) (i) an assertion of the form $C(a)$ if $a^{\mathcal{I}} \in C^{\mathcal{I}}$; (ii) an assertion of the form $R(a, b)$ if $(a^{\mathcal{I}}, b^{\mathcal{I}}) \in R^{\mathcal{I}}$. Given a KB=$(\mathcal{T}, \mathcal{A})$, we say that a model $\mathcal{M}$ satisfies KB if it satisfies all the inclusions in $\mathcal{T}$ and all the assertions in $\mathcal{A}$.*

Definition 6 (Eligible extension $\widehat{\mathcal{A}}$). *Given an $\mathcal{ALC} + \mathbf{T}_{\mathbf{R}}^{exp}$ KB=$(\mathcal{T}, \mathcal{A})$ and an extension $\widehat{\mathcal{A}}$ of $\mathcal{A}$ as in Definition 4, we say that $\widehat{\mathcal{A}}$ is eligible if there exists an $\mathcal{ALC} + \mathbf{T}_{\mathbf{R}}^{exp}$ model $\mathcal{M}$ as in Definition 5 that satisfies KB'=$(\mathcal{T}, \mathcal{A} \cup \widehat{\mathcal{A}})$.*

Definition 7 (Order between eligible extensions). *Given KB=(T, A) and the set $\mathbb{S}$ of Definition 3, let $s = [s_1, s_2, \ldots, s_n]$ and $r = [r_1, r_2, \ldots, r_n]$, with $s, r \in \mathbb{S}$. Let $\widehat{A_s}$ and $\widehat{A_r}$ be two eligible extensions of A corresponding to s and r (Definition 4). We say that $s < r$ if there exists a bijection δ between s and r such that, for each $(s_i, r_j) \in \delta$, it holds that $s_i \leq r_j$, and there is at least one $(s_i, r_j) \in \delta$ such that $s_i < r_j$. We say that $\widehat{A_s}$ is more surprising (or less trivial) than $\widehat{A_r}$ if $s < r$.*

Intuitively, a string s whose elements are "lower" than the ones of another string r corresponds to a less trivial ABox. For instance, let us consider a KB whose typicality inclusions are $\mathbf{T}(C) \sqsubseteq_1 D$ and $\mathbf{T}(E) \sqsubseteq_2 F$, and such that $\mathbf{T}(C)(a)$, $\mathbf{T}(C)(b)$, and $\mathbf{T}(E)(b)$ are entailed in $\mathcal{ALC} + \mathbf{T}_\mathbf{R}^{RaCl}$. Given the strings $s = [1, 1, 0]$ and $r = [1, 0, 2]$, we have that $s < r$, because there exists a bijection $\{(1, 1), (0, 0), (1, 2)\}$. The assumptions $\mathbf{T}(C)(a)$ and $\mathbf{T}(C)(b)$ corresponding to s are then considered less trivial than $\mathbf{T}(C)(a)$ and $\mathbf{T}(E)(b)$ corresponding to r. It is worth noticing that the order of Definition 7 is partial: as an example, the strings $[1, 1, 0]$ and $[0, 0, 2]$ are not comparable, in the sense that $[1, 1, 0] \not< [0, 0, 2]$ and $[0, 0, 2] \not< [1, 1, 0]$. In order to choose between two incomparable situations, we introduce the following notion of weak order: intuitively, given two incomparable extensions $\widehat{A_s}$ and $\widehat{A_r}$, we assume that $\widehat{A_s}$ is weakly less trivial than $\widehat{A_r}$ if $\widehat{A_r}$ is strictly included in another *eligible* extension $\widehat{A_u}$ more trivial than $\widehat{A_s}$, i.e. $\widehat{A_r} \subset \widehat{A_u}$ and $s < u$:

Definition 8 (Weak preference). *Given a KB=(T, A), let $\widehat{A_s}$ and $\widehat{A_r}$ be two eligible extensions of A such that neither $\widehat{A_s}$ is more surprising than $\widehat{A_r}$ nor $\widehat{A_r}$ is more surprising than $\widehat{A_s}$. We say that $\widehat{A_s}$ is (weakly) more surprising (or (weakly) less trivial) than $\widehat{A_r}$ if there exists an eligible extension $\widehat{A_u}$ of A such that (i) $\widehat{A_s}$ is more surprising than $\widehat{A_u}$ (Definition 7) and (ii) $\widehat{A_r} \subset \widehat{A_u}$.*

Definition 9 (Minimal (perfect) extensions). *Given a KB=(T, A) and the set $\mathbb{S}$ of strings of plausible assumptions (Definition 3), we say that an eligible extension $\widehat{A_s}$ is minimal if there is no other eligible extension $\widehat{A_r}$ which is (weakly) more surprising (or (weakly) less trivial) than $\widehat{A_s}$.*

Given the above definitions, we can define a notion of entailment in $\mathcal{ALC} + \mathbf{T}_\mathbf{R}^{exp}$. Intuitively, given a query F, we check whether F follows in the monotonic logic $\mathcal{ALC} + \mathbf{T}_\mathbf{R}$ from a given KB, whose ABox is augmented with extensions that are minimal (perfect) as in Definition 9. We can reason either in a skeptical way, by asking that F is entailed if it follows in *all* KBs, obtained by considering each minimal extension of the ABox, or in a credulous way, by assuming that F is entailed if there exists at least one extension of the ABox allowing such inference. This is stated in a rigorous manner by the following definition:

Definition 10 (Entailment in $\mathcal{ALC} + \mathbf{T}_\mathbf{R}^{exp}$). *Given a KB=$(T, A)$ and given $\mathbb{C}$ a set of concepts, let $\mathcal{E}$ be the set of all extensions of A that are minimal as in Definition 9. Given a query F, we say that (i) F is skeptically entailed from*

KB in $\mathcal{ALC} + \mathbf{T_R^{exp}}$, written $KB \models^{sk}_{\mathcal{ALC}+\mathbf{T_R^{exp}}} F$, if $(\mathcal{T}, \mathcal{A} \cup \widehat{\mathcal{A}}) \models_{\mathcal{ALC}+\mathbf{T_R}} F$ for all $\widehat{\mathcal{A}} \in \mathcal{E}$; (ii) F is credulously entailed from KB in $\mathcal{ALC} + \mathbf{T_R^{exp}}$, written $KB \models^{cr}_{\mathcal{ALC}+\mathbf{T_R^{exp}}} F$, if there exists $\widehat{\mathcal{A}} \in \mathcal{E}$ such that $(\mathcal{T}, \mathcal{A} \cup \widehat{\mathcal{A}}) \models_{\mathcal{ALC}+\mathbf{T_R}} F$.

At a first glance, one could have the impression that the notions of rank in the semantics of $\mathcal{ALC} + \mathbf{T_R^{RaCl}}$, where elements with lowest rank are the most typical ones, and the semantics of expectedness of Definitions 7 and 9, where lower ranks correspond to more surprising scenarios, are in conflict. However, this is not the case: ranks in the semantics are introduced in order to define extensions of typicality concepts, and this is also considered in the expectation semantics to select plausible typicality assumptions. The rank among extensions is rather used in order to choose surprising scenarios, to restrict the number of typicality assumptions to satisfy cardinality restrictions: the *unexpectedness* is the additional ingredient to select surprising scenarios by fixing cardinality restrictions, where all candidates try to maximize the typicality of individuals.

4   A Decision Procedure for $\mathcal{ALC} + \mathbf{T_R^{exp}}$

In this section we describe a decision procedure for reasoning in the logic $\mathcal{ALC} + \mathbf{T_R^{exp}}$. We consider skeptical and credulous entailment. In both cases, we exploit the decision procedure to show that the problem of entailment in the logic $\mathcal{ALC} + \mathbf{T_R^{exp}}$ is in ExpTime. This allows us to conclude that reasoning about typicality and defeasible inheritance in surprising scenarios is essentially inexpensive, in the sense that reasoning retains the same complexity class of the underlying standard Description Logic $\mathcal{ALC}$, which is known to be ExpTime-complete [19].

We define a procedure computing the following steps: 1. compute the set $\mathbb{C}_a$ of all typicality assumptions that are minimally entailed from the KB in the nonmonotonic logic $\mathcal{ALC}+\mathbf{T_R^{RaCl}}$; 2. compute all possible extensions of the ABox and select perfect extensions; 3. check whether the query F is entailed from at least one extension/all the extensions of KB in the monotonic logic $\mathcal{ALC} + \mathbf{T_R}$ plus cardinality restrictions.

Step 3 is based on reasoning in the monotonic logic $\mathcal{ALC} + \mathbf{T_R}$: to this aim, the procedure relies on a polynomial encoding of $\mathcal{ALC}+\mathbf{T_R}$ into $\mathcal{ALC}$ introduced in [20] and then on reasoning with cardinality restrictions. Step 1 is based on reasoning in the nonmonotonic logic $\mathcal{ALC} + \mathbf{T_R^{RaCl}}$: in this case, the procedure computes the rational closure of an $\mathcal{ALC} + \mathbf{T_R}$ knowledge base by means of the algorithm introduced in [13], which is sound and complete with respect to the minimal model semantics recalled in Sect. 2.2. Also the algorithm computing the rational closure relies on reasoning in the monotonic logic $\mathcal{ALC}+\mathbf{T_R}$, then on the above mentioned polynomial encoding in $\mathcal{ALC}$. We assume unary encoding of numbers in cardinality restrictions in order to exploit the results in [21], namely that reasoning in $\mathcal{ALCO}$, extending $\mathcal{ALC}$ with qualified number restrictions, is ExpTime-complete also with cardinality restrictions. Due to space limitations, here we only introduce the overall procedure for reasoning in $\mathcal{ALC} + \mathbf{T_R^{exp}}$ and we analyze its complexity, whereas we remind to [20] for the procedures for reasoning in $\mathcal{ALC} + \mathbf{T_R}$ and $\mathcal{ALC} + \mathbf{T_R}^{RaCl}$.

Let KB=$(\mathcal{T} \cup \mathcal{T}_{card}, \mathcal{A})$ be an $\mathcal{ALC} + \mathbf{T_R^{exp}}$ knowledge base, where $\mathcal{T}_{card}$ is a set of cardinality restrictions and $\mathcal{T}$ does not contain cardinality restrictions. Let $\mathcal{T}'$ be the set of inclusions of $\mathcal{T}$ without the degrees of expectedness: $\mathcal{T}' = \{\mathbf{T}(C) \sqsubseteq D \mid \mathbf{T}(C) \sqsubseteq_d D \in \mathcal{T}\}$, that the procedure will take into account in order to reason in $\mathcal{ALC} + \mathbf{T_R}$ and $\mathcal{ALC} + \mathbf{T_R}^{RaCl}$ for checking query entailment and finding all plausible typicality assumptions, respectively. Other inputs of the procedure are a finite set of concepts $\mathbb{C}$ and a query F. Algorithm 1 checks whether F is skeptically entailed from the KB in the logic $\mathcal{ALC} + \mathbf{T_R^{exp}}$, namely whether KB $\models^{sk}_{\mathcal{ALC}+\mathbf{T_R^{exp}}} F$. For credulous entailment (KB $\models^{cr}_{\mathcal{ALC}+\mathbf{T_R^{exp}}} F$), just replace lines 20–23 in Algorithm 1 with:

20 :   **for each** $\widehat{\mathcal{A}}_i \in \mathcal{E}$ **do**

21 :   **if**$(\mathcal{T}' \cup \mathcal{T}_{card}, \mathcal{A} \cup \widehat{\mathcal{A}}_i) \models_{\mathcal{ALC}+\mathbf{T_R}} F$ **then**

22 :   **return** KB $\models^{cr}_{\mathcal{ALC}+\mathbf{T_R^{exp}}} F$

23 :   **return** KB $\not\models^{cr}_{\mathcal{ALC}+\mathbf{T_R^{exp}}} F$

By exploiting the procedures above, we can show that (the proof is omitted in order to save space):

Theorem 1 (Complexity of entailment). *Given a KB in $\mathcal{ALC} + \mathbf{T_R^{exp}}$ and a query F whose size is polynomial in the size of KB, assuming the unary encoding of numbers in cardinality restrictions of KB, the problem of checking skeptically (resp. credulously) whether KB $\models^{sk}_{\mathcal{ALC}+\mathbf{T_R^{exp}}} F$ (resp. KB $\models^{cr}_{\mathcal{ALC}+\mathbf{T_R^{exp}}} F$) is* ExpTime-*complete.*

Since reasoning in the underlying standard $\mathcal{ALC}$ is ExpTime-complete, we can conclude that reasoning about typicality in surprising scenarios is essentially inexpensive.

5   Reasoning About Scenarios "in the Middle"

The approach described in the previous sections could be extended in order to take also into account *all* the extensions of the ABox satisfying cardinality restrictions, i.e. the eligible extensions of Definition 6. The idea is to reason about *all* plausible scenarios, each one equipped with a degree of expectedness, representing a sort of probability, allowing the user to choose the one he considers more adequate for his application, ranging from the most trivial scenario to the most surprising one.

We iteratively define sets of extensions representing scenarios "in the middle": intuitively, at each step, we consider the extensions of ABox that are minimal w.r.t. the weak preference among extensions of ABox in Definition 8. At the next step, only remaining eligible extensions are considered, and so on. This is formally stated as follows:

Algorithm 1. Skeptical entailment in $\mathcal{ALC} + \mathbf{T}_\mathbf{R}^{exp}$: $KB \models_{\mathcal{ALC}+\mathbf{T}_\mathbf{R}^{exp}}^{sk} F$

```
 1: procedure SKEPTICALENTAILMENT((𝒯 ∪ 𝒯_card, 𝒜), 𝒯', F, ℂ)
 2:    ℂ_𝒜 ← ∅                                              ▷ build the set 𝕊 of plausible assumptions
 3:    for each C ∈ ℂ do
 4:       for each individual a ∈ 𝒜 do                     ▷ Reasoning in 𝒜ℒ𝒞 + 𝐓_𝐑^{RaCl}
 5:          if (𝒯', 𝒜) ⊨_𝒜ℒ𝒞 +𝐓_𝐑^{RaCl}𝐓(C)(a) then ℂ_𝒜 ← ℂ_𝒜 ∪ {𝐓(C)(a)}
 6:    d_𝒜 ← build the ordered multiset of avg degrees of Definition 2 given 𝒯 and ℂ_𝒜
 7:    𝕊 ← build strings of plausible extensions as in Definition 3 given ℂ_𝒜 and d_𝒜
 8:    𝒜_pl ← ∅                                              ▷ build plausible extensions of 𝒜
 9:    for each d_i ∈ 𝕊 do
10:       build the extension 𝒜̂_i corresponding to d_i
11:       𝒜_pl ← 𝒜_pl ∪ 𝒜̂_i
12:    𝒜_el ← ∅                             ▷ select eligible extensions checking cardinality restrictions
13:    for each 𝒜̂_i ∈ 𝒜_pl do               ▷ Reasoning in 𝒜ℒ𝒞 + 𝐓_𝐑 plus cardinality restrictions
14:       if (𝒯' ∪ 𝒯_card, 𝒜 ∪ 𝒜̂_i) is satisfiable in 𝒜ℒ𝒞 + 𝐓_𝐑 then
15:          𝒜_el ← 𝒜_el ∪ 𝒜̂_i
16:    for each 𝒜̂_i ∈ 𝒜_el do                      ▷ check preference among extensions of 𝒜
17:       for each 𝒜̂_j ∈ 𝒜_el do
18:          if d_i < d_j then let 𝒜̂_i < 𝒜̂_j
19:    ℰ ← {𝒜̂_i | ∄𝒜̂_j ∈ 𝒜_el  such that  𝒜̂_j < 𝒜̂_i}              ▷ select perfect extensions
20:    for each 𝒜̂_i ∈ ℰ do        ▷ query entailment in 𝒜ℒ𝒞 + 𝐓_𝐑 plus cardinality restrictions
21:       if (𝒯' ∪ 𝒯_card, 𝒜 ∪ 𝒜̂_i) ⊭_𝒜ℒ𝒞+𝐓_𝐑 F then
22:          return KB ⊭_{𝒜ℒ𝒞+𝐓_𝐑^{exp}}^{sk} F              ▷ a perfect extension not entailing F
23:    return KB ⊨_{𝒜ℒ𝒞+𝐓_𝐑^{exp}}^{sk} F                  ▷ F is entailed in all perfect extensions
```

Definition 11 (Extensions "in the middle"). *Given a KB=*$(\mathcal{T}, \mathcal{A})$*, let* $\mathbb{E}$ *be the set of all eligible extensions as in Definition 6. We define extensions "in the middle" as follows: - we let* $\mathcal{E}_0 = \mathcal{E}$*, where* $\mathcal{E} \subseteq \mathbb{E}$ *is the set of eligible extensions that are minimal as in Definition 9; - while* $\mathcal{E}_i$ *is not empty, let* $\mathcal{E}_{i+1}$ *be the extensions in* $\mathbb{E} - (\mathcal{E}_0 \cup \mathcal{E}_1 \cup \ldots \cup \mathcal{E}_i)$ *that are minimal with respect to the order relation of Definition 8.*

Definition 11 describes a sequence of sets of eligible extensions $\mathcal{E}_0, \mathcal{E}_1, \ldots, \mathcal{E}_n$ with a degree of expectedness i associated to each one. We can formally define what we mean for reasoning in more or less surprising scenarios:

Definition 12. (Entailment in $\mathcal{ALC}+\mathbf{T}_\mathbf{R}^{exp}$**).** *Given a KB=*$(\mathcal{T}, \mathcal{A})$ *and a query* F*, we say that (i)* F *is skeptically entailed from KB in* $\mathcal{ALC} + \mathbf{T}_\mathbf{R}^{exp}$ *at degree* i*, for* $i \in \mathbb{N}$*, written* $KB \models_{\mathcal{ALC}+\mathbf{T}_\mathbf{R}^{exp}}^{sk^i} F$*, if* $(\mathcal{T}, \mathcal{A} \cup \widehat{\mathcal{A}}) \models_{\mathcal{ALC}+\mathbf{T}_\mathbf{R}} F$ *for all* $\widehat{\mathcal{A}} \in \mathcal{E}_i$*; (ii)* F *is credulously entailed from KB in* $\mathcal{ALC}+\mathbf{T}_\mathbf{R}^{exp}$*, written* $KB \models_{\mathcal{ALC}+\mathbf{T}_\mathbf{R}^{exp}}^{cr^i} F$*, if there exists* $\widehat{\mathcal{A}} \in \mathcal{E}_i$ *such that* $(\mathcal{T}, \mathcal{A} \cup \widehat{\mathcal{A}}) \models_{\mathcal{ALC}+\mathbf{T}_\mathbf{R}} F$*.*

It can be observed that, since Algorithm 1 computes all the extensions of an ABox, it can be adapted in order to compute the sets of extensions $\mathcal{E}_0, \mathcal{E}_1, \ldots, \mathcal{E}_n$, so that we can conclude that also reasoning about scenarios "in the middle" is EXPTIME-complete. We can also think of reasoning, credulously or skeptically, in combinations of plausible scenarios, for instance in all scenarios $\mathcal{E}_i$ such that $i < k$ for a given and fixed k.

6 Discussion and Future Issues

In this work we have provided a nonmonotonic procedure for preferential Description Logics in order to reason about surprising scenarios in presence of cardinality restrictions on concepts. In future work we aim at extending this approach to more expressive Description Logics, in particular the logics underlying the standard language for ontology engineering OWL. As a first step, in [22] the logic with the typicality operator and the rational closure construction have been applied to the description logic $\mathcal{SHIQ}$.

A comparison with probabilistic approaches will be also object of further investigations. To the best of our knowledge, the literature lacks a formalization of surprising scenarios in probabilistic formalizations of knowledge, however it is worth observing that a surprising scenario could be defined as a set of facts with a low probability, then one can think of restricting the attention to less probable outcomes.

Acknowledgements. The author is partially supported by the project "ExceptionOWL: Nonmonotonic Extensions of Description Logics and OWL for defeasible inheritance with exceptions" by Università degli Studi di Torino and Compagnia di San Paolo, call 2014 "Excellent (young) PI".

References

1. Bonatti, P.A., Lutz, C., Wolter, F.: The complexity of circumscription in DLs. J. Artif. Intell. Res.(JAIR) **35**, 717–773 (2009)
2. Baader, F., Hollunder, B.: Priorities on defaults with prerequisites, and their application in treating specificity in terminological default logic. J. Autom. Reasoning (JAR) **15**(1), 41–68 (1995)
3. Bonatti, P.A., Faella, M., Sauro, L.: Defeasible inclusions in low-complexity DLs. J. Artif. Intell. Res. (JAIR) **42**, 719–764 (2011)
4. Donini, F.M., Nardi, D., Rosati, R.: Description logics of minimal knowledge and negation as failure. ACM Trans. Comput. Logics (ToCL) **3**(2), 177–225 (2002)
5. Casini, G., Straccia, U.: Rational closure for defeasible description logics. In: Janhunen, T., Niemelä, I. (eds.) JELIA 2010. LNCS (LNAI), vol. 6341, pp. 77–90. Springer, Heidelberg (2010). doi:10.1007/978-3-642-15675-5_9
6. Casini, G., Straccia, U.: Defeasible inheritance-based description logics. J. Artif. Intell. Res. (JAIR) **48**, 415–473 (2013)
7. Straccia, U.: Default inheritance reasoning in hybrid kl-one-style logics. In: Proceedings of the 13th International Joint Conference on Artificial Intelligence (IJCAI 1993), pp. 676–681. Morgan Kaufmann (1993)
8. Bonatti, P.A., Faella, M., Petrova, I., Sauro, L.: A new semantics for overriding in description logics. Artif. Intell. **222**, 1–48 (2015)
9. Giordano, L., Gliozzi, V., Olivetti, N., Pozzato, G.L.: ALC+T: a preferential extension of description logics. Fundam. Inform. **96**, 341–372 (2009)
10. Giordano, L., Gliozzi, V., Olivetti, N., Pozzato, G.L.: A nonmonotonic description logic for reasoning about typicality. Artif. Intell. **195**, 165–202 (2013)

11. Giordano, L., Gliozzi, V., Olivetti, N., Pozzato, G.L.: Preferential vs rational description logics: which one for reasoning about typicality? In: Coelho, H., Studer, R., Wooldridge, M. (eds.) Proceedings of the 19th European Conference on Artificial Intelligence (ECAI 2010). FAIA (Frontiers in Artificial Intelligence and Applications), Lisbon, Portugal, vol. 215, pp. 1069–1070. IOS Press (August (2010)

12. Giordano, L., Gliozzi, V., Olivetti, N., Pozzato, G.L.: Reasoning about typicality in low complexity DLs: the logics $\mathcal{EL}^\perp\mathbf{T}_{min}$ and $DL - Lite_c\mathbf{T}_{min}$. In: Walsh, T. (ed.) Proceedings of the 22nd International Joint Conference on Artificial Intelligence (IJCAI 2011), Spain, pp. 894–899. IOS Press, Barcelona (2011)

13. Giordano, L., Gliozzi, V., Olivetti, N., Pozzato, G.L.: Semantic characterization of rational closure: from propositional logic to description logics. Artif. Intell. **226**, 1–33 (2015)

14. Kraus, S., Lehmann, D., Magidor, M.: Nonmonotonic reasoning, preferential models and cumulative logics. Artif. Intell. **44**(1–2), 167–207 (1990)

15. Lehmann, D., Magidor, M.: What does a conditional knowledge base entail? Artif. Intell. **55**(1), 1–60 (1992)

16. Baader, F., Buchheit, M., Hollunder, B.: Cardinality restrictions on concepts. Artif. Intell. **88**(1–2), 195–213 (1996)

17. Pozzato, G.L.: Preferential description logics meet sports entertainment: cardinality restrictions and perfect extensions for a better royal rumble match. In: Ancona, D., Maratea, M., Mascardi, V. (eds.) Proceedings of the 30th Italian Conference on Computational Logic CILC 2015. CEUR Workshop Proceedings, Genova, 1–3 July 2015, Italy, vol. 1459, pp. 159–174. CEUR-WS.org (2015)

18. Bordino, I., Mejova, Y., Lalmas, M.: Penguins in sweaters, or serendipitous entity search on user-generated content. In: He, Q., Iyengar, A., Nejdl, W., Pei, J., Rastogi, R. (eds.) 22nd ACM International Conference on Information and Knowledge Management, CIKM 2013, San Francisco, CA, USA, 27 October–1 November 2013, pp. 109–118 (2013)

19. Baader, F., Calvanese, D., McGuinness, D., Nardi, D., Patel-Schneider, P.: The Description Logic Handbook - Theory, Implementation, and Applications, 2nd edn. Cambridge University Press, Cambridge (2010)

20. Giordano, L., Gliozzi, V., Olivetti, N., Pozzato, G.L.: Minimal model semantics and rational closure in description logics. In: Eiter, T., Glimm, B., Kazakov, Y., Krötzsch, M. (eds.) DL2016 Informal Proceedings of the 26th International Workshop on Description Logics. CEUR Workshop Proceedings, vol. 1014, pp. 168–180. CEUR-WS.org (2013)

21. Tobies, S.: The complexity of reasoning with cardinality restrictions and nominals in expressive description logics. J. Artif. Intell. Res. (JAIR) **12**, 199–217 (2000)

22. Giordano, L., Gliozzi, V., Olivetti, N., Pozzato, G.L.: Rational closure in $\mathcal{SHIQ}$. In: DL 2014, 27th International Workshop on Description Logics. CEUR Workshop Proceedings, vol. 1193, pp. 543–555. CEUR-WS.org (2014)

Natural Language Processing

A Resource-Driven Approach for Anchoring Linguistic Resources to Conceptual Spaces

Antonio Lieto, Enrico Mensa, and Daniele P. Radicioni[(✉)]

Dipartimento di Informatica, Università degli Studi di Torino, Turin, Italy
{lieto,mensa,radicion}@di.unito.it

Abstract. In this paper we introduce the TTCS system, so named after Terms To Conceptual Spaces, that exploits a resource-driven approach relying on BabelNet, NASARI and ConceptNet. TTCS takes in input a term and its context of usage and produces as output a specific type of vector-based semantic representation, where conceptual information is encoded through the Conceptual Spaces (a geometric framework for common-sense knowledge representation and reasoning). The system has been evaluated in a twofold experimentation. In the first case we assessed the quality of the extracted common-sense conceptual information with respect to human judgments with an online questionnaire. In the second one we compared the performances of a conceptual categorization system that was run twice, once fed with extracted annotations and once with hand-crafted annotations. In both cases the results are encouraging and provide precious insights to make substantial improvements.

Keywords: NLP · Lexical semantics · Lexical resources integration

1 Introduction

The development of reliable knowledge sources to use in different scenarios (such as automatic reasoning, recognition, categorization, *etc.*) represents an active area of research in the AI community. In this paper we face the problem of automatically generating a Conceptual Space representation starting from text and passing through a pipeline involving the integrated use of different linguistic resources: BabelNet [1], NASARI [2] and ConceptNet [3]. The resulting representation enjoys the interesting property of being anchored to such resources, thus providing a uniform interface between the linguistic and the conceptual level.

Conceptual Spaces (CSs) can be thought of as a particular class of vector representations where knowledge is represented as a set of quality dimensions, and where a geometrical structure is associated to each quality dimension. For example, the concept *color* is characterized by 3 quality dimensions: brightness, saturation and hue. Brightness varies from white to black, so it can be represented as a linear dimension with two endpoints; saturation ranges from grey to full intensity and it is, therefore, isomorphic to an interval of the real line; hue can

© Springer International Publishing AG 2016
G. Adorni et al. (Eds.): AI*IA 2016, LNAI 10037, pp. 435–449, 2016.
DOI: 10.1007/978-3-319-49130-1_32

be arranged in a circle where complementary colors (e.g. red-green) lie opposite one another. Then, a possible CS for representing colors is a three-dimensional space with a structure resembling the color spindle.

In this setting, concepts correspond to convex regions,[1] and regions with different geometrical properties correspond to different sorts of concepts [5]. Here common-sense conceptual representation and reasoning have a natural geometrical interpretation, since prototypes (the most relevant representatives of a category from a cognitive point of view, see [6]) correspond to the geometrical centre of a convex region (the centroid). Also exemplars-based representations can be mapped onto points in a multidimensional space, and their similarity can be computed as the intervening distance between each two points, based on some suitable metrics such as Euclidean and Manhattan distance, or standard cosine similarity. Of course, exemplars can be used to calculate the centroid (i.e. the prototype) of each conceptual region.

It is widely accepted that knowledge acquisition is a severe and long-standing bottleneck in many applications [7]. However, while it is possible to ingest existing broad coverage resources such as the formal ontology OpenCyc or domain ontologies, unfortunately no broad coverage resources exist containing common-sense knowledge compliant to the CSs framework.[2] Also, wide-coverage semantic resources such as DBPedia and ConceptNet, in fact, mostly fail to represent the sort of common-sense information based on prototypical and default information which is usually required to perform forms of plausible reasoning.

Focus of this work is precisely how to extract common-sense information, suitable to be encoded through CSs.

This work has the following main strengths: the TTCS can be used to build a broad-coverage, basically domain-independent knowledge base implementing the geometrical representational tenets of CSs; additionally, the TTCS can be used to integrate different recent, state-of-the-art, lexical and semantic resources to operate them in a novel fashion (so to exploit both BabelNet concepts and ConceptNet relations). Finally, the TTCS builds on the relations harvested from ConceptNet to design a procedure to fill the appropriate dimensions in the CS representation thus producing Lexicalized Conceptual Spaces (i.e., CSs whose representations are fully endowed with BabelSynsetIds). On the whole, the TTCS is part of a broader effort aimed at collecting common-sense knowledge to overcome some of the limitations proper to most symbolic-oriented resources (like formal ontologies) in handling forms of non-monotonic reasoning [8].

---

[1] Recently the convexity constraint of conceptual spaces has been argued as a plausible but not necessary condition for the characterisation of concepts within this framework in the case, for example, of the adoption of non-euclidean metrics (see [4]). In our case-study we considered such constraint as proposed in the original theory since we didn't consider non-euclidean metrics.

[2] In fact, we remark that differently from CSs, formal ontologies are not suited for representing defeasible, prototypical knowledge and for dealing with the corresponding typicality-based conceptual reasoning (e.g., non-monotonic inference). For example, for the concept *dog*, OpenCyc does not represent that "typically" dogs bark and woof because common-sense traits are not necessary/sufficient for defining this category.

The paper is structured as follows: we first survey related literature, elaborate on the existing approaches and about the differences with the current proposal (Sect. 2); we then illustrate in full detail the strategy implemented by the TTCS system and report some elements of the build of a resource where information is encoded based on the CSs framework (Sect. 3). The experimentation to assess the obtained results is described in Sect. 4, where we show how the obtained resource has been evaluated through an on-line questionnaire, and employed as the knowledge base used by a conceptual categorization system. The final remarks on future work conclude the paper (Sect. 5).

2   Related Work

Automatically extracting semantic information and annotating texts is an open problem in various fields of AI, and especially for the NLP community [9]. In the last few years many different methodologies and systems for the construction of unified lexical and semantic resources have been proposed.

Some of them are directly referred to the extraction of Conceptual Spaces representations. Existing approaches, for example, try to induce Conceptual Spaces based on distributional semantics by directly accessing huge amounts of textual documents to extract the multidimensional feature vectors that describe the Conceptual Spaces. In particular, [10] try to learn a different vector space representation for each semantic type (e.g. movies), given a textual description of the entities in that domain (e.g., movie reviews). Specifically, they use multidimensional scaling (MDS) to construct the space and identify directions corresponding to salient properties of the considered domain in a *post-hoc* analysis.

Other approaches that show some similarities with this proposal aim at learning word embeddings from text corpora. Word embeddings [11–13] represent the meaning of words as points in a high-dimensional Euclidean space, and are in this sense reminescent of Conceptual Spaces. However, they differ from Conceptual Spaces in at least two crucial ways that limit their usefulness for applications in knowledge representation, e.g., in automatically dealing with inconsistencies. First, word embedding models are mainly aimed at modelling similarity (and notions such as analogy), and are not aimed at providing a geometric representation of the conceptual information (e.g., by representing concepts as convex regions where prototypical effects are naturally modelled). Moreover, the dimensions of a word embedding space are essentially meaningless, while quality dimensions in Conceptual Spaces directly reflect salient cognitive properties of the underlying domain.

Differently from such approaches that aim at extracting Conceptual Spaces (or similar multidimensional vector representations) from text or textual corpora, we use an approach that explicitly relies on existing linguistic resources. We assume that such resources already represent an intermediate step between the lexical level and the conceptual level of the representation, which we are targeting.

Existing resources, in general, can be arranged into two main classes: handcrafted resources –created either by expert annotators, such as WordNet [14],

FrameNet [15] and VerbNet [16], or through collaborative initiatives, such as ConceptNet [17]–; and resources built by automatically combining the above ones, like in the case of BabelNet [1]. Recently, great efforts have been invested to make such resources interoperable, such as the UBY platform [18,19] and the LEMON model [20]. Specifically, UBY is a lexical resource that combines a wide range of lexica (*WordNet*, *Wiktionary*, *Wikipedia*, FrameNet, VerbNet and *OmegaWiki*),[3] by converting them into lexica compliant to the ISO standard Lexical Markup Framework (LMF). Similar to the UBY-LMF project, also the LEMON project relies on the adoption of the LMF for standardisation purposes: LEMON builds on the LexInfo project [21], and it has the purpose of mapping lexical information onto symbolic ontologies. Ontologies, in turn, record the linguistic realizations for classes, properties and individuals. Our system does not directly compare with approaches based on formal ontologies (and standard logic-oriented symbolic representations in general), since the notion of meaning we are currently considering is complementary to ontological information that, on the other hand, is not explicitly committed to represent and reason on common-sense information. In our case, meaning is associated to terms that are mapped onto CSs via the identifiers provided by BabelNet, which is used as a reference framework for concept identifiers.

Some similarities can be drawn with works aiming at aligning WordNet (WN) and FrameNet (FN) [22,23]. The TTCS system shares some traits with the latter approaches, in that we provide a method to put together different linguistic resources. However, at the current stage of development, we do not align the exploited resources, but we rather provide a method for the intelligent multi-resource integration and exploitation, aimed at extracting relevant common-sense information that can be useful to fill Conceptual Spaces dimensions.

None of the mentioned proposals addresses the issue of integrating resources and extracting information to the ends of providing common-sense conceptual representations. The rationale underlying the TTCS is to extract the conceptual information hosted in BabelNet (and its vectorial counterpart, NASARI) and to exploit the relations in ConceptNet so to rearrange BabelNet concepts in a semantic network enriched with ConceptNet relations. Differently from the surveyed works, this is done by leveraging the lexical-semantic interface provided by such resources. In the next Section we illustrate our strategy.

3   TTCS: Terms to Conceptual Spaces

The TTCS system takes in input a pair $\langle t, ctx_t \rangle$ where t is a term and ctx_t is the context in which t occurs,[4] and produces as output a set of attribute-value pairs:

$$\bigcup_{d \in \mathcal{D}} \{ \langle ID_d, \{v_1, \cdots, v_n\} \rangle \} \tag{1}$$

---

[3] Resources marked with emphasized fonts are harmonized in UBY in both the English and the German version.

[4] Typically, the context is composed by one or more sentences; without loss of generality, in the present setting the context has been retrieved by accessing the DBPedia page associated to t.

Algorithm 1. The control strategy of the TTCS system.

---
input : the pair $\langle t, ctx_t \rangle$
output : $\bigcup_{d \in \mathcal{D}} \{\langle ID_d, \{v_1, \cdots, v_n\}\rangle\}$
1: /* Associate t to a NASARI vector v_i */
2: /* and yield the lexicalized concept c^t*/
3: $c^t \leftarrow \arg\max_i (\text{WO}(ctx_t, v_i))$

4: **for each** edge $e_i \in E$ in ConceptNet, such that $t \xrightarrow{e_i} t'_i$ **do**
5:     retrieve the terms t'_i related to t
6:     **if** t'_i isRelevant to t for the meaning conveyed by c^t **then**
7:         $C \leftarrow C \cup c^t_i$ /*C set of concepts related to c^t*/
8:     **end if**
9: **end for**
10: **for each** pair $c^t_i \in C, d \in \mathcal{D}$ **do**
11:     **if** d is filled by ConceptNet edges E^d and $c^t \xrightarrow{e_i} c^t_i \mid e_i \in E^d$ **then**
12:         /* ConceptNet-driven mapping */
13:         fill the dimension d identified by ID_d with c^t_i as value
14:     **else if** d is filled by dictionary Dic and $c^t_i \in Dic_d$ **then**
15:         /* Dictionary-driven mapping */
16:         fill the dimension d identified by ID_d with c^t_i as value
17:     **end if**
18: **end for**
19: **return** $\bigcup_{d \in \mathcal{D}} \{\langle ID_d, \{v_1, \cdots, v_n\}\rangle\}$

---

where ID_d is the identifier of the d-th quality dimension, and $\{v_1, \cdots, v_n\}$ is the set of values chosen for d. Such values will be used as fillers for Conceptual Space dimensions $d \in \mathcal{D}$. The output of the system is then a Conceptual Space representation of the input term.

The control strategy implemented by the TTCS is described in Algorithm 1, and it includes two main steps:

– **semantic extraction** phase (lines 1–9): starting from the input term and its context, we access NASARI to provide that term with a BabelSynsetId (simply ID in the following) so to identify the correct concept. This step corresponds to a simple though effective form of word-sense disambiguation, which relies on NASARI vectors. Once the concept underlying t has been identified, we explore its ConceptNet connections and extract a bag-of-concepts semantically related to the seed term;
– **semantic matching** phase (lines 10–18): a new empty exemplar is created, corresponding to an empty vector in the CSs; we then use the bag-of-concepts extracted in the previous phase to identify the values suitable as fillers for the Conceptual Space quality dimensions.

3.1   Semantic Extraction

The semantic extraction creates a bag-of-concepts C containing a set of values for the conceptual representation of a given concept. We can distinguish two steps: the *concept identification* and the *extraction*.

Concept identification. As regards as the concept identification (Algorithm 1, line 3), given the pair $\langle t, ctx_t \rangle$, we employ NASARI to acquire an ID id_t for the

Table 1. The list of the considered ConceptNet relations.

IsA	PartOf	MemberOf	HasA
CapableOf	AtLocation	HasProperty	Attribute
MadeOf	SymbolOf	UsedFor	InstanceOf

input term. At the end of this step we obtain a lexicalized concept c^t, which is referred to as *seed concept*.

The problem of assigning an ID to the term t can be cast to the problem of associating a NASARI vector to t. A set of *candidate vectors* V is individuated; a NASARI vector v is a candidate for the meaning of t iff t is contained in the synset (the set of synonyms) associated to the head of the vector v.[5] We distinguish three cases:

- V is empty: we cannot identify any concept for t, and the process stops;
- V contains exactly one element v_1: t is identified by the ID of v_1;
- V has more than one element: for each $v_i \in V$ we compute the weighted overlap between ctx_t and all the synsets that appear as body of v_i. The vector v_k that has maximum weighted overlap [24] is then chosen as best candidate, so t is identified by the ID of v_k. If the weight of all the candidates is zero, we cannot identify any concept for t and the process stops.

Once the concept identification task has terminated, we have a lexicalized concept c^t that represents the semantics of t by means of a NASARI vector.

Extraction. In the extraction step (Algorithm 1, lines 4–9), we access the ConceptNet node associated with t, scan t's outgoing edges $e_i \in E$, and retrieve the related terms $\{t'_1, \ldots, t'_n\}$, such that $t \xrightarrow{e_1} t'_1, \ldots, t \xrightarrow{e_n} t'_n$. The list of 12 relations that are presently considered –out of the 57 relations available in ConceptNet– is provided in Table 1.

Since ConceptNet does not provide any anchoring mechanism to associate its terms to meaning identifiers (the BabelSynsetIds), it is necessary to determine which edges are relevant for the concept associated to t, that is in the meaning conveyed by c^t. In particular, when we access the ConceptNet page for t, we find not only the edges regarding t intended as c^t, but also all the edges regarding t in any possible meaning. Ultimately, in this phase we look for the set of concepts related to c^t, that is the set $C = \{c_1^t, \cdots, c_k^t\}$, with $k \leq n$.

---

[5] NASARI *unified* vectors are composed by a *head* concept (represented by its ID in the first position) and a *body*, that is a list of synsets related to the head concept. Each synset ID is followed by a number that grasps its correlation with the head concept. It is worth noting that in order to reduce the number of required accesses to BabelNet we built an all-in-one resource that maps each ID referred in NASARI vectors onto its synset terms.

To select only (and possibly all) the edges that concern c^t we introduce the notion of *relevance*. The devised algorithm is as follows:

1. Access the ConceptNet node regarding t and consider its set E of edges.
2. For each $e_i \in E$, we call t'_i the term linked to t via e_i, and verify that t'_i is relevant to t in the meaning intended by c^t. The term t'_i is relevant if either it appears within the first (highest weighted) α synsets[6] in the NASARI vector of c^t, or if the set of nodes directly linked to the node t'_i in ConceptNet shares at least β terms[7] with the NASARI vector of c^t.
3. If t'_i is relevant, we then instantiate a concept c^t_i, that we identify through the ID of the first synset in the c^t NASARI vector. Finally, c^t_i is added to the result set C.[8]

For example, given in input the term 'bank' and a usage context such as 'A bank is a financial institution that creates credit by lending money to a borrower', we first disambiguate the term to identify the concept c^{bank}. Then, we inspect the edges of the ConceptNet node 'bank' and thanks to the relevance notion we get rid of sentences such as 'bank isA flight maneuver' (which has nothing to do with the sense bank-financial institution) since the term 'flight maneuver' is not present in the vector associated to concept c^{bank}; conversely, we'll accept sentences such as 'bank HasA branch', as related to bank-financial institution. Finally, 'branch' will be identified as a concept and then added to C.

3.2   Semantic Matching

The semantic matching phase consists in generating a new exemplar ex in the CS representation, and in filling it with the information previously extracted. An exemplar is a list of sets of IDs, where each set corresponds to a quality dimension; it is named and identified in accordance with the seed term t and its meaning c^t.

However, only in some cases C will be rich enough to completely fill the exemplar, which thus can be partially empty; interestingly, Conceptual Spaces are robust to lacking and/or noisy information.

Dimension anchoring. The process of assigning a certain value to a quality dimension is called *dimension anchoring*, and it is carried out for every pair $c^t_i \in C$ and $d \in \mathcal{D}$, where $\mathcal{D}$ is the set of quality dimensions of the Conceptual Space. Quality dimensions can be either directly filled based on ConceptNet edges or checked through a dictionary (please also refer to Algorithm 1, lines 13 and 16).

---

[6] α is presently set to 100.

[7] β is presently set to 3.

[8] We note that the presence of t'_i in the vector of c is guaranteed only if t'_i was detected as relevant through the first relevance condition. So, if t'_i does not appear in the vector of c^t, the identification process fails, and the term will not be added to the result set C.

Table 2. List of the considered quality dimensions; the last two columns indicate respectively whether each dimension is filled in a dictionary-driven (DD column) or in a ConceptNet-driven (CND column) way.

Name	BabelSynsetId	Metric	DD	CND
class	00016733n	no	-	IsA
family	00032896n	no	✓	-
shape	00021751n	no	✓	-
color	00020726n	yes	✓	-
locationEnv	00057017n	yes	✓	-
atLocation	00051760n	no	-	AtLocation
feeding	00029546n	yes	✓	-
hasPart	00021395n	no	-	HasA
partOf	00021394n	no	-	PartOf
locomotion	00051798n	yes	✓	-
symbol	00075653n	no	-	SymbolOf
function	00036822n	no	-	UsedFor

In the former case (ConceptNet-driven approach) the process of extracting values to fill d leverages the set of edges E^d: if c_i^t is related to c^t by an edge included in E^d (the set of edges that are relevant to dimension d), then c_i^t is a valid value and it is added to ex as value for the quality dimension d, identified by ID_d like indicated in Eq. 1. In the latter case (dictionary-driven approach) we exploit the dictionary associated with d: if c_i^t is included in the dictionary, then it is a valid value and it is added to ex as value for the quality dimension d, like indicated in Eq. 1.

Additionally, every quality dimension can be metric or not (the whole picture is provided in Table 2). For metric quality dimensions we devised a set of translation maps (e.g., in the aforementioned case of *color*, we directly translate the *red* color into its L*a*b color space: $\langle 53, 80, 67 \rangle$).

Translation of metric quality dimensions. In the Conceptual Spaces theory metrical values are fundamental to carry out forms of common-sense reasoning by exploiting the distances between exemplars in the resulting geometrical framework. After the new exemplar ex is filled with the values extracted through the above mentioned procedure, we translate the values of the *metric* quality dimensions by exploiting the related translation maps.

Translation maps have been devised to map the extracted values onto the corresponding set of metric values in the Conceptual Space. For example, the *locomotion* dimension is used to account for the type of movement (1:swim, 2:dig, 3:crawl, 4:walk, 5:run, 6:roll, 7:jump, 8:fly). In the Conceptual Space representation the above mentioned values are translated into a numerically ordered scale such that the distance between indexes of values mirrors the semantic distance between the different types of locomotion: e.g., in this setting 'dig' and 'crawl' are assumed to be closer than 'swim' and 'fly' [25].

3.3   Building the CSs Representation

In order to build an actual Conceptual Space, the TTCS took in input a set of 593 cross-domain pairs term-context; such contexts have been obtained by crawling DBPedia, and by extracting the abstracts therein. To briefly account for the computational effort required in the concept identification step, the TTCS handled over 2.8 M NASARI vectors (restricting to consider the first 100 features). By and large, 592 terms out of the initial 593 were associated to a NASARI vector: the failure was due to the fact that no NASARI vector was found containing the given term in the synset associated to its head.

In the extraction step, the TTCS accessed around 10 M of ConceptNet assertions, linking about 3 M concepts. In this step 28 K ConceptNet nodes (on average 47.6 per concept) were extracted; 2.3 K out of 28 K concepts were selected, by finally retaining the relevant and correctly identified ones.[9]

The semantic extraction phase ended up with 516 success cases (the 76 failures were caused by the lack of the ConceptNet node, rather than by the extraction of irrelevant concepts); where the bag-of-concepts C contains at least one extracted concept. For 30 lexicalized concepts the resulting bag-of-concept did not contain any suitable value to fill the exemplars. This led to a total of 486 correctly extracted exemplars, and producing overall 2, 388 dimension values (on average 4.9 per exemplar).

Whether and to what extent the information extracted by the TTCS approaches human common-sense knowledge is the object of the next Section.

4   Evaluation

We devised a twofold experimentation aimed at assessing (i) the quality of the extracted common-sense conceptual information via human assessment; and (ii) the usefulness of the obtained representations in the context of a specific conceptual categorization task.

4.1   Human Evaluation of the Extracted Conceptual Information

The first evaluation regards the assessment of the quality of the extracted conceptual space representations through the TTCS, based on human common-sense judgement. More precisely, the evaluation is intended to assess both correctness (does TTCS output reasonable information for the considered concept?) and completeness (does TTCS output all relevant information for the considered concept?) of the extracted information.[10]

Twenty-seven volunteers, 17–52 years of age (average 37) were recruited for this experiment, mostly from the Department of Computer Science from the authors' University (11 females and 16 males), all naïve to the experiment.

---

[9] Correctly identified concepts are those for which the whole procedure produces an output.

[10] Questionnaires are available at: http://goo.gl/am0S2f.

Table 3. The accuracy results on the deletions (Table 3-a) and insertions (Table 3-b).

a. Analysis of the deletions suggested by human subjects.

	accuracy	agreement
all deletions	83.12%	83%
relevant deletions	68.21%	57%

b. Analysis of the insertions suggested by human subjects.

	# insertions	agreement
all insertions	149	3%
relevant insertions	7	58%

The human subjects have been provided with a concept (e.g., *dog*) and some related common-sense statements obtained from the representations extracted by the TTCS (e.g., "Dogs have fur", "Dogs are animals", *etc.*). In this setting, participants had to assess each statement by indicating (i) whether it was appropriate or not for the concept at hand; and (ii) any further statement reputed essential in order to complete the common-sense description of the considered concept. Participants were randomly split into 2 groups (respectively composed by 12 and 15 participants); subjects had to provide their assessment through an on-line questionnaire containing statements about 15 concepts randomly extracted from the obtained Conceptual Space resource.[11] Each piece of information available in the conceptual space has been used to automatically generate a statement, in such a way that all information collected by the TTCS has undergone this experimentation. Overall 173 statements have been proposed to human subjects for evaluation.

According to the experimental design, for each concept we recorded the number of statements that were found inappropriate (*deletions*), and the number of added statements (*insertions*). In particular, we recorded two distinct metrics: one considering *all* answers, and one examining *relevant* answers. As '*all* answers' metrics we recorded all of the responses, whilst the second metrics has been designed to tame the sparsity of human answers. As regards as the '*relevant* answers' metrics, we defined as *relevant* a deletion (or an insertion) that occurs when a statement about a given concept is not accepted (or felt to be lacking, but necessary) by at least 3 participants. For example, given the concept *mouse* and the statements 'Mouse lives in desert' and 'Mouse is snake food', we recorded a relevant deletion for the former statement which was refused by 3 participants, but not for the latter one, which was refused by only 1 participant.

---

[11] We acknowledge that compared to similar experiments (such as [26, 27]) such data is rather small, and defer to future work an extensive evaluation.

Results. The final figures of our results are reported in Table 3. Let us start by considering the result on deletions (Table 3-a). The participants produced on the whole 2, 340 judgements; the *all answers* raw datum is that in 1, 945 cases they accepted the considered statement (thereby determining a 83.12 % of correct results). As regards as *relevant deletions*, 55 statements were refused by at least 3 participants, thus leading to 32 % *relevant deletions* (and to its complement, 68 % of accuracy). The relevant deletions occurred mainly for statements rather obscure or incorrect, such as 'A bullet is spherical', 'Banana is a dessert', or 'Singer is at location show'. Besides, in order to evaluate how reliable are the collected judgements, we also measured the agreement in participants' answers. Agreement values were measured about the *all deletions* metrics.[12] The average agreement concerning the *deletions* amounts to 83 %; this datum grasps that there is a neat consensus on which statements are acceptable and which ones are not (be them counter-intuitive, or explicitly incorrect, and irrespective of individual preferences and experiences).

On the other hand, as regards as insertions (please refer to Table 3-b), we registered a clear data sparsity which resulted in a low agreement (3 %). The few cases where the participants provided *relevant insertions* (that is, proposing at least 3 times the same statement for insertion), point out an information lacking from the set of statements extracted for that concept: e.g., the TTCS missed to extract that 'Airplanes fly', that 'Camels have two humps' and that 'Chameleons change color'. Despite there is a consistent difference in the agreement between the *all insertions* and *all deletions* metrics, there is on the other hand a similar agreement rate on relevant deletions and insertions. Or, equivalently, there is a prominent analogous accord on both mistaken and missing information.

A general insight emerging from the collected data is that the output obtained by the TTCS system mostly corresponds to the characterization of the conceptual information in terms of prototypical, common-sense knowledge. This element is crucial, since most of the available KBs are not equipped with this sort of distilled but salient information. On the other hand, common sense knowledge represents exactly the type of knowledge crucially used by humans for efficient heuristic reasoning, and it could be adopted by artificial systems aiming at providing forms of plausible automatic reasoning. In next Section we show a concrete application requiring this sort of knowledge.

4.2   A Case Study in Conceptual Reasoning and Categorization

The obtained Conceptual Space resource produced by the TTCS has been additionally evaluated in a practical case study involving a basic conceptual categorization task. In particular, a target concept illustrated by a simple common-sense linguistic description had to be identified; for this experiment we exploited

---

[12] As the ratio between the number of deletions expressed for a given statement and the number of assessments obtained by that statement: e.g., the statement 'Soap has function of scent' has been questioned by 2 participants out of 12. The agreement on such deletion was computed as $2/12 = 16.7\%$.

Table 4. Results in a conceptual categorization task, where the output of the system fed with information extracted by the TTCS system is compared against the results obtained by the categorization system fed with information annotated by hand.

CSs annotation	Number of input descriptions	Correct categorization
Manual	60	52 (87.7%)
TTCS	60	41 (68.3%)

an existing categorization system that relies on a hybrid knowledge base coupling annotated Conceptual Spaces encoding common-sense information, and an external ontological component, represented by the OpenCyc ontology [28–30]. In this evaluation we compared the output provided by this system in two different executions: in the former case the categorization system made use of manually annotated Conceptual Spaces, whilst in the latter one it was fed with the Conceptual Spaces extracted by the TTCS system. A set of 60 common-sense textual descriptions has been given in input to the categorization system in both conditions (with manual or automatically obtained CSs); these stimuli have been built by a multidisciplinary team composed of neuro-psychologists, linguists and philosophers in the frame of a project aimed at investigating the brain activation of visual areas in tasks of lexical processing.

The whole categorization pipeline works as follows. The input to the system is a simple sentence, like 'The feline with mane and big jaws', and the correct output is the category evoked by the description (the category *lion* in this case).[13] Correctly identified categories represent a gold standard which has been individuated based on the results of an experiment involving human subjects [28]; both outputs have thus been compared to human answers.

Results. The obtained results are reported in Table 4, that provides a comparison between the accuracy of the categorization system adopting the automatically obtained conceptual representations and the accuracy of the same system endowed with manually annotated representations. Although the obtained accuracy is lower than that obtained when using hand-crafted knowledge, the performance of the system using data extracted by the TTCS is still acceptable, especially if we consider the increase in the coverage. In fact, the extracted KB includes around 500 conceptual representations, while only 300 conceptual representations were present in the manually annotated one. Also, we are now able to extract salient information to fill CSs representations with virtually no domain restriction, thus attaining a much broader resource representing common-sense knowledge in terms of Lexicalized Conceptual Spaces (i.e. CSs whose representations are fully endowed with BabelSynsetIds).

---

[13] In essence, the employed system executes a two-steps categorization process: it first computes a result based on Conceptual Spaces, and it then checks the validity of the obtained result against an ontological knowledge base.

5 Conclusions

In this paper we have presented TTCS, a system that takes a textual input and returns the corresponding common-sense conceptual representation encoded in terms of Lexicalized Conceptual Spaces. This representation is obtained through a novel method leveraging different linguistic resources such as Babel-Net, NASARI and ConceptNet. The results obtained through the human assessment are encouraging for what concerns the acceptability of the extracted representations; also, the TTCS output has been fed as input to a broader cognitively-inspired categorization system, with an interesting outcome.

The ongoing and future work is represented by the attempt to build a wider, general, common-sense resource in term of Lexicalized Conceptual Space. Such knowledge resource is complementary, and easily integrable onto existing encyclopedic knowledge resources such as BabelNet, since the interface between the lexical and conceptual level would be grounded on the BabelNet synsets. All of the TTCS IDs are actually anchored to BabelNet synset IDs, so the output of TTCS is, *de facto*, already connected to the semantic network of BabelNet. In addition, such a resource would benefit from the geometrical features proper to Conceptual Spaces representations, that are especially helpful in applications that mix different types of reasoning strategies for tasks such as conceptual categorization, question answering, *etc.*. The obtained resource will also enable us to extend the present evaluation towards a larger coverage and more quantitative scenario which will furnish further insights for iteratively refining the TTCS.

References

1. Navigli, R., Ponzetto, S.P.: BabelNet: the automatic construction, evaluation and application of a wide-coverage multilingual semantic network. Artif. Intell. **193**, 217–250 (2012)
2. Camacho-Collados, J., Pilehvar, M.T., Navigli, R.: NASARI: a novel approach to a semantically-aware representation of items. In: Proceedings of NAACL, pp. 567–577 (2015)
3. Speer, R., Havasi, C.: Representing general relational knowledge in ConceptNet 5. In: LREC, pp. 3679–3686 (2012)
4. Hernández-Conde, J.V.: A case against convexity in conceptual spaces. Synthese, 1–27 (2016)
5. Gärdenfors, P.: The Geometry of Meaning: Semantics Based on Conceptual Spaces. MIT Press, Cambridge (2014)
6. Rosch, E.: Cognitive representations of semantic categories. J. Exp. Psychol. Gen. **104**, 192–233 (1975)
7. Lenat, D.B., Prakash, M., Shepherd, M.: CYC: using common sense knowledge to overcome brittleness and knowledge acquisition bottlenecks. AI Mag. **6**, 65 (1985)
8. Frixione, M., Lieto, A.: Representing concepts in formal ontologies: compositionality vs. typicality effects. Logic Logical Philos. **21**, 391–414 (2012)
9. Reeve, L., Han, H.: Survey of semantic annotation platforms. In: Proceedings of the 2005 ACM symposium on Applied computing, pp. 1634–1638. ACM (2005)

10. Derrac, J., Schockaert, S.: Inducing semantic relations from conceptual spaces: a data-driven approach to plausible reasoning. Artif. Intell. **228**, 66–94 (2015)
11. Pennington, J., Socher, R., Manning, C.D.: Glove: global vectors for word representation. EMNLP **14**, 1532–1543 (2014)
12. Mikolov, T., Sutskever, I., Chen, K., Corrado, G.S., Dean, J.: Distributed representations of words and phrases and their compositionality. In: Advances in Neural Information Processing Systems, pp. 3111–3119 (2013)
13. Turney, P.D., Pantel, P., et al.: From frequency to meaning: vector space models of semantics. J. Artif. Intell. Res. **37**(1), 141–188 (2010)
14. Miller, G.A.: WordNet: a lexical database for English. Commun. ACM **38**(11), 39–41 (1995)
15. Baker, C.F., Fillmore, C.J., Lowe, J.B.: The Berkeley framenet project. In: Proceedings of the 17th International Conference on Computational Linguistics. Association for Computational Linguistics, vol. 1, pp. 86–90 (1998)
16. Levin, B.: English Verb Classes and Alternations: A Preliminary Investigation. University of Chicago Press, Chicago (1993)
17. Havasi, C., Speer, R., Alonso, J.: ConceptNet: a lexical resource for common sense knowledge. Recent Adv. Nat. Lang. Process. V Sel. Pap. RANLP **309**, 269 (2007)
18. Eckle-Kohler, J., Gurevych, I., Hartmann, S., Matuschek, M., Meyer, C.M.: UBY-LMF-a uniform model for standardizing heterogeneous lexical-semantic resources in ISO-LMF. In: LREC, pp. 275–282 (2012)
19. Gurevych, I., Eckle-Kohler, J., Hartmann, S., Matuschek, M., Meyer, C.M., Wirth, C.: UBY: a large-scale unified lexical-semantic resource based on LMF. In: Proceedings of the 13th Conference of the European Chapter of the Association for Computational Linguistics. Association for Computational Linguistics, pp. 580–590 (2012)
20. McCrae, J., Montiel-Ponsoda, E., Cimiano, P.: Integrating WordNet and Wiktionary with lemon. In: Chiarcos, C., Nordhoff, S., Hellmann, S. (eds.) Linked Data in Linguistics, pp. 25–34. Springer, Heidelberg (2012)
21. Buitelaar, P., Cimiano, P., Haase, P., Sintek, M.: Towards linguistically grounded ontologies. In: Aroyo, L., et al. (eds.) ESWC 2009. LNCS, vol. 5554, pp. 111–125. Springer, Heidelberg (2009). doi:10.1007/978-3-642-02121-3_12
22. Tonelli, S., Pianta, E.: A novel approach to mapping framenet lexical units to Wordnet synsets. In: Proceedings of the Eighth International Conference on Computational Semantics. Association for Computational Linguistics, pp. 342–345 (2009)
23. Ferrández, O., Ellsworth, M., Munoz, R., Baker, C.F.: Aligning FrameNet and WordNet based on semantic neighborhoods. LREC **10**, 310–314 (2010)
24. Pilehvar, M.T., Jurgens, D., Navigli, R.: Align, disambiguate and walk: a unified approach for measuring semantic similarity. In: ACL, vol. 1, pp. 1341–1351 (2013)
25. Bejan, A., Marden, J.H.: Constructing animal locomotion from new thermadynamics theory. Am. Sci. **94**, 342–349 (2006)
26. Tonelli, S., Pighin, D.: New features for framenet: Wordnet mapping. In: Proceedings of the Thirteenth Conference on Computational Natural Language Learning. Association for Computational Linguistics, pp. 219–227 (2009)
27. De Cao, D., Croce, D., Basili, R.: Extensive evaluation of a framenet-wordnet mapping resource. In: LREC (2010)
28. Lieto, A., Minieri, A., Piana, A., Radicioni, D.P.: A knowledge-based system for prototypical reasoning. Connection Sci. **27**, 137–152 (2015)

29. Lieto, A., Radicioni, D.P., Rho, V.: A common-sense conceptual categorization system integrating heterogeneous proxytypes and the dual process of reasoning. In: Proceedings of the International Joint Conference on Artificial Intelligence (IJCAI), pp. 875–881. AAAI Press, Buenos Aires (2015)
30. Lieto, A., Radicioni, D.P., Rho, V.: Dual PECCS: a cognitive system for conceptual representation and categorization. J. Exp. Theor. Artif. Intell., 1–20 (2016)

Analysis of the Impact of Machine Translation Evaluation Metrics for Semantic Textual Similarity

Simone Magnolini[1,2(✉)], Ngoc Phuoc An Vo[3], and Octavian Popescu[4]

[1] University of Brescia, Brescia, Italy
`magnolini@fbk.eu`
[2] FBK, Trento, Italy
[3] Xerox Research Centre Europe, Meylan, France
`an.vo@xrce.xerox.com`
[4] IBM T.J. Watson Research, Yorktown, USA
`o.popescu@us.ibm.com`

Abstract. We present a work to evaluate the hypothesis that automatic evaluation metrics developed for Machine Translation (MT) systems have significant impact on predicting semantic similarity scores in Semantic Textual Similarity (STS) task, in light of their usage for paraphrase identification. We show that different metrics may have different behaviors and significance along the semantic scale [0–5] of the STS task. In addition, we compare several classification algorithms using a combination of different MT metrics to build an STS system; consequently, we show that although this approach obtains remarkable result in paraphrase identification task, it is insufficient to achieve the same result in STS. We show that this problem is due to an excessive adaptation of some algorithms to dataset domain and at the end a way to mitigate or avoid this issue.

Keywords: Semantic textual similarity · Machine translation evaluation metrics · Paraphrase recognition

1 Introduction

Semantic related tasks have become a noticed trend in Natural Language Processing (NLP) community. Particularly, the Semantic Textual Similarity (STS) task has captured a huge attention in the NLP community despite being recently introduced since SemEval 2012 and continuing in 2013, 2014 and 2015 [1–4]. Basically, the task requires to build systems which can compute the similarity degree between two given sentences. The similarity degree is scaled as a real score from 0 (no relevance) to 5 (semantic equivalence). The evaluation is done by computing the correlation between human judgment scores and systems' predictions by the mean of Pearson correlation method.

In contrast, Machine Translation (MT) evaluation metrics are designed to assess if the output of a MT system is semantically equivalent to a set of reference

© Springer International Publishing AG 2016
G. Adorni et al. (Eds.): AI*IA 2016, LNAI 10037, pp. 450–463, 2016.
DOI: 10.1007/978-3-319-49130-1_33

translations. In SemEval 2012, the system made by de Souza et al. [5] and then the system Barron Cedeo et al. [6] introduced the approach of using a set of MT evaluation metrics together with other lexical and syntactic features to predict the semantic similarity scores in STS. Although this approach shows promising results, there was no in-depth analysis on the impact of the evaluation metrics to the overall performance and how each metric behaves on STS data. Moreover, as being inspired by the literature [7] for paraphrase recognition, we decide to analyze the impact of MT evaluation metrics in STS; doing this we implicitly analyze the relationship among three tasks: MT evaluation, paraphrase identification and STS.

The contribution of this paper consists of four folds, (1) to obtain a clear idea of how each individual metric behaves and correlates with the human-judgment semantic similarity, (2) to examine the approach of combining a set of chosen metrics to build regression models for predicting the semantic similarity scores and analyze the incorporation of these metrics in regarding to the overall performance of the system, (3) to analyze the behaviour of individual metric in the different domains from the STS domain-independent datasets, and (4) to propose a strategy to use the MT metrics inside specific context.

The remainder of this paper is organized as follows: Sect. 2 presents the description of different MT evaluation metrics, Sect. 4 reports the experimental settings, Sect. 4 is the evaluation and discussion, Sect. 5 proposes a way to use MT metrics to achieve promising result, and finally, Sect. 6 is conclusions and future work.

2   Related Work

In the literature [5] is the very first system adopted a set of MT evaluation metrics for STS task. These metrics are implemented in the Asiya Open Toolkit for Automatic Machine Translation (Meta-) Evaluation [8] extracting features at different linguistic levels: lexical, syntactic and semantic. At the lexical level, they exploited different n-gram (BLEU, NIST, ROUGE, GTM, and METEOR) and edit distance based metrics (WER, PER, TER, and TER-Plus) which return different lexical similarity results. At the syntactic level, they use this Asiya tool to compute the similarity between two given texts using constituency parse tree, dependency tree, and shallow syntax. Next, they considered three different types of information, i.e., discourse representation, named entities, and semantic roles to compute the semantic similarity between two texts.

At STS 2013, the system UPC-CORE [6] built a Radial Basis Functions (RBS) regression model using features extracted from 13 machine translation metrics (also obtained by Asiya toolkit), an Explicit Semantic Analysis model (ESA) built from opening paragraphs of Wikipedia 2010 articles, a data predictor, and a subsets of featured obtained from the Takelab system [9] in STS 2012. In particular, the MT evaluation metrics have been used at three levels: lexical, syntactic and semantic. At lexical level, they used two metrics of Translation Error Rate: TER and TER_{pA}, two measures of lexical precision:

BLEU and NIST, one measure of lexical recall: $ROUGE_W$, and four variants of METEOR: exact, stemming, synonyms, and paraphrasing, and a lexical metric accounting for F-Measure. At syntactic level, they used three metrics to learn similarity from dependency parse trees [10]: $DP\text{-}HWCMi_c\text{-}4$ for grammatical categories chains, $DP\text{-}HWCMi_r\text{-}4$ for grammatical relations, and $DP\text{-}O_r$ for words ruled by non-terminal nodes; one metric $CP\text{-}STM_4$ [10] estimating the similarity using constituency parse trees. Finally, at semantic level, they used three metrics measuring the similarities by semantic roles: $SR\text{-}O_r$, $SR\text{-}M_r$, $SR\text{-}O_r$, and two metrics estimating by discourse representation: $DR\text{-}O_r$ and $DR\text{-}O_{rp}$.

Though these two system adopted features from the MT evaluation metrics together with other features to build a regression model to compute the semantic similarity, they did not have any in-dept analysis on the behavior and significance of each metrics towards the overall accuracy. There was also no analysis on the combination of these metrics to see if this combination leads to a good model for computing semantic similarity.

On the other hand, the work in literature [7] used eight MT evaluation metrics BLEU, NIST, TER, TER_p, METEOR, SEPIA, BADGER, and MAXSIM to obtain a remarkable result in paraphrase identification on the Microsoft Research paraphrase corpus (MSRP) [11] and Plagiarism Detection Corpus (PAN). They evaluated the significance of each metric in the binary classification, subsequently, they speculated the three metrics having best performance on both MSRP and PAN datasets (BLEU, NIST and TER), and then they combined the eight metrics to build a binary classifier. However, the MSRP and PAN datasets are very different from STS datasets because STS consists of several domain-independent datasets for both training and testing. Hence, the analysis on the significance of individual metric and the combination of all metrics towards overall performance is very different from what have been done in this literature.

3   Machine Translation Evaluation Metrics

Technically, the MT evaluation metric assesses the semantic equivalence between the translation hypothesis produced by a MT system and the reference translation. In STS task, the idea of using MT evaluation metrics is adopted to improve the word alignment between two given sentences which consequently leads to better prediction of semantic similarity scores. In this study, we employ eight commonly used metrics from three different groups of MT evaluation metrics, (1) the n-gram based metrics (METEOR, BLEU, SEPIA and TESLA), (2) the edit-distance based metrics (TER and TERp), and (3) the information theory based metric (BADGER). We create a system similar to the one described in [7] that use all these types of MT metrics for paraphrase identification, but we substitute MAXSIM with TESLA a newer system based on MAXSIM.

METEOR (Metric for Evaluation of Translation with Explicit ORdering). We use the latest version (1.5) of METEOR [12] that finds alignments between sentences based on exact, stem, synonym and paraphrase matches between words and phrases. Segment and system level metric scores are calculated based on the alignments between sentence pairs.

BLEU (BiLingual Evaluation Understudy). We use BLEU [13] because it is one of the most commonly used metrics and it has a high reliability. The BLEU metric computes as the amount of n-gram overlap, for different values of n = 1,2,3 and 4, between the system output and the reference translation, in our case between sentence pairs. The score is tempered by a penalty for translations that might be too short. BLEU relies on exact matching and has no concept of synonymy or paraphrasing.

NIST (National Institute of Standards and Technology). We use the NIST score [14] that is a variant of BLEU that uses the arithmetic mean of n-gram overlaps, rather than the geometric mean. It also weights each n-gram according to its informativeness as indicated by its frequency. We use the single output of the script that is based on n-grams with different values of n = 1,2,3,4,5,6,7,8 and 9.

SEPIA. We use the latest version (0.2) of [15], it is a syntactically-aware metric designed to focus on 184 structural n-grams with long surface spans that cannot be captured efficiently with surface n-gram metrics. Like BLEU, it is a precision-based metric and requires a length penalty to minimize the effects of length.

BADGER 2.0. This metric is based on the original BADGER [16] architecture but utilizes an extended version of the Smith Waterman Similarity Metric most often used for DNA sequence analysis. The substitution cost has been modified to use a multilingual knowledge base that supports English, Czech, Spanish, German and French. Word similarity is a combination of t-measures for association and the Dice Coefficient for relations. This version is much faster then the original Burrows Wheeler Transformation (BWT)/SpatterMap based system due to pre-computation of the z-scores for word to word substitution costs.

TESLA (Translation Evaluation of Sentences with Linear-programming-based Analysis). This metric based on MAXSIM [17] was first proposed in [18]. The simplest variant, TESLA-M (M stands for minimal), is based on N-gram matching, and utilizes light-weight linguistic analysis including lemmatization, part-of-speech tagging, and WordNet synonym relations.

Ter (Translation Error Rate). We use the 0.7.25 version of TER [19]. TER computes the number of edits needed to fix the translation output so that it matches the reference. TER differs from word error rate (WER) in which it includes a heuristic algorithm to deal with shifts in addition to insertions, deletions and substitutions.

TERp (TER-Plus). The last metric that we use is TERp [20] building upon the core TER algorithm and providing additional edit operations based on stemming, synonymy and paraphrase.

4   Experiments

4.1   Datasets

The STS dataset consists of several datasets: STS 2012, STS 2013, STS 2014 and STS 2015 [1–4]. Each sentence pair is annotated with the semantic similarity score in the scale [0–5]. Table 1 shows the summary of STS datasets and sources over the years. For training, we use all data in STS 2012, 2013 and 2014; and for testing, we use STS 2015 datasets.

Table 1. Summary of STS datasets in 2012, 2013, 2014, 2015.

Year	Dataset	Pairs	Source
2012	MSRpar	1500	newswire
2012	MSRvid	1500	video descriptions
2012	OnWN	750	OntoNotes, WordNet glosses
2012	SMTnews	750	Machine Translation evaluation
2012	SMTeuroparl	750	Machine Translation evaluation
2013	Headlines	750	newswire headlines
2013	FNWN	189	FrameNet, WordNet glosses
2013	OnWN	561	OntoNotes, WordNet glosses
2013	SMT	750	Machine Translation evaluation
2014	Headlines	750	newswire headlines
2014	OnWN	750	OntoNotes, WordNet glosses
2014	Deft-forum	450	forum posts
2014	Deft-news	300	news summary
2014	Images	750	image descriptions
2014	Tweet-news	750	tweet-news pairs
2015	Images	750	image description
2015	Headlines	750	news headlines
2015	Answers-students	750	student answers, reference answers
2015	Answers-forum	375	answers in stack exchange forums
2015	Belief	375	forum data exhibiting committed belief

4.2   Evaluation Methods

We use two different evaluation methods to evaluate the impact of the metrics on our training dataset, (1) the Pearson correlation between the metric outputs and the gold standards which is the official evaluation method used in STS task; and (2) the RELIEF [21] analysis implemented in WEKA [22] to estimate the quality of MT evaluation metric output in regression.

4.3 Settings

Firstly, we employ the eight metrics to compute the semantic similarity between given sentences on the training dataset. We use the default configuration for all metrics, except the "-norm" option for METEOR that tokenizes and normalizes punctuation and lowercase, as suggested in its documentation; and the "-c" option for TER and TERp that roofs the score to 100. Then we normalize all the output results to the scale [0–1]. We use the script given with SemEval 2015, Task #2 data to calculate the Pearson correlation between MT metrics and gold standards and the ReliefFAttributeEval attribute selector implemented in Weka [22] to compute RELIEF.

Next, we combine the outputs of these eight metrics to build seven different regression models using different classification algorithms in WEKA (i.e. IsotonicRegression, LeastMedSq, MultilayerPerceptron, SimpleLinearRegression, M5Rules, M5 Model Trees, and DecisionTable). We only use the default settings of each algorithm without tuning any parameter because our goal is to compare the results of different approaches, not to obtain high performance. We evaluate each model twice, (i) by a 10-fold cross validation on training data, and (ii) we evaluate the model on the test data (STS 2015 dataset). For the comparison, we use the official baseline of STS task which uses the bag-of-words approach to represent each sentence as a vector in the multidimensional token space (each dimension has 1 if the token is present in the sentence, 0 otherwise) and computes the cosine similarity between vectors.

5 Evaluations and Discussions

In this section we describe all the analysis we did, showing how the results drive our research from a type of analysis to the other. We take into consideration for every step the impact of every single metric and their combination. We start with a general analysis, then we focus on the behaviour of the metrics inside Score Bracket after that we conclude with an analysis of every dataset.

5.1 Evaluation of Individual Metric in the Score Bracket

The Pearson correlation and RELIEF analysis of each single metric compared to the human-annotation scores are presented in Table 2. According to both

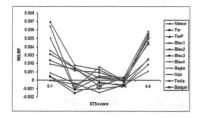

Fig. 1. RELIEF analysis.

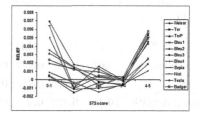

Fig. 2. Pearson correlation.

methods, the METEOR tends to be the superior metric, while in contrast TESLA has low values in both. We split the BLEU metric into four values for 1-gram, 2-gram, 3-gram and 4-gram. The Pearson correlation shows that the smaller size of n-gram overlap, the more correlation with the human judgment obtained. In overall, except TER that has inverse correlation, other metrics have reasonable direct correlation with the human annotation scores.

We also investigate the behaviour of each metric deeper inside each score bracket in the STS semantic scale. We plot the output of each metric in corresponding to each score bracket [0–1], [1–2], [2–3], [3–4] and [4–5] to see how each MT metric behaves on each score bracket. The results of RELIEF analysis and Pearson correlation in Figs. 1 and 2 show that most of the metrics perform well in two particular score brackets [0–1] and [4–5]. This means that by deploying MT evaluation metrics for STS task, the system will be able to obtain a high precision of predicting the semantic similarity for two cases "not/almost not relevant" and "equivalent/almost equivalent". This investigation can help to significantly improve the overall performance of a STS system by increasing the accuracy of predicting the scores in brackets [0–1] and [4–5]. In contrast, both figures have a central region where the correlation scores decrease significantly.

Table 2. Evaluation of the different features on the training dataset.

	RELIEF	Pearson
METEOR	0.007419	0.56065
TER	−0.000417	−0.25673
TERp	0.00017	0.21047
BLEU-1	0.000431	0.368
BLEU-2	−0.001846	0.31801
BLEU-3	−0.002737	0.27074
BLEU-4	−0.00346	0.27233
SEPIA	0.000446	0.37073
NIST	0.002711	0.48416
TESLA	−0.000111	0.19984
BADGER	0.001812	0.27824

This enlightens an important difference between the impact of these metrics on STS task and on the paraphrase recognition task: while MT metrics show acceptable performance distinguishing the border regions, i.e. the most similar (almost paraphrase) and the most dissimilar, they have worse performance in the middle regions. This is compatible with some conclusions in [7] *"not every part of a sentence is equally important for recognizing semantic equivalence or nonequivalence"*, but in this task this issue has a wider impact on the performance.

Table 3. Evaluation of the different algorithms (IR: IsotonicRegression, LMS: Least-MedSq, MLP: MultilayerPerceptron, SLR: SimpleLinearRegression, M5R: M5Rules, M5P: M5 Model Trees, DT: DecisionTable, Baseline: STS Baseline, BestSys: 1st ranked system in STS 2015).

	IR	LMS	MLP	SLR	M5R	M5P	DT	Baseline	BestSys
Cross-validation	0.610	0.659	0.622	0.560	0.753	0.753	0.697	0.382	-
Test set	0.739	0.725	0.718	0.714	0.714	0.711	0.632	0.587	**0.801**
Standard deviation	0.429	0.422	0.498	0.443	0.353	0.350	0.381	0.579	-

5.2   Evaluation of Metric Combination on the Entire Dataset

We examine the impact of the combination of all metrics to the overall performance in STS by building several regression models using all the metric outputs as features. The results of these analysis are reported in Table 3 which shows, (i) the average of the 10-Folds cross-validation on the training data, (ii) the overall performance on the test data, and (iii) to better describe the different algorithms, we also report the standard deviation (SD) of the ten standard deviations from ten folds; we use this measure as an index to evaluate if the performances of the classifier during cross-validation are uniform or present some instability due to specific fold.

We group the models into two groups by a threshold of the standard deviation (SD = 0.41) in which the lower SD, the more reliable model is and vice versa. It is interesting to notice that more stable models (on the right hand side) perform well the cross-validation on the training dataset, but obtain low performance on the test dataset, in a margin of 4 %. Nevertheless, the less stable models (on the left hand side) obtain better results on the test dataset and low performance on the cross-validation, in a margin of 7–15 %. From our observation, another important aspect is that not all the algorithms use all given features in the same way, but during the training phase Isotonic Regression (IR) and Simple Linear Regression (SLR) discard other features and use only METEOR metric.

Another interesting observation is how different learning approaches of different algorithms taking advantage from MT metrics. Some algorithms can learn more information from the combination of these metrics and perform well the cross validation on training data, but when being evaluated on the test data, the model is strongly penalized by the domain-independence datasets in STS. In our case the STS 2012, 2013 and 2014 datasets are different from the STS 2015, which leads to an overfitting of the systems that builds the model using all these features. On the other hand, algorithms which are not so optimized can use MT metrics in a more flexible way to obtain good result on the test dataset. The 1st ranked system in STS 2015 [23] is a ridge regression model that uses two types of features: the proportion of content aligned words on the total content words and the cosine similarity of the component wise average of 400-d vectors obtained using the word2vec[1].

5.3   Evaluation of Individual Metric in the Datasets

We examine the impact of the single MT metrics on the different subsets that compose our dataset to better understand the overfitting of the systems that take advantage of the different metrics. We conduct our analysis using again the Pearson correlation and RELIEF analysis of each single metric compared to the human-annotation scores. We calculate these two indexes for every metric and for every subsets (i.e. METEOR score correlation with human-annotation score for the headlines 2013 dataset etc.), then to compare these correlations

---

[1] https://code.google.com/p/word2vec/.

Table 4. Mean (M), Standard Deviation (SD) and Coefficient of variation (CV) of Pearson and RELIEF of the MT metrics in the datasets STS 2012, STS 2013, STS 2014 and STS 2015.

	Pearson			RELIEF		
2012	M	SD	CV	M	SD	CV
METEOR	0.524	0.115	0.220	0.0120	0.0108	0.897
TER	−0.337	0.111	**0.331**	0.0028	0.0075	2.711
TERp	−0.037	0.194	**5.265**	0.0049	0.0082	1.665
BLEU-1	0.406	0.097	0.238	0.0053	0.0072	1.347
BLEU-2	0.350	0.078	0.222	0.0025	0.0063	2.469
BLEU-3	0.283	0.100	**0.353**	−0.0023	0.0054	2.293
BLEU-4	0.259	0.078	**0.302**	−0.0052	0.0052	1.004
SEPIA	0.418	0.068	0.162	0.0018	0.0078	4.390
NIST	0.522	0.057	0.110	0.0086	0.0109	1.273
TESLA	0.268	0.242	**0.903**	0.0063	0.0104	1.643
BADGER	0.335	0.115	**0.343**	0.0067	0.0074	1.104
2013						
METEOR	0.469	0.154	**0.329**	0.0036	0.0085	2.371
TER	−0.251	0.203	**0.808**	0.0005	0.0024	4.923
TERp	0.001	0.265	**263.466**	0.0053	0.0044	0.833
BLEU-1	0.305	0.153	**0.500**	0.0002	0.0018	10.375
BLEU-2	0.277	0.135	**0.487**	−0.0023	0.0024	1.056
BLEU-3	0.253	0.122	**0.481**	−0.0033	0.0016	0.478
BLEU-4	0.237	0.164	**0.692**	−0.0034	0.0030	0.886
SEPIA	0.326	0.143	**0.439**	−0.0027	0.0019	0.708
NIST	0.417	0.181	**0.433**	0.0034	0.0062	1.795
TESLA	0.267	0.116	**0.435**	−0.0026	0.0054	2.078
BADGER	0.315	0.195	**0.617**	0.0020	0.0010	0.518
2014						
METEOR	0.632	0.084	0.133	0.0082	0.0036	0.445
TER	−0.412	0.154	**0.375**	−0.0011	0.0031	2.863
TERp	−0.081	0.164	**2.041**	−0.0033	0.0034	1.034
BLEU-1	0.479	0.104	0.218	0.0006	0.0033	5.850
BLEU-2	0.424	0.054	0.127	−0.0023	0.0029	1.227
BLEU-3	0.356	0.064	0.179	−0.0031	0.0028	0.910
BLEU-4	0.319	0.077	0.240	−0.0048	0.0030	0.633
SEPIA	0.462	0.086	0.187	−0.0009	0.0039	4.396
NIST	0.535	0.083	0.155	0.0020	0.0051	2.567
TESLA	0.384	0.083	0.217	−0.0013	0.0056	4.171
BADGER	0.467	0.087	0.186	0.0008	0.0041	5.102
2015						
METEOR	0.690	0.067	0.098	0.0063	0.0047	0.739
TER	−0.523	0.097	0.185	0.0007	0.0034	4.754
TERp	−0.204	0.094	**0.462**	0.0017	0.0016	0.963
BLEU-1	0.605	0.070	0.115	0.0015	0.0021	1.432
BLEU-2	0.531	0.086	0.162	0.0012	0.0020	1.674
BLEU-3	0.418	0.110	0.262	−0.0002	0.0017	8.815
BLEU-4	0.332	0.112	**0.337**	−0.0010	0.0012	1.139
SEPIA	0.582	0.063	0.109	0.0003	0.0021	8.090
NIST	0.648	0.047	0.072	0.0042	0.0024	0.579
TESLA	0.449	0.132	0.293	0.0008	0.0038	4.721
BADGER	0.538	0.071	0.133	0.0014	0.0026	1.901

we calculate Mean (M), Standard Deviation (SD) and Coefficient of variation (CV), that is $SD/|M|$, of two correlation indexes in the datasets STS 2012, STS 2013, STS 2014 and STS 2015. The results in Table 4 show interesting aspects of the metrics in the datasets. The Mean stress, again, that not all the metrics are correlated in the same way with the human judgment, this is in line with our previous experiment and confirm our assumptions. The Coefficient of variation of Pearson shows that in some year the dataset is composed by subsets with different correlation with the metrics, we, arbitrary, set a threshold of 0.3 (in bold the Pearson CVs that exceed), that according to our expertise seems tolerant enough, to evaluate if, for a certain metric, the dataset is uniform or not. We notice that for every year at least two metrics exceed our threshold and in STS 2012 and STS 2013 the behaviour of the metrics changes significantly from a subset to the other, in particular in STS 2013 every metric exceed the threshold. The RELIEF CV enlightens that the impact of the single metric when they are combined is even more susceptible to the change of subset, hence of domain.

5.4   Evaluation of Metric Combination in the Datasets

The RELIEF analysis of the MT metrics in the different data subset suggests that the relative importance of the metrics changes significantly from a domain to the other. This idea combined with the different degree of correlation that the different metrics have with the human-judgment of similarity rises some questions: (i) is possible to overcome this issue combining the metrics with a classification algorithm or (ii) is the difficulty to classify the dataset connected to the dataset itself? We evaluate the different algorithms already taken into consideration for the previous experiment on the different data subset, the Mean of the results are in Table 5. The cross-validation of the algorithms perform quite well in every data subset, but is possible to identify three subsets with low performances (i.e. under 50 %). These results give an explanation to the phenomena in the previous experiments: this training data, hence these domains, are not possible to solve using the MT metrics as features, this causes the high variability of correlation between MT metrics and STS score and the low performance of the algorithms that better adapt their model on the training data.

Table 5. Mean of the performance of the different algorithms on the data subset.

2012	M	2013	M	2014	M	2015	M
MSRpar Training	0.579	Headlines	0.705	Headlines	0.691	Images	0.753
MSRvid Training	0.729	**FNWN**	**0.215**	OnWN	0.727	Headlines	0.729
SMTeuroparl Training	0.634	OnWN	0.701	Deft-forum	0.513	Answers-students	0.759
MSRpar Testing	0.559	**SMT**	**0.384**	Deft-news	0.647	Answers-forum	0.559
MSRvid Testing	0.774			Images	0.712	Belief	0.704
SMTeuroparl Testing	**0.484**			Tweet-news	0.750		
OnWN	0.667						
SMTnews	0.559						

6   A Domain Driven Approach to the Task

We evaluate again the two most promising algorithms, IsotonicRegression (IR) best performance on test set and M5 Model Trees (M5P) best performance on cross-validation, removing from the training set the subsets that the previous analysis shows to be not useful, i.e. all the ones with mean of the performance under 50 %. The result are in Table 6, for both algorithms there is an improvement of the performances.

The results in Table 7 are the output of the systems using only domain related training set. We train the classifiers using 2013 and 2014 Headlines datasets to classify the 2015 Headlines dataset and 2012 MSRvid and 2014 Images dataset to classify the 2015 Images dataset. In this case the performance of the M5P classifier exceeds the performance of the IR classifier achieving promising result.

Table 6. Performance of the algorithms (IR: IsotonicRegression, M5P: M5 Model Trees) with a reduced training set.

	IR	M5P	BestSys
Cross-validation	0.624	0.755	
Test set	0.740	0.731	**0.801**

In overall, all the regression models using combination of MT metrics outperform the task baseline in both cross validation on training dataset (by a large margin of 18–37 %) and performance on test dataset (by a margin of 5–15 %). However, none of these models can compare to the best system on the test dataset, the difference between the best model and the best system is a large margin of 7 %. The performance of this approach is strongly influenced by the domain similarity of the training dataset with the test dataset. With a careful selection of the training datasets the classifiers can reduce the difference between the best model and the best system to less than 6 %. This proves that using only MT metric is not sufficient and efficient enough to solve the STS task. But combining MT metrics with other linguistic features may return promising result.

Table 7. Performance of the algorithms (IR: IsotonicRegression, M5P: M5 Model Trees) with a reduced training set.

		IR	M5P	BestSys
Headlines	Cross-validation	0.701	0.753	
	Test set	0.734	0.770	**0.825**
Images	Cross-validation	0.698	0.814	
	Test set	0.761	0.808	**0.864**

7   Conclusions and Future Work

In this study, we show the notable characteristic of the MT metrics as features for the STS task. The distribution of correlation between MT metrics and STS human judgment indicates that this feature is reliable only in the border regions of the [0–5] scale, in particular in [0–1] and [4–5]. This result means that, MT metrics have interesting degrees of correlation with STS, like paraphrase identification, but weaker. So they are useful features for the task, but from the other side it means that they can not be used alone, because their performances are very low in the [1–4] range. Among the different metrics, METEOR has superior property compared to others and it proves to be an useful feature, even alone, to build acceptable STS systems. We enlighten the susceptibility of the metrics to domain changes, in particular we have noticed that algorithms that build models that fit the training data have a problem of overfitting. If training data can be selected to fit the test data classifier like M5P can take great advantage and obtain promising result using MT metrics, otherwise if the training dataset and the test one are different is better to use a more general algorithm like Isotonic Regeression and limit the use of the metrics only to the most promising ones, in our case METEOR. We show two issues show the difference between the tasks (paraphrase identification and STS): the [1–4] rage is usually considered inside the non paraphrase class and the paraphrase identification datasets are without specific domains.

In future we want to investigate more on the impact of other MT metrics on STS task. We have taken into consideration only some metrics with general model; it would be interesting to study also the performances of this approach training the metric on the specific data of the task. Our aim is to find the most useful MT metric or the best combination of metrics among others, and the most reliable and effective algorithm to obtain better performance on the STS task. Another interesting aspect is to study how MT metrics can be used as features for other similar task like Recognizing Textual Entailment (RTE) or Contradiction Detection.

References

1. Agirre, E., Diab, M., Cer, D., Gonzalez-Agirre, A.: SemEval-2012 task 6: a pilot on semantic textual similarity. In: Proceedings of the First Joint Conference on Lexical and Computational Semantics-Volume 1: Proceedings of the Main Conference and the Shared Task, and Volume 2: Proceedings of the Sixth International Workshop on Semantic Evaluation, pp. 385–393. Association for Computational Linguistics (2012)
2. Agirre, E., Cer, D., Diab, M., Gonzalez-Agirre, A., Guo, W.: Shared task: semantic textual similarity, including a pilot on typed-similarity. In: The Second Joint Conference on Lexical and Computational Semantics, *SEM 2013. Association for Computational Linguistics, Citeseer (2013)
3. Agirre, E., Baneab, C., Cardiec, C., Cerd, D., Diabe, M., Gonzalez-Agirre, A., Guof, W., Mihalceab, R., Rigaua, G., Wiebeg, J.: SemEval-2014 task 10: multilingual semantic textual similarity. In: SemEval 2014, p. 81 (2014)

4. Agirre, E., Banea, C., Cardie, C., Cer, D., Diab, M., Gonzalez-Agirre, A., Guo, W., Lopez-Gazpio, I., Maritxalar, M., Mihalcea, R., Rigau, G., Uria, L., Wiebe, J.: SemEval-2015 task 2: semantic textual similarity, English, Spanish and pilot on interpretability. In: Proceedings of the 9th International Workshop on Semantic Evaluation (SemEval 2015). Association for Computational Linguistics, Denver, CO, June 2015

5. de Souza, J.G.C., Negri, M., Mehdad, Y.: FBK: machine translation evaluation and word similarity metrics for semantic textual similarity. In: Proceedings of the First Joint Conference on Lexical and Computational Semantics-Volume 1: Proceedings of the Main Conference and the Shared Task, and Volume 2: Proceedings of the Sixth International Workshop on Semantic Evaluation, pp. 624–630. Association for Computational Linguistics (2012)

6. Barrón-Cedeño, A., Màrquez Villodre, L., Fuentes Fort, M., Rodríguez Hontoria, H., Turmo Borras, J.: UPC-core: what can machine translation evaluation metrics and Wikipedia do for estimating semantic textual similarity?. In: The Second Joint Conference on Lexical and Computational Semantics, *SEM 2013, pp. 1–5 (2013)

7. Madnani, N., Tetreault, J., Chodorow, M.: Re-examining machine translation metrics for paraphrase identification. In: Proceedings of the 2012 Conference of the North American Chapter of the Association for Computational Linguistics: Human Language Technologies, pp. 182–190. Association for Computational Linguistics (2012)

8. Giménez, J., Màrquez, L.: Asiya: an open toolkit for automatic machine translation (meta-)evaluation. Prague Bull. Math. Linguist. 94, 77–86 (2010)

9. Šarić, F., Glavaš, G., Karan, M., Šnajder, J., Bašić, B.D.: TakeLab: systems for measuring semantic text similarity. In: Proceedings of the First Joint Conference on Lexical and Computational Semantics-Volume 1: Proceedings of the Main Conference and the Shared Task, and Volume 2: Proceedings of the Sixth International Workshop on Semantic Evaluation, pp. 441–448. Association for Computational Linguistics (2012)

10. Liu, D., Gildea, D.: Syntactic features for evaluation of machine translation. In: Proceedings of the ACL Workshop on Intrinsic and Extrinsic Evaluation Measures for Machine Translation and/or Summarization, pp. 25–32. Citeseer (2005)

11. Dolan, B., Quirk, C., Brockett, C.: Unsupervised construction of large paraphrase corpora: exploiting massively parallel news sources. In: Proceedings of the 20th International Conference on Computational Linguistics, p. 350. Association for Computational Linguistics (2004)

12. Denkowski, M., Lavie, A.: Meteor universal: language specific translation evaluation for any target language. In: Proceedings of the EACL 2014 Workshop on Statistical Machine Translation (2014)

13. Papineni, K., Roukos, S., Ward, T., Zhu, W.J.: BLEU: a method for automatic evaluation of machine translation. In: Proceedings of the 40th Annual Meeting on Association for Computational Linguistics, pp. 311–318. Association for Computational Linguistics (2002)

14. Doddington, G.: Automatic evaluation of machine translation quality using n-gram co-occurrence statistics. In: Proceedings of the Second International Conference on Human Language Technology Research, pp. 138–145. Morgan Kaufmann Publishers Inc. (2002)

15. Habash, N., Elkholy, A.: SEPIA: surface span extension to syntactic dependency precision-based MT evaluation. In: Proceedings of the NIST Metrics for Machine Tanslation Workshop at the Association for Machine Translation in the Americas Conference, AMTA-2008. Citeseer, Waikiki, HI (2008)

16. Parker, S.: Badger: A new machine translation metric. Metrics for Machine Translation Challenge (2008)
17. Chan, Y.S., Ng, H.T.: Maxsim: A maximum similarity metric for machine translation evaluation. In: ACL, pp. 55–62. Citeseer (2008)
18. Liu, C., Dahlmeier, D., Ng, H.T.: Tesla: translation evaluation of sentences with linear-programming-based analysis. In: Proceedings of the Joint Fifth Workshop on Statistical Machine Translation and Metrics MATR, pp. 354–359. Association for Computational Linguistics (2010)
19. Snover, M., Dorr, B., Schwartz, R., Micciulla, L., Makhoul, J.: A study of translation edit rate with targeted human annotation. In: Proceedings of Association for Machine Translation in the Americas, pp. 223–231 (2006)
20. Snover, M.G., Madnani, N., Dorr, B., Schwartz, R.: TER-Plus: paraphrase, semantic, and alignment enhancements to translation edit rate. Mach. Transl. **23**(2–3), 117–127 (2009)
21. Robnik-Sikonja, M., Kononenko, I.: An adaptation of relief for attribute estimation in regression. In: Fisher, D.H. (ed.) Fourteenth International Conference on Machine Learning, pp. 296–304. Morgan Kaufmann (1997)
22. Hall, M., Frank, E., Holmes, G., Pfahringer, B., Reutemann, P., Witten, I.H.: The weka data mining software: an update. ACM SIGKDD Explor. Newsl. **11**(1), 10–18 (2009)
23. Sultan, M.A., Bethard, S., Sumner, T.: DLS@CU: sentence similarity from word alignment and semantic vector composition. In: SemEval 2015, p. 148 (2015)

QuASIt: A Cognitive Inspired Approach to Question Answering for the Italian Language

Arianna Pipitone$^{(\boxtimes)}$, Giuseppe Tirone, and Roberto Pirrone

Dipartimento di Ingegneria Chimica, Gestionale, Informatica e Meccanica (DICGIM),
Università degli Studi di Palermo, Palermo, Italy
{arianna.pipitone,giuseppe.tirone,roberto.pirrone}@unipa.it

Abstract. In this paper we present QuASIt, a Question Answering System for the Italian language, and the underlying cognitive architecture. The term *cognitive* is meant in the procedural semantics perspective, which states that the interpretation and/or production of a sentence requires the execution of some cognitive processes over both a perceptually grounded model of the world, and a linguistic knowledge acquired previously. We attempted to model these cognitive processes with the aim to make an artificial agent able both to understand and produce natural language sentences. The agent runs these processes on its inner domain representation using the linguistic knowledge also. In this sense, QuASIt is both a rule-based and ontology-based question answering system.

In the model, rules are aimed at understanding the query in terms of the linguistic typology of the question, and enabling its semantic processing as regards the search for the answer in the structured knowledge from DBPedia Italian project. Also the free explicative text in support of the query is analyzed if available. QuASIt attempts to answer for both multiple choice and essay questions. The model is presented, the implementation of the system is detailed, and some experiments are discussed.

Keywords: Question answering · Cognitive architecture · Linguistic typology

1 Introduction

This paper presents a new cognitive model for Question Answering (QA); the model attempts to solve some limitations of both traditional ontology-based and statistical approaches, such as the small scale of the underlying natural language models. Ontology-based QA systems have more advantages than statistical ones [1]: they offer additional information about the answer, provide reliability measures and can motivate how the answer was produced. However, the linguistic models on which these methods are based are poorly sensible to the evolution of language that is its *fluidity*: this is a typical aspect of human interactions. Moreover, such models fail when lexical resources (WordNet [2], FrameNet [3]

© Springer International Publishing AG 2016
G. Adorni et al. (Eds.): AI*IA 2016, LNAI 10037, pp. 464–476, 2016.
DOI: 10.1007/978-3-319-49130-1_34

or MultiWordnet [4]) are no exhaustively available for the language; this is the case of the Italian.

The model we propose keeps the ontology-based QA advantages, attempts to be open-domain like statistical approaches, and it is sensible to the fluidity of the languages. These objectives are achieved by the underlying cognitive architecture we defined: the term *cognitive* derives from the procedural semantics theory, which states that the cognitive processes related to natural language are executed on two kinds of knowledge, that is the perceptually grounded knowledge of the world, and the linguistic one. Our approach attempts to reproduce such processes in an artificial agent to make it able both to understand the query and to produce the answer.

The knowledge about the world is modeled by the domain ontology. As our model implementation is currently under development, no perceptual channels have been considered apart from the query input, and the answer output. So the agent has all its knowledge coded in advance by the ontology. Considering that the ontology can be replaced (our agent makes inferences based on the mere ontological structures) QuASIt can be considered an open-domain system.

The linguistic knowledge is the grammar of the language, and it is modeled by the usage patterns in that language, such as a word, a combination of words, an idiom, and so on. The generalization of the linguistic model, that is separated from the domain knowledge, is one of the main focuses of this work; it is obtained by the *construction grammar* (CxG). The language model abstraction is the result of both the *continuum* and the *abstract categorizations* of constructions. The continuum is between quite abstract grammatical constructions and the *item-based* constructions. The abstract categorizations relate semantics to syntax and allow to conceptualize the meaning and the function of a language.

Modeling the linguistic knowledge by construction grammar, allow us to define general usage patterns of queries that represent the question's linguistic typology. The cognitive processes of query understanding are related to the operations performed by the agent for handling the pattern in the comprehension of questions. The production of the correct answer is obtained by other cognitive processes that, considering the content fitted to the query pattern, extract a subgraph from DBPedia Italian[1] and make reasoning on the nodes of this subgraph for retrieving the correct information; such processes are based on a set of correspondences we defined purposely between the query's typology and some specific DBPedia ontology properties. Moreover, the rules for answer production consider two cases: essay questions and multiple choice ones. In the first case the paragraphs extracted from the text contained in the nodes of the subgraph are analyzed for producing the correct answer by matching them against the retrieved properties in the DBPedia ontology. In case of multiple choice questions, the candidate answers are used both to condition the pruning of the subgraph from DBPedia and to guide the search inside the text contained in its nodes. Candidate answers are used also to search inside the free text in support of the query, if this is available.

---

[1] http://it.dbpedia.org/.

The rest of the paper is arranged as follows. Section 2 reports theoretical backgrounds and motivations for this work, along with a survey of the QA literature related to the systems for the Italian language. Our cognitive model is detailed in Sect. 3, and a simple case study is developed. In Sect. 4, a brief survey on the system implementation is reported, and the experiments are presented; experiments were carried out using the datasets released in the Question Answering For Machine Reading Evaluation (QA4MRE) challenges[2] in 2011 and 2012 respectively. Finally, in Sect. 5 some conclusions are drawn, and future work is outlined.

2   Background and Motivations

Human natural languages are open systems; they exhibit a high degree of evolution in short time. Evolution depends on speakers, which tend to change expressiveness in each situation of real life, rapidly breaking the linguistics conventions. When artificial systems (like QA systems) interact with humans, these variations represent a significant problem; often QA systems fail because their linguistic model is not able to deal with either new meanings or new expressions emerging during a single dialogue session. In statistical approaches (all the tools presented to the well-known competition TREC LiveQA [5] and those described in [6]), such changes might be not sensibly observable despite frequent training, and the QA system might not adapt to new sentences or catch users' attention with correct interactions. Other approaches try very hard to separate the issues concerned to efficiency from issues concerned to grammar representation [7]. Also, if the linguistic sources are not exhaustive for a language, such approaches might not be used; this challenge was taken up in [8] where a new ontology-based QA system is defined. Such a system extracts data from a federation of websites, developing a multilingual environment, which implies the ability to manage several languages and conceptualizations. However, in this approach a large use of linguistic sources is made, because they are linked to the domain ontology by [9]. Undeniably, the underlying computational linguistic model of an artificial agent should handle the *fluidity* of language to face the problem outlined above. QuASIt tries to do this by inspiring its model to both the *Constructions Grammar* (CxG) [10] and the cognitive processes, which are the basis of procedural semantics [11].

CxG is a "symbolic grammar" because all elements have a surface form that is the symbolic representation of the element in the human's mind. Grammatical structures have symbolic representations too, which are the conjunctions of elementary items. All these elements (both structures and items) are considered as tied and related intrinsically to other knowledge structures in the mind. The basic units of CxG are the *constructions*; a construction is a regular pattern which has a conventionalized meaning and function [12]. The meaning side of a construction is captured in a *semantic pole*, while all the aspects related to form, as the syntax, are captured in a *syntactic pole*.

---

[2] http://nlp.uned.es/clef-qa/repository/qa4mre.php.

In [13] a psychologically plausible account of language was made, by investigating some general cognitive principles. Differently from our approach, authors keep into account the linguistic problems only. Instead, we want to integrate both linguistic and world knowledge, according to cognitive linguistics mainstream. With the aim of generalizing our linguistic model, the main properties we referred to are the *continuum* and the *abstract categorizations* of constructions. The continuum is the result of the taxonomy of constructions on which the grammar relies on; quite general constructions subsume the so-called *item-based* ones, that are built out of lexical materials and frozen syntactic patterns, according to such a taxonomy. The continuum is realized through the constructions' open slots where sentences with specific semantic and syntactic structures can fit. The semantic and syntactic categorizations are the means by which constructions relate meaning to form, and allow the *conceptualization of meaning*. As an example, many languages categorize the specific roles of the participants in an event represented by the verb in terms of abstract semantic categories like agent, patient and so on, before mapping them into abstract syntactic categories like the nominative case. Syntactic categories translate further into surface forms.

Categorizations allowed us to abstract the linguistic typology of a query, which is next fitted to the user questions. Such an approach is obviously much more efficient than having an idiosyncratic way to express each question because fewer constructions are needed, and new queries can be understood even if the meaning of the whole question is unknown; this represents also the solution of the fluidity issue.

The only existent computational version of CxG is the Fluid Construction Grammar (FCG) [14] that is an engine that implements both parsing and production using the same set of constructions. FCG is based on two mechanisms: *unification* and *merging*. In parsing, a *transient structure* owning only the syntactic pole is fitted with the set of constructions; when a construction is unified, the transient structure is merged with its slots, and new slots are added in the semantic pole. Production uses the same mechanism by swapping semantic and syntactic pole as the initiator of the process; the transient structure starts owning only the semantic pole that is unified with constructions, and the syntactic one is next merged.

In our system, we use parsing as explained before, while the production of the answer is different, and it will be discussed next.

3   Model Description

The proposed cognitive architecture is depicted in Fig. 1. As already explained, the domain ontology and the CxG represent the knowledge of the world and the linguistic knowledge respectively. We referred to the semantic and syntactic categorizations for defining the cognitive processes of the agent. Two kinds of processes have been devised: the first is related to the *conceptualization of meaning* that associates a perceived external entity (i.e. a word, a visual percept, and so on) to an internal concept. The conceptualization of meaning allows to associate a sense to a perceived form; in our case, the forms are the words of the query, and the internal concepts are the nodes of the domain knowledge. The second process is related to the *conceptualization of form*, that associates a form or a syntactic expression to a meaning; it is the well-known *lexical access* process. In our case the lexical access is implemented by the strategies for producing the correct form of the answer; such a form depends on the way QuASIt can be used, that is in both *multiple choice questions* and *essay questions*. Generally, the form is the value of a property's range in the conceptualized nodes that are inferred by the system. In the specific case of multiple choice questions the form must be one of the proposed answers, that is inferred in the same way of the essay questions (i.e. using the values of the ranges of the involved properties). If no answer can be inferred in this way, the support text is used, which is derived from both the text associated to the nodes, such as an abstract, and the text associated to the questions, if available. In the figure, the knowledge about the world and the linguistic knowledge are located respectively in the *Domain Ontology Base* and the *Linguistic Base*. The *Mapping to Meanings* (MtM) and the *Mapping to Forms* (MtF) modules are the components that model the cognitive processes related to the conceptualization of meaning and form respectively. The *Unification Merging* module is essentially the FCG engine used to perform query comprehension. All the components are detailed in the following subsections.

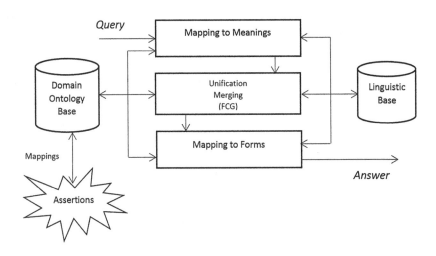

Fig. 1. The QuASIt cognitive architecture

3.1   Domain Knowledge

The ontology forms the structural backbone of the domain, and it represents the terminological box on which the assertions are mapped; assertions are the facts of the domain, and they can be derived from text as in the case of the Wikipedia pages that are mapped to the DBPedia OWL ontology. Assertions can be derived also from a database, or they can be included in the ontology directly. Some of the authors implemented various mapping strategies of assertions from databases [15,16] to be included in the model presented in this paper, but the description of such strategies is out of the scope of this work.

Formally, the domain ontology is the tuple $O = \langle C_o, P_o, T_s, L, P_d \rangle$ defined according to the W3C technical report specification[3], where:

- $C_o = \{cl_i\}$ is the set of type 1 classes;
- P_o is the set of the object properties, so that:

$$P_o = \{o_i \mid o_i = (cl_j, cl_k) \ \ cl_j, cl_k \in C_o\};$$

- $T_s = \{t_i\}$ is the set of literal datatypes;
- $L = \{l_i\}$ is the set of literal strings used in the ontology as values of t_i;
- P_d is the set of the datatype properties, so that:

$$P_d = \{d_i \mid d_i = (cl_j, l_k) \ \ cl_j \in C_o, \ l_k \in L\}.$$

The ontology formal definition provided here does not include individuals, that are the so-called *facts* or *instances*. In our work we considered the case where facts are obtained from a set of strategies for mapping assertions to the terminological structures, that are formalized by the *map* function so that $map : C_o \cup P_o \cup P_d \cup L \rightarrow I$, and $map(oe)$ returns the set I containing the instances for the ontological element oe defined by the assertion mapping.

3.2   Linguistic Knowledge

The linguistic base contains the set of constructions we defined for representing the linguistic typologies of a query in Italian. In particular, considering that the objective of QuASIt is to answer to general questions about the domain, the typologies we referred to are the *direct real* interrogative sentences, which are related to something that is really unknown, and not to the *direct rhetoric* ones. Such a set is grouped in a taxonomy of constructions for implementing the corresponding continuum. The more general construction is the *direct interrogative*, which includes in order a particle, a verb, and what we called the *question topic*, that can be either a syntagma or a dependent clause. The following code in a simplified FCG syntax shows the more abstract construction representing the direct interrogative, that is the top unit:

```
((?Top (sem-subunits (== ?particle ?verb ?questopic)))
(?particle (sem-cat (== (particle ?x))))
(?verb (sem-cat (== (verb ?x ?y ?z))))
(?questopic (sem-cat (== (questopic ?z))))
<->
((?Top (syn-subunits (== ?particle ?verb ?questopic)))
(syn-cat (==1 (pos (DI))))
(?form (form (== (meets ?particle ?verb)(meets ?verb ?questopic)))))
```

---

[3] https://www.w3.org/TR/owl-ref/.

The double arrow separates the semantic pole and the syntactic one, which are written respectively over and under the arrow. The `meets` operator establishes an order between sub-units: in this case the particle comes before the verb, and the verb before the question topic. The question mark identifies variables that can be unified: `?particle`, `?verb` and `?questopic`. The `sem-cat` slot contains the semantic category for each variable. We defined slots purposely by trivial significance: `particle`, `verb` and `questopic`. Similarly, the `syn-cat` slot contains the syntactic category that is, in the case of the direct interrogative, the `DI` value. Some subsumed constructions might look as:

```
((?particle-unit  (meaning ((particle ?particle))))
<->
(?particle-unit (form ((syn-cat ADV))((string "Quando")))))
((label "annomorte" "annonascita" ...)))))
```

and

```
((?questopic-unit  (sem-subunits (== ?noun ?pre ?npr)))
(?pre (sem-cat (== (noun ?x))))
(?npr (sem-cat (== (npr ?y))))
(?pre (sem-cat (== (pre ?z))))
<->
((?questopic-unit (syn-subunits (== ?npr ?pre ?npr)))
(syn-cat (==1 (pos (SUBORDINATE))))
(?form (form (== (meets ?npr ?pre)(meets ?pre ?npr)))))
```

The first construction represents the particle that is the Italian adverb "*Quando*"; it is an item-based construction because the specific string form is indicated. In such a construction, the `label` slot is used to indicate the properties to be searched for in DBPedia to disambiguate the user request, as explained in Sect. 3.3. These properties convey information about the question topic, and are linguistically related to the specified item. There can be different `label` slots for each item. For the sake of clarity, only one `label` is reported in the example.

The second example is another abstract construction, representing a question topic. Obviously, there are many kinds of question topic abstractions. In the example, the topic is the ordered conjunction of a proper noun, a preposition and another proper noun because this conjunction generally identifies the entity the user is questioning about. The particle and the verb that are unified next, define what piece of information is requested for the entity. For example, the phrase *Torre di Pisa*, which is POS tagged as (*Torre.NP di.PREP Pisa.NP*), is unified by means of the POS tags with both the generic constructions of the proper noun, and the item-based construction related to the preposition "*di*". Unification is then used to put constructions into conjunction to the more abstract question topic in the example. This strategy will be detailed in Subsect. 3.3. Summarily, we defined the sub-units of both general constructions and item-based ones where a string is associated, as in the case of adverbs, prepositions, conjunctions, and so on. These constructions are unified with the transient structure of the user question; if the user question does not contain any slot of the item-based constructions, more general constructions are unified. In this way, the model allows the comprehension of the query even if some of its part are unknown or incomplete.

3.3 Mapping to Meanings

The *Mapping to Meanings* module (MtM) implements the set of processes used by the agent to access the meanings of the natural language query formulated by the user. The query is first chunked by a POS tagger. A chunk is a set of consecutive tokens with the same POS tag; it is a n-gram of query words having the same syntactic category. Chunks are used for filling the transient structure that will unify with the constructions in the linguistic base; the semantic pole of this structure will be empty, while the syntactic one will contain two slots; the former will be related to the syntactic category of the chunk, the latter will contain its string form. The set of operations implementing the cognitive processes of agent's comprehension, are formalized in what follows.

Let consider the query $Q = \{q_1, q_2, ..., q_n\}$, where q_i is the i-th token. Being T the set of all POS tags, and $t \in T$ a specific tag, the chunks set C is the partition of Q so that:

$$C = \{c_i | c_i = \bigcup_l^k q_j, \ pos(q_j) = t_i \forall j \in [l \dots k], l, k \in [1 \dots n], \ t_i \in T, \ q_j \in Q\}$$

where the function $pos : Q \to T$ returns the POS tag of a token. The transient structure represents a sentence that has been understood partially; the MtM builds such a structure considering the chunks set of the query. The base process claims that *for each chunk a couple of slots is built in the transient structure, containing respectively the POS tag and the tokens of the chunk.* As an example, if the question is "*Quando è morto Riccardo Zandonai?*" the module outputs the following chunks expressed as set of bindings: {*(Quando . ADV), (è . VP), (morto . ADJ), (Riccardo Zandonai . NPR)*}.

Formally, given the set C, the initial transient structure t_s is a couple $t_s = \langle sem, syn \rangle$ where $sem = \emptyset$, while syn is a set of couples $syn = \{(\text{syn-cat } t_i)(\text{string}c_i), c_i \in C, t_i \in T\}$. The complete transient structure for the example is represented in Fig. 2.

The structure is then unified with the linguistic base. Through the unified constructions, a general query pattern emerges, and the semantic side is filled. In the example, the unified pattern is:

```
((?Top (sem-subunits (== ?particle ?verb ?questopic)))
(?particle (sem-cat (== (particle Quando))))
(?verb (sem-cat (== (verb morto))))
(?questopic (sem-cat (== (questopic Riccardo Zandonai))))
<->
```

```
((?Top (syn-subunits (== ?particle ?verb ?questopic)))
(syn-cat (==1 (pos (DI))))
(?form (form (== (meets ?particle ?verb)(meets ?verb ?questopic))
((syn-cat ADV) (string "Quando"))
((syn-cat V)(string "morto"))
((syn-cat NPR)(string "Riccardo Zandonai"))
))
```

where all query components emerge as result of a sequences of unification steps. Formally, we call F the store collecting all these components. In particular, $F = F_p \cup F_v \cup F_q$, where F_p contains the particle slots, F_v contains the verb slots, and F_q contains the question topic. In the example, $F_p = \{\text{Quando}\}$, $F_v = \{\text{morto}\}$ and $F_q = \{\text{Riccardo Zandonai}\}$. The question topic is what the user wants to know, and it is then searched for into the domain ontology to extract the corresponding *assertion subgraph*.

Extraction of the assertion subgraph corresponds exactly to conceptualization of the perceived words. In this work, extraction is simply implemented by rough intersections between the stems of the words in the unified query construction, and the labels of the domain ontology. Being *stem* the function that returns the stem of the words in its argument, *couple* the function that returns both elements of a couple, and f_i a generic element belonging to either F_q or F_v, the assertion extraction process is modeled by the following functions:

- $a_c : F_q \rightarrow C_o$ that returns the set $a_c(f_i) = \{cl_k \mid stem(f_i) = stem(i_j), i_j = map(cl_k), cl_k \in C_o\}$ composed by the ontological classes whose instances label's stems map to the stems of question topic words contained in F_q. The set $I = \{i_j\}$ will contain all such instances;
- $a_p : F_v \rightarrow P_o \cup P_d$ returning the set $a_p(f_i) = \{p_j \mid stem(f_i) = stem(p_j), p_j \in P_o \cup P_d \wedge couple(p_j) \supset a_c(f_i)\}$ composed by the properties of the classes in $a_c(f_i)$ whose label's stems map to the stems of the verb in F_v.

The result is a sub-ontology $A = \langle C_a, P_a, I \rangle$ where $C_a = \bigcup_{F_q} a_c(f_i)$, $P_a = \bigcup_{F_v} a_p(f_i)$, and I the set of individuals retrieved by a_c. Its worth noting that in this case the ontology definition includes the instances. In the proposed example, $C_a = \{\text{Person}\}$, $I = \{\text{Riccardo Zandonai}\}$ (the map returned by a_c is exactly this instance). Properties result to be $P_a = \{\text{luogomorte, annomorte}\}$ that are connected to nodes with the same names in DBPedia; these are the properties whose stems match to the stem of the verb specified in the unified query pattern ($F_v = \{\text{morto}\}$). To complete the comprehension process, QuASIt prunes further the properties in the subgraph by the label slot annotated in F_p. In the example, QuASIt prunes the luogomorte property because it is not annotated in the label slot of the adverb *Quando*.

3.4 Mapping to Forms

Once the assertion subgraph has been retrieved QuASIt run the processes aimed at simulating lexical access in human mind. Here the correct expressions related

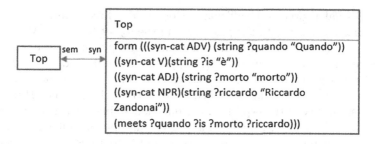

Fig. 2. The initial transient structure corresponding to the query in the example.

to the comprehended concepts must be formulated. The structure of the answer depends on the type of question; QuASIt accepts both essay questions and multiple choice ones. In the first case the correct answer has a free structure, while in the second case it must be one of the proposed candidates. We will describe both cases separately, but all questions share the same strategy to retrieve information from the support text, if available.

Search in the support text represents the way according to which the agent attempts to retrieve an information that is not in its knowledge base, and it learns a possible answer by comprehending a plain text dealing with the question topic. We implemented such process by extracting from the text the sentence with the highest syntactic similarity to the one owned by the agent; such a sentence can be either the query itself or one of the multiple answers. In fact, the query and the multiple choices represent the only source of information the agent owns to devise the correct answer when it has no coded knowledge about the question topic. Let $S = \{s_1, s_2...s_n\}$ be the sentence owned by the agent, and $P = \{p_1, p_2, ...p_m\}$ a sentence in the support text. The most similar sentence $\widehat{P}$ to S maximizes the following similarity measure m:

$$m = n - (\alpha l + \beta u)$$

where $n = |S \cap P|$ is the number of tokens that are both in S and P, $l = 1 - \frac{|S \cap P|}{|S|}$ is the number of "lacking tokens" that are tokens belonging to S that are not in $S \cap P$, while $u = 1 - \frac{o(P,S)}{|S \cap P|}$ is the number of "unordered tokens" that is the number of tokens in P that do not have the same order in S; here $o(a,b)$ is the function returning maximum number of ordered tokens in a with respect to b. Both l and u are normalized in the range $[0 \dots 1]$; they are penalty values representing syntactical differences among the sentences. The higher u and l are, the lower is the sentences similarity. The α and β parameters weight the penalty, and they have been evaluated empirically through experimentation.

Essay Questions. In case of essay questions, QuASIt produces the assertions that are contained in the range of the properties belonging to the assertion subgraph, which are separated by comma and conjunctions. Right now, we do not produce a formal answer as the affirmative form of the direct interrogative. Assertions are those connected to the properties in the a_p set by the MtM strategies. If a_p is empty, QuASIt does not know the question topic, and it starts searching for external information sources. The support text contained in the DBPedia `Abstract` property related to the question topic, and the one contained in the question itself, if available, are searched for to extract the sentence $\widehat{P}$ that maximizes the similarity with respect to the question string. In this case the plain n value is used because the question string has an obvious different order with respect to the sentences in the support text, and it contains much less tokens.

Multiple Choice Questions. QuASIt can be used for choosing the correct answer to a question in a set of candidates. These candidates represent the sentences owned by the system, and they are managed according to the strategies explained above. In particular, two steps are executed: first QuASIt searches for each candidate in the assertion subgraph, filtering the best matching answer according to the metric m. If no candidates match to any assertion, QuASIt refers to the available support text, that is the concatenation of the DBPedia `Abstract` property related to the question topic and the support text enclosed in the question itself, if available. Again QuASIt searches for the best matching candidate to some sentence in the support text according to the similarity value m. The form of the answer is directly the winning candidate.

4   Experiments and Discussions

Due to the poor number of tests and tools for QA in the Italian language, it was hard to find a way for benchmarking our system. As already mentioned we used the QA4MRE2011 and QA4MRE2012 datasets, that were created to test the systems submitted in the related tasks at CLEF 2011 and CLEF 2012 conferences, providing multiple-choice questions. In this way it is possible to make a quantitative performance measure. The task focuses on reading a support text, and answering to a set of questions about the information that is stated or implied in the text. More in detail, the datasets consist respectively of 120 and 160 questions, each with 5 possible answers and just one correct answer. Questions are grouped by topic and, for each topic, support texts are provided containing information about the relative group of questions (Table 1).

Table 1. Test results

Test dataset	Correct	NoA	Total	Accuracy	c@1
QA4MRE2011	40	2	120	0,33	0,33
QA4MRE2012	46	1	160	0,29	0,29

5   Conclusions

QuASIt has been presented in this work, that is a cognitive model for an artificial agent performing question answering in Italian, along with its implementation. QuASIt is able to answer both multiple choice and essay questions using an ontology-based approach. Our cognitive model is inspired to the procedural semantics perspective, so it runs different cognitive processes on two sources of knowledge: domain knowledge, and linguistic one. Domain knowledge has been formalized in the implementation using the DBPedia ontology, while Construction Grammar (CxG) has been used to represent linguistic knowledge. Particularly, language fluidity has been addressed using the FCG implementation of CxG. QuASIt analyses the question through FCG, isolates the question topic, searches for the properties' ranges in the DBPedia ontology that are linguistically related to the question topic by the particle and the verb in the questions, and outputs the answer. In case of multiple choice questions, all candidate answers are matched against both all the retrieved information, and the support text, if available. Experimentation has been carried out using the datasets provided in the QA4MRE 2011 and 2012 challenges. Results are satisfactory, and QuASIt can be no doubt considered as a state-of-the art system with respect to the performance values reported by the QA literature. QuASIt implementation is being refined continuously. Current work is aimed mainly at deepening the FCG linguistic analysis of both the question and the answer candidates to refine the similarity measure, and to produce formal affirmative answers in any case, through FCG the production.

References

1. McGuinness, D.L.: Question answering on the semantic web. IEEE Intell. Syst. **1**, 82–85 (2004)
2. Fellbaum, C.: WordNet: An Electronic Lexical Database. Bradford Books, Cambridge (1998)
3. Baker, C.F., Fillmore, C.J., Lowe, J.B.: The Berkeley framenet project. In: Proceedings of the 36th Annual Meeting of the Association for Computational Linguistics and 17th International Conference on Computational Linguistics, ACL 1998, vol. 1, pp. 86–90. Association for Computational Linguistics, Stroudsburg, PA, USA (1998)
4. Pianta, E., Bentivogli, L., Girardi, C.: MultiWordNet: developing an aligned multilingual database. In: Proceedings of the First International Conference on Global WordNet, January 2002
5. Dean-Hall, A., Clarke, C.L.A., Kamps, J., Kiseleva, J., Voorhees, E.M.: Overview of the TREC 2015 contextual suggestion track. In: Voorhees, E.M., Ellis, A. (eds.) TREC, Volume Special Publication, pp. 500–319, National Institute of Standards and Technology (NIST) (2015)
6. Boubiche, D.E., Hidoussi, F., Cruz, H.T., Bouziane, A., Bouchiha, D., Doumi, N., Malki, M.: Question answering systems: survey and trends. Procedia Comput. Sci. **73**(2015), 366–375 (2015). International Conference on Advanced Wireless Information and Communication Technologies (AWICT 2015)

7. Sag, I.A., Wasow, T., Bender, E.: Syntactic Theory: A Formal Introduction, 2nd edn. Center for the Study of Language and Information, Stanford (2003)
8. Basili, R., Hansen, D.H., Paggio, P., Pazienza, M.T., Zanzotto, F.M.: Ontological resources and question answering
9. Basili, R., Pazienza, M.T., Zanzotto, F.M.: Exploiting the feature vector model for learning linguistic representations of relational concepts. In: Workshop on Adaptive Text Extraction and Mining (ATEM 2003) Held in Conjuction with European Conference on Machine Learning (ECML 2003) (2003)
10. Hoffmann, T., Trousdale, G., Hoffmann, T., Trousdale, G.: Construction grammar introduction
11. Spranger, M., Pauw, S., Loetzsch, M., Steels, L.: Open-ended procedural semantics. In: Steels, L., Hild, M. (eds.) Language Grounding in Robots, pp. 153–172. Springer, Boston (2012)
12. Goldberg, A., Suttle, L.: Construction grammar. Wiley Interdisc. Rev. Cogn. Sci. **1**(4), 468–477 (2010)
13. Hoffmann, T., Trousdale, G., Boas, H.C.: Cognitive construction grammar
14. Steels, L.: Introducing fluid construction grammar (2011)
15. Pipitone, A., Anastasio, F., Pirrone, R.: HOWERD: a hidden Markov model for automatic OWL-ERD alignment. In: ICSC, pp. 477–482. IEEE Computer Society (2016)
16. Pipitone, A., Pirrone, R.: A hidden Markov model for automatic generation of ER diagrams from owl ontology. In: ICSC 2014, pp. 135–142. IEEE Computer Society (2014)

Spoken Language Understanding for Service Robotics in Italian

Andrea Vanzo[1], Danilo Croce[2], Giuseppe Castellucci[3(✉)], Roberto Basili[2], and Daniele Nardi[1]

[1] Department of Computer, Control and Management Engineering
"Antonio Ruberti", Sapienza University of Rome, Rome, Italy
{vanzo,nardi}@dis.uniroma1.it
[2] Department of Enterprise Engineering,
University of Roma Tor Vergata, Rome, Italy
{croce,basili}@info.uniroma2.it
[3] Department of Electronic Engineering,
University of Roma Tor Vergata, Rome, Italy
castellucci@ing.uniroma2.it

Abstract. Robots operate in specific environments and the correct interpretation of linguistic interactions depends on physical, cognitive and language-dependent aspects triggered by the environment. In this work, we describe a Spoken Language Understanding chain for the semantic parsing of robotic commands, designed according to a Client/Server architecture. This work also reports a first evaluation of the proposed architecture in the automatic interpretation of commands expressed in Italian for a robot in a Service Robotics domain. The experimental results show that the proposed solution can be easily extended to other languages for a robust Spoken Language Understanding in Human-Robot Interaction.

Keywords: Spoken language understanding · Automatic interpretation of robotic commands · Grounded language learning · Human robot interaction

1 Introduction

End-to-end communication in natural language between humans and robots is challenging for the deep interaction of different cognitive abilities. For a robot to react to a user command like *"porta il libro sul tavolo nel laboratorio"*[1], a number of implicit assumptions should be met. First, at least three entities, libro (book), tavolo (table) and laboratorio (laboratory), must exist in the environment and the speaker must be aware of such entities. Hence, the robot must have access to an inner representation of the objects, e.g., an explicit map of the environment. Second, mappings from lexical references to real world entities

---

[1] In English, *"bring the book on the table in the laboratory"*.

© Springer International Publishing AG 2016
G. Adorni et al. (Eds.): AI*IA 2016, LNAI 10037, pp. 477–489, 2016.
DOI: 10.1007/978-3-319-49130-1_35

must be made available. *Grounding* [1], here, should correspond to the explicit linking of symbols (e.g., words) to the information perceived about the context. Spoken Language Understanding (SLU) for interactive dialogue systems acquires a specific nature when applied to Interactive Robotics. Linguistic interactions are context-aware in the sense that both the user and the robot access and make references to the environment (i.e., entities of the real world). In the above example, whenever a table is actually in the laboratory, the GOAL of the action referred by the verb *"portare"* (*"to bring"*) is [*sul tavolo nel laboratorio*], i.e., the book has to be brought on the table in the laboratory. On the contrary, if there are no tables in the laboratory, [*sul tavolo*] is needed to locate the book nearby the robot and the GOAL refers to [*nel laboratorio*], i.e., the book is on a table and it has to be brought in the laboratory. Hence, robot interactions need to be *grounded*, as meaning depends on the state of the physical world and interpretation crucially interacts with perception, as pointed out by psycho-linguistic theories [2]. The integration of perceptual information derived from the robot's sensors with an ontologically motivated description of the world provides an augmented representation of the environment, called *semantic map* in [3]. In this map, the existence of real world objects can be associated to *lexical* information, in the form of entity names given by a knowledge engineer or uttered by a user, as in Human-Augmented Mapping [4]. While SLU for Interactive Robotics has been mostly carried out over the evidences specific to the linguistic level, e.g., in [5–7], we argue that such process should be accomplished in a harmonized and coherent manner. In fact, SLU has been already addressed in other works (see, for example, [8,9]) where perceptual knowledge is neglected in disambiguating among the structures produced by a linguistic parser.

This paper presents a processing chain for the interpretation of spoken commands. This chain is based on the approach proposed in [10] that integrates both linguistic and perceptual information to realize a context-aware interpretation of robotic commands. In particular, the interpretations coherently express constraints about the world (with all the entities composing it), the Robotic Platform (with all its inner representations and capabilities) and the pure linguistic level. Moreover, we present an experimental evaluation of the proposed chain over a dataset of commands in Italian, to validate its effectiveness with respect to different languages. To the best of our knowledge this is the first work addressing SLU of robotic commands in Italian Language. Preliminary results confirm the effectiveness of the adopted approach even in Italian: a first processing chain in Italian can be in fact obtained by annotating about 10 sentences representing typical ways to express a robotic command in a domestic environment.

In Sect. 2, the overall processing work-flow is introduced. In Sect. 3, we provide an architectural description of the chain, as well as an introduction about its integration with a generic robot. In Sect. 4, we present the experimental results of the proposed system over a dataset of Italian commands. Finally, in Sect. 5 we derive the conclusions.

2   The Language Understanding Cascade

A command interpretation system for a robotic platform must produce interpretations of user utterances. In this paper, the understanding process is based on the theory of the Frame Semantics [11]; in this way, we aim at giving a linguistic and cognitive basis to the interpretations. In particular, we consider the formalization promoted in the FrameNet [12] project, where actions expressed in user utterances can be modeled as *semantic frames*. Each frame represents a micro-theory about a real world situation, e.g., the actions of *bringing* or *motion*. Such micro-theories encode all the relevant information needed for their correct interpretation. This information is represented in FrameNet via the so-called *frame elements*, whose role is to specify the participating entities in a frame, e.g., the THEME frame element represents the object that is taken in a *bringing* action.

As an example, let us consider the following sentence: *"porta il libro sul tavolo"*. This sentence can be intended as a command (in Italian), whose effect is to instruct a robot to bring a book on a table. The language understanding cascade should produce its FrameNet-annotated version, that is:

$$[porta]_{Bringing} \ [il \ libro]_{\text{THEME}} \ [sul \ tavolo]_{\text{GOAL}} \tag{1}$$

Semantic frames can thus provide a cognitively sound bridge between the actions expressed in the language and the implementation of such actions in the robot world, namely plans and behaviors.

The whole SLU process has been designed as a cascade of reusable components, as shown in Fig. 1. As we deal with vocal commands, their (possibly multiple) hypothesized transcriptions derived from an Automatic Speech Recognition (ASR) engine constitute the input of this process. It is composed by four

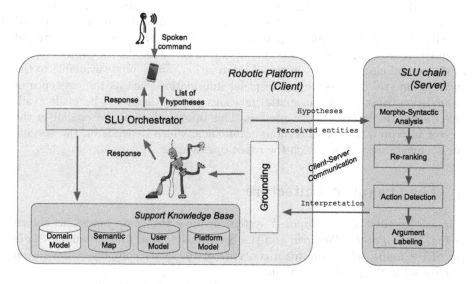

Fig. 1. Overall architecture of the SLU chain

modules, whose final output is the interpretation, later adopted to implement the corresponding robotic actions. First, **Morpho-syntactic analysis** is performed over the available utterance transcriptions by applying morphological analysis and Part-of-Speech tagging. In our evaluations, an off-the-shelf tool is adopted for this module, i.e., the *Chaos* parser [13]. Then, if more than one transcription hypothesis is available, the **Re-ranking** module can be activated to compute a new ranking of the hypotheses, in order to get the best transcription out of the initial ranking. This module is realized through a learn-to-rank approach, where a Support Vector Machine exploiting a combination of linguistic kernels is applied, according to [14]. Third, the best transcription is the input of the **Action Detection** (AD) component. The evoked frames in a sentence are detected, along with the corresponding evoking words, the so-called lexical units. Let us consider the one recurring sentence: the AD should produce the following interpretation $[porta]_{Bringing}$ *il libro sul tavolo*. The AD step is realized through a sequential labeling approach: each token of a sentence is labeled through a Markovian Support Vector Machine [15] with respect to the possible frames evoked by the token, according to [7]. The final step is the **Argument Labeling**, where a set of frame elements is retrieved for each frame. This process is realized in two sub-steps. First, the *Argument Identification* (AI) finds the spans of all the possible frame elements. Then, the *Argument Classification* (AC) assigns the suitable label (i.e. the frame element) to each span thus producing the final tagging shown in the Eq. 1. The Argument Labeling phase is realized through a sequential labeling algorithm similar to the one of the previous phase. Here, each token of a sentence is associated to one (or none) frame element of the detected frame.

Notice that both the re-ranking and the semantic parsing phases can be realized in two different settings. They can either exploit only linguistic information to solve the given task, or they can embed also perceptual knowledge coming from a semantic map into the process. In the first case, the information used to solve the task comes from linguistic inputs, as the sentence itself or external linguistic resources. These models correspond to the methods discussed in [7,14]. In the second case, robot's *perceptual* information can be made available to the chain, as in [10]. In this way, perceptual information such as the existence of grounded entities, as well as spatial relations among them, can be made available during the interpretation process. This information can be crucial in the correct interpretation of ambiguous commands, which depends on the specific environmental setting in which the robot operates.

3   The Overall Architecture

The architecture of the proposed system involves two main actors, as shown in Fig. 1: the *Robotic Platform* and the *Spoken Language Understanding Chain* (or SLU Chain), where the main concepts of the latter component have been introduced in the previous section.

A Client-Server communication schema between the SLU chain and the Robot allows maintaining a perspective on the SLU Chain, which strictly emphasizes the independence from the Robotic Platform, in order to maximize the reusability and integration in heterogeneous robotic settings. The SLU process we propose exhibits semantic capabilities (e.g., disambiguation, predicate detection or grounding into robotic actions and environments), that are designed to be general enough to be representative of a large set of application scenarios.

It is obvious that an interpretation process must be achieved even when no information about the domain/environment is available, i.e., a scenario involving a *blind* but speaking robot, or when the actions a robot can perform are not made explicit, that we would call an *unaware* linguistic robot. This is the case when the command *"porta il libro sul tavolo nel laboratorio"* is not paired with any additional information and the ambiguity with respect to the argument spans, i.e., [*il libro sul tavolo*]$_{\text{THEME}}$ [*nel laboratorio*]$_{\text{GOAL}}$ vs. [*il libro*]$_{\text{THEME}}$ [*sul tavolo nel laboratorio*]$_{\text{GOAL}}$, cannot be resolved. At the same time, the platform makes available methods to specialize its semantic interpretation process to individual situations where more information is available about goals, environment and capabilities of the robot. These methods are expected to support the optimization of the core SLU platform against a specific interactive robotics setting, in a cost-effective manner. In fact, whenever more information about the environment perceived by the robot (e.g., a semantic map) or about its capabilities is provided, the interpretation of a command can be improved by exploiting a more focused context. It means that whenever the sentence *"porta il libro sul tavolo nel laboratorio"* is provided along with information about the presence and possible positions of a `table` referred by the word *tavolo* in a `laboratory` (*laboratorio*) the system should be able to detect and disambiguate the intended action. In order to better describe the different operating modalities of the proposed SLU Chain, some assumptions about the Robotic Platform must be made explicit: this will allow to precisely establish functionalities and resources that the robot needs to provide to unlock the more complex processes. These information will be used to express the experience that the robot is able to share with the user (i.e., the perceptual knowledge about the environment, where the linguistic communication occurs and some lexical information and properties about objects in the environment) and some level of awareness about its own capabilities (e.g., the primitive actions that the robot is able to perform, given its hardware components).

In the following, each component of the architecture in Fig. 1 will be discussed and analyzed[2].

3.1   The Robotic Platform

The SLU Chain contemplates a generic Robotic Platform, whose task, domain and physical setting are not necessarily specified. In order to make the SLU Chain independent from the above specific aspects, we will assume that the platform requires at least the following modules:

---

[2] A more detailed description of the proposed SLU Chain along with usage instructions can be found at http://sag.art.uniroma2.it/sluchain.html.

– an Automatic Speech Recognition (ASR) system;
– a SLU Orchestrator;
– a Grounding and Command Execution;
– a Physical Robot.

Additionally, an optional component, i.e., the *Support Knowledge Base*, is expected to maintain and provide the contextual information discussed above. While the discussion about the Robotic Platform is out of the scope of this work, all the other components are hereafter shortly summarized.

ASR system. An ASR engine allows to transcribe a spoken utterance into one or more possible transcriptions. In the actual release, the ASR is here performed through an *ad-hoc* Android application. In fact, it relies on the official *Google ASR API*[3] and offers valuable performances for an off-the-shelf solution. The main requirement of this solution is that the device hosting the software must feature an Internet connection in order to provide transcriptions for the spoken utterance. This App can be deployed on both Android smartphones and tablets.

Once a new sentence is uttered by the user, this component outputs a list of candidate hypothesis transcriptions. The communication with the entire system is realized through TCP Sockets. In this setting, the Android ASR App implements a TCP Client, feeding the SLU Chain with lists of hypotheses.

SLU Orchestrator. The SLU Orchestrator implements a TCP Server for the Android App, here coded as a ROS[4] node waiting for Client requests. Once a new request arrives (a list of transcriptions for a given spoken sentence), this module is in charge of extracting the perceived entities from a structured representation of the environment (here, a sub-component of the Support Knowledge Base) and of sending the list of hypothesized transcriptions to the SLU Chain along with the list of the perceived entities.

The communication protocol requires the serialization of such information in two different JSON objects. In order to obtain the desired interpretation, only the list of transcription is mandatory. In fact, even though environment information is essential for the perception-driven chain, whenever it is not provided, the chain operates in a blind setting. The SLU orchestrator has been decoupled from the SLU Chain as it can be employed for other purposes, such as tele-operating the robot by means of a virtual joypad coded into the Android App. To this end, it can be personalized (or even replaced with a new one), by adding further functionalities and features, provided that the communication protocol is respected. The orchestrator, managing the communication between the Android App, the SLU Chain and the Robotic Platform, is provided along with the SLU Chain, so that robustness in the communication is guaranteed. In this way, the robotic developers are in charge of: (i) the ROS node of the target Robotic System; (ii) the definition of the policies for the acquisition of perceptual knowledge; and (iii) the manipulation of the structure representing the interpretation returned

---

[3] http://goo.gl/4ZkdU.
[4] http://www.ros.org/.

by the SLU Chain. In fact, the SLU orchestrator, besides acting as TCP Server for the Android App, represents also the Client interface toward the SLU Chain.

Grounding and Command Execution. Even though the grounding process is placed at the end of the SLU processing chain, it is discussed here as it represents part of the Robotic Platform. In fact, grounding has been completely decoupled from the SLU Chain, as it may involve perception capabilities and information unavailable to the SLU Chain or, in general, out of the linguistic dimension. Nevertheless, this situation can be partially compensated by defining mechanisms to exchange some of the grounding information with the linguistic reasoning component. However, grounding is always carried out on board of the robot, as it represents the most general situation.

The grounding carried out by the robot is triggered by a logical form expressing one or more actions through logic predicates, which potentially correspond to specific frames. The output of the whole SLU process embodies the produced logic form. This latter exposes: the recognized actions that are thus linked to specific robotic operations (primitive actions or plans); the predicate arguments (e.g., objects and location involved in the targeted action) detected and linguistically linked to the objects/entities of the current environment. A fully grounded command is thus obtained where possible through the complete instantiation of the robot action (or plan) and its final execution.

3.2 The SLU Chain

The SLU Chain component implements the language understanding cascade described in Sect. 2. It realizes the SLU service as a black-box component, so that the complexity of each inner sub-task is hidden to the robotic engineer. The service is realized through a server accepting connections on a predefined port. It is entirely coded in Java and released as a single Jar file, along with the required folders containing linguistic models, configurations files and other resources. Hence, it can be run through command line, so that it is easier to integrate it within any architecture. Operationally, the chain takes three parameters as inputs: *type* of the chain (`basic` or `simple`), *output format* (`XDG`, `AMR` or `TAB`) and *listening port* (e.g., `9090`). The first parameter defines the type of the chain going to be initialized. While `basic` refers to a setting where only linguistic information is employed, i.e., the *blind* situation, `simple` refers to the more complex chain, where perceptual features are taken into account in the interpretation process.

The second parameter specifies the desired output format. The type `XDG` refers to a Java data structure specifically devoted to the overall linguistic analysis of a command, called *eXtendend Dependency Graph*, whose details can be found in [13]. The type `AMR` refers to the *Abstract Meaning Representation*, a semantic representation language proposed in [16]. This formalism allows to express semantics, neglecting both the original sentence and its syntactic structure. Given the sentence *"porta il libro sul tavolo"*, the corresponding AMR format is:

```
(t1 / porta-Bringing
   : Theme (l1 / il libro)
   : Goal (t2 / sul tavolo)
)
```

4   Experimental Evaluation

In this section, we report a preliminary experimental evaluation of the Spoken Language Understanding (SLU) Chain presented in this paper applied in the interpretation of commands in Italian. The experiments have been designed in order to verify the robustness of the adopted SLU solution in the robotic context with different languages. The evaluation reported here extends the experiments already carried out in [10], where the above SLU chain has been evaluated against commands in English.

We produced an Italian dataset by translating a significant subset of English commands from the HuRIC corpus [17] already used in [10]. Each translated command is also manually labeled according to the Frame Semantics theory, that provides a semantic layer for the command interpretation process, as discussed in Sect. 2: semantic frames and frame elements are here used to represent the meaning of commands, reflecting the actions a robot can accomplish in a home environment. To this end, we considered the same set of FrameNet-inspired semantic frames adopted in [10], that act as language independent primitives for the robot's possible actions. Linguistic information required for each processing step has been extracted by using the *Chaos* parser [13]. The dataset is composed of 188 different commands, whose actions are represented by 14 different frames. It contains 211 annotated frames (i.e., almost $1, 12$ annotated frame per sentence) and 304 annotated roles (i.e., 1.62 role per sentence). The composition of the dataset in terms of number of sentences evoking each frame and number of annotated examples for each role is reported in Table 1.

In the following experiments, we first evaluated each sub-module in the chain separately, then we focused in the overall processing chain, thus considering the error propagated during the analysis.

4.1   Evaluation of the Individual Modules in the SLU Chain

The proposed SLU Chain has been first evaluated by considering in isolation each sub-modules discussed in Sect. 2, i.e., the Action Detection (AD), Argument Identification (AI) and Argument Classification (AC) sub-modules. To this end, we invoke each module by assuming that the information provided by the previous step in the chain is always correct. Moreover, the evaluation has been carried out considering the correct transcriptions, i.e., not contemplating the error introduced by the Automatic Speech Recognition (ASR) system. In this way, we focus on the errors of the SLU Chain and avoid the bias introduced by the ASR system. Given the limited size of the training material, experiments have been performed

Table 1. Distribution of frames and frame elements in the Italian dataset

Frame	Examples	Frame	Examples
Being_in_category	2	*Inspecting*	4
CATEGORY	1	DESIRED_STATE_OF_AFFAIRS	3
ITEM	2	PURPOSE	2
		INSTRUMENT	2
Being_located	13	*Following*	12
THEME	12	COTHEME	6
LOCATION	11	GOAL	3
COTHEME	1	MANNER	7
TIME	1	SOURCE	1
Bringing	39	*Motion*	34
GOAL	12	GOAL	28
BENEFICIARY	2	SOURCE	1
THEME	34	PATH	2
SOURCE	6	MANNER	1
PLACE	2	THEME	1
		DIRECTION	2
		DEGREE	2
Change_direction	6	*Entering*	1
DIRECTION	6	GOAL	1
Change_operational_state	15	*Releasing*	2
DEVICE	15	THEME	2
Closure	5	*Searching*	29
CONTAINING_OBJECT	1	MANNER	2
PLACE	2	PHENOMENON	29
INSTRUMENT	4	GROUND	8
MANNER	1	DEGREE	1
Placing	21	*Taking*	28
THEME	18	THEME	28
GOAL	20	SOURCE	16
SOURCE	1	PLACE	4

in a leave-one-out setting, i.e., each example is in turn removed from the dataset and it is adopted as test example: the remaining examples are adopted to train the chain while performances are derived by averaging results across the entire dataset. In these experiments, we do not consider perceptual information derived from the environment where the command has been pronounced, as these new commands in Italian are not completely aligned with the maps used in [10], yet.

Table 2. Experimental evaluation over the Italian dataset of each single sub-module in terms of Precision (P), Recall (R) and F1-Measure (F1)

Action detection			Argument identification			Argument classification		
P	R	F1	P	R	F1	P	R	F1
86.39 %	78.57 %	82.29 %	81.82 %	77.23 %	79.46 %	84.49 %	84.49 %	84.49 %

Results reported here are thus comparable with those obtained in [10], when a pure linguistic approach is addressed. We report the performance measures, in terms of Micro Precision (P), Recall (R) and F1-Measure (F1), with respect to the single sub-module.

In the AD phase, P, R and F1 measure the system effectiveness in correctly recognizing the frame(s) for each sentence, i.e., the robotic action(s) in our scenario. In the AI phase, they quantify the system ability in recognizing the exact boundaries of each argument in the frame. This means that *every* token (i.e., span) of every argument must be properly detected. In the AC phase, they are a measure of the correctness of the role label assignment to each span.

The results for the three phases are reported in Table 2. Even though this setting does not reflect a real operating scenario, where the performance drop is due to the error propagation during the semantic understanding process, this experiment provides an interesting food for thought about the complexity of each task. First, the most challenging task seems to be the AI, whose F1 is the lowest among the three phases, i.e., 79.46 % of F1 is obtained. On the contrary, AD and AC obtain higher F1 scores (82.28 % and 84.49 %, respectively).

These results are in general lower with respect to the results obtained over the entire HuRIC, in [10]. In fact, while in the AD task over the English dataset the system obtains a F1 score of 94.67 %, the same evaluation over Italian commands achieve a F1 score of 82.29 %. A similar drop of performances is observed both in AI and AC: in the AI task the English dataset allows the system to reach 90.74 % in the F1 score, while in AC the score is 94.93 %. These results are compared, respectively, with 79.46 % and 84.49 % obtained over Italian commands. We speculate that such drop is mainly due to the size of the involved dataset. In fact, while Italian data count a total of 188 commands, the evaluation of the system over the English language has been carried out over 527 commands.

However, this empirical investigation confirms the overall trend of performances, with the Argument Identification task the most complex one, and proves that the proposed system can be robustly extended to other languages.

4.2   Evaluation of the Whole Process

In a second experiment, we analyze the error propagation through the whole SLU Chain. To this end, the performances measured in each step take into account the errors made by the previous one. As an example, let us consider the AI sub-module, where the identification of the frame elements does depend on the

Table 3. Experimental evaluation over the Italian dataset of the whole processing chain in terms of Precision (P), Recall (R) and F1-Measure (F1)

Action detection			Argument identification			Argument classification		
P	R	F1	P	R	F1	P	R	F1
86.39 %	78.57 %	82.29 %	85.27 %	63.04 %	72.49 %	79.02 %	58.61 %	67.30 %

frames assigned in the previous step, i.e. the AD sub-module. If an action is not detected, its corresponding argument will be not identified neither. This issue is considered in the evaluation of the next steps, while it has been neglected in the previous evaluations. This setting thus reflects a more realistic operating scenario, where the performance drop is due to the error propagation. Again, we report the performance measures in terms of Micro Precision (P), Recall (R) and F1-Measure (F1), with respect to each single sub-module.

The results for the three phases are reported in Table 3. As expected, a performance drop across the SRL steps has been obtained, when possible incorrect information has been provided at each step. While the AD phase gets the same results (i.e., the non-gold setting for the AD is not provided as it is the first step in the proposed chain) if we consider the AI phase, the F1 score of 79.46 % in Table 2 measured with gold-standard information drops to 72.49 % of Table 3, when enabling error propagation by feeding non-gold information through modules. A performance drop is observed in the AC phase when compared with the gold setting, where we measure a F1 score of 67.30 % against the 84.49 % of the gold setting. However, the overall chain seems to be quite robust to the error propagation as, given the previous measurement made in isolation, a lower result was expected. In fact, the coarse multiplication of the F1 scores obtained in the single steps, i.e., 82.29 %, 79, 46 % and 84, 49 %, corresponds to about 55 % of F1. Such experimental results suggest that the proposed solution is promising for the development of SLU chains in different languages, as these results have been obtained only labeling about 13 sentences per frame, i.e., robot command.

5   Conclusion

In this paper, we presented a SLU processing chain focused on the problem of interpreting commands in the Mobile Service Robotics domain. The proposed solution relies on well-known theories, such as Frame Semantics and Distributional Semantics and leverages Machine Learning algorithms to support the interpretation of commands. These characteristics enabled for a more robust interpretation of the sentences against language variability. Moreover, even though the SLU Chain is completely decoupled from the Robotic Platform, the final interpretation has been tied to the environment surrounding the robot: it will allow to inject perceptual knowledge into the feature modeling process, as foreseen by the English chain ([10]). In order to prove the effectiveness of the proposed tool, we conducted some experiments on a real robotic scenario by addressing a new language, i.e., interpreting commands in Italian. Preliminary evaluations show promising results that can be obtained by only labeling a very limited set of examples, i.e., about 10 sentences for each robot action, and by relying only on pure linguistic information. Further evaluations will take into consideration both an extended version of the Italian dataset and the alignment of its commands with perceptual knowledge. We expect that the proposed SLU chain can support the development of natural language interfaces for Human Robot Interaction for further languages than English and Italian.

References

1. Harnad, S.: The symbol grounding problem. Physica D Nonlinear Phenom. **42**(1–3), 335–346 (1990)
2. Tanenhaus, M., Spivey-Knowlton, M., Eberhard, K., Sedivy, J.: Integration of visual and linguistic information during spoken language comprehension. Science **268**, 1632–1634 (1995)
3. Nüchter, A., Hertzberg, J.: Towards semantic maps for mobile robots. Robot. Auton. Syst. **56**(11), 915–926 (2008)
4. Diosi, A., Taylor, G.R., Kleeman, L.: Interactive SLAM using laser and advanced sonar. In: Proceedings of the 2005 International Conference on Robotics and Automation, pp. 1103–1108 (2005)
5. Chen, D.L., Mooney, R.J.: Learning to interpret natural language navigation instructions from observations. In: Proceedings of the 25th AAAI Conference, pp. 859–865 (2011)
6. Matuszek, C., Herbst, E., Zettlemoyer, L.S., Fox, D.: Learning to parse natural language commands to a robot control system. In: Desai, J.P., Dudek, G., Khatib, O., Kumar, V. (eds.) Experimental Robotics. STAR, vol. 88, pp. 403–415. Springer, Heidelberg (2012)
7. Bastianelli, E., Castellucci, G., Croce, D., Basili, R., Nardi, D.: Effective and robust natural language understanding for human-robot interaction. In: Proceedings of ECAI 2014. IOS Press (2014)
8. Tellex, S., Kollar, T., Dickerson, S., Walter, M., Banerjee, A., Teller, S., Roy, N.: Approaching the symbol grounding problem with probabilistic graphical models. AI Mag. **32**(4), 64 (2011)
9. Matuszek, C., FitzGerald, N., Zettlemoyer, L.S., Bo, L., Fox, D.: A joint model of language and perception for grounded attribute learning. In: ICML. icml.cc/Omnipress (2012)
10. Bastianelli, E., Croce, D., Vanzo, A., Basili, R., Nardi, D.: A discriminative approach to grounded spoken language understanding in interactive robotics. In: Proceedings of the 25th IJCAI, New York (2016)
11. Fillmore, C.J.: Frames and the semantics of understanding. Quad. Semantica **6**(2), 222–254 (1985)
12. Baker, C.F., Fillmore, C.J., Lowe, J.B.: The berkeley framenet project. In: Proceedings of ACL and COLING, pp. 86–90 (1998)
13. Basili, R., Zanzotto, F.M.: Parsing engineering and empirical robustness. Nat. Lang. Eng. **8**(3), 97–120 (2002)
14. Basili, R., Bastianelli, E., Castellucci, G., Nardi, D., Perera, V.: Kernel-based discriminative re-ranking for spoken command understanding in HRI. In: Baldoni, M., Baroglio, C., Boella, G., Micalizio, R. (eds.) AI*IA 2013. LNCS (LNAI), vol. 8249, pp. 169–180. Springer, Heidelberg (2013). doi:10.1007/978-3-319-03524-6_15

15. Altun, Y., Tsochantaridis, I., Hofmann, T.: Hidden markov support vector machines. In: Proceedings of ICML (2003)
16. Banarescu, L., Bonial, C., Cai, S., Georgescu, M., Griffitt, K., Hermjakob, U., Knight, K., Koehn, P., Palmer, M., Schneider, N.: Abstract meaning representation for sembanking. In: Proceedings of the 7th Linguistic Annotation Workshop and Interoperability with Discourse, Sofia, Bulgaria, ACL, pp. 178–186, August 2013
17. Bastianelli, E., Castellucci, G., Croce, D., Basili, R., Nardi, D.: HuRIC: a human robot interaction corpus. In: Proceedings of LREC 2014, Reykjavik, Iceland, May 2014

Planning and Scheduling

DARDIS: Distributed And Randomized DIspatching and Scheduling

Thomas Bridi[1,4(✉)], Michele Lombardi[1,4], Andrea Bartolini[2,3,4],
Luca Benini[2,3,4], and Michela Milano[1,4]

[1] DISI, University of Bologna, Viale Risorgimento 2, 40123 Bologna, Italy
{thomas.bridi,michele.lombardi2,michela.milano}@unibo.it
[2] DEI, University of Bologna, Viale Risorgimento 2, 40123 Bologna, Italy
{a.bartolini,luca.benin}@unibo.it
[3] Integrated Systems Laboratory, ETH Zurich, Zurich, Switzerland
[4] University of Bologna, Viale del Risorgimento, 2, 40136 Bologna, Italy
http://www.informatica.unibo.it

Abstract. Scheduling and dispatching are critical enabling technologies in supercomputing and grid computing. In these contexts, scalability is an issue: we have to allocate and schedule up to tens of thousands of tasks on tens of thousands of resources. This problem scale is out of reach for complete and centralized scheduling approaches. We propose a distributed allocation and scheduling paradigm called DARDIS that is lightweight, scalable and fully customizable in many domains. In DARDIS each task offloads to the available resources the computation of a probability index associated with each possible start time for the given task on the specific resource. The task then selects the proper resource and start time on the basis of the above probability. The scheduler can be customized with different policies to fit several objective functions like load balancing or makespan. We evaluate our approach in the domain of grids and supercomputers. We compare DARDIS with the most widely used algorithms used in these specific domains to show that this approach can reach better solutions in several cases.

Keywords: Distributed scheduling · Variable constraints · Soft/Hard constraints

1 Introduction

Computing is everywhere in modern society and large-scale computing services are at the foundation of mega-trends like smart grids infrastructures and the Internet of Things (IoT). Large-scale computing infrastructure like grids and High-Performance Computing (HPC) facilities require efficient workload scheduling and dispatching.

Consider for example the number of computational nodes a scheduler has to manage for high performance computers like the top 1 HPC in 2015 [1] or the future HPC machine planned in the USA [2]. This machine features a number

© Springer International Publishing AG 2016
G. Adorni et al. (Eds.): AI*IA 2016, LNAI 10037, pp. 493–507, 2016.
DOI: 10.1007/978-3-319-49130-1_36

of nodes estimated between 50'000 and 1'000'000 [3,4]. Classical job schedulers are rule-based [5]. These are heuristic schedulers that use rules to prioritize jobs. In these scheduling systems a job requests a set of resources on which the job will execute. The scheduler checks for each job if it can execute on a node while respecting the capacity of the target resources. If the job can use the requested amount of resources, the job is executed. To enforce fairness, supercomputing centers use different queues or partitions, targeting jobs with different resource request and different expected waiting time. This expected waiting time is viewed by the final user as a sort of soft-deadline and measures the Quality of Service (QoS). It is quite clear that for these large scale machines a centralized, optimization-based scheduler [6,7], is not a feasible option. Hence, scalable, distributed schedulers are needed to handle thousands of nodes while at the same time optimizing efficiency metrics (e.g., reducing operating cost, maximizing utilization).

This work takes inspiration from Randomized Load Control proposed in [8]. In Randomized Load Control, the algorithm schedules appliances using a probability distribution based on a desired total energy consumption profile. In this work, it is shown that the approach obtains results that match well the desired profile. We substantially extend that work to the case of multiple resources and introduce new start times generators and dispatching policies.

In this work we present a Distributed And Randomized DIspatching and Scheduling (DARDIS) approach that is:

- **Distributed** to scale to an ultra-large system. The scheduler and dispatcher basically leave the dispatching choice to the task and each resource then schedules its own tasks.
- **Supporting variable resources utilization profile.** Each resource, besides its capacity, exhibits a (variable) desired utilization profile.
- **Randomized.** The scheduler can choose the proper probability distribution for selecting resources and start times to optimize different objective functions.
- **Customizable.** We will show 12 different setups of DARDIS each one obtained by combining different scheduling and dispatching policies.
- **Deadlines aware.** Each activity can specify a time window in which it should start. This can be used i.e., by facility administrators to create activities with different priorities (the smaller the window the higher the priority). In these system, the administrator could also decide to apply a pricing model inversely proportional size of the user-specified window.

Results show that this approach does not only obtain benefits derived by the reduction of the overhead w.r.t. rule-based schedulers, but also we show that DARDIS can improve waiting and makespan w.r.t. rule-based schedulers. Moreover, the introduction of a desired utilization profile can be used to minimize costs derived by the resource utilization. In addition, DARDIS can be tailored to produce a utilization profile that never exceeds the desired profile and follows its shape. This constraint on the desired profile can also be used as soft constraint (as we do in the presented test). We, also, show that in this case the overutilization over the desired profile obtained by DARDIS is created only to minimize metrics with a higher specified importance.

We have chosen to apply our approach to two domains (Grid computing and HPC) whose instances are large scale both in the number of tasks and in the number of resources.

1.1 High Performance Computing

Today's commercial schedulers are centralized and based on rule-based policies. In [4], it is advocated that this approach will fail in computing an entire schedule in future large scale HPC installations. In addition, the intrinsic computational power of future HPC installations is bounded by their total consumed power which has a practical limit of 20 MWatt due to constraint in the energy provisioning system [9]. While it is easy to predict the worst nominal power consumption of an HPC infrastructure, its real utilization depends on workload properties and on the current system utilization. Many works such as [10], demonstrated that a job scheduler can be proactively used to constrain the power consumption at run-time by setting a desired power profile and schedule on the machine only the jobs which satisfy this constraint. This has the potential of reducing system over-provisioning. Moreover, the cooling power and cost required to cool down the heat generated by the system activity depends on the overall power and environmental conditions [11]. Borghesi [10] show that by dynamically modulating the power profile according to the environmental temperature, it is possible to improve the overall energy efficiency. As a matter of fact, this scenario requires to schedule a set of large number of activities (jobs) in a large number of resources (nodes) while satisfying a desired profile (power budget) which is variable in time.

1.2 Grid Computing

In Grid domains, the scheduler has to deal with fewer resources w.r.t the HPC domain, but a larger number of jobs. Usually a number between tens to hundreds of resources (clusters) have to manage tens of thousands of activities (jobs).

The work in [12] shows that usual grid installations are composed by several clusters which share the same power supply source. Moreover, in this scenario high utilization peaks can create problems to the energy distribution network. The same problem is present when a heavily loaded cluster faces an abrupt workload shift which leads to a negative power peak. These two situations have been shown to be dangerous for the power supply network and usually are translated into a penalty from the energy distributor.

This problem can be mitigated by imposing a desired utilization profile on the resources which forbids abrupt changes in the utilization profile.

2 Approach

The workload dispatching and scheduling problem can be modeled by a set of resources res_r, with $r \in R$ and a set of activities a_i with $i \in A$, with R the number of resources and A the number of activities. Each resource has a capacity c_r, a desired profile $dp_r(t)$ and a utilization profile $up_r(t)$, with $t \in [0, \ldots, Eoh]$ (*Eoh* End of horizon). The desired profile is a profile decided by the administrator that shows how the resource should be used (in term of amount of resource used by activities) in time. The utilization profile is the amount of resource already used and reserved to scheduled activities. This profile is a cyclic profile repeated in time. As in example for HPC and grid computing this could be a daily utilization profile.

Each activity is submitted to the system at a time instant q_i. At the submission it specifies its earliest start time est_i, the latest start time lst_i, its duration wt_i, and the amount of resource required req_i.

The scheduling problem consists of allocating each activity to a given resource and assigning it a start time st_i and a resource ur_i such that

$$st_i :: [est_i, \ldots, lst_i] \; \forall i \in A$$
$$ur_i \in res_r \; \forall i \in A, r \in R$$
$$\sum_i req_i \le c_r \; \forall i \in A | st_i \le t \wedge st_i + wt_i > t, \forall t \in [0, \ldots, Eoh] \qquad (1)$$
$$\sum_i req_i \le dp_r(t) \; \forall i \in A | st_i \le t \wedge st_i + wt_i > t, \forall t \in [0, \ldots, Eoh]$$

The main idea of the proposed scheduler is to partition the decision process in two main phases performed by two separate software entities: the task agent and the resource manager. The *task agent* is responsible for the activity submission and the dispatching. This agent resides into the user-space (e.g., in HPC and grid computing the task agent is localized into the user workstation/PC). The *resource manager* is responsible for the scheduling. This agent resides into the resources host (e.g., in HPC it is the host node and in grid computing it is the cluster interfaced to the web).

Figure 1 shows the different phases of our approach. This phases are subdivided in:

Job Submission (1) - Our approach starts with a task agent submitting an activity to all the resource managers of the system. At submission time, it specifies the activity ID, the amount of required resources, the maximum execution time (referred to as walltime), the earliest start time and the latest start time for its execution. After the submission to all the resource managers, the task agent waits for the responses.

Start time probability generation (2) - Each resource manager receives the submitted activity and starts the start time probability generation phase in which the manager generates a start time for the activity according to an internal rule (Sect. 2.1).

Start time response (3) - After the start time generation, the resource manager sends a generated start time to the task agent.

Resource selection (4) - The task agent, after receiving the responses from all the resources, applies a policy (Sect. 2.2) to select the resources for the activity execution. In case of error (like communication fault) a time out on the task

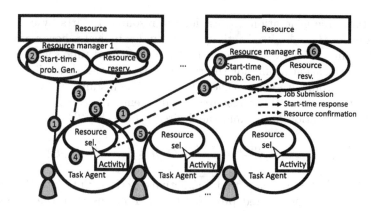

Fig. 1. DARDIS architecture and phases (number ordering corresponds to time progression)

agent let the protocol to continue with the responses obtained until that time. A number of missing responses from the resource managers could increase the probability to not find a feasible schedule for the activity before its deadline. This probability depends on the state of the used resources. However, if an allocation is not found this case is ascribable to the case of an activity that cannot be scheduled within its deadline.

Resource confirmation (5) - The task agent sends the result to all the resource managers involved in the submission, namely, the one selected and those not selected.

Resource reservation (6) - the resource managers in which the activity has to execute, reserves the proper capacity for the execution, by modifying the utilization profile.

DARDIS can handle situations like unexpected activity termination. This event simply triggers the resource manager in which the activity executed to update the profile. However, different solutions can be implemented like i.e., the rescheduling of the queued jobs. This is left for future works.

It must be noted that the number of exchanged messages for scheduling a single activity depends linearly on the number of the resources of the system. In particular, for each activity we exchange $3 * M$ messages where M is the number of resources involved in the submission.

In the following sections, we detail the start time probability generation process and the resource selection one.

2.1 Start Time Probability Generation

As described above, the agent responsible for the start time generation is the resource manager. This agent manages the resources and the activities allocated on the resource itself.

The start time generation process for the resource j starts by computing a fitting index for the submitted activity i. This index indicates how many parallel runs of the same activity could be executed at a given start time s while satisfying the desired utilization profile for the resource. Due to the variability over time of the desired profile, we have to check for each time instant $t \in \{s, \ldots, s + wt_i\}$ how many times the activity resource requirement req_i can fit the space left between the utilization profile and the desired profile (Eq. 2).

$$I'(s) = min_t(\frac{dp_j(t) - up_j(t)}{req_i}) \; \forall t \in \{s, \ldots, s + wt_i\} \tag{2}$$

Note that $I'(s) = 1$ means that the activity perfectly fits into the resource without exceeding the desired profile. $I'(s) > 1$ means that the activity fits the desired profile and leaves some resource for other activities. If $I'(s) < 1$, it means that the activity exceeds the desired profile. The capacity instead cannot be exceeded by definition. To handle this case, we use Eq. 3. Where \ represents integer division.

$$I(s) = min(I'(s), min_t((c_j - up_j(t)) \setminus req_i)), t \in \{s, \ldots, s + wt_i\} \tag{3}$$

The index distribution I is calculated for each possible start time between the earliest start time est_i and the latest start time lst_i of the activity: $I = \{I(est_i), \ldots, I(lst_i)\}$. In this way we obtain the fitting profile for the activity.

We defined three generators for the start time selection:

- **First:** the goal of this start-time generation procedure is to maximize the throughput of the entire system. This deterministic selection, works by picking up the first start time st_i with $I(st_i) \geq 1$.
- **Uniform:** the goal of this generator is to produce a scheduler that allocate resources following the shape of the desired profile for its entire window. This is a probabilistic selection that chooses a random number rnd in the range $[0, \ldots, \sum_{s=est_i}^{lst_i} I(s)]$. The start time st_i is then obtained by imposing the conditions $\sum_{t=0}^{st_i} I(t) \geq rnd$ and $\sum_{t=0}^{st_i - 1} I(t) < rnd$. If the selected start time has a fitting index $I(t) < 1$, t is increased until the condition $I(t) \geq 1$ is verified. If a start time is not found in this range, the search is repeated starting from $I(est_i)$ and the first $I(t) \geq 1$ is chosen. If a start time cannot be found in the entire range, the allocation is infeasible on the specific resource and the generator fails returning a null value.
- **Exponential:** this generator has been designed to reach a trade-off between throughput and profile chase. This is a probabilistic generator that chooses a random number rnd following the distribution $(lst_i - est_i)e^{-(lst_i - est_i)x}$. Then it computes the start time by imposing the conditions $\sum_{t=0}^{st_i} I(t) \geq rnd$ and $\sum_{t=0}^{st_i - 1} I(t) < rnd$. If the selected start time has a fitting index $I(t) < 1$, it increases t until the condition $I(t) \geq 1$ is verified. If a start time is not found in this range, the search is repeated from $I(est_i)$ and the first $I(t) \geq 1$ is chosen. If a start time cannot be found in the entire range, the allocation is infeasible on the specific resource and the generator fails returning a null value.

Complexity. The complexity of the resource manager resides into the start time generation algorithm. The algorithm calculates a fitting profile for a submitted activity. This fitting profile is composed by $k = lst - est$ fitting indexes. Finally, a fitting index is derived by checking the desired profile and the resource required by the activity for each time instant of its execution wt_i. The complexity is given by $O(k * wt_i)$. Being k and wt_i dependent from the size of the input variables, this algorithm is pseudo-polynomial.

2.2 Resource Selection

After the start time generation process, the task agent receives the responses from all the resource managers involved in the submission. This algorithm selects the resources for the activity execution.

The designed policies are:

- **MIN_START:** it selects the resource that will execute the activity first. If more than one resource returns the same start time, the resource is selected randomly among these. This approach goes in the direction of optimizing the activity throughput.
- **MAX_PROB:** it selects the resource that gives the highest fitting index. This means that it selects the most unloaded resource. This approach is designed to minimize the standard deviation from the desired profile. As for the previous policy if more than one resource has the maximum fitting index, the resource is randomly selected among these.
- **MIN_PROB:** it selects the resource that gives the lowest fitting index. This policy is designed to ensure the best fitting for the desired profile. This is useful when we have to prefer solutions that saturate one resource before starting filling another one. If more than one resource has the minimum fitting index, the resource is selected randomly among these.
- **RANDOM:** it selects randomly the resources using a uniform distribution. This policy is designed to enforce each resource to have the same probability to host an activity.

Complexity. The complexity of the task agent resides into the resource selection algorithm. The algorithm searches through the responses from all resource managers R for (usually) a minimum or a maximum. Being the response composed by 0 or 1 values, the complexity is $O(R)$. For this reason, the complexity of the task agent is polynomial.

3 Evaluation

In this section we first define the performance metrics we will use to perform our evaluation. We show two different tests: one on the Grid domain and the other in the HPC domain. For each test, we describe the experimental setup we choose to evaluate our solution against state-of-the-art schedulers and we show two sets of results: a performance and an overhead evaluation.

We used the following metrics for the comparison:

- Makespan: the completion time of the set of activities.
- Total waiting time of activities: the sum of all the activities waits.
- Tardiness: the sum of the delay w.r.t. the latest start time of the activity.
- Overutilization: volume of resource utilization in time, that exceeds the desired profile.
- Number of late jobs: computed as number of jobs exceeding their deadline.
- Dissimilarity: quantifies the dissimilarity of the final utilization profile and the desired utilization profile. It is designed to indicate if two functions have a different shape without considering the difference of volumes. The metric uses as input two functions: the desired profile $g(t)$ and the utilization profile to measure $f(t)$. These two profiles are considered from the time instant 0 to the maximum deadline. First, the function $f(t)$ is multiplied by factor ψ, computed as $\psi = min_t(\frac{g(t)}{f(t)})$. This multiplication of the utilization profile with ψ is used to compare the Dissimilarity of two utilization profiles with the same desired profile without the bias introduced by high differences in the length of the two utilization profile. Than we compute the discrete auto-correlation for $g(t)$ R_{gg} and the cross-correlation between $g(t)$ and $f(t)$ R_{gf}. The metric is obtained as: $Dissimilarity = \frac{R_{gg}(0)-R_{gf}(0)}{R_{gg}(0)}$. This metric is not affected by the makespan of the profile: other metrics (as for example the squared difference between the desired and the utilization profiles) cannot compare utilization profiles with different durations. Obviously, this metric is meaningful only in presence of jobs deadlines.

The different setups of DARDIS have been tested and compared against the KDistr scheduler [13] and different setups for the rule-based scheduler. The rule-based scheduler has been implemented to obtain a fair comparison: the scheduler avoids creating overutilization on the desired profile by limiting the resource capacity at schedule time to the amount of resource in the desired profile at the current time stamp. DARDIS has been implemented using the MPI library. Tests have been executed on a server with 2xIntel Xeon DP 12 Core E5-2670v3 and 128GB of RAM on instances coming from the domain of HPC and Grid computing, and compared with state of the art approaches in both domains.

The KDistr scheduler has been presented in [13]. The system is composed by a hierarchy of meta schedulers with one root. All the jobs are submitted to the root, then the root sends the job to K meta schedulers. The first scheduler that executes the job informs the other schedulers that the job is already in execution. Due to the fact that two different schedulers can schedule the same job at the same time, the authors use an atomic scheduling cycle. As reported by [13], the best result is obtained with $K = 4$. For this reason, we choose this value. As for our variable-profile-aware rule-based scheduler, in KDistr we have avoided overutilization on the desired profile, to make a fairer comparison.

The rule-based scheduling setups are:

- RB-FCFS: the jobs are ordered by increasing earliest start time. The algorithm checks to not exceed the desired profile at scheduling time.
- RB-DF: the jobs are ordered by increasing latest start time. The algorithm checks to not exceed the desired profile at scheduling time.
- RB-WT: the jobs are ordered by increasing walltime. The algorithm checks to not exceed the desired profile at scheduling time.

Each DARDIS set-up takes its name by the names of the policy and the generator that uses.

We performed two different test. The first test with 2000 jobs and 672 cores to compare the results obtained by a rule-based and a distributed scheduler and to obtain a baseline for comparison on scalability. The second test with 35583 jobs and 6900 cores to test the scalability and to investigate deeply the results compared to different rule-based schedulers.

3.1   Test in Grid Domain

All the 12 (4*3) different set-up of the scheduler have been tested and compared against the RB-FCFS and the KDistr scheduler [13]. To evaluate the performance of the proposed solution we considered a test bed composed by 2000 activity submissions in 1440 time instants. The system consists of 42 nodes with a capacity c_j of 16. The desired profile $dp_j(t)$ is randomly generated for each node with a distribution that simulates a system for which we want to enforce a utilization profile that is maximized during the night (e.g., to exploit lower temperature and lower electric power prices in the night).

The activity submission q_i follows a normal distribution (which is customary in HPC and Grid computing) with $\mu = 720, \sigma = 270$. However, the submissions are bounded by the interval $[0, \ldots, 1440]$. The amount of required resource req_i are randomly generated following a normal distribution with $\mu = 12, \sigma = 3$. Also, for the required resource we have the bound $[0, \ldots, 16]$. In this test the est_i is equal to the submission time q_i, and the lst_i is equal to $est_i + 144 \cdot k$, where k is randomly generated in the interval from 5 to 5 plus the amount of resources required by the activity. Finally, wt_i is generated following a normal distribution with $\mu = 30, \sigma = 60$.

Table 1 shows the first set of results in absolute values DARDIS set-up for each metric. We can notice that the DARDIS approach avoids both overutilization and deadline violations while RB-FCFS and KDistr fail to do that. Obtaining 1329 and 16587 respectively in Tardiness and Overutilization for the RB-FCFS and 670 and 216408 for the KDistr. From the Total waiting time and Makespan metrics, the MIN_START policies with the first and exp generators obtain better results of the RB-FCFS scheduler (which is better than the KDistr): we obtain 18 % and 2.6 % of improvement for Total waiting time and Makespan of the MIN_START first, and 17.9 % and 2.6 % of improvement for the MIN_START exp w.r.t. the RB-FCFS scheduler. This surprisingly good result

Table 1. Schedulers comparison on Grid domain test

	Makespan	Total wait	Tardiness	Late jobs	Over-util	Dissimilarity
RB-FCFS	2573	829192	1329	12	16587	0.698
KDistr	2594	787418	670	8	216408	0.695
MIN_START first	2506	679852	0	0	0	0.680
MIN_START exp	2506	680878	0	0	0	0.684
MIN_START unif.	2918	880735	0	0	0	0.675
MIN_PROB first	3266	1484953	0	0	0	0.618
MIN_PROB exp	3889	1233828	0	0	0	0.628
MIN_PROB unif.	5866	2467005	0	0	0	0.559
MAX_PROB first	2961	901531	0	0	0	0.649
MAX_PROB exp	3023	927965	0	0	0	0.644
MAX_PROB unif.	5193	1915255	0	0	0	0.598
RANDOM first	3150	1014540	0	0	0	0.657
RANDOM exp	3022	987353	0	0	0	0.653
RANDOM unif.	5333	1903513	0	0	0	0.587

can be explained in two ways. The first is the randomness in the dispatching process contributes to decrease resource fragmentation and this is translated in a better resource utilization. The second that: the computational overhead for a scheduling cycle is much lower in our scheduler w.r.t. the RB-FCFS scheduler, thus leading to lower waiting time and better resource utilization. From the Dissimilarity metrics, all the DARDIS set-ups obtain better result w.r.t. the RB-FCFS and KDistr. This is due not only to the fact that the majority of the utilization is fitted on the left side of the time axis (also done by the MIN_START first and exp set-up), but also to the fact that the RB-FCFS scheduler creates a considerable amount of Overutilization, exceeding the desired profile. Another important consideration, extrapolated from the table, is that usually the Total waiting time and Makespan are strictly correlated metrics. Conversely, a scheduler that obtains good results in the Dissimilarity metric, usually obtains bad results in Total waiting time and Makespan.

3.2   Test in HPC Domain

For this test we choose three DARDIS setups: MIN_START first, MAX_PROB exp and RANDOM uniform. We compared our approach to the three rule-based schedulers: RB-FCFS, RB-DF, and RB-WT. The KDistr scheduler is not a feasible approach in this domain due to the fact that it can not schedule parallel jobs in nodes managed by different sub-scheduler.

The test is based on the parallel workload archive of the CEA Curie system [14]. This system originally was composed by 360 nodes with four 8-core processors and 128 GB of RAM. The scheduler in use is Slurm [15], and the system is subdivided in 33 partitions. The schedulers have been tested on 300 out of 360 of the fat nodes of the system for a total of 9600 cores and 38 TB of RAM. A total of 35538 jobs submitted in 22 days of regular workload have been extracted from the trace log and used for the benchmark. This trace log does not consider explicit deadlines. However, there are implicit soft deadlines in the setup which define the user satisfaction. For this reason, we use the arrival time as est_i. After that, we extrapolated the average waiting time for each partition $aqt(partition)$ and used it to compute the $lst_i = est_i + aqt(partition)$.

Table 2 shows the first set of results in absolute values.

For the makespan, total waiting and tardiness metrics the best results are obtained by the MIN_START first. MIN_START first outperforms all the rule-based schedulers of the 41–42% in Makespan, and 31–50% in total waiting and tardiness. The other two versions of DARDIS obtain poorer results for all these metrics. This is due to the fact that their optimization goals are in contraposition to these metrics. For the number of late jobs metric, the best result is obtained by the MAX_PROB exp DARDIS. MAX_PROB exp DARDIS outperforms all the rule-based schedulers of the 54–57% in this metric. But also the others two versions of DARDIS obtain good result under this metric: MIN_START first outperforms the rule-based schedulers of the 47–51% while the RANDOM uniform outperforms the rule-based schedulers of the 53–56%. From the table we can notice that rule-based schedulers obtain better results in overutilization. It is important to note the motivation why both these scheduling approaches create overutilization. In these three setups, DARDIS is configured to use the desired profile as a soft constraint. Moreover, by configuration this soft-constraint has lower priority than the deadline soft-constraint. Under this consideration we can motivate the high overutilization as result of the strict deadline. For the rule-based schedulers, the variable profile is a hard constraint which is checked only at submission time. This means that the overutilization obtained by DARDIS has been caused by a decrease in the number of late jobs while the overutilization obtained by the rule-based scheduler derives from the architecture of the scheduler itself. For the dissimilarity, RANDOM uniform outperforms the rule-based schedulers by the 0.2–2.4 % while MIN_START first and the MAX_PROB exp DARDIS behave similarly to the rule-based approaches.

Table 2. Scheduler comparison on HPC domain test

	Makespan	Total Wait	Tardiness	Late jobs	Over-util.	Dissimilarity
MIN_START first	1010311	2521427205	2509717507	14722	781886	0.990
MAX_PROB exp	15384108	20868955885	20857350093	12725	1227779	0.981
RANDOM uniform	17317960	23966165379	23950189604	13141	785639	0.974
RB-FCFS	1702691	4993205642	5004811604	29102	257856	0.976
RB-DF	1725376	4082239604	4051655803	30098	368234	0.980
RB-WT	1721744	3673981393	3637368656	27908	395657	0.998

Table 3. Overhead comparison for an activity scheduling on the grid domain of DARDIS, RB-FCFS and KDistr in seconds

	Mean	Std. dev.
DARDIS Scheduling	0.021	0.111
DARDIS Dispatching	2.13E-04	1.38E-004
Rule-based	0.895	1.030
KDistr	0.750	0.711

Table 4. Overhead comparison for an activity scheduling on the HPC domain of DARDIS and rule-based scheduler in seconds

	Mean	Std. dev.
DARDIS Scheduling	0.018	0.129
DARDIS Dispatching	0.006	0.025
Rule-based total	5.356	5.077

3.3 Overhead Comparison

Tables 3 and 4 show the overhead for the computation of an entire job scheduling respectively for the test on the Grid domain and the HPC domain. The overhead of DARDIS is subdivided in scheduling and dispatching while for the rule-based and KDistr we have only total overhead of a scheduling cycle. In the Grid domain, comparing the sum of scheduling and dispatching overhead of our approach to the rule-based scheduler we can evince that our approach is 42 times faster. Comparing the KDistr, DARDIS is 35 times faster. In the HPC domain, in which the number of resources is drastically increased and the execution interval of each job $[est_i, \dots, lst_i]$ is in general shorter w.r.t. the intervals used in the Grid test, our approach turns out to be 214 times faster than the rule-based scheduler. These results give us the confirmation that this approach is more scalable w.r.t. one of the most reactive scheduler at the state-of-the-art.

4   Related Work

The problem studied in this work is a resource-constrained project scheduling problem (RCPSP) [16]. In the literature a plethora of works on this subject can be found [17–19]. However, real world and real-time instances are usually solved with heuristic algorithms, e.g., rule based schedulers [20].

Ramamritham et al. [21] present a distributed scheduler. The proposed approach is based on bids for the dispatching. These bids can be random or based on estimations. This could lead to the condition in which a job has to migrate to avoid exceeding its deadline. In our work we do not use estimations, and the dispatching phase considers all the resources of the system. For this reason, our work does not need the job migration, and if an activity exceeds its deadline it is due to the high utilization of all the resources of the system.

A number of works using Particle Swarm Optimization for scheduling can be found in literature [22–24]. These algorithms are optimization algorithm that explore a set of feasible solutions. The problem with these algorithms is the computational overhead. The best result obtained in this paper on a number of nodes and jobs halved w.r.t. our tests, show a computational overhead 6 times higher than ours. Distributed implementations of this approach have been studied [25] for different kinds of problems but never applied to scheduling.

The work presented by Montresor [26] shows the application of an ant colony algorithm to the problem of the scheduling in peer-to-peer systems. In this scheduler, the resources are nests, the ants have the duty to migrate jobs from highly loaded resources to low loaded resources. The starting assumption of this work is that a job can be migrated even during its execution. This assumption is not true in the majority of the domains studied by our work. Moreover, the authors consider only the load balancing objective. Finally, the ant colony approach does not consider the scheduling horizon for further optimization.

Optimization techniques have been applied to the problem of distributed scheduling [27–29]. However, as demonstrated in [30], centralized optimization approach cannot scale up to large-size systems. These distributed approach add to the overhead of a centralized approach also an overhead due to communications between agents. For this reason, these approach are unfeasible in a real-time HPC schedulers.

5 Conclusion

In conclusion, we presented a new scheduling approach for large scale systems where the number of resources and the number of activities make a centralized approach infeasible. The approach is highly scalable due to its distributed nature. We have shown that the approach could obtain better result w.r.t. rule-based schedulers and the KDistr scheduler. Moreover, the possibility to specify a variable profile of desired utilization increases the possibility of customization. We have shown 4 different policies for the dispatching and 3 different approaches for the scheduling. This leads to 12 different set-ups of the scheduler. These 12 set-ups have been evaluated w.r.t. both RB-FCFS and KDistr schedulers in the Grid domain, and the best setups obtained from the Grid test have been evaluated in the HPC domain w.r.t. three different rule-based schedulers. The tests show impressive results in the Tardiness and Overutilization metrics. Furthermore, we chose three different set-ups: one that overcomes the FIFO and KDistr schedulers in Makespan and Total waiting time, one that obtains the utilization form factor most similar to the desired profile and one trade-offs between these two results. Moreover, results on the overhead confirmed our hypothesis that this approach can overcome the problem of the scalability.

Future work will explore several directions. We will introduce the concept of resource cluster with heterogeneous resources. We will introduce the concept of activity unit for an activity: two activities units can ask different resources but they have to be synchronized. In some cases, the activity duration is not given but a min and max values are known. Finally, we will introduce new techniques to reduce the overhead, as for example a timer for the activity dispatching.

Acknowledgments. This work was partially supported by the FP7 ERC Advance project MULTITHERMAN (g.a. 291125), by the YINS RTD project (no. 20NA21

150939), evaluated by the Swiss NSF and funded by Nano-Tera.ch with Swiss Confederation financing and by CINECA.

References

1. NSCC: Tianhe-2 service page (2015). http://www.nscc-gz.cn/Product/ HighPerformanceComputingService/ServiceCharacteristics.html#Page_1
2. BBC: Supercomputers: Obama orders world's fastest computer (2015). http:// www.bbc.com/news/technology-33718311
3. Attig, N., Gibbon, P., Lippert, T.: Trends in supercomputing: the european path to exascale. Comput. Phys. Commun. **182**(9), 2041–2046 (2011)
4. Lavignon, J., et al.: Etp4hpc strategic research agenda achieving hpc leadership in europe (2013). http://www.etp4hpc.eu/wp-content/uploads/2013/06/ETP4HPC_ book_singlePage.pdf
5. Salot, P.: A survey of various scheduling algorithm in cloud computing environment. Int. J. Res. Eng. Technol. (IJRET) (2013). ISSN 2319-1163
6. Bartolini, A., Borghesi, A., Bridi, T., Lombardi, M., Milano, M.: Proactive workload dispatching on the EURORA supercomputer. In: O'Sullivan, B. (ed.) CP 2014. LNCS, vol. 8656, pp. 765–780. Springer, Heidelberg (2014). doi:10.1007/ 978-3-319-10428-7_55
7. Borghesi, A., Collina, F., Lombardi, M., Milano, M., Benini, L.: Power capping in high performance computing systems. In: Pesant, G. (ed.) CP 2015. LNCS, vol. 9255, pp. 524–540. Springer, Heidelberg (2015). doi:10.1007/978-3-319-23219-5_37
8. Van Den Briel, M., Scott, P., Thiébaux, S.: Randomized load control: A simple distributed approach for scheduling smart appliances. In: Proceedings of the 23th International Joint Conference on Artificial Intelligence, pp. 2915–2922. AAAI Press (2013)
9. Bergman, K., Borkar, S., Campbell, D., Carlson, W., Dally, W., Denneau, M., et al.: Exascale computing study: technology challenges in achieving exascale systems. Defense Advanced Research Projects Agency Information Processing Techniques Office (DARPA IPTO), Technical report 15 (2008)
10. Borghesi, A., Conficoni, C., Lombardi, M., Bartolini, A.: Ms3: A mediterraneanstile job scheduler for supercomputers-do less when it's too hot!. In: 2015 International Conference on High Performance Computing & Simulation (HPCS), pp. 88–95. IEEE (2015)
11. Feng, X., Ge, R., Cameron, K.W.: Power and energy profiling of scientific applications on distributed systems. In: 19th IEEE International, Parallel and Distributed Processing Symposium, 2005, Proceedings, p. 34. IEEE (2005)
12. Mehta, V.K.: Variable load on power station (2005). http://www.nct-tech.edu.lk/ Download/Technology%20Zone/Variable%20Load%20on%20Power%20Station.. pdf
13. Subramani, V., Kettimuthu, R., Srinivasan, S., Sadayappan, P.: Distributed job scheduling on computational grids using multiple simultaneous requests. In: 11th IEEE International Symposium on High Performance Distributed Computing, HPDC-11 2002, Proceedings, pp. 359–366. IEEE (2002)
14. Feitelson, D.: The cea curie log (2012). http://www.cs.huji.ac.il/labs/parallel/ workload/l_cea_curie/index.html
15. Yoo, A.B., Jette, M.A., Grondona, M.: SLURM: simple linux utility for resource management. In: Feitelson, D., Rudolph, L., Schwiegelshohn, U. (eds.) JSSPP 2003. LNCS, vol. 2862, pp. 44–60. Springer, Heidelberg (2003). doi:10.1007/10968987_3

16. Blazewicz, J., Lenstra, J.K., Kan, A.R.: Scheduling subject to resource constraints: classification and complexity. Discrete Appl. Math. **5**(1), 11–24 (1983)
17. Hartmann, S.: A self-adapting genetic algorithm for project scheduling under resource constraints. NRL **49**(5), 433–448 (2002)
18. Damay, J., Quilliot, A., Sanlaville, E.: Linear programming based algorithms for preemptive and non-preemptive rcpsp. Eur. J. Oper. Res. **182**(3), 1012–1022 (2007)
19. Bhaskar, T., Pal, M.N., Pal, A.K.: A heuristic method for rcpsp with fuzzy activity times. Eur. J. Oper. Res. **208**(1), 57–66 (2011)
20. Haupt, R.: A survey of priority rule-based scheduling. Oper. Res. Spektrum **11**(1), 3–16 (1989)
21. Ramamritham, K., Stankovic, J., Zhao, W., et al.: Distributed scheduling of tasks with deadlines and resource requirements. IEEE Trans. Comput. **38**(8), 1110–1123 (1989)
22. Izakian, H., Tork Ladani, B., Zamanifar, K., Abraham, A.: A novel particle swarm optimization approach for grid job scheduling. In: Prasad, S.K., Routray, S., Khurana, R., Sahni, S. (eds.) ICISTM 2009. CCIS, vol. 31, pp. 100–109. Springer, Heidelberg (2009). doi:10.1007/978-3-642-00405-6_14
23. Zhan, S., Huo, H.: Improved pso-based task scheduling algorithm in cloud computing. J. Inform. Comput. Sci. **9**(13), 3821–3829 (2012)
24. Izakian, H., Ladani, B.T., Abraham, A., Snasel, V.: A discrete particle swarm optimization approach for grid job scheduling. Int. J. Innovative Comput. Inform. Control **6**(9), 4219–4233 (2010)
25. Vanneschi, L., Codecasa, D., Mauri, G.: A comparative study of four parallel and distributed pso methods. New Gener. Comput. **29**(2), 129–161 (2011)
26. Montresor, A., Meling, H., Babaoğlu, Ö.: Messor: load-balancing through a swarm of autonomous agents. In: Moro, G., Koubarakis, M. (eds.) AP2PC 2002. LNCS (LNAI), vol. 2530, pp. 125–137. Springer, Heidelberg (2003). doi:10.1007/3-540-45074-2_12
27. Benhamou, F. (ed.): CP 2006. LNCS, vol. 4204. Springer, Heidelberg (2006)
28. Gomes, C.P., van Hoeve, W.J., Selman, B.: Constraint programming for distributed planning and scheduling. AAAI Spring Symposium: Distributed Plan and Schedule Management, vol. 1, pp. 157–158 (2006)
29. Rolf, C.C., Kuchcinski, K.: Distributed constraint programming with agents. In: Bouchachia, A. (ed.) ICAIS 2011. LNCS (LNAI), vol. 6943, pp. 320–331. Springer, Heidelberg (2011). doi:10.1007/978-3-642-23857-4_32
30. Bridi, T., Bartolini, A., Lombardi, M., Milano, M., Benini, L.: A constraint programming scheduler for heterogeneous high-performance computing machines. IEEE Trans. Parallel Distrib. Syst. **27**(10), 2781–2794 (2016). doi:10.1109/TPDS.2016.2516997. ISSN:1045-9219

Steps in Assessing a Timeline-Based Planner

Alessandro Umbrico[1], Amedeo Cesta[2], Marta Cialdea Mayer[1],
and Andrea Orlandini[2(✉)]

[1] Dipartimento di Ingegneria, Università degli Studi Roma TRE, Rome, Italy
[2] Consiglio Nazionale delle Ricerche,
Istituto di Scienze e Tecnologie della Cognizione, Rome, Italy
andrea.orlandini@istc.cnr.it

Abstract. The "timeline-based" is a particular paradigm of temporal planning that has been successfully applied in many real-world scenarios. Different timeline-based planning systems have been developed, each using its own planning specification language and solving techniques. An analysis of the differences between such kind systems has not been addressed yet. In previous work we have developed EPSL a planning tool successfully applied in real-world manufacturing scenarios. During subsequent projects our tool achieved a level of stability and a relative maturity. In this paper we start addressing the problem of comparison with other timeline-based planners and presents an analysis that concerns the EUROPA2 framework which can be considered the *de-facto* standard for timeline-based planning. In the present work we analyze the modeling and solving capabilities of the two frameworks. This phase of our study identifies differences and discusses strengths and weaknesses when solving the same problem.

Keywords: Timeline-based planning · Planning and Scheduling · Constraint-based planning

1 Introduction

Timeline-based planning is an approach to temporal planning research which has been successfully applied to real-world problems [1–4] where time constitutes a crucial factor to deploy effective planning applications. The main feature of the approach stems in the capacity of modeling and dealing with temporal constraints and in the capability of integrating planning and scheduling (P&S) in a unified solving approach. Indeed, a lot of the reasons for its success stay in the modeling capability of the systems that support such applications. Several applications are supported by various timeline-based general purpose architectures, some of the most known are EUROPA2 [5], IxTeT [6], ASPEN [7] and APSI [8]. Despite the practical success there is not a uniform shared view of what timeline-based planning is. Thus each existing framework applies its own *interpretation* of the planning approach. In contrast to action-based planning, theoretical aspects of timeline-based planning were not investigated until very

© Springer International Publishing AG 2016
G. Adorni et al. (Eds.): AI*IA 2016, LNAI 10037, pp. 508–522, 2016.
DOI: 10.1007/978-3-319-49130-1_37

recently. A formal description of the problem has been proposed in [9], while a formalization in terms of *flexible* timelines appeared in [10], later extended in [11] to account also for plan controllability issues. Meanwhile, the connection between timelines and Timed Game Automata has been investigated for the purpose of plan verification [12,13] and robust plan execution [14]. Also, initial steps for a complexity-theoretic characterization of the planning problem has been recently proposed in [15].

As a counterpart of the formal work presented in [11], we have developed a general purpose timeline-based planning architecture, called EPSL (Extensible Planning and Scheduling Library), proposing a hierarchy-based approach for modeling and solving timeline-based problems [16]. Such system has been successfully tested during subsequent research projects to support a manufacturing plant (see [4,17,18]) and an industrial robotics scenario [19]. Through the use in these projects the EPSL tool has achieved both a level of stability and a relative maturity.

In this paper, we start addressing the problem of comparing EPSL with other timeline-based planners and present an analysis that concerns EUROPA2, a planning framework developed at NASA [5] which, given also the wide spectrum of missions that have used it, can be considered a *de-facto* state of the art for timeline-based planning systems. The goal of the paper is to provide the reader with an initial report about the differences between the two frameworks taking into account their modeling and solving capabilities. Namely, rather than focusing on a comparison of performances, we aim at understanding the different features of the frameworks in order to highlight weak and strength points of the approach we are pursuing. A general interesting result is that the development of EUROPA2 seems to have led the NASA's framework to assimilate some features of "classical" PDDL-like approach to planning. Conversely, EPSL framework maintains a modeling and solving approach inspired by the original idea of timeline-based planning as introduced in [1].

Plan of the Paper. The next section of the paper provides a brief description of the timeline-based approach also introducing a general formal framework. Then, we describe the features of the EPSL planning tool and the EUROPA2 framework. We also introduce a ROVER planning domain exploited as the case study for the comparison. The following section goes into details for the comparison considering the planning domain models, the solving approaches and some features of the generated plans. Finally, the paper ends with some conclusions concerning the overall work and future developments.

2   Timeline-Based Planning

The modeling assumption underlying the timeline-based planning approach is inspired by classical Control Theory: the problem is modeled by identifying a set of *relevant features* whose behavior over time is to be controlled in order to obtain a desired set of goals. In this respect, the domain features under control are modeled as a set of temporal functions whose values have to be decided over a temporal horizon. Such functions are synthesized during problem solving by posting planning decisions. The evolution of a single temporal feature over a time horizon is called the *timeline* of that feature. The temporal behavior of an element of the domain is described in shape of a timeline, which is a sequence of values (states or actions) the related element of the domain can assume over time. Thus the set of the timelines of a domain (called the "timeline-based plan") describes the behavior of the overall system. A planner synthesizes timeline-based plans by posting temporal constraints between states or actions according to some domain rules that model the physical and logical constraints of the system and its elements.

2.1   Planning with Flexible Timelines

Despite the practical success of timeline-based approach to solve real world problems, a shared view and a well-defined formalization of the main planning concepts were missing (but see pointers in the introduction). For the purpose of this paper, we refer to the generic timeline-based planning framework presented by Cialdea Mayer et al. [11]. A timeline-based planning domain is composed by a set of features to be controlled over time. These features are modeled by means of *multi-valued state variables* that specify causal and temporal constraints characterizing the allowed temporal behavior of such domain features. A state variable describes the set of values $v \in V$ the related feature may assume over time with their flexible duration. For each value $v_i \in V$ the state variable describes also the set of values $v_j \in V$ (where $i \neq j$) that are allowed to follow v_i and the related *controllability property*. If a value $v \in V$ is tagged as *controllable* then the system can decide the actual duration of the value. If a value $v \in V$ is tagged as *uncontrollable* instead, the system cannot decide the duration of the value, the value is under the control of the *environment*. The behavior of state variables may be further restricted by means of *synchronization rules* that allow to specify temporal constraints between different values. Namely, while state variables specify *local* rules for the single features of the domain, synchronizations represent *global* rules specifying how different features of the domain must behave together. A *planning domain* is composed by a set of state variables and a set of synchronization rules. Specifically, there are two types of state variable in a planning domain. The *planned variables* that model the domain features the system can control (or partially control). The *external variables* that model domain features completely outside the control of the system. External state variables model features of the *environment* the system cannot control but that must care about in order to successfully carry out activities.

Planning with timelines usually entails considering sequence of valued intervals and time flexibility is taken into account by requiring that the durations of valued intervals, called *tokens*, range within given bounds. In this regard, a flexible plan represents a whole set of non-flexible timelines, whose tokens respect the (flexible) duration constraints. However a set of flexible timelines do not convey enough information to represent a flexible plan. The representation of a flexible plan must include also information about the relations that must hold between tokens in order to satisfy the synchronization rules of the planning domain. A flexible plan Π over the horizon H is defined by a set of flexible timelines FTL and a set of temporal relations R representing a possible choice to satisfy the synchronization rules. Given the concepts above, a *planning problem* is defined by a temporal horizon H, a planning domain D, a planning goal G which specifies a set of tokens and constraints to satisfy and the *observations* O which completely describes the flexible timelines for all the external variables of the domain. Consequently a flexible plan Π is a solution for a planning problem if it satisfies the planning goal and if it does not make any hypothesis on the behavior of the external variables (i.e. the plan does not change the *observation* of the problem).

Moreover, given a solution plan Π, it is important to check the controllability properties of Π in order to verify the executability of the plan. The controllability problem aims at verifying if there exists a way to execute a plan according to the possible (temporal) evolutions of the environment [20]. Namely controllability properties are to define a set of decisions (i.e., a feasible temporal allocation of all plan's controllable events/intervals - *schedule*) that guarantee the execution of a plan according to the known possible evolutions of the environment (i.e., uncontrollable events/intervals of a plan). Different controllability properties have been introduced differing for the assumptions made on the known evolutions of the environment [20]. The most relevant property is the *dynamic controllability* of a plan which entails that a *dynamic execution strategy* exists to dynamically decide the schedule of the controllable intervals/events of a plan by reasoning on the perceived evolution of the environment. From a planning point of view, it is also important to check that partial plans maintain such property overall the solving process.

2.2   EPSL: A General-Purpose P&S Framework

The Extensible Planning and Scheduling Library (EPSL) [16] is the result of a research effort started after the analysis of different timeline-based systems (e.g., [5,6,21]) as well as some previous experiences in deploying timeline-based solvers for real world domains [22]. EPSL relies on the APSI framework [8] which provides the modeling capabilities to represent timeline-based domain. In particular, EPSL extends the APSI framework in order to comply with the semantics proposed in [11]. EPSL provides a modular software library which allows users to easily define timeline-based planners by specifying strategies and heuristics to be applied in a specific application.

The key point of EPSL modularity is the *planner interpretation*. A planner is a compound element whose solving process is affected by the specific set of *modules* applied. Indeed, a EPSL-based planner implements a plan refinement search by combining several modules responsible for managing different aspects of the search: (i) a *search strategy* for managing the fringe of the search space; (ii) a set of *resolvers* for detecting and solving different types of flaws on the plan; (iii) a *selection heuristics* for analyzing the flaws detected on the current plan and selecting the *best* flaw to solve for plan refinement. Thus the actual behavior of the solving process of a EPSL-based planner is determined by the particular strategy, the heuristics and the resolvers set in the configuration of the P&S application. In this regard, the set of resolvers determines the expressivity of the framework. Adding new resolvers allows the framework to manage new types of feature of the domain (e.g., different type of resources) and detect/solve a wider range of flaws of a plan. Heuristics affect the performances of the solving process by encapsulating a *flaw-selection criteria* which guides the plan-refinement procedure to select the most *promising* flaw to solve at each iteration. Finally, the strategies may affect the *qualities* of the generated plans by encapsulating plan evaluation criteria that estimate some desired properties of the (partial) plans (e.g. plan cost).

Several EPSL-based solvers have been deployed in the real-world scenarios mentioned before [4,18,19]. Specifically, we have applied a hierarchy-based modeling and solving approach which identifies two types of state variables (in addition to the external ones described in the formalization) [16]. The *primitive variables* model the set of *low-level tasks* that can be directly executed by the system to control. The *functional variables* model the *complex tasks* that can be performed by combining the available primitive ones. Namely functional variables abstract the behavior of the system by modeling the functional capabilities it can perform. Synchronization rules define a *hierarchical task decomposition* which decomposes the values of functional variables in terms of temporal constraints between values of primitive variables (complex domains may have several *functional layers* between the top of the hierarchy and the primitive layer). The resulting hierarchical structure of the domain is then exploited by the solving process by means of a domain independent heuristics which allows to improve the performance of the framework as shown in [16]. During its solving process, EPSL also performs a *pseudo-controllability* check of partial plans as a necessary but not sufficient requirement for guaranteeing dynamic controllability.

The modular architecture of EPSL allows to easily extend the solving capabilities of the framework by adding new modules (e.g. heuristics or resolvers). Thus EPSL provides an enhanced framework for developing applications in which designers may focus on single aspects of the solving process without dealing with all the details related to timeline-based planner implementation.

2.3   The EUROPA2 Framework

With this paper, we start addressing the problem of comparing EPSL with other timeline-based planners. The natural starting point is to focus on the EUROPA2 framework which can be considered the *de-facto* standard for timeline-based planning. EUROPA2 is one of the most known timeline-based tools in the literature, its most known incarnation has been in the DS1 mission [2], attempts of formalization are given in [23,24], the more recent, public domain version is described in [5].

Similarly to EPSL, EUROPA2 models a planning domain by identifying a set of features to control over time. The modeling approach in EUROPA2 relies on an object-oriented language, the *New Domain Description Language* – NDDL. An *object* models a specific feature of the domain. The behavior of an object can be described by means of *predicates* and/or *actions* that represent respectively states or operations the related feature can assume or perform over time. However not all objects behave in the same way. Indeed some types of objects may have specific rules that constrain their temporal evolutions. A *timeline* is a particular type of object whose temporal behavior is constrained to be a sequence of not overlapping values (i.e. predicates) over time.

EUROPA2 allows to model also *renewable resources* and *consumable resources* in addition to *general objects* and *timeline objects*. Renewable resources represent *shared* features of the domain with a limited capacity that can be *consumed* over time (e.g., a pool of workers in a manufacturing environment, or a communication channel). Namely no *production* of the resource is needed because the amount of resource consumed by an activity is restored as soon as the activity ends. Consumable resources represent *shared* features of the domain with a limited capacity that can be either *consumed* or *produced* over time (e.g., a battery). In this case *production* activities are needed in order to restore the capacity of the resource after *consumptions*. Objects of the domain declare the predicates or actions they can assume over time. The temporal behaviors of domain objects are constrained by means of *compatibilities*. Broadly speaking a compatibility represents a general rule composed by the *head* which represents the predicate or action the rule applies to, and a *body* which specifies a set of predicates/values of other objects together with a set of temporal and parameter constraints that must hold between the *head* of the rule and the target of the constraint.

Given a domain specification, a planning problem is composed by a *temporal horizon* and an *initial configuration*. The initial configuration describes the initial (partial) plan in terms of a set of predicates on the timelines and a set of goals that can be either actions to perform or predicates to achieve (i.e., a desired final state). Note that the term *timeline* is used to refer either to the domain objects or the temporal evolutions of all the objects of the domain. EUROPA2 applies an action-based modeling approach where *general objects* specify actions to constrain the predicates of the timelines of the domain. Thus given a set of domain objects and compatibilities, the solving process generates the temporal behaviors of the timelines according to the compatibilities of the planning domain and the specified goals. The EUROPA2 solving process relies on a constraint-based engine to encode a partial-plan refinement procedure. As described in [25], the planner implements a plan refinement search which starts from the initial configuration and incrementally refines the related plan by adding and ordering predicates or actions to the timelines until a final consistent configuration is found. Namely the refinement process consists of detecting and solving *flaws* on the current partial plan. EUROPA2 is able to manage three types of flaw during the solving process: (i) *open condition flaws* represent (sub)goals generated during the planning process; (ii) *ordering flaws* represent overlapping predicates of a timeline; (iii) *unbound variable flaws* represent variables of the underlying CSP engine that must be instantiated.

A feature of EUROPA2 solver worth being mentioned is that a planning goal can be either an action to perform or a desired *state* to reach. In the latter case the solving process applies all the actions defined in the model to *support* the desired state. Namely the solver acts like in PDDL planning where operators are applied to preconditions according to the desired effects.

3   Comparing EPSL and EUROPA2

In order to assess the actual maturity of EPSL, here an initial comparison with EUROPA2 is performed evaluating different aspects. Indeed, the objective is to compare the two timeline-based frameworks by taking into account aspects concerning the modeling approach, the expressiveness, the solving capabilities and the features of the generated plans. In this regard, the comparison between EPSL and EUROPA2 entails two steps: first, an analysis of different modeling capabilities considering the features of the planning domain models exploited by the systems; then, a comparison of the differences in the solving approaches shown by the two planners. To this aim, we consider a well known planning domain, i.e., a ROVER planning domain, as a reference domain to analyze the different modeling features as well as to discuss the main differences on the planning processes also considering planning problems of growing complexity.

The ROVER planning domain has been extracted from the scenario described on the EUROPA2's web site concerning an autonomous exploration rover[1]. This scenario represents a well known application context in AI [26,27]. Specifically, a rover is a robotic device endowed with a wheeled base to explore the environment, an instrument to sample rocks and collect scientific data that can be communicated back to Earth. The domain consists of a rover which must navigate between known points of interest and collect scientific data by means of payload instruments (e.g., camera) and communicate such data to Earth. Usually, some requirements are to be satisfied during the execution of a mission in order to successfully carry out tasks and guarantee the operational requirements of the rover. In the ROVER domain, the mission plans must satisfy the following requirements: (i) the instruments of the rover must be set in a safe position (i.e., *stowed*) while the rover is moving; (ii) the rover must be still at a requested location and place the instrument accordingly in order to take a sample of the target (e.g., a rock); (iii) the rover must not move when communicating data to Earth.

3.1   Comparing Modeling Capabilities

Although following the same planning approach, APSI and EUROPA2 use different ways for modeling the domain features. They rely on two different domain specification languages to model planning domains, i.e., DDL and NDDL respectively. And, given the ROVER planning domain, two different models are then

---

[1] https://github.com/nasa/europa/tree/master/examples/Rover.

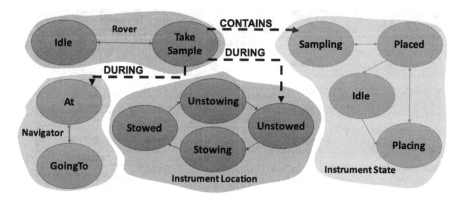

Fig. 1. EPSL model of the ROVER planning domain

analyzed according to the features of the specific languages. A preliminary analysis of the modeling capabilities suggests some main relevant differences concerning the *expressiveness* of the two frameworks. Indeed, EUROPA2 allows to model a wider range of domain features than EPSL. EUROPA2 can model *consumable resources* while EPSL cannot. For instance, EPSL is not able to model the *battery* of the ROVER planning domain. On the other hand, EUROPA2 does not allow to model *uncontrollable* features in a planning domain while EPSL allows to specify *external variables* to model features of the environment to *monitor*. Namely, EPSL can model a *visibility window* with a ground station on Earth in order to allow the rover control system to plan scientific data communication tasks within given time periods. For comparison purposes, a revised version of the ROVER scenario is defined in order to obtain planning models of *equivalent complexity* for the two planners. Thus, the original ROVER planning domain has been *simplified* by not considering *battery management* and *communication activities*[2].

A model of the ROVER *domain in* EPSL. EPSL allows to define planning domain by defining a set of state variables to be controlled over time and a set of synchronization rules to coordinate their temporal behaviors. Figure 1 shows the set of state variables defined to model the ROVER domain. The figure shows also temporal constraints entailed by the synchronization rules of the domain that allow to satisfy the goals (i.e., take samples). In general, EPSL allows to follow a *hierarchical approach* to domain modeling which starts by identifying a set of *primitive variables* that model the primitive/atomic tasks the system may directly execute. Then, *functional variables* are defined to model complex tasks, called *functions*, the system can perform over time by composing the primitive ones. Namely, functions represent complex tasks that cannot be directly

---

[2] Examples of the DDL and NDDL planning specification files for the ROVER domain considered in this paper can be found at the following link: http://tinyurl.com/TLRoverDomains-zip.

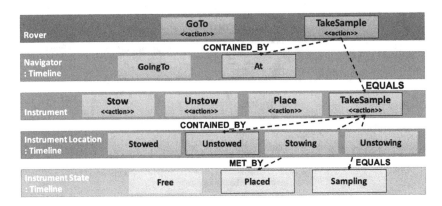

Fig. 2. EUROPA2-based model of the ROVER planning domain

performed by a single component of the system. Rather functions entail a coordination among system's (internal) devices (i.e., the primitive variables of the planning domain). Synchronization rules describe the *hierarchical decomposition* of the agent's functions in terms of *primitive tasks* the system's devices can directly handle. Complex domains may require different hierarchical levels. Then, a *hierarchical decomposition* may involve both primitive tasks and other functions. Thus, the EPSL modeling approach uses synchronizations to perform hierarchical task decomposition similarly to classical HTN planning.

In Fig. 1, the *Navigator* is a *primitive variable* which models the navigation facility of the rover. The *At* and *GoingTo* values represent that the rover can either be still at a known location or moving towards another location. Similarly, the *Instrument State* is a *primitive variable* which models the operating state of the rover's instrument. The *Rover* component of the domain is a *functional variable* which models the set of functions the rover may perform. The associated *TakeSample* value represents a high-level task (i.e., a function) the rover must perform by coordinating the behaviors of *Navigation*, *Instrument Location* and *Instrument State* variables. A dedicated synchronization rule specifies the set of temporal constraints that must hold to perform the *TakeSample* function (see the black dotted arrows in Fig. 1). Then, a consistent (temporal) behavior for taking a sample (i.e., *Sampling* value of *Instrument State* variable) requires that the rover is located at the target's location (i.e., *At*), the instrument is active (i.e., *Unstowed* value of *Instrument Location* variable).

A model of the ROVER *domain in* EUROPA2. Figure 2 depicts the EUROPA2 model generated for the ROVER planning domain. EUROPA2 uses an object-oriented modeling language to represent the features of a planning domain. The *objects* of the domain are described in terms of *predicates* and *actions*. Predicates represent the states that objects can assume over time. Actions represents the operations that objects can perform over time. There are two types of objects that compose a planning domain. *Timeline Objects* model the features of a domain that may change over time, e.g. the physical position of the rover or the state of an instrument. *Objects* model the set of *actions* that can be performed to change the state of one or a set of Timeline Objects, e.g. the action for moving the rover from an initial position to a destination position. Then, for each action, *compatibilities* specifies the set of constraints that allow to build the plan. Namely, compatibilities specify the constraints affecting the temporal evolutions of one (or more) Timeline Object(s), e.g. a compatibility for a move action specifies that the rover must be at the initial position *before* the action start, and that must be at the destination position *after* the execution of the action.

Considering the model in Fig. 2, the instrument facility is modeled by means of two timelines, i.e., the *InstrumentState* and the *InstrumentLocation* timelines. They model the set of states and positions the instrument may assume over time. The *Instrument* object provides actions for controlling the device. The predicates of the *InstrumentState* timeline model the "operational status" of the device over time. The predicate *Placed* means that the instrument is placed on target. The predicate *Sampling* means that the instrument is sampling a particular target. Similarly the predicates of *InstrumentLocation* timeline model the position of the device over time. The predicate *Stowed* means that the instrument is stowed and it cannot perform sampling operations. The predicate *Unstowed* means that the instrument is ready to use. The actions of *Instrument* object model operations needed to properly manage the device. The action *Unstow* represents the operation which allows to "activate" the device by changing the position of the instrument from *Stowed* to *Unstowed*. Similarly the action *TakeSample* represents the operation which allows the device to actually take a sample of a desired target. The black dotted arrows in Fig. 2 represent some of the temporal constraints required by the compatibilities defined on the corresponding actions of the domain. In this regard it is important to point out that actions have *conditions* and *effects* that must hold to apply and *execute* the action. *Effects* represent predicates that the execution of the action adds to the plan. *Conditions* represent predicates that must be part of the plan in order to "execute" the action. This is an important aspect to take into account while modeling planning problems with EUROPA2. Indeed, an action-based planning perspective is actually pursued while solving problems (see next section for further considerations). That is, the EUROPA2 planner checks conditions and effects of actions in order to find a suitable sequence of actions that allow to build timelines and satisfy the desired goals.

Also, some differences can be noted between the formalization of EUROPA2 given in [24] and its actual implementation. These differences mainly concern the compatibility specification and their expressiveness w.r.t. synchronizations of EPSL. Indeed, from a theoretical point of view, compatibilities are less expressive than synchronizations because they do not allow to specify constraints between tokens of the rule's body. Namely, a compatibility can only specify constraints between the head of the rule and tokens in the rule's body. Moreover, unlike EPSL, EUROPA2 does not use quantified Allen's temporal relations. However, the "concrete implementation" of the framework overcomes all these (theoretical) limitations by exploiting the underlying CSP engine. Indeed, it is possible to explicitly constrain compatibilities' tokens by specifying CSP's linear constraints between the tokens' temporal variables (i.e., tokens time point and duration variables). For instance, it is possible to specify a *before* temporal constraint between two tokens in the body of a compatibility by adding a linear constraint between the end time of the first token and the start time of the second token (e.g. $endTime(a) < startTime(b)$ where a and b are tokens declared into the body of a compatibility).

3.2 Comparing Solving Capabilities and Generated Plans

Both EPSL and EUROPA2 apply a plan refinement procedure which starts with an initial partial plan and some goals to plan for. The solving process iteratively refines the plan by solving *flaws* until a complete and valid plan is found [16,25]. However there are some relevant differences worth to be underscored (also related to the different modeling approaches). Given the ROVER planning models described in the previous section, we have defined several problem instances by considering an increasing number of planning goals (i.e., the number of targets to sample) to compare the solving capabilities of EPSL and EUROPA2. On these problems, the collected results show that EPSL performs rather better than EUROPA2 in terms of deliberation time. Obviously, this is not sufficient to support any general claim about the actual effectiveness of the planning systems. Providing a complete analysis entails to refer a set of benchmark domains with multiple problem instances for timeline-based planning and, to the best of our knowledge, such benchmark is still missing. Thus, a thorough comparison of the solving performances provided by the two frameworks is kept outside the scope of this work and left as future work. In this paper, the main objective of the experiments is to check the suitability of the defined models and assess the features of the plans generated by the two frameworks. Therefore, the most important result elicited from the comparison concerns, again, the different interpretation of timeline-based planning in EUROPA2 and EPSL frameworks. Indeed, despite they share the same conceptual origin (see [1]), they have developed two different ways of handling timeline-based problems.

The experimental campaign shows a first difference between EPSL and EUROPA2 concerning the *interpretation* of a solution plan. Namely, EPSL generates plans providing a set of timelines as a continuous sequences of (temporally) ordered tokens and a set of temporal constraints that relate their start and end times, while EUROPA2 interprets timelines as "discrete" sequence of ordered values. Namely, timelines may contain *gaps* according to EUROPA2's interpretation. The actual presence of gaps on the plans generated by EUROPA2 depends from the planning model specification provided as input. The user is then responsible for specifying a set of compatibilities that avoid gaps in the solution plans. On the contrary, the responsibility of filling gaps in EPSL is on the planning algorithm. Then, EPSL exploits state variables of the planning domain to guarantee consistent behaviors of the domain features. And every time a gap (i.e., a temporal interval with no value) is detected between two tokens, the EPSL solving algorithm checks the state variable specification to extract the allowed transition between the values involved. This interpretation relies on the assumption that a gap represent an *uncertainty* about the actual behavior of the feature because it could assume any value in the related *unbounded* temporal intervals. A key point of this aspect is that the planner is not only responsible for applying the constraints specified by the user but it is also responsible for ensuring consistent transitions between the values according to state variable specification. In EUROPA2 such a behavior must be achieved by specifying a set of actions and related compatibilities that model the possible transitions of the objects. Such an operation is not always simple and, thus, the correctness of the generated plans strongly relies on the expertise of the user actually modeling the planning domain.

An interesting feature of the EUROPA2 modeling approach is the use of general objects to model actions of the domain. Objects can be seen as *relaxed* timelines where tokens are allowed to overlap in time. Thus, EUROPA2 planner can generate plans with parallel actions if they do not violate the related compatibilities' constraints. On the contrary, EPSL relies on state variables to model functional variables and the related tasks. Thus, *complex tasks* cannot overlap in time even if the related synchronization rules would allow parallelism. Let us suppose, for example, that the rover of the planning domain is endowed with two instruments and that the rover must sample two targets at the same location. In such a scenario, the rover should be able to perform the two *TakeSample* tasks in parallel. EUROPA2 models the *TakeSample* as an action of an object. Thus, the planner can generate plans where the rover performs the two planning goals (i.e., two *Take-Sample* actions) in parallel by allowing the related tokens to overlap. Conversely, EPSL models *TakeSample* as a value of a functional variable (i.e., a function of the rover) and the related tokens of the timeline are not allowed to overlap. Thus, the planner can only generate plans where the two goals are in sequence.

3.3   Easy of Use

As discussed above, a main general comment is that EUROPA2 seems to have been influenced by "classical" PDDL-like planning techniques that lead the framework move towards an action-based approach rather than a "behavior-based" approach to planning with timelines. This mainly affect the modeling approach of EUROPA2 framework. The user must be aware of the solving process of the framework defining a suitable set of actions (and compatibilities) that allow the planner to build a valid plan. As a consequence, the user is supposed to completely specify how the planner can build the timelines of the plan. Conversely EPSL approach is compliant with the original idea of timeline-based planning where the focus is on the temporal behavior of the domain features to control. Indeed, state variables model the features of the domain by describing how they can *autonomously* evolve over time. Synchronizations allow the user to constrain these possible evolutions (i.e., temporal behaviors) to coordinate components and realize some complex operations. Thus the user is supposed to "simply" declare the values the domain components must assume over time in order to realize the desired (complex) behavior. The EPSL solving process is then responsible for building the timelines according to the requirements of the domain and the desired goals.

As an example, in the ROVER domain, the EUROPA2 user is supposed to provide a complete specification while the EPSL user can simply declare that the rover must be at the target's location *during* sampling operations in order to successfully perform a *TakeSample* task. Then, in EPSL it is not necessary to declare the "rule" which allows the rover to reach the target's locations. Such a rule is encoded by the *Navigator* state variable. And the EPSL planner is responsible for checking whether the rover must move or not towards the desired location, and building the related timeline accordingly. In other words, as envisioned in the original idea of timeline-based planning [1], the crucial point is that users of the planning framework are not supposed to be aware of the internal functioning of the particular planning algorithm/technique adopted to deploy effective P&S applications. Users can be an expert of the particular application domain without being *forced* to know the details of the solving mechanism of the planner. This is a very important feature of a planning framework and it represent a *long-term* goal we are pursuing in order to design a general purpose tool which can be easily exploited by *end-users*. Thus most of the EUROPA2 modeling and solving capabilities rely on the expertise of the user and, as a consequence, a deep knowledge of the solving mechanism of the EUROPA2 framework is requested in order to develop effective P&S applications. Consequently we may argue that EUROPA2 approach to timeline-based planning seems to be harder to apply than the EPSL approach. However it is important to point out that EUROPA2 gives to the (skilled) users a total control of the plan generation process.

4   Conclusions

In this paper we have summarized some recent results in the development of a general-purpose timeline-based planning framework, called EPSL. Then we have described the approach followed to compare EPSL with EUROPA2, the most known timeline-based planning framework in the literature. We have analyzed the modeling and solving capabilities of the frameworks by taking into account a ROVER domain, which represents a "classical" planning domain extracted from a real-world application scenario. Despite EPSL and EUROPA2 share the same origin of timeline-based planning [1], the evaluation has pointed out some relevant differences between the two frameworks. Indeed, we have found significant differences in terms of both the modeling capabilities and the solving approach. After the assessment, we can conclude that the most relevant difference between the frameworks concerns the *usage* and the *level-of-expertise* needed to develop effective P&S applications. In our opinion, EPSL unlike EUROPA2, does not require a deep understanding of the solving process to design P&S applications. Thus EPSL seems to be easier to use than EUROPA2 for *end-users* that may have a deep knowledge of the specific application domain but not a good background in AI planning and the related planner applied. However the assessment has also pointed out some deficiencies in EPSL that we are going to address in the next future to improve the framework (e.g., introducing *consumable resources* and taking into account *functional variables* concurrency issue). This paper provides an initial report that aims at starting an evaluation of the features and capabilities of the EPSL framework we are developing. Future work will extend the evaluation by addressing performance features with other timeline-based planning systems and taking into account also the "new generation" of planning frameworks like CHIMP [28] and FAPE [29].

References

1. Muscettola, N.: HSTS: Integrating planning and scheduling. In: Zweben, M., Fox, M.S. (eds.) Intelligent Scheduling. Morgan Kauffmann (1994)
2. Jonsson, A., Morris, P., Muscettola, N., Rajan, K., Smith, B.: Planning in interplanetary space: theory and practice. In: Proceedings of the Fifth International Conference on AI Planning and Scheduling. AIPS-00 (2000)
3. Py, F., Rajan, K., McGann, C.: A systematic agent framework for situated autonomous systems. In: Proceedings of the 9th International Conference on Autonomous Agents and Multiagent Systems. AAMAS-10 (2010)
4. Borgo, S., Cesta, A., Orlandini, A., Umbrico, A.: A planning-based architecture for a reconfigurable manufacturing system. In: The 26th International Conference on Automated Planning and Scheduling (ICAPS) (2016)
5. Barreiro, J., Boyce, M., Do, M., Frank, J., Iatauro, M., Kichkaylo, T., Morris, P., Ong, J., Remolina, E., Smith, T., Smith, D.: EUROPA: a platform for AI planning, scheduling, constraint programming, and optimization. In: The 4th International Competition on Knowledge Engineering for Planning and Scheduling. ICKEPS 2012 (2012)
6. Ghallab, M., Laruelle, H.: Representation and control in IxTeT, a temporal planner. In: 2nd International Conference on Artificial Intelligence Planning and Scheduling (AIPS), pp. 61–67 (1994)
7. Chien, S., Rabideau, G., Knight, R., Sherwood, R., Engelhardt, B., Mutz, D., Estlin, T., Smith, B., Fisher, F., Barrett, T., Stebbins, G., Tran, D.: ASPEN - automated planning and scheduling for space mission operations. In: Proceedings of Space Ops 2000 (2000)
8. Cesta, A., Fratini, S.: The timeline representation framework as a planning and scheduling software development environment. In: Proceedings of the 27th Workshop of the UK Planning and Scheduling Special Interest Group. PlanSIG-08, Edinburgh, 11–12 December 2008
9. Cimatti, A., Micheli, A., Roveri, M.: Timelines with temporal uncertainty. In: 27th AAAI Conference on Artificial Intelligence (AAAI) (2013)
10. Cialdea Mayer, M., Orlandini, A., Umbrico, A.: A formal account of planning with flexible timelines. In: The 21st International Symposium on Temporal Representation and Reasoning (TIME), pp. 37–46. IEEE (2014)
11. Mayer, Cialdea: M., Orlandini, A., Umbrico, A.: Planning and execution with flexible timelines: a formal account. Acta Informatica 53(6), 649–680 (2016)
12. Cesta, A., Finzi, A., Fratini, S., Orlandini, A., Tronci, E.: Flexible timeline-based plan verification. In: Mertsching, B., Hund, M., Aziz, Z. (eds.) KI 2009. LNCS (LNAI), vol. 5803, pp. 49–56. Springer, Heidelberg (2009). doi:10.1007/978-3-642-04617-9_7
13. Cesta, A., Finzi, A., Fratini, S., Orlandini, A., Tronci, E.: Analyzing flexible timeline plan. In: Proceedings of the 19th European Conference on Artificial Intelligence. ECAI 2010, vol. 215. IOS Press (2010)

14. Cialdea Mayer, M., Orlandini, A.: An executable semantics of flexible plans in terms of timed game automata. In: The 22nd International Symposium on Temporal Representation and Reasoning (TIME). IEEE (2015)
15. Gigante, N., Montanari, A., Cialdea Mayer, M., Orlandini, A.: Timelines are expressive enough to capture action-based temporal planning. In: The 23rd International Symposium on Temporal Representation and Reasoning (TIME). IEEE (2016)
16. Umbrico, A., Orlandini, A., Mayer, M.C.: Enriching a temporal planner with resources and a hierarchy-based heuristic. In: Gavanelli, M., Lamma, E., Riguzzi, F. (eds.) AI*IA 2015. LNCS (LNAI), vol. 9336, pp. 410–423. Springer, Heidelberg (2015). doi:10.1007/978-3-319-24309-2_31
17. Carpanzano, E., Cesta, A., Orlandini, A., Rasconi, R., Suriano, M., Umbrico, A., Valente, A.: Design and implementation of a distributed part routing algorithm for reconfigurable transportation systems. Int. J. Comput. Integr. Manuf. (2015) http://www.tandfonline.com/doi/full/10.1080/0951192X.2015.1067911
18. Borgo, S., Cesta, A., Orlandini, A., Rasconi, R., Suriano, M., Umbrico, A.: Towards a cooperative knowledge-based control architecture for a reconfigurable manufacturing plant. In: 19th IEEE International Conference on Emerging Technologies and Factory Automation (ETFA). IEEE (2014)
19. Cesta, A., Orlandini, A., Bernardi, G., Umbrico, A.: Towards a planning-based framework for symbiotic human-robot collaboration. In: 21th IEEE International Conference on Emerging Technologies and Factory Automation (ETFA). IEEE (2016)
20. Morris, P.H., Muscettola, N., Vidal, T.: Dynamic control of plans with temporal uncertainty. In: International Joint Conference on Artificial Intelligence (IJCAI), pp. 494–502 (2001)
21. Fratini, S., Pecora, F., Cesta, A.: Unifying planning and scheduling as timelines in a component-based perspective. Arch. Control Sci. **18**(2), 231–271 (2008)
22. Cesta, A., Cortellessa, G., Fratini, S., Oddi, A.: MrSPOCK: steps in developing an end-to-end space application. Comput. Intell. **27**(1), 83–102 (2011)
23. Frank, J., Jonsson, A.: Constraint based attribute and interval planning. J. Constraints **8**(4), 339–364 (2003)
24. Bernardini, S.: Constraint-based temporal planning: issues in domain modelling and search control. PhD thesis, Università degli Studi di Trento (2008)
25. Bernardini, S., Smith, D.E.: Towards search control via dependency graphs in Europa2. In: ICAPS Workshop on Heuristics for Domain Independent Planning (HDIP) (2009)
26. Bresina, J.L., Jónsson, A.K., Morris, P.H., Rajan, K.: Activity planning for the mars exploration rovers. In: International Conference on Automated Planning and Scheduling (ICAPS), pp. 40–49 (2005)
27. Fratini, S., Cesta, A., De Benidictis, R., Orlandini, A., Rasconi, R.: APSI-based deliberation in Goal Oriented Autonomous Controllers. In: 11th Symposium on Advanced Space Technologies in Robotics and Automation. ASTRA-11 (2011)
28. Stock, S., Mansouri, M., Pecora, F., Hertzberg, J.: Online task merging with a hierarchical hybrid task planner for mobile service robots. In: 2015 IEEE/RSJ International Conference on Intelligent Robots and Systems (IROS), pp. 6459–6464, September 2015
29. Dvořák, F., Barták, R., Bit-Monnot, A., Ingrand, F., Ghallab, M.: Planning and acting with temporal and hierarchical decomposition models. In: 2014 IEEE 26th International Conference on Tools with Artificial Intelligence (ICTAI), pp. 115–121, November 2014

Formal Verification

Learning for Verification in Embedded Systems: A Case Study

Ali Khalili, Massimo Narizzano, and Armando Tacchella[(✉)]

DIBRIS, Via Opera Pia 13, 16145 Genova, Italy
khalili.ir@gmail.com,
{massimo.narizzano,armando.tacchella}@unige.it

Abstract. Verification of embedded systems is challenging whenever control programs rely on black-box hardware components. Unless precise specifications of such components are fully available, learning their structured models is a powerful enabler for verification, but it can be inefficient when the system to be learned is data-intensive rather than control-intensive. We contribute a methodology to attack this problem based on a specific class of automata which are well suited to model systems wherein data paths are known to be decoupled from control paths. We test our approach by combining learning and verification to assess the correctness of grey-box programs relying on FIFO register circuitry to control an elevator system.

Keywords: Automata learning · Formal verification · Knowledge-based software engineering

1 Introduction

One of the main hurdles on the path towards extensive adoption of formal verification techniques is that implementation details are not always available and structured specifications are notoriously hard to come by. While several reasons contribute to this state of affairs, the consequence is that many systems in use today lack adequate specifications or make use of under-specified components [1]. This situation is all but infrequent in embedded systems where third-party hardware components are used as parts and only their interface and some informal description about their behavior is available. We call these systems grey-box, as they are composed by some parts for which implementation details or models are available (white-box) and other parts for which neither implementation details nor models are available (black-box). The research question we consider is thus how to enable verification on such grey-box systems.

Among the wide choice of formal verification techniques available, we focus on Model checking (MC), a technique that aims to establish that properties specified in some temporal logic hold in all allowable executions of the system — see, e.g., [2]. In case of MC, black-box components pose an obstacle to the application of all known techniques. To overcome such obstacle, several authors considered

© Springer International Publishing AG 2016
G. Adorni et al. (Eds.): AI*IA 2016, LNAI 10037, pp. 525–538, 2016.
DOI: 10.1007/978-3-319-49130-1_38

automata learning techniques — see, e.g., [3,4]. In spite of many success stories, and the availability of effective tools like LEARNLIB [5], in cases where components cannot be modeled as having small-sized input alphabets, learning is hardly applicable [6]. This can be an issue when dealing with data-intensive, rather than control-intensive (sub)systems, because classical Angluin-style [7] methods are not well suited for such systems.

We showcase an application of automata learning to enable MC techniques for checking requirements of embedded grey-box systems. The main idea is to inductively infer abstract models of black-box hardware components as finite-state machines (FSMs) and use such models in place of the actual components to verify the whole grey-box system. Models of black-box components are learned using AIDE [8], our tool suite comprising several algorithms for learning automata. The system we consider is an elevator-control program deployed on a third-party embedded platform endowed with some FIFO registers that can be used for scheduling elevator calls. The first task we accomplish is to learn models of such registers by interacting with a hardware simulator. With respect to other works in the literature [6,9] we make a simplifying, but realistic assumption, i.e., that the data path and the control path are separable. By this we mean that the specific values exchanged between the embedding context and the system will have no effect on its behavior. On the other hand, the actual values exchanged are important to assess the correctness of the component. In this sense, we regard this as a "data-intensive" rather than "control-intensive" system. FIFO register models are encoded into the language of the model checker SPIN [10]; albeit the example we show is specific, the encoding generalizes to all sorts of models learned by AIDE. Finally, we verify the whole system using SPIN; the model of the control system is based on previous contributions by Nagafuji and Yamaguchi [11] and Attie et al. [12].

Our experiments show that learning models from black-box hardware parts is an effective path to increase reliance in the system as a whole. At least in our experience, the scalability challenge is still mainly on the side of model checking, i.e., AIDE can learn models larger than those verifiable with SPIN, and verification with stubs is bound to generate up to millions of spurious counterexamples that can make verification results useless in practice. The rest of the paper is structured as follows. In Sect. 2 we introduce basic definitions and notations regarding automata learning and model checking. In Sect. 3 we present the elevator system including the FIFO register, and in Sect. 4 the experimental results related to automata learning and model checking of various properties on the overall system. Finally, we conclude the paper in Sect. 5 with some remarks and an agenda for future works.

2   Background

2.1   Automata Learning

Automata learning — also known as automata-based identification or grammatical inference — is a set of techniques that enables the inference of formal models of systems considering examples of their execution. Automata learning can be divided into two wide categories, i.e., passive and active learning. In passive learning, there is no control over the observations received to learn the model, whereas in active learning, the target system can be experimented with, and experimental results are collected to learn a model. In this paper, we focus on the latter kind of techniques, thereby assuming that the system under learning (SUL) is always available for controlled experimentation. Active automata learning is pioneered by Gold in [13], and later refined by Angluin in [7], where she proposed a polynomial-time algorithm, called L^*, for learning deterministic finite state automatons (DFAs). L^* algorithm has been extended to many different settings.

Since we are interested in systems in which there is a clear separation between user-provided inputs and system-generated output, we consider Mealy machines as reference models for black-box systems. Adaptation of L^* for identifying Mealy machines, where there is a clear separation between user-provided inputs and system-generated outputs, was first developed by Niese [14] and it was further extended by Shahbaz's [15] L_M^+ algorithm — for more details, see [16]. A further extension, the N^* algorithms described in [8], caters for *non-deterministic Mealy machines (NFM)*. From a formal point of view, a NFM is a quintuple $M = (Q, \Sigma_I, \Sigma_O, q_0, \tau)$ where Σ_I is the input alphabet and Σ_O is the output alphabet; Q is a set of *states*, $q_0 \in Q$ is the *initial state*, and $\tau : Q \times \Sigma_I \to 2^{Q \times \Sigma_O}$ is the transition relation. N^* performs identification of NFMs in an iterative fashion wherein two phases alternate. In the first phase, the algorithm asks output queries to a procedure that simulates an oracle knowing the internals of the black-box system perfectly. Given that the oracle is just a theoretical device, in practice, such procedure answers the query by experimenting with the actual black-box component. Moreover, since the component is assumed to be non-deterministic, a single input may yield different outputs. In theory, an oracle should answer with the set of outputs for any given input. In practice, such a set is obtained by repeating the input query under a no-rare-event assumption. In the second phase, given the results of the output queries, N^* builds a "conjecture automata" and passes it to another oracle-like procedure to test equivalence with the black-box components. While in theory such procedure should answer positively or negatively based on its knowledge of the component, in practice the equivalence check must be approximated using testing techniques — see [8] for further details. Here we wish to remark that our system AIDE [8] includes all the algorithms mentioned above, including an extension which applies to the case study presented in Sect. 3.

Since modeling real-life systems often requires a finite number of interaction primitives (methods, operations, commands, protocol messages), but actual interactions often carry additional data values (parameters, resource identifiers, authentication credentials), standard Mealy machines might be not expressive enough. *Register Mealy Machines* (RMMs) are an extension that equips the structural skeleton of Mealy machines with a finite set of registers. The increase in expressiveness of RMMs makes learning such models intrinsically more complex. In [9] an approach based on counterexample guided abstraction refinement (CEGAR) is proposed to construct models of black-box RMMs. This approach is implemented in a tool called TOMTE[1] which, together with the learners provided by LEARNLIB [5], enables identification of RMMs. Another approach for inference of RMMs is presented in [6], where a dedicated learning algorithm is proposed. At the time of this writing, an implementation of this approach is also available in the LEARNLIB public repository, whereas our system AIDE does not include algorithms for RMM learning.

2.2   Model Checking

Model checking — see, e.g., [2] — is a prominent formal verification technique for assessing functional properties of information and communication systems. The prerequisites of model checking are (i) a model of the system under consideration and (ii) a list of properties that the system must fulfill expressed is some formal logic. While not essential in theory, the availability of a system that can automate the check is taken for granted in practical applications. The task of such system is to perform an exploration of the state space of the system, until either a violation of the stated property is found, or no more new states can be explored. While there are various tools that support model checking for a variety of modeling and property-specification languages, they can be divided into two broad categories, namely explicit-state and symbolic-state model checkers. The former category encompasses tools that maintain the set of explored states using an explicit data structure, i.e., one in which the main elements stored are descriptions of the explored states. The latter category encompasses tools that represent the set of explored states as a logical formula on state variables, such that the formula is satisfied only when the variables are evaluated to explored states. The details of model checking algorithms are beyond the scope of this paper — see, e.g., [17] for further details. Here it is sufficient to say that the crucial problem is that, while many model checking algorithms are polynomial in the size of the state space, the state space size is huge for all but the simplest models. At present, this is the main limit for the applicability of model checking techniques, which makes scalability the main parameter of evaluation in our experimental analysis.

In our experiments, we use the explicit-state model checker SPIN [10] to evaluate correctness of properties. The reason of our choice is that SPIN is a mature and well-maintained tool which has been successfully deployed to verify a wide variety of industrial-size applications, from operating systems software and communications protocols to railway signaling systems [10]. The modeling language of SPIN is PROMELA (PROcess MEta LAnguage), a formalism to describe communicating finite-state machines. The basic building blocks of PROMELA are *Process, Data Objects* and *Message Channels*. A *Process* defines the behavior of a (sub)system, and is defined by the keyword `Proctype` followed by the process name, the list of input parameters, and the body of the process which consists of data declarations and statements. *Data objects* are declared in a C-language

---

[1] http://tomte.cs.ru.nl/.

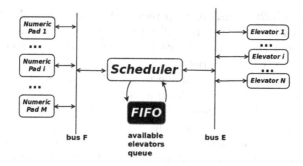

Fig. 1. Functional diagram of the elevator control system

style. Finally, *Message Channels* admit two operations, *send* and *receive*, where each channel has associated a *message* type, and only messages of that type can be sent and received. The channel declaration allows for the specification of a capacity. When the capacity is zero, the channel implements a *rendezvous* communication, i.e., the sender cannot proceed after a send unless the receiver reads the message, and symmetrically, the reader cannot proceed after a read unless the sender sends a message. When the capacity is at least one, and unless the buffer is either full or empty, the reader and the sender can process messages without the need of synchronizing. This is called a *buffered* channel in PROMELA.

Correctness properties expressing requirements about the behavior of a system can be specified in different ways. In our experiments we consider the *never claim* construct of SPIN. Intuitively, a never claim can be used to define a system behavior that, for whatever reason, is of special interest. It is most commonly used to specify behavior that should never happen (invariants, or safety properties) but it can also be used to specify behavior that should happen (liveness properties). Formally, the claim is defined as a series of propositions, or Boolean expressions, on the system state that must become true in the sequence specified for the behavior of interest to be matched. A never claim can be used to match either a finite or an infinite behavior. Finite behavior is matched if the claim can reach its final state. Infinite behavior is matched if the claim permits an ω-acceptance cycle. We do not present the syntax of never claims, but we rely on standard notation from automata theory instead, and present never claims as automata.

3   The Elevator Case Study

The behavior of the system under consideration is schematized in Fig. 1. The system is composed by N elevators moving trough M different floors. On each floor, users have access to a number pad from which requests to go to specific floors can be issued. Each elevator, when free, sends a message to the scheduler which puts the elevator into a queue. Each numeric pad sends user requests to the scheduler which assigns to each request the first available elevator. The scheduler uses a FIFO policy when allocating free elevators, i.e., the elevator that was freed first should be allocated first. From an implementation point of view, we assume that the control system (scheduler) is implemented as a program whose source code is available to us, whereas the FIFO register used to queue elevators' requests is available as a part of an off-the-shelf embedded computing platform whereon the control system runs. We assume that only the FIFO register simulator is available to us.

The model of the system presented in Fig. 1, is composed by three sub-models, namely number pad, available elevators queue and elevator. In more details:

- A number pad is a PROMELA process which non-deterministically generates a request — thereby simulating user input to the system — and sends its request to the scheduler. As a convention, each number pad process is named as $Pad[f]$, where f is the floor associated to the number pad. For each floor, user requests are stored by an internal variable which can be accessed with the syntax $f.request$.
- The available elevators queue maintains data about elevators availability and corresponds to the black-box FIFO register available in the embedded computing platform. The process model is inferred as described in the next subsection. Here we just mention that the only operations supported by the queue are *push* and *pop* primitives with the usual semantics.
- Each elevator is also a PROMELA process named as $Elevator[e]$, where e is the unique identification number of the elevator. An elevator has three main state variables, namely *floor* representing the current floor of the elevator, *state* indicating if the elevator is moving($MOVE$) or stopping ($STOP$); finally, *request* stores the floors that must served and has the value $NONE$ if no request has to be served. The internal state of an elevator e can be accessed using the syntax $e.<variable>$ where $<variable>$ is one of *floor*, *state* and *request*. Each elevator sends its availability to the scheduler using a dedicated channel ($busE$) and waits for a request. Once the request has been satisfied, it notifies to the scheduler its availability again.
- The *scheduler* coordinates user requests with elevator availability. Each available elevator is pushed into a queue (the available elevators queue) and is popped when a user request should be served. The scheduler communicates with the Numeric Pad trough a dedicated bus ($busF$).

An elevator system like the one described above should fulfill a number of safety and liveness requirements. Considering the literature [11,12], we were able to find a number of typical constraints that we describe as SPIN never claims. Instead of using SPIN syntax, we describe them in a graphical notation — see Fig. 2 — such that each never claim corresponds to an automaton where states are represented with labeled circles, initial states are marked by a short entry arrow while unwanted (acceptance) states are marked with a double circle. Transitions are marked by a guard, and a transition occurs when the guard is true — for a complete reference see [18]. In our experimental setting we verify 4 different properties, three of which are presented in Fig. 2.

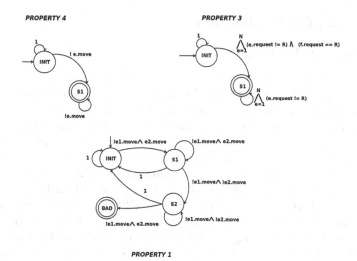

Fig. 2. Never Claims of the Properties 1, 3 and 4. Notice that $e_i.move$ is a syntactic sugar for $e_i.state == MOVE$.

Property 1: "*The scheduler should respect the FIFO policy*", i.e., if an elevator e_1 stops before the elevator e_2, then e_1 should be scheduled before e_2. This property is described in Fig. 2 (bottom) where *init* is the initial state and *bad* is a state that we do not want to reach. Each path that leads to the bad state shows that e_1 stops before e_2 but restarts after it. This property should be checked for each pair of elevators (e_i, e_j) where e_i is not equal to e_j.

Property 2: "*An elevator cannot be scheduled while it is moving*". This safety property can be checked by adding an assertion into each elevator process; the assertion should guard the fact that the scheduler can send a new load request only if the elevator is not moving.

Property 3: "*Each user request should always be satisfied*" — a typical *liveness* property. In Fig. 2 (top-right) we show the structure of the never claim for a generic request R. Starting from the init state, assuming that the request R arrives at some point, if the system cannot serve it then looping on state $S1$ shows that the property is not satisfied. This property should be written and checked for each specific request.

Property 4. The last property states that "*It can never happen that an elevator is not scheduled*". In other words if an elevator stops, it cannot be stopped forever. The never claim in Fig. 2 accepts this behavior: starting from the initial state when an elevator stops (!e.move) then a cycle on $S1$ violates the property. Since elevators are subjected to periodic maintenance, this property guarantees that, over a sufficiently long period, all the elevators will have the same service time.

4   Learning for Verification

4.1   Learning FIFO Registers

While RMMs could fit our purposes, our case study requires learning models of black-box systems wherein the control component is independent from the data component. This is a substantial simplification over RMMs, one that allows for simpler learning algorithms and a more efficient identification process with respect to RMMs. To reap these benefits, we have introduced *Parametrized Mealy Machines* (PMMs), a restricted class of RMMs. In the following, we briefly describe PMMs to the extent required to understand the case study and the experiments in this paper. For lack of space, we do not describe the inference algorithm for PMMs which is implemented in AIDE as a relatively straightforward extension of Shahbaz's [15] L_M^+ algorithm to infer Mealy automata.

PMMs are defined assuming an unbounded domain D of data values, a finite set of input symbols Σ_I, and a finite set of output symbols Σ_O where each input or output symbol is parameterized and takes a single formal parameter from D^2. The set Σ_I (Σ_O) is called the input (output) alphabet of the machine. Let further $X = \{x_1, \ldots, x_m\}$ be a finite set of *registers*. An *assignment* is a partial mapping $\rho : X \to X \cup \Sigma_I$. A *Parametrized Mealy Machine* (PMM) is a tuple $(Q, q_0, \Sigma_I, \Sigma_O, D, X, \tau)$ where Q is a finite set of *locations*; $q_0 \in Q$ is the *initial location*; Σ_I and Σ_O are the finite sets of *parametrized input* and *output symbols*, respectively; D is the *data domain* of input and output parameters; X is the set of *registers*; and $\tau : Q \times Q \times (\Sigma_I \times D) \times (\Sigma_O \times D) \times (X \cup \Sigma_I)^X$ is a finite set of *transitions* in the form $\langle q, q', (i, d_i), (o, d_o), \rho \rangle$ where q and q' are the source and destination locations of the transition, $i \in \Sigma_I$ is the input symbol, $o \in \Sigma_O$ is the output symbol, $d_i \in D$ is the user-provided input data, $d_o \in D$ is the generated output data, and ρ is an assignment.

To characterize the semantics of PMMs, we first define a *valuation* as a partial mapping $\nu : X \to D$ which determines the values of *active* registers. A *state* is a pair (q, ν) where q is a location and ν is a valuation. The initial state of the machine is always $(q, \emptyset)$, i.e., it has an empty valuation and no register is active. One *step* of a PMM takes it from state (q, ν) to state (q', ν') by input (i, d) and emits the output (o, d') if there is a transition $\langle q, q', (i, d), (o, d'), \rho \rangle$ such that ν' is the updated valuation, where $\nu'(x_j) = \nu(x_k)$ whenever $\rho(x_j) = x_k$ and $\nu'(x_j) = d$ whenever $\rho(x_j) = i$. In each step, (i) an input i with its parameter d is given to the machine, (ii) the machine may assign the value of the input parameter d to one of its registers x_j, provided that $\rho(x_j) = i$, (iii) registers may be copied, if there is some j, k such that $\rho(x_j) = x_k$, (iv) an output action o with its parameter d' is generated, and finally (v) the current location of machine changes from q to q'. Notice that in (ii) the PMM may change the value of a register, and in (iii) active registers may change and/or their values can be copied. The generated output parameter d' may come from a register or it can be some constant in D. Similarly to RMMs, the *execution* of machine is defined

---

[2] Notice that we consider only input and output symbols of arity one. This can be extended for arbitrary, but fixed a priori, number of parameters.

as a finite alternating sequence of states and steps $u_0, s_0, u_1, ..., u_n$ such that u_i is a concrete state and s_i is one step for all $i < n$.

A model of the available elevators queue learned by AIDE with its PMM inference algorithm is shown in Fig. 3. In this case we have assumed that the queue has 3 elements at most, which correspond to three PMM registers R0, R1 and R2. The system has only two interaction primitives (input symbols), namely PUSH and POP. The identified system has a total of four states and **0** is the initial state. Every transition is labeled as "$i/o/r$" where i is the concrete input symbol, o is the output symbol and r are the register operations. For instance from state **0** to state **1** the action "PUSH, d/NONE/R0 = d" means that data item d is pushed on the queue, no output is given and the data item is stored in R0. Notice that subsequent PUSH operations use registers in increasing order, and corresponding POP operations "shift" the registers to maintain queue ordering.

In Fig. 3 (right) we show the encoding of the PMM in Fig. 3 (left) into PROMELA. The encoding procedure is standard and works for any model inferred by AIDE. In particular every PMM is translated into a PROMELA process with two input channels, namely *inCH* and *outCH*, both with capacity 0. In the learned FIFO model, *inCH* is used to communicate PUSH and POP operations from some external process, and *outCH* is used to return the output of the request, i.e., the first request to be served in case of a POP request. The translation also caters for some local variables, namely "S", "d" and an array "R". Variable "S" is used to store the current state of the PMM, variable "d" is used as a temporary storage for incoming data, and array "R" corresponds to the registers. In the learned FIFO model, "S" takes values in $\{0, 1, 2, 3\}$, corresponding to the states of the PMM in Fig. 3 (left) and "R" is an array of three elements — indexed from 0 to 2 — corresponding to the three registers of the PMM. The main body of the process corresponding to a PMM is just a loop which updates the state according to the current state and the input channel value, thereby implementing the PMM computation semantics. For instance, the transition from state 2 to state 3 in Fig. 3 (left) is coded into lines 5–8 in Fig. 3

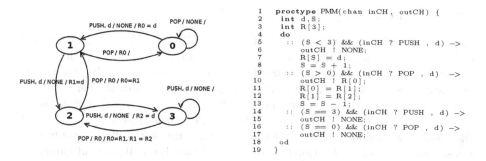

Fig. 3. FIFO queue model as learned by AIDE (left); translation into a PROMELA Process (right)

(right): when a $PUSH, d$ is received as input, the guard at line 5 becomes true, the result of the operation — NONE in this case — is given as output (line 6), then the value of the input is stored into the first empty register (line 7), and the state is updated (line 8). It is easy to see how the example given in Fig. 3 can be generalized to the same FIFO model with a different number of places in the queue, and it is also straightforward to see how the construction is apt to simulate any PMM learned by AIDE.

Table 1. Running time of AIDE to learn a model of the available elevators queue.

Size of queue	OQ	EQ	Time (s)
5	1768	2	46
10	11406	4	150
15	43496	6	466
20	106650	7	1257

The running time[3] spent by AIDE to learn FIFO queues of various sizes is presented in Table 1. Here we report the size of the queue, the number of output queries (OQ) and equivalence queries (EQ) performed by AIDE, as well as the total run-time (in CPU seconds). Output queries correspond to experiments in which AIDE asks the simulator to provide output on a specific input. The answer is used by AIDE to construct a conjecture about the structure of the SUL. Equivalence queries correspond to sets of experiments in which AIDE tries to understand whether its current conjecture could be a model of the SUL or not. As it could be expected, by increasing the number of positions, the number of output queries and the number of steps to obtain the right conjecture increases. Indeed, the PMM inference algorithm built in AIDE is able to learn FIFO registers of up to 20 places in about 20 min of CPU time. As we will see in the following, verification turns out to be unfeasible already for much smaller queue sizes.

4.2   Model Checking Elevator-Control

In the verification experiments we compile the PROMELA code without any optimization technique. Liveness properties (3 and 4) are verified with (weak) fairness conditions by adding -DNFAIR=n as a flag, where n is the number of processes fired in this case. We consider different instances of the elevator system — including the model inferred by AIDE — varying the number P of places in the queue with $P \in \{2, 4, 8\}$, the number E of elevators with $E \in \{3, 4, 5\}$ and the number F of floors with $F \in \{4, 5, 6\}$, for a total of 27 different configurations. The results of verification for properties (1) to (4) on all the configurations

---

[3] All the experiments reported in this paper ran on an Intel i7 3.4 GHz PC equipped with 32 GB of RAM and running Ubuntu 14.04. The inference of the FIFO queue models is performed by AIDE using a simulator.

Table 2. Results of property verification on the elevator control system. The group "System" defines the parameters of the system: #P is the number of places in the queue, #E is the number of elevators and #F is the number of floors. Pi stands for *Property* (i) in Sect. 3. The group "Stub" is for property verification with stubs, whereas "Model" is for verification based on the learnt model. For each property, we report the result (column "V") which is either "S" for a successfull verification (the property holds), "U" for an unsuccessfull verification (the property does not hold), and "M" for memory out. When using stubs, an integer value (column "R") reporting the number of (spurious) errors is given; the column "T" reports the CPU time (in seconds) used by SPIN to verify the properties.

| System | | | Stub | | | | | | | | Model | | | | | | | |
| | | | P1 | | P2 | | P3 | | P4 | | P1 | | P2 | | P3 | | P4 | |
#P	#E	#F	$R(\times10^3)$	T	$R(\times10^3)$	T	$R(\times10^3)$	T	$R(\times10^3)$	T	V	T	V	T	V	T	V	T
2	3	4	274	8	309	1	1425	172	301	6	U	1	S	1	S	90	U	1
2	3	5	510	14	572	1	5495	672	541	10	U	1	S	2	S	364	U	1
2	3	6	850	23	952	2	7000	428	885	16	U	1	S	2	S	1070	U	1
2	4	4	1714	128	3164	5	5000	416	2791	50	U	1	S	5	M	435	U	1
2	4	5	5116	236	5789	8	6000	292	5000	76	U	1	S	9	M	393	U	1
2	4	6	8435	392	9435	12	7000	242	5000	27	U	1	S	14	M	370	U	2
2	5	4	16979	1437	17979	12	5000	462	6000	30	U	1	S	33	M	370	U	4
2	5	5	23405	2645	24405	14	6000	351	6000	30	U	1	S	63	M	372	U	5
2	5	6	26000	2858	27000	14	7000	201	6000	27	U	1	S	102	M	356	U	7
4	3	4	274	8	309	1	1425	171	301	5	S	11	S	1	S	114	S	6
4	3	5	510	14	572	1	5495	672	541	10	S	19	S	2	S	460	S	10
4	3	6	850	23	952	2	7000	424	885	16	S	30	S	3	S	1464	S	16
4	4	4	2795	128	3164	5	5000	415	2791	51	S	184	S	7	M	421	S	62
4	4	5	5116	236	5789	8	6000	292	5000	76	S	328	S	13	M	379	S	116
4	4	6	8435	392	9435	12	7000	241	5000	27	S	554	S	21	M	392	S	199
4	5	4	16979	1437	17979	12	5000	460	6000	30	U	19	S	63	M	363	U	41
4	5	5	23405	2646	24405	14	6000	353	6000	30	U	33	S	112	M	331	U	76
4	5	6	26000	2857	27000	15	7000	199	6000	27	U	55	S	214	M	329	U	117
8	3	4	274	8	309	1	1425	171	301	5	S	11	S	1	S	119	S	6
8	3	5	510	14	572	1	5495	672	541	10	S	20	S	2	S	478	S	10
8	3	6	850	23	952	2	7000	425	885	16	S	32	S	3	S	1509	S	17
8	4	4	2795	128	3164	5	5000	415	2791	50	S	188	S	7	M	421	S	64
8	4	5	5116	236	5789	8	6000	292	5000	76	S	346	S	13	M	386	S	121
8	4	6	8435	392	9435	12	7000	242	5000	27	S	578	S	22	M	395	S	208
8	5	4	16979	1438	17979	12	5000	460	6000	30	S	2645	S	64	M	356	S	680
8	5	5	23405	2645	24405	14	6000	351	6000	30	S	4745	S	113	M	373	S	1344
8	5	6	26000	2858	27000	14	7000	197	6000	26	S	8156	S	210	M	373	S	2153

are presented in Table 2, where each line of the table represents a different configuration of the elevator system. Observing Table 2, we can see that using the learnt model all the configuration, except for property *P*3, can be verified by SPIN. Moreover the time spent to verify the system is always less than one hour, except for the last two configurations shown. We can also notice that the verification result of each property depends on the parameters of the system. For instance, when the system has a queue with size 2, property (4) is always unsatisfied, while with size 8 it is always satisfied. Indeed when the size of the queue is less than the number of elevators, it could happen that the scheduler tries to add an elevator to the queue but the corresponding message is dropped (the queue is full). When this happens, the elevator is stopped forever, since the scheduler cannot pop it from the queue. This phenomenon happens also for property (1) while properties (2) and (3) are unaffected.

If we compare the results of verification with and without model, we can see that in both cases properties that do not hold are reported correctly. — e.g., property (1) with a queue of size 2. However, when using a learned model, the counter-example provided by SPIN corresponds to a concrete failed execution, whereas when using stubs the concrete counter-example must be searched among a huge number of spurious ones — no less than hundreds of thousands, event in the simplest configurations. If we consider properties that hold, then the two methodologies are simply incomparable. When using the model, SPIN correctly reports that no counter-example can be found, while using stubs we have again a huge number of spurious counter-examples. While all of them could be in principle checked on the real systems, it is clearly inefficient to do so (e.g., there are 27 millions of counter-examples when checking property 2 in configuration $P = 4$, $E = 5$ and $F = 6$). Another thing to consider is that usually stubs are used to replace parts of a system in order to speed up verification, based on the assumption that complete systems may be more difficult to verify than incomplete ones. However, we can see that in this case incomplete systems may generate millions of spurious counter-examples which take a lot of time to generate and check on the real system, making the original assumption untenable. For instance, looking at the configuration with $P = 8$, $E = 5$ and $F = 4$ the time spent to generate more than 16 millions of counter-examples for property (1) is 1438 s, while the time to check correctness of the same property using a model is within a $2\times$ factor. Under this perspective, while the time spent for learning the FIFO model is not negligible with respect to the time spent for verification, we observe that (i) the learning time of AIDE for a queue with a given number of places is amortized over several configurations, (ii) SPIN can check the majority of the properties in the majority of the configurations and (iii) the advantage of using a model is clear since there is no need to deal with a huge number of counter-examples.

5   Conclusions

In this paper, we contributed a new methodology to improve model checking in grey-box systems by leveraging automated inference of black-box ones. In particular, learning models of black-box components as finite-state machines becomes a pre-requisite of model checking. In the case of model checking, our methodology, albeit restricted to learning a FIFO register for an elevator control system, witnesses that learning models of black-box components is effective. Indeed, scalability in the verification of the whole design was only limited by SPIN results before we could consider the largest FIFO register learned by AIDE as a component. The experimental analysis shows that our methodology can avoid the generation of a large number of spurious counter-examples when verifying a grey-box system and thus it can be more efficient in all the cases where the time needed to generate and verify spurious errors is higher than the time required to learn black-box component (amortized on several checks) and to verify the overall system.

References

1. Howar, F., Steffen, B.: Learning models for verification and testing — special track at ISoLA 2014 track introduction. In: Margaria, T., Steffen, B. (eds.) ISoLA 2014. LNCS, vol. 8802, pp. 199–201. Springer, Heidelberg (2014). doi:10.1007/978-3-662-45234-9_14

2. Baier, C., Katoen, J.P.: Principles of Model Checking. MIT press, Cambridge (2008)

3. Peled, D., Vardi, M.Y., Yannakakis, M.: Black box checking. In: Wu, J., Chanson, S.T., Gao, Q. (eds.) Formal Methods for Protocol Engineering and Distributed Systems. IFIP, vol. 28, pp. 225–240. Springer, Heidelberg (1999). doi:10.1007/978-0-387-35578-8_13

4. Groce, A., Peled, D., Yannakakis, M.: Adaptive model checking. Logic J. IGPL 14(5), 729–744 (2006)

5. Raffelt, H., Steffen, B., Berg, T., Margaria, T.: LearnLib: a framework for extrapolating behavioral models. Int. J. Softw. Tools Technol. Transfer (STTT) 11(5), 393–407 (2009)

6. Howar, F., Isberner, M., Steffen, B., Bauer, O., Jonsson, B.: Inferring semantic interfaces of data structures. In: Margaria, T., Steffen, B. (eds.) ISoLA 2012. LNCS, vol. 7609, pp. 554–571. Springer, Heidelberg (2012). doi:10.1007/978-3-642-34026-0_41

7. Angluin, D.: Learning regular sets from queries and counterexamples. Inf. Comput. 75(2), 87–106 (1987)

8. Khalili, A., Tacchella, A.: Learning nondeterministic Mealy machines. In: Proceedings of the 12th International Conference on Grammatical Inference (ICGI), pp. 109–123 (2014)

9. Aarts, F.: Tomte: Bridging the Gap between Active Learning and Real-World Systems. Ph.D. thesis, Radboud University Nijmegen (2014)

10. Holzmann, G.J.: The model checker Spin. IEEE Trans. Softw. Eng. 23(5), 279–295 (1997)

11. Nagafuji, K., Yamaguchi, S.: Éclair: An elevator group controller model checking system based on s-ring and spin. In: 2014 IEEE 3rd Global Conference on Consumer Electronics (GCCE), pp. 178–181. IEEE (2014)

12. Attie, P.C., Lorenz, D.H., Portnova, A., Chockler, H.: Behavioral compatibility without state explosion: design and verification of a component-based elevator control system. In: Gorton, I., Heineman, G.T., Crnković, I., Schmidt, H.W., Stafford, J.A., Szyperski, C., Wallnau, K. (eds.) CBSE 2006. LNCS, vol. 4063, pp. 33–49. Springer, Heidelberg (2006). doi:10.1007/11783565_3

13. Gold, E.M.: System identification via state characterization. Automatica 8(5), 621–636 (1972)

14. Niese, O.: An Integrated Approach to Testing Complex Systems. Ph.D. thesis, Universität Dortmund, Dortmund, Germany (2003)

15. Shahbaz, M.: Reverse Engineering Enhanced State Models of Black Box Software Components to Support Integration Testing. Ph.D. thesis, Institut Polytechnique de Grenoble, Grenoble, France (2008)
16. Steffen, B., Howar, F., Merten, M.: Introduction to active automata learning from a practical perspective. In: Bernardo, M., Issarny, V. (eds.) SFM 2011. LNCS, vol. 6659, pp. 256–296. Springer, Heidelberg (2011). doi:10.1007/978-3-642-21455-4_8
17. Clarke, E.M., Grumberg, O., Peled, D.: Model checking. MIT press, Cambridge (1999)
18. Holzmann, G.J.: The SPIN Model Checker: Primer and Reference Manual, vol. 1003. Addison-Wesley, Reading (2004)

Learning in Physical Domains: Mating Safety Requirements and Costly Sampling

Francesco Leofante and Armando Tacchella[✉]

DIBRIS, Università Degli Studi di Genova,
Via All'Opera Pia, 13, 16145 Genova, Italy
francesco.leofante@edu.unige.it, armando.tacchella@unige.it

Abstract. Agents learning in physical domains face two problems: they must meet safety requirements because their behaviour must not cause damage to the environment, and they should learn with as few samples as possible because acquiring new data requires costly interactions. Active learning strategies reduce sampling costs, as new data are requested only when and where they are deemed most useful to improve on agent's accuracy, but safety remains a standing challenge. In this paper we focus on active learning with support vector regression and introduce a methodology based on satisfiability modulo theory to prove that predictions are bounded as long as input patterns satisfy some preconditions. We present experimental results showing the feasibility of our approach, and compare our results with Gaussian processes, another class of kernel methods which natively provide bounds on predictions.

Keywords: Machine learning · Automated reasoning · Formal verification

1 Introduction

Artificial agents that act in physical domains can be equipped with learning capabilities in order to acquire the dynamics of interest, thus saving developers from the burden of devising explicit models — see [1–3] for examples. However, learning in physical domains poses two challenges. The first one is safety, i.e., freedom from unacceptable risk to the outside from the functional and physical units considered. In practice, guaranteeing safety amounts to controlling undesirable behaviours when the agent acts based on a model of the environment.[1] The second challenge relates to the cost of sampling, i.e., acquiring new data for learning. In physical domains, this requires non-negligible time and energy to be accomplished, causes wear of parts and usually calls for qualified human supervision. For these reasons, agents acting in physical domains should learn with as little experimentation as possible.

---

[1] One may also consider safety issues occurring at the time of learning. While this is an important aspect in the development of autonomous agents, we assume that learning is carried out safely, e.g., in a simulator or within a controlled physical environment.

© Springer International Publishing AG 2016
G. Adorni et al. (Eds.): AI*IA 2016, LNAI 10037, pp. 539–552, 2016.
DOI: 10.1007/978-3-319-49130-1_39

To minimize costs, we would like to query only for potentially informative samples, i.e., the ones that we expect to improve the accuracy of the model to the greatest extent. An effective approach to reduce sampling effort is Active Learning (AL) — see, e.g., [4] for a recent survey. In the context of physical domains, active learning amounts to iteratively choose the actions to be performed that are deemed most informative, and acquire the corresponding response from the environment. Actions and related responses are added to the sample set, and the agent is trained again using the augmented set. This procedure can be repeated until a desired level of accuracy is reached, and the final sample set is expected to contain the smallest number of elements needed. However, to the best of our knowledge, there is no AL methodology which takes into account even very basic safety requirements.

In this work, our goal is to exploit the sampling paucity of AL techniques by supplementing them with safety guarantees in order to tackle learning in physical domains. We focus on active learning using Support Vector Regression (SVR) — see, e.g., [5] for a tutorial introduction. The choice of SVR is motivated by the popularity of kernel-based methods, and the fact that identification of unknown environment functions mapping stimuli to responses is a fairly common task in physical domains. Different active learning criteria have been proposed for Support Vector Regression — see [4] for a comprehensive survey. In our study we focus on the *query by committee* algorithm [6] and on a recent proposal by Demir and Bruzzone [7] wherein kernel-based clustering algorithms are used to infer whether a new data sample can be informative.

To provide safety guarantees on SVRs we propose a methodology based on Satisfiability Modulo Theory (SMT) solvers [8]. Our approach is inspired by similar works on neural networks [9], and it assumes that the main requirement is to bound the response of the learning agent, given that supplied stimuli are within acceptable operational parameters. Intuitively, verification of SVRs via SMT involves abstraction of the learned model into a set of constraints expressed in linear arithmetic over real numbers. The abstraction process is guaranteed to provide an over-approximation of the concrete SVR. In this way, when the SMT solver proves that the response of the abstract SVR cannot exceed a stated bound as long as the input values satisfy given preconditions, the concrete SVR has the same property. On the other hand, if a counterexample is found, then it is either an artefact of the abstraction, or a true counterexample proving that the SVR is not satisfactory in terms of safety. The former case calls for a refinement of the abstraction, whereas the second may entail further training of the SVR.

Based on an artificial physical domain simulated with V-REP [10], we present experimental results to demonstrate the feasibility of our approach. Furthermore, we compare it with Gaussian Processes (GPs) for regression [11], another class of kernel methods which natively provide *probabilistic* bounds on their predictions. The remainder of this paper is organized as follows. Section 2 gives background notions on SVRs, SMT solvers and GPs. Section 3 defines the details of the proposed methodology, while experimental results are discussed in Sect. 4. Concluding remarks are presented in Sect. 5.

2 Preliminaries

2.1 Kernel-Based Methods for Regression

Support Vector Regression (SVR) is a supervised learning paradigm, whose mathematical foundations are derived from Support Vector Machine (SVM) [12, 13]. Suppose we are given a training set $T = \{(x_1, y_1) \dots (x_l, y_l)\} \subset \mathcal{X} \times \mathbb{R}$, where $\mathcal{X}$ is the input space with $\mathcal{X} = \mathbb{R}^d$. In the context of ε-SVR [14] for the linear case, our aim is to compute a function $f(x)$ which takes the form

$$f(x) = w \cdot x + b \text{ with } w \in \mathcal{X}, b \in \mathbb{R} \tag{1}$$

such that $f(x)$ deviates at most ε from the targets $y_i \in T$ with $\|w\|$ as small as possible. A solution to this problem is to introduce slack variables ξ_i, ξ_i^* to handle otherwise infeasible constraints. The formulation of the optimization problem stated above then becomes:

$$
\begin{aligned}
\text{minimize} \quad & \tfrac{1}{2}\|w\|^2 + C \sum_{i=1}^{l} (\xi_i + \xi_i^*) \\
\text{subject to} \quad & y_i - w \cdot x_i - b \leq \varepsilon + \xi_i \\
& w \cdot x_i + b - y_i \leq \varepsilon + \xi_i^* \\
& \xi_i, \xi_i^* \geq 0
\end{aligned}
\tag{2}
$$

The constant $C > 0$ determines the trade-off between the "flatness" of f, i.e., how small is $\|w\|$, and how much we want to tolerate deviations greater than ε.

In order to solve the above stated optimization problem one can use the standard dualization method using Lagrange multipliers. The Lagrange function writes

$$
\begin{aligned}
L := \frac{1}{2}\|w\|^2 + C \sum_{i=1}^{l} (\xi_i + \xi_i^*) - \sum_{i=1}^{l} (\eta_i \xi_i + \eta_i^* \xi_i^*) \\
- \sum_{i=1}^{l} \alpha_i (\varepsilon + \xi_i - y_i + w \cdot x_i + b) - \sum_{i=1}^{l} \alpha_i^* (\varepsilon + \xi_i^* - y_i - w \cdot x_i - b)
\end{aligned}
\tag{3}
$$

where $\eta_i, \eta_i^*, \alpha_i, \alpha_i^*$ are Lagrange multipliers. Solving the dual problem allows to rewrite (1) as

$$f(x) = \sum_{i=1}^{l} (\alpha_i - \alpha_i^*) x_i \cdot x + b \tag{4}$$

The algorithm can be modified to handle non-linear cases. This could be done by mapping the input samples x_i into some feature space $\mathcal{F}$ by means of a map $\Phi : \mathcal{X} \to \mathcal{F}$. However this approach can easily become computationally infeasible for high dimensionality of the input data. A better solution can be obtained by applying the so-called *kernel trick*. Observing Eq. (4) it is possible to see that the SVR algorithm only depends on dot products between patterns x_i. Hence it suffices to know the function $K(x, x') = \Phi(x) \cdot \Phi(x')$ rather than Φ explicitly.

By introducing this idea the optimization problem can be rewritten, leading to the non-linear version of (4):

$$f(x) = \sum_{i=1}^{l}(\alpha_i - \alpha_i^*)K(x_i, x) + b \tag{5}$$

where $K(\cdot)$ is called *kernel function* as long as it fulfils some additional conditions — see, e.g., [5] for a full characterization of K.

Another kind of kernel-based regression can be obtained if we view the function to be estimated from the training set T as a *stochastic process*, i.e., a collection of random variables $\{f(x) \mid x \in \mathcal{X}\}$, where $\mathcal{X} \subset \mathbb{R}^d$ is the input space. In particular, we consider *Gaussian processes* [11], i.e., stochastic processes which can be fully specified by their mean and covariance functions

$$\mu(x) = E\{f(x)\} \qquad C(x, x') = E\{(f(x) - \mu(x)(f(x') - \mu(x'))\} \tag{6}$$

In a Gaussian process, any finite subset of random variables has a joint Gaussian distribution. Without loss of generality, here we always choose the mean function to be identically zero. There are many reasonable choices for the covariance function as long as a semi-positive definite covariance matrix is generated for any set of points. A common choice for C is the so-called *squared exponential* covariance function, defined as

$$C(x, x') = e^{-\frac{1}{2}\frac{\|x - x'\|^2}{l^2}} \tag{7}$$

with l being the characteristic length-scale of the process. This parameter can be informally thought of as the distance between two points in input space for which the function value changes significantly.

If we consider a training set $T = \{(x_1, y_1) \ldots (x_l, y_l)\} \subset \mathcal{X} \times \mathbb{R}$, the predictive distribution for an unseen test case x^* can be obtained from the $l+1$ dimensional joint Gaussian distribution for the outputs of the l training cases and the test case, by conditioning on the observed targets in the training set. The predictive distribution is defined as

$$\begin{aligned} \hat{f}(x^*) &= \mathbf{k}^T(x^*)K^{-1}\mathbf{y} \\ \sigma_{\hat{f}}^2(x^*) &= C(x^*, x^*) - \mathbf{k}^T(x^*)K^{-1}\mathbf{k}(x^*)\} \end{aligned} \tag{8}$$

where $\mathbf{k}(x^*) = (C(x^*, x_1) \ldots C(x^*, x_l))^T$, K is the covariance matrix for the training data $K_{ij} = C(x_i, x_j)$ and $\mathbf{y} = (y_1, \ldots, y_l)^T$. Intuitively, learning with Gaussian processes amounts to maintain a distribution over potential regression functions consistent with T and choosing among their predictions the one which is a-posteriori more consistent with the observation — see [15] for a complete characterization of Gaussian Process Regression. The key point of interest to us is that Gaussian Process Regression (GPR) provide s both a prediction $\hat{f}(x^*)$ and a probabilistic bound $\sigma_{\hat{f}}^2(x^*)$ which can be used to quantify how much the actual prediction floats around the expected value. Notice that such a bound is not natively provided by ε-SVR predictions.

2.2   Satisfiability Modulo Theory

A Constraint Satisfaction Problem (CSP) [16] is defined over a *constraint network*, i.e., a finite set of *variables*, each ranging over a given domain, and a finite set of *constraints*. Intuitively, a constraint represents a combination of values that a certain subset of variables is allowed to take. In this paper we are concerned with CSPs involving linear arithmetic constraints over real-valued variables. From a syntactical point of view, the constraint sets that we consider are built according to the following grammar:

$$
\begin{aligned}
set &\rightarrow \{\ constraint\ \}^*\ constraint \\
constraint &\rightarrow (\{\ atom\ \}^*\ atom) \\
atom &\rightarrow bound \mid equation \\
bound &\rightarrow value\ relop\ value \\
relop &\rightarrow <\ \mid\ \leq\ \mid\ =\ \mid\ \geq\ \mid\ >
\end{aligned}
\qquad
\begin{aligned}
value &\rightarrow var \mid \text{-}\ var \mid const \\
equation &\rightarrow var = value \text{ - } value \\
&\mid\ var = value + value \\
&\mid\ var = const \cdot value
\end{aligned}
$$

where *const* is a real value, and *var* is the name of a (real-valued) variable. Formally, this grammar defines a fragment of linear arithmetic over real numbers known as Quantifier-Free Linear Arithmetic over Reals (QF_LRA) [8].

From a semantical point of view, let X be a set of real-valued variables, and μ be an *assignment* of values to the variables in X, i.e., a partial function $\mu : X \rightarrow \mathbb{R}$. For all μ, we assume that $\mu(c) = c$ for every $c \in \mathbb{R}$, and that $relop^\mu$, $+^\mu$, $-^\mu$, $\cdot^\mu$ denote the standard interpretations of relational and arithmetic operators over real numbers. Linear constraints are thus interpreted as follows:

- A constraint $(a_1 \ldots a_n)$ is true exactly when at least one of the atoms a_i with $i \in \{1, \ldots, n\}$ is true.
- Given $x, y \in X \cup \mathbb{R}$, an atom $x\ relop\ y$ is true exactly when $\mu(x)\ relop^\mu\ \mu(y)$ holds.
- Given $x \in X$, $y, z \in X \cup \mathbb{R}$, and $c \in \mathbb{R}$
  - $x = y \odot z$ with $\odot \in \{+, -\}$ is true exactly when $\mu(x) =^\mu \mu(y) \odot^\mu \mu(z)$,
  - $x = c \cdot y$ is true exactly when $\mu(x) =^\mu \mu(c) \cdot^\mu \mu(y)$,

Given an assignment, a constraint is thus a function that returns a Boolean value. A constraint is *satisfied* if it is true under a given assignment, and it is *violated* otherwise. A constraint network is *satisfiable* if it exists an assignment that satisfies all the constraints in the network, and *unsatisfiable* otherwise.

Different approaches are possible to solve the CSPs, this work is confined to Satisfiability Modulo Theories (SMT) [8]. SMT is the problem of deciding the satisfiability of a first-order formula with respect to some decidable theory $\mathcal{T}$. In particular, SMT generalizes the Boolean satisfiability problem (SAT) by adding background theories such as the theory of real numbers, the theory of integers, and the theories of data structures (*e.g.*, lists, arrays and bit vectors). The idea behind SMT is that the satisfiability of a constraint network can be decided in two steps. The first one is to convert each arithmetic constraint to a Boolean constraint, e.g., the constraint network $x \geq y$, $(y > 0, x > 0)$, $y \leq 0$ is converted to the Boolean CSP A, (B, C), $\neg B$ where $A, B, C \in \{0, 1\}$ and "$\neg$" is the symbol for Boolean negation. A specialized Boolean satisfiability solver can now be invoked to compute an assignment that satisfies all the constraints. In the example above, $A = 1, B = 0, C = 1$ satisfies all the constraints. Notice that if no such assignment exists, then the set of arithmetic constraints is trivially unsatisfiable. The second step is to check the consistency of the assignment in the corresponding background theory. Considering the assignment $A = 1, B = 0, C = 1$ we need to check whether the system of linear inequalities

$$\begin{cases} x \geq y \ (A) \\ y \leq 0 \ (\neg B) \\ x > 0 \ (C) \end{cases}$$

is feasible. This is clearly the case for the example at hand, and we thus conclude that the original CSP admits a solution. Checking that the Boolean assignment is feasible in the underlying mathematical theory can be performed by a specialized reasoning procedure (SMT solvers) – any procedure that computes feasibility of a set of linear inequalities in the case of QF_LRA. If the consistency check fails, then a new Boolean assignment is requested and the SMT solver goes on until either an arithmetically consistent Boolean assignment is found, or no more Boolean assignments can be found.

3   Verifying SVR via Abstraction

The main objective of our research is to verify SVR models, i.e., given a trained SVR ς and a suitable formal logic L, show that

$$\models_L \forall x (pre(x) \rightarrow post(\varsigma(x))) \tag{9}$$

holds, where pre and $post$ are formulas expressed in L defining preconditions on inputs and postconditions on output predictions. However, in most practical applications SVRs utilize non-linear and/or transcendental kernel functions which make the problem undecidable if the logic L is expressive enough to model them. This problem has been tackled in previous works [9,17] introducing an abstraction framework based on interval arithmetic. Given an SVR ς and properties pre, $post$, we can define a corresponding abstract SVR $\widehat{\varsigma}$ and properties $\widehat{pre}$, $\widehat{post}$ such that

$$\models_{L'} \forall x' (\widehat{pre}(x') \rightarrow \widehat{post}(\varsigma(x'))) \tag{10}$$

is decidable, and (9) holds whenever (10) does. If (10) does not hold, then there is an abstract input x' such that the preconditions are satisfied, but the output of the abstract SVR $\widehat{\varsigma}$ violates the postconditions. This counterexample may have two possible explanations: if a concrete input x can be extracted from x' such that $\varsigma(x)$ violates $post$, then (9) does not uphold; on the other hand, if no concrete input can be found, then x' is a spurious counterexample and a refinement of the abstraction is needed.

In the spirit of [17], we show how to define $\widehat{\varsigma}$ from any SVR ς such that L' is QF_LRA. In particular, our experimental analysis focuses on radial basis function (RBF) kernels

$$K(x_i, x) = e^{-\gamma\|x-x_i\|^2} \text{ with } \gamma = \frac{1}{2\sigma^2} \tag{11}$$

Given a SVR, we consider its input domain to be defined as $\mathcal{I} = D_1 \times \cdots \times D_n$, where $D_i = [a_i, b_i]$ with $a_i, b_i \in \mathbb{R}$, and the output domain to be defined as $\mathcal{O} = [c, d]$ with $c, d \in \mathbb{R}$. A concrete domain $D = [a, b]$ is abstracted to $[D] = \{[x, y] \mid a \leq x \leq y \leq b\}$, i.e., the set of intervals inside D, where $[x]$ is a generic element.

To abstract the RBF kernel K, we observe that (11) corresponds to a Gaussian distribution $\mathcal{G}(\mu, \sigma)$ with mean $\mu = x_i$ — the i-th support vector of ς — and variance $\sigma^2 = \frac{1}{2\gamma}$. Given a real-valued abstraction parameter p, let us consider the interval $[x, x + p]$. It is possible to show that the maximum and minimum of $\mathcal{G}(\mu, \sigma)$ restricted to $[x, x + p]$ lie in the external points of such interval, unless $\mu \in [x, x + p]$ in which case the maximum of $\mathcal{G}$ lies in $x = \mu$. The abstraction $\widehat{K}_p$ is complete once we consider two points x_0, x_1 with $x_0 < \mu < x_1$ defined as the points in which $\mathcal{G}(x_i) \ll G(\mu)$ with $i \in 0, 1$, where by "$a \ll b$" we mean that a is at least one order of magnitude smaller than b^2. We define the *abstract kernel function for the i-th support vector* as

$$\widehat{K}_p([x, x+p]) = \begin{cases} [\min(\mathcal{G}(x), \mathcal{G}(x+p)), \mathcal{G}(\mu)] & \text{if } \mu \in [x, x+p] \subset [x_0, x_1] \\ [\min(\mathcal{G}(x), \mathcal{G}(x+p)), \max(\mathcal{G}(x), \mathcal{G}(x+p))] & \text{if } \mu \notin [x, x+p] \subset [x_0, x_1] \\ [0, \mathcal{G}(x_0)] & \text{otherwise} \end{cases} \tag{12}$$

According to Eq. (12), p controls to what extent $\widehat{K}_p$ over-approximates K since large values of p correspond to coarse-grained abstractions and small values of p to fine-grained ones. A pictorial representation of Eq. (12) is given in Fig. 1. The *abstract SVR* $\widehat{\varsigma}_p$ is defined rewriting Eq. (5) as

$$\widehat{\varsigma}_p = \sum_i^l (\alpha_i - \alpha_i^*)\widehat{K}_p(x_i, x) + b \tag{13}$$

where we abuse notation and use conventional arithmetic symbols to denote their interval arithmetic counterparts.

It is easy to see that every abstract SVR $\widehat{\varsigma}_p$ can be modelled in QF_LRA. Let us consider a concrete SVR $\varsigma : [0, 1] \rightarrow (0, 1)$, i.e., a SVR trained on $T = \{(x_1, y_1) \ldots (x_l, y_l)\}$ with $x_i \in [0, 1] \subset \mathbb{R}$ and $y_i \in (0, 1) \subset \mathbb{R}$. The abstract input domain is defined by:

$$x_i \geq 0 \qquad x_i \leq 1 \tag{14}$$

---

[2] Given a Gaussian distribution, 99.7 % of its samples lie within $\pm 3\sigma$. In this case however, our RBF formulation does not include any normalization factor and stopping at $\pm 3\sigma$ would not be sufficient to include all relevant samples. The heuristics described in the text was therefore adopted to solve the problem.

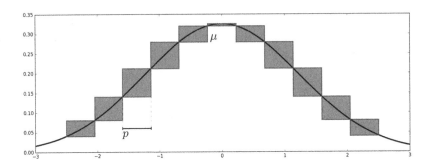

Fig. 1. Interval abstraction of a Gaussian kernel.

The coefficients and the intercept seen in Eq. (13) carry over to $\widehat{\varsigma}_p$ and are defined by constraints of the form

$$\alpha_i - \alpha_i* = \langle \alpha_i - \alpha_i* \rangle \qquad b = \langle b \rangle \tag{15}$$

where the notation $\langle \cdot \rangle$ represents the actual values obtained by training ς. To complete the encoding, we must provide constraints that correspond to (12) for all the support vectors of ς. Assuming $p = 0.5$, the encoding for the i-th kernel writes

> **if** $(x \leq x_0)$
>   **then** $(K_i \geq 0)$ $(K_i \leq \mathcal{G}(x_0))$
> ...
> **if** $(0 < x)(x \leq 0.5)$
>   **then** $(K_i \geq \min(\mathcal{G}(0), \mathcal{G}(0.5)))$ $(K_i \leq \max(\mathcal{G}(0), \mathcal{G}(0.5)))$
> ...
> **if** $(x_1 < x)$
>   **then** $(K_i \geq 0)$ $(K_i \leq \mathcal{G}(x_0))$

The expression "**if** $t_1 \ldots t_m$ **then** $a_1 \ldots a_n$" is an abbreviation for the set of constraints

$$(\neg t_1 \ldots \neg t_m a_1)$$
$$\ldots \tag{16}$$
$$(\neg t_1 \ldots \neg t_m a_n)$$

where $t_1, \ldots, t_m$ and $a_1, \ldots a_n$ are atoms expressing bound on variables, and $\neg$ is Boolean negation (e.g., $\neg(x < y)$ is $(x \geq y)$). Finally, the output of the SVR is just

$$f(x) = \langle \alpha_1 - \alpha_1^* \rangle K_i(x_1, x) + \ldots + \langle \alpha_n - \alpha_n^* \rangle K_n(x_n, x) + \langle b \rangle \tag{17}$$

with $x_1, \ldots, x_n$ being the support vectors of ς

4   Experimental Setup

A pictorial representation of the case study we consider is shown in Fig. 2. In this domain setup, a ball has to be thrown in such a way that it goes through a goal, possibly avoiding obstacles present in the environment. The learning agent interacting with the domain is allowed to control the force $f = (f_x, f_y)$ applied to the ball when it is thrown. Given that the coordinate origin is placed at the center of the wall opposing the goal — see Fig. 2 — we chose $f_x \in [-20, 20]N$ and $f_y \in [10, 20]N$, respectively. The range for f_x is chosen in such a way that the whole field, including side walls, can be targeted, and the range for f_y is chosen so that the ball can arrive at the goal winning pavement's friction both in straight and kick shots. The collision between the side walls and the ball is completely elastic. In principle, the obstacle can be placed everywhere in the field, but we consider only three configurations, namely (a) no obstacle is present, (b) the obstacle occludes straight line trajectories — the case depicted in Fig. 2 — and (c) the obstacle partially occludes the goal, so that not all straight shots may go through it.

The domain of Fig. 2 is implemented using the V-REP simulator [10]. Concrete SVRs are trained using the Python library SCIKIT-LEARN, which in turn is based on LIBSVM [18]. Verification of abstract SVRs is carried out using Mathsat [19]. The Python library GPy [20] is used to train Gaussian process regression. All experiments were performed on a pc running Ubuntu Trusty 14.04 LTS, Intel Core i7 CPU at 2.40 GHz and 8 GB of RAM. In the remainder of this section, we present two sets of experiments. The first one in Subsect. 4.1 is aimed to understand which active learning method is best on the domain described above. The second one in Subsect. 4.2 is aimed to evaluate SMT-based verification of SVRs to see whether prediction bounds can be obtained by means of the techniques described in Sect. 3, and to compare such deterministic bounds with the probabilistic ones that can be obtained through GPR.

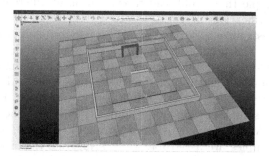

Fig. 2. Experimental setup in V-REP. In a field delimited by fences, a green ball is to be thrown from one side into the goal (brown door) on the opposite side. An obstacle (grey wall) may be placed inside the field in such a position that all or some straight shots will always be ineffective. (Color figure online)

Table 1. Preliminary evaluation of the active learning algorithms. Statistics on the number of AL runs needed when no obstacle is present (case a), using different subsets of the feasible trajectories. "Mean" and "STD" are the sample mean and standard deviation, respectively; "Median" is the 50 %-percentile, and IQR is the difference between 75 %-percentile and 25 %-percentile. When the median coincides with the maximum number of allowable runs, the IQR value cannot be computed (denoted with "–").

Algorithm	Mean	STD	Median	IQR
Straight and Kick shots				
Random	97.6	13.68	100.0	–
QBC	88.27	24.28	100.0	–
DB	95.53	19.49	100.0	–
Straight shots				
Random	40.06	11.7364	40.0	5.0
QBC	15.05	3.7275	15.0	3.0
DB	14.75	0.8275	15.0	1.0
Kick shots				
Random	16.3	18.29	12.0	8.25
QBC	15.9	25.12	3.0	19.25
DB	9.28	13.51	7.0	5.0

4.1 Assessing Active Learning with SVR

We train SVRs on a dataset of n samples built as $T = \{(f_i, y_i), \ldots, (f_n, y_n)\}$ where $f_i = (f_{xi}, f_{yi})$ is the force vector applied to the ball to throw it, and y_i is the observed distance from the center of the gate, which is also the target of the regression. Obviously, the regression problem could be extended, e.g., by including the initial position of the ball, but our focus here is on getting the simplest example that could lead to evaluation of different active learning paradigms and verification thereof.

Preliminary tests were performed in an obstacle-free environment, using specific subsets of all the feasible trajectories. Three algorithms were compared, namely a random strategy for adding new samples to the learning set, the *query by committee* algorithm and the one proposed by Demir and Bruzzone [7] — listed in Table 1 as "Random", "QBC" and "DB", respectively. Each test episode consists in adding samples to the model, and it ends when either the Root Mean Squared Error (RMSE) is smaller than a threshold (here 0.2), or when at most 100 additional samples are asked for. Table 1 suggests that when the learning problem involves one single class of shots (e.g., only straight shots), the algorithms provide better results. Indeed, when both straight and kick shots are possible, all three algorithms require more than 100 additional samples in at least 50 % of the runs in order to reach the required RMSE.

We conducted further experiments, this time considering the three cases (a, b, c) described before, and the full set of feasible trajectories in each case.

Table 2. Preliminary evaluation of the active learning algorithms. Statistics on the number of AL runs needed for cases a, b and c, legend as in Table 1.

Algorithm	Mean	STD	Median	IQR
Case a				
Random	97.6	13.68	100.0	–
QBC	88.27	24.28	100.0	–
DB	95.53	19.49	100.0	–
Case b				
Random	64.07	36.18	69.0	71.0
QBC	67.22	43.44	100.0	–
DB	63.63	39.49	100.0	–
Case c				
Random	91.94	23.26	100.0	–
QBC	85.46	24.67	100.0	–
DB	93.13	23.35	100.0	–

The results obtained, shown in Table 2, confirm the observation made on Table 1. The best results are indeed obtained for case b, wherein the obstacle is placed in such a way that only kick shots are feasible, thus turning the learning problem into a single-concept one. Overall, as a result of this preliminary analysis, we opted to choose Demir and Bruzzone for applying SMT-based verification and the comparative analysis with GPR.

4.2   Bounding Learning Results with SVR and GPR

In order to prove that predictions are bounded as long as input patterns satisfy some preconditions, we verify a property called *stability* in [9] which is formally defined as follows

$$\forall x_1, x_2 \in \mathcal{I} : ||x_1 - x_2|| \leq \delta \rightarrow ||\varsigma(x_1) - \varsigma(x_2)|| \leq \epsilon \tag{18}$$

where $\delta, \epsilon \in \mathbb{R}^+$ are two arbitrary constants defining the amount of variation that we allow in the input and the corresponding perturbation of the output. Stability is used to verify if, for a given input point x_1, there exist another input x_2 such that the condition of Eq. (18) does not hold (i.e., $||\varsigma(x_1) - \varsigma(x_2)|| \geq \epsilon$). The values of δ and ϵ are set as follows:

- $\delta \geq p$, where p is the abstraction parameter of Sect. 3. This is because asking for a δ smaller than the abstraction parameter would not allow to obtain meaningful results.
- ϵ is chosen to be arbitrarily small, and then incremented if the verification manages to find non-spurious counterexamples showing that condition (18) does not hold.

Table 3. Comparison between output given by the two methodologies. All values are referred to an input pair $f = [-17.8, 10.3]$. Real values for input forces in a neighbourhood δ of f are in the range $[-0.508, 0.196]$, with the real output corresponding to f being equal to -0.051. "MAE" is the mean absolute error from the correct prediction, "p" is the abstraction parameter, ϵ is the required confidence as per Eq. (18) and σ is the uncertainty associated with the GP prediction.

	Samples	MAE	p	ε	Output interval	CPU time (s)
SVR + SMT	10 samples	0.036	0.09	0.38	$[-0.078, -0.049]$	465.201
	20 samples	0.032	0.09	0.40	$[-0.087, -0.066]$	2041.645
	40 samples	0.030	0.06	0.42	$[-0.053, -0.033]$	7480.110
	Samples	MAE	σ		Output interval	Time
GP	10 samples	0.106	0.291		$[-0.164, -0.140]$	3.336
	20 samples	0.095	0.303		$[-0.152, -0.129]$	6.805
	40 samples	0.099	0.280		$[-0.162, -0.137]$	14.035

Running verification on the abstract SVR will therefore produce as output lower and upper bounds on the input and on the predictions, together with the current value of the abstraction parameter for which the value has been produced (i.e., δ, ϵ and p respectively). We also trained a GPR using an AL strategy based on uncertainty reduction. At each iteration, a new point is added where the variance of the GP was the highest among some test points. After the GP is trained on the same input space of the SVR, we can use it to obtain a prediction for a given input, together with the associated variance.

The tests carried out to compare SVR + SMT with GPR are presented in Table 3. While the prediction performances of the two approaches do not differ substantially, it is possible to observe that the more samples we provide the SVR with, the closer are the deterministic bounds computed by the SMT to the real ones. GPR tends to produce predictions with variances that even exceed the range in which real values lie. As it is evident from table 2, GPR outperforms SVR + SMT solvers in terms of CPU time needed to certify a given bound. This result is not surprising however, as constraint satisfaction problems are computationally hard. However, in applications where certifying the bounds beyond a mere statistical guarantee is important, it could be worth paying the additional price charged by SMT solvers.

5   Conclusion

In this work we presented an abstraction-refinement approach that enables the application of formal methods to verify regression models learned in physical domains using SVR. The feasibility of our approach has been demonstrated with experimental results. Furthermore, an analysis has been carried out to compare the performances of our approach with GPR. Preliminary results are promising: while GPR is undoubtedly faster than SVR + SMT, the additional guarantees provided by the latter approach far exceed those of GPRs. Furthermore, our experiments gave interesting insights into both the challenges as well as the potentials coming from the application of SMT solvers to (active) learning that add on previous results in the literature. We expect that future extensions of our work might have an impact in fields where active learning is the standard method for training adaptive agents, e.g., when learning through reinforcement with continuous rather than discrete state-action-space representations — see [21] for a discussion. Our future works will focus on whether the proposed methodology can scale to real-world problems, possibly posing verification of abstract SVR models as a challenge problem for the SMT research community.

References

1. Argall, B.D., Chernova, S., Veloso, M., Browning, B.: A survey of robot learning from demonstration. Robot. Auton. Syst. **57**(5), 469–483 (2009)
2. Nguyen-Tuong, D., Peters, J.: Model learning for robot control: a survey. Cogn. process. **12**(4), 319–340 (2011)
3. Kober, J., Peters, J.: Reinforcement learning in robotics: a survey. In: Wiering, M., van Otterlo, M. (eds.) Reinforcement Learning. ALO, vol. 12, pp. 579–610. Springer, Heidelberg (2012)
4. Settles, B.: Active learning literature survey. Univ. Wis. Madison **52**(55–66), 11 (2010)
5. Smola, A.J., Schölkopf, B.: A tutorial on support vector regression. Stat. Comput. **14**(3), 199–222 (2004)
6. Seung, H.S., Opper, M., Sompolinsky, H.: Query by committee. In: Proceedings of the Fifth Annual Workshop on Computational Learning Theory, pp. 287–294. ACM (1992)
7. Demir, B., Bruzzone, L.: A multiple criteria active learning method for support vector regression. Pattern Recogn. **47**(7), 2558–2567 (2014)
8. Barrett, C.W., Sebastiani, R., Seshia, S.A., Tinelli, C.: Satisfiability modulo theories. Handb. Satisfiability **185**, 825–885 (2009)
9. Pulina, L., Tacchella, A.: Challenging smt solvers to verify neural networks. AI Commun. **25**(2), 117–135 (2012)
10. Rohmer, E., Singh, S.P., Freese, M.: V-rep: a versatile and scalable robot simulation framework. In: 2013 IEEE/RSJ International Conference on Intelligent Robots and Systems (IROS), pp. 1321–1326. IEEE (2013)
11. Williams, C.K.I., Rasmussen, C.E.: Gaussian processes for regression. In: Advances in Neural Information Processing Systems 8, NIPS, Denver, CO, November 27–30, 1995, pp. 514–520 (1995)
12. Boser, B.E., Guyon, I.M., Vapnik, V.N.: A training algorithm for optimal margin classifiers. In: Proceedings of the Fifth Annual Workshop on Computational Learning Theory, pp. 144–152. ACM (1992)
13. Guyon, I., Boser, B., Vapnik, V.: Automatic capacity tuning of very large VC-dimension classifiers. In: Advances in Neural Information Processing Systems, pp. 147–147 (1993)
14. Vapnik, V.N.: The Nature of Statistical Learning Theory. Springer, New York (1995)

15. Rasmussen, C.E.: Gaussian processes in machine learning. In: Bousquet, O., Luxburg, U., Rätsch, G. (eds.) ML -2003. LNCS (LNAI), vol. 3176, pp. 63–71. Springer, Heidelberg (2004). doi:10.1007/978-3-540-28650-9_4

16. Mackworth, A.K.: Consistency in networks of relations. Artif. Intell. **8**(1), 99–118 (1977)

17. Pulina, L., Tacchella, A.: An abstraction-refinement approach to verification of artificial neural networks. In: Touili, T., Cook, B., Jackson, P. (eds.) CAV 2010. LNCS, vol. 6174, pp. 243–257. Springer, Heidelberg (2010). doi:10.1007/978-3-642-14295-6_24

18. Chang, C.C., Lin, C.J.: Libsvm: a library for support vector machines. ACM Trans. Intell. Syst. Technol. (TIST) **2**(3), 27 (2011)

19. Cimatti, A., Griggio, A., Schaafsma, B.J., Sebastiani, R.: The MathSAT5 SMT solver. In: Piterman, N., Smolka, S.A. (eds.) TACAS 2013. LNCS, vol. 7795, pp. 93–107. Springer, Heidelberg (2013). doi:10.1007/978-3-642-36742-7_7

20. GPy: GPy: a gaussian process framework in python, since 2012. http://github.com/SheffieldML/GPy

21. Pathak, S., Pulina, L., Metta, G., Tacchella, A.: How to abstract intelligence? (if verification is in order). In: Proceedings of 2013 AAAI Fall Symp. How Should Intelligence Be Abstracted in AI Research (2013)

Author Index

Adorni, Giovanni 135, 406
Alberti, Marco 351, 364
Alviano, Mario 149
Amendola, Giovanni 164
Amoretti, Michele 35
An Vo, Ngoc Phuoc 450
Angelastro, Sergio 308
Angiani, Giulio 51

Bandini, Stefania 118
Bartolini, Andrea 493
Barushka, Aliaksandr 65
Basili, Roberto 76, 477
Belahcen, Anas 283
Benini, Luca 493
Bergenti, Federico 105, 179
Bianchini, Monica 283
Bossek, Jakob 3
Bridi, Thomas 493

Cagnoni, Stefano 35, 51
Calimeri, Francesco 192, 223
Castellucci, Giuseppe 76, 477
Cesta, Amedeo 508
Chesani, Federico 208
Chuzhikova, Natalia 51
Cota, Giuseppe 351
Cottone, Pietro 294
Croce, Danilo 76, 477
Crociani, Luca 118

d'Avila Garcez, Artur S. 334
De Cao, Diego 76
De Masellis, Riccardo 208
De Paola, Alessandra 377
Di Francescomarino, Chiara 208
Dodaro, Carmine 149, 164

Esposito, Floriana 308

Ferilli, Stefano 308
Ferraro, Pierluca 377
Fornacciari, Paolo 51
Fuscà, Davide 192, 223

Gaglio, Salvatore 294, 377
Germano, Stefano 223
Ghidini, Chiara 208
Gliozzi, Valentina 392
Guarino, Nicola 237
Guizzardi, Giancarlo 237

Hájek, Petr 65

Khalili, Ali 525
Khoshkangini, Reza 250
Koceva, Frosina 135

Lamma, Evelina 364
Leofante, Francesco 539
Leone, Enrico 89
Leone, Nicola 164
Lieto, Antonio 435
Lisi, Francesca Alessandra 266
Lo Re, Giuseppe 294, 377
Lombardi, Michele 493

Maggini, Marco 321
Magnolini, Simone 450
Manna, Carlo 13
Mayer, Marta Cialdea 508
Mello, Paola 208
Mensa, Enrico 435
Milano, Michela 493
Monica, Stefania 105, 179
Montali, Marco 208
Mordonini, Monica 35, 51
Muhammad Fuad, Muhammad Marwan 26

Nardi, Daniele 477
Narizzano, Massimo 525

Orlandini, Andrea 508
Ortolani, Marco 294

Pandolfo, Laura 406
Pecori, Riccardo 35
Pergola, Gabriele 294

Perri, Simona   192, 223
Pini, Maria Silvia   250
Pipitone, Arianna   464
Pirrone, Roberto   464
Popescu, Octavian   450
Pozzato, Gian Luca   418
Pulina, Luca   406

Radicioni, Daniele P.   435
Redavid, Domenico   308
Ricca, Francesco   164
Riguzzi, Fabrizio   351, 364
Roli, Andrea   35
Rossi, Alessandro   321
Rossi, Francesca   250
Rossi, Gianfranco   179
Rossi, Silvia   89

Sani, Laura   35
Scarselli, Franco   283

Serafini, Luciano   334
Serra, Roberto   35
Staffa, Mariacarla   89

Tacchella, Armando   525, 539
Tessaris, Sergio   208
Tirone, Giuseppe   464
Tomaiuolo, Michele   51
Trautmann, Heike   3

Umbrico, Alessandro   508

Vanzo, Andrea   477
Vicari, Emilio   35
Villani, Marco   35
Vizzari, Giuseppe   118

Zangari, Jessica   192, 223
Zese, Riccardo   351, 364

Printed in the United States
by Bo...ters

Printed in the United States
By Bookmasters